Advancing the Science of Climate Change

America's Climate Choices:
Panel on Advancing the Science of Climate Change

Board on Atmospheric Sciences and Climate

Division on Earth and Life Studies

NATIONAL RESEARCH COUNCIL
OF THE NATIONAL ACADEMIES

THE NATIONAL ACADEMIES PRESS
Washington, D.C.
www.nap.edu

THE NATIONAL ACADEMIES PRESS • 500 Fifth Street, N.W. • Washington, DC 20001

NOTICE: The project that is the subject of this report was approved by the Governing Board of the National Research Council, whose members are drawn from the councils of the National Academy of Sciences, the National Academy of Engineering, and the Institute of Medicine. The members of the committee responsible for the report were chosen for their special competences and with regard for appropriate balance.

This study was supported by the National Oceanic and Atmospheric Administration under contract number DG133R08CQ0062. Any opinions, findings, conclusions, or recommendations expressed in this material are those of the author(s) and do not necessarily reflect the views of the sponsoring agency.

International Standard Book Number-13: 978-0-309-14588-6 (Book)
International Standard Book Number-10: 0-309-14588-0 (Book)
International Standard Book Number-13: 978-0-309-14589-3 (PDF)
International Standard Book Number-10: 0-309-14589-9 (PDF)
Library of Congress Catalog Control Number: 2010940606

Additional copies of this report are available from the National Academies Press, 500 Fifth Street, N.W., Lockbox 285, Washington, DC 20055; (800) 624-6242 or (202) 334-3313 (in the Washington metropolitan area); Internet, http://www.nap.edu

Cover images:

Far left: courtesy of Department of Agriculture/Agricultural Research Service. Photograph by Scott Bauer.

Middle left: Borden, K., and S. Cutter. 2008. Spatial patterns of natural hazards mortality in the United States. *International Journal of Health Geographics* 7 (1):64.

Middle right: Courtesy of DOE/NREL; Credit - Sandia National Laboratories.

Far right: Courtesy of National Oceanic and Atmospheric Administration/Department of Commerce. Photograph by Commander John Bortniak, NOAA Corps, August 1991.

Printed in the United States of America

THE NATIONAL ACADEMIES

Advisers to the Nation on Science, Engineering, and Medicine

The **National Academy of Sciences** is a private, nonprofit, self-perpetuating society of distinguished scholars engaged in scientific and engineering research, dedicated to the furtherance of science and technology and to their use for the general welfare. Upon the authority of the charter granted to it by the Congress in 1863, the Academy has a mandate that requires it to advise the federal government on scientific and technical matters. Dr. Ralph J. Cicerone is president of the National Academy of Sciences.

The **National Academy of Engineering** was established in 1964, under the charter of the National Academy of Sciences, as a parallel organization of outstanding engineers. It is autonomous in its administration and in the selection of its members, sharing with the National Academy of Sciences the responsibility for advising the federal government. The National Academy of Engineering also sponsors engineering programs aimed at meeting national needs, encourages education and research, and recognizes the superior achievements of engineers. Dr. Charles M. Vest is president of the National Academy of Engineering.

The **Institute of Medicine** was established in 1970 by the National Academy of Sciences to secure the services of eminent members of appropriate professions in the examination of policy matters pertaining to the health of the public. The Institute acts under the responsibility given to the National Academy of Sciences by its congressional charter to be an adviser to the federal government and, upon its own initiative, to identify issues of medical care, research, and education. Dr. Harvey V. Fineberg is president of the Institute of Medicine.

The **National Research Council** was organized by the National Academy of Sciences in 1916 to associate the broad community of science and technology with the Academy's purposes of furthering knowledge and advising the federal government. Functioning in accordance with general policies determined by the Academy, the Council has become the principal operating agency of both the National Academy of Sciences and the National Academy of Engineering in providing services to the government, the public, and the scientific and engineering communities. The Council is administered jointly by both Academies and the Institute of Medicine. Dr. Ralph J. Cicerone and Dr. Charles M. Vest are chair and vice chair, respectively, of the National Research Council.

www.national-academies.org

AMERICA'S CLIMATE CHOICES: PANEL ON ADVANCING THE SCIENCE OF CLIMATE CHANGE

PAMELA A. MATSON (Chair), Stanford University, California
THOMAS DIETZ (Vice Chair), Michigan State University, East Lansing
WALEED ABDALATI, University of Colorado at Boulder
ANTONIO J. BUSALACCHI, JR., University of Maryland, College Park
KEN CALDEIRA, Carnegie Institution of Washington, Stanford, California
ROBERT W. CORELL, H. John Heinz III Center for Science, Economics, and the Environment, Washington, D.C.
RUTH S. DEFRIES, Columbia University, New York, New York
INEZ Y. FUNG, University of California, Berkeley
STEVEN GAINES, University of California, Santa Barbara
GEORGE M. HORNBERGER, Vanderbilt University, Nashville, Tennessee
MARIA CARMEN LEMOS, University of Michigan, Ann Arbor
SUSANNE C. MOSER, Susanne Moser Research & Consulting, Santa Cruz, California
RICHARD H. MOSS, Joint Global Change Research Institute (Pacific Northwest National Laboratory/University of Maryland), College Park, Maryland
EDWARD A. PARSON, University of Michigan, Ann Arbor
A. R. RAVISHANKARA, National Oceanic and Atmospheric Administration, Boulder, Colorado
RAYMOND W. SCHMITT, Woods Hole Oceanographic Institution, Massachusetts
B. L. TURNER, II, Arizona State University, Tempe
WARREN M. WASHINGTON, National Center for Atmospheric Research, Boulder, Colorado
JOHN P. WEYANT, Stanford University, California
DAVID A. WHELAN, The Boeing Company, Seal Beach, California

NRC Staff

IAN KRAUCUNAS, Study Director
PAUL STERN, Director, Committee on the Human Dimensions of Global Change
ART CHARO, Senior Program Officer, Space Studies Board
MAGGIE WALSER, Associate Program Officer
KATHERINE WELLER, Research Associate
GYAMI SHRESTHA, Christine Mirzayan Science and Policy Fellow
ROB GREENWAY, Program Associate

Foreword: About America's Climate Choices

Convened by the National Research Council in response to a request from Congress (P.L. 110-161), *America's Climate Choices* is a suite of five coordinated activities designed to study the serious and sweeping issues associated with global climate change, including the science and technology challenges involved, and to provide advice on the most effective steps and most promising strategies that can be taken to respond.

The Committee on America's Climate Choices is responsible for providing overall direction, coordination, and integration of the *America's Climate Choices* suite of activities and ensuring that these activities provide well-supported, action-oriented, and useful advice to the nation. The committee convened a Summit on America's Climate Choices on March 30–31, 2009, to help frame the study and provide an opportunity for high-level input on key issues. The committee is also charged with writing a final report that builds on four panel reports and other sources to answer the following four overarching questions:

- What short-term actions can be taken to respond effectively to climate change?
- What promising long-term strategies, investments, and opportunities could be pursued to respond to climate change?
- What are the major scientific and technological advances needed to better understand and respond to climate change?
- What are the major impediments (e.g., practical, institutional, economic, ethical, intergenerational, etc.) to responding effectively to climate change, and what can be done to overcome these impediments?

The Panel on Limiting the Magnitude of Future Climate Change was charged to describe, analyze, and assess strategies for reducing the net future human influence on climate. This report focuses on actions to reduce domestic greenhouse gas emissions and other human drivers of climate change, such as changes in land use, but also considers the international dimensions of climate stabilization.

The Panel on Adapting to the Impacts of Climate Change was charged to describe, analyze, and assess actions and strategies to reduce vulnerability, increase adaptive capacity, improve resiliency, and promote successful adaptation to climate change in different regions, sectors, systems, and populations. The panel's report draws on a wide range of sources and case studies to identify lessons learned from past experiences, promising current approaches, and potential new directions.

The Panel on Advancing the Science of Climate Change was charged to provide a concise overview of past, present, and future climate change, including its causes and its impacts, and to recommend steps to advance our current understanding, including new observations, research programs, next-generation models, and the physical and human assets needed to support these and other activities. This report focuses on the scientific advances needed both to improve our understanding of the integrated human-climate system and to devise more effective responses to climate change.

The Panel on Informing Effective Decisions and Actions Related to Climate Change was charged to describe and assess different activities, products, strategies, and tools for informing decision makers about climate change and helping them plan and execute effective, integrated responses. The panel's report describes the different types of climate change-related decisions and actions being taken at various levels and in different sectors and regions; it develops a framework, tools, and practical advice for ensuring that the best available technical knowledge about climate change is used to inform these decisions and actions.

America's Climate Choices builds on an extensive foundation of previous and ongoing work, including National Research Council reports, assessments from other national and international organizations, the current scientific literature, climate action plans by various entities, and other sources. More than a dozen boards and standing committees of the National Research Council were involved in developing the study, and many additional groups and individuals provided additional input during the study process. Outside viewpoints were also obtained via public events and workshops (including the Summit), invited presentations at committee and panel meetings, and comments received through the study website, *http://americasclimatechoices.org*.

Collectively, the *America's Climate Choices* suite of activities involves more than 90 volunteers from a range of communities including academia, various levels of government, business and industry, other nongovernmental organizations, and the international community. Responsibility for the final content of each report rests solely with the authoring panel and the National Research Council. However, the development of each report included input from and interactions with members of all five study groups; the membership of each group is listed in Appendix A.

Preface

The Panel on Advancing the Science of Climate Change is one of four panels convened under the *America's Climate Choices* suite of activities, which is collectively responsible for providing advice on the most effective steps and most promising strategies that the nation can take to respond to climate change (see Foreword). Our charge was to provide a concise overview of past, present, and future climate change, including its causes and its impacts, and to recommend steps to advance our current understanding of climate change and the effectiveness of responses to it (see Appendix B).

The panel's first challenge was to decide how to summarize the large volume of excellent peer-reviewed research by the national and international community to produce a concise overview of what is known. We recognize that this report is not brief; we decided that comprehensiveness was essential to the report's credibility. In addition to drawing on the new scientific results being published nearly every week, we were aided in this task by the final U.S. Global Change Research Program (USGCRP) Synthesis and Assessment Product *Global Climate Change Impacts in the United States* (USGCRP, 2009a), the recent National Research Council (NRC) report *Restructuring Federal Climate Research to Meet the Challenges of Climate Change* (NRC, 2009k), and the four volumes of the fourth assessment report by the Intergovernmental Panel on Climate Change (IPCC, 2007a-d). In keeping with the overarching goals of the *America's Climate Choices* study, we focus on the scientific knowledge that we thought would be of greatest interest to decision makers facing crucial choices about how to respond to climate change. Likewise, in looking to the future, we emphasize the scientific advances that could help decision makers identify, evaluate, and implement effective actions to limit its magnitude and adapt to its impacts.

The body of science reviewed by the Panel on Advancing the Science of Climate Change makes a compelling case that climate change is occurring and suggests that it threatens not just the environment and ecosystems of the world but the well-being of people today and in future generations. Climate change is thus a sustainability challenge. We hope that, for those who are skeptical or uncertain about what the body of scientific evidence tells us, our report will be informative. The scientific process is never "closed"—new ideas are always part of scientific debate, and there is always more to be learned—but scientific understanding does advance over time as some ideas are supported by multiple lines of evidence while others prove inconsistent with the data or basic principles. Our understanding of climate change and its causes and consequences have advanced in this way.

The panel also examined the adequacy of the science base needed to improve the effectiveness of actions taken to limit the magnitude of future climate change and adapt to its inevitable impacts. Decision makers in the federal government, state governments, tribes, corporations, municipalities, and nongovernmental organizations, as well as citizen decision makers, are beginning to act. Climate research over the past three decades, however, has been driven largely by a need to better understand rather than to explicitly respond to climate change. Until recently, there has been relatively little research focused on the development and implementation of climate-friendly energy sources or land use practices, socioeconomic and behavioral processes that affect responses, adaptation strategies, analytical approaches to evaluate trade-offs and unintended consequences of actions, policy mechanisms, and other response issues. To address the need for new kinds of knowledge, we recommend some significant changes to the nation's climate change research enterprise.

Our report covers a great deal of scientific territory and has been accomplished over a relatively short time period. For this, we thank our tremendously dedicated panel members and remarkably talented NRC study director Ian Kraucunas. The report also benefitted from the insights and assistance of several members of our sister panels and the Committee on America's Climate Choices; in particular, we thank Kris Ebi, George Eads, Bob Fri, Linda Mearns, and Susan Solomon. In addition, we thank Mike Behrenfeld, Bill Nordhaus, Michele Betsill, Peter Schultz, Chris Field, and others who contributed written materials or spoke at panel meetings. We also benefitted from many one-on-one discussions throughout the study process and from the comments and perspectives contributed through the *America's Climate Choices* website.[1]

The report also would not have been possible without the dedication and contributions of the NRC staff. In addition to study director Ian Kraucunas, we thank Paul Stern, who provided many good ideas and written contributions throughout the study; Art Charo, who staffed the workshop on geoengineering held in June 2009; Maggie Walser, who assisted with the panel's response to external review comments; Madeline Woodruff and Joe Casola, who contributed to several chapters; Katie Weller, who compiled the references for the report—a huge job; our science writers/editors Lisa Palmer and Yvonne Baskin; Rob Greenway, who provided logistical support; and Chris Elfring, who provided wise advice at several points in the process.

[1] *http://americasclimatechoices.org.*

There is still much to learn about the physical phenomenon of global climate change and its social, economic, and ecological drivers and consequences. There is also a great deal to learn about how to respond effectively without creating serious unintended consequences and, where possible, creating multiple co-benefits. If the scientific progress of the past few decades is any indication, we can expect amazing progress, but only if there is adequate demand, support, and organization for the nation's new era of climate change research.

Pamela Matson, *Chair*, and Thomas Dietz, *Vice Chair*
Panel on Advancing the Science of Climate Change

Acknowledgments

This report has been reviewed in draft form by individuals chosen for their diverse perspectives and technical expertise, in accordance with procedures approved by the NRC's Report Review Committee. The purpose of this independent review is to provide candid and critical comments that will assist the institution in making its published report as sound as possible and to ensure that the report meets institutional standards for objectivity, evidence, and responsiveness to the study charge. The review comments and draft manuscript remain confidential to protect the integrity of the deliberative process. We wish to thank the following individuals for their participation in their review of this report:

DOUG ARENT, National Renewable Energy Laboratory
DONALD F. BOESCH, University of Maryland
VIRGINIA BURKETT, U.S. Geological Survey
ROBERT DICKINSON, The University of Texas at Austin
DAVID GOLDSTON, Natural Resources Defense Council
DENNIS HARTMANN, University of Washington
JEANINE A. JONES, California Department of Water Resources
THOMAS R. KARL, National Oceanic and Atmospheric Administration
ARTHUR LEE, ChevronTexaco Corporation, San Ramon
GERALD A. MEEHL, National Center for Atmospheric Research
JERRY M. MELILLO, Marine Biological Laboratory
WILLIAM D. NORDHAUS, Yale University
ARISTIDES A.N. PATRINOS, Synthetic Genomics, Inc.
ORTWIN RENN, Institute of Management and Technology
RICHARD RICHELS, Electric Power Research Institute, Inc.
THOMAS C. SCHELLING, University of Maryland
ROBERT H. SOCOLOW, Princeton University
AMANDA STAUDT, National Wildlife Federation
MICHAEL TOMAN, The World Bank
JOHN M. WALLACE, University of Washington

Although the reviewers listed above have provided many constructive comments and suggestions, they were not asked to endorse the conclusions or recommendations nor did they see the final draft of the report before its release. The review of this report was overseen by **Andrew Solow** (Marine Policy Center) and **Robert Frosch** (Harvard

University). Appointed by the National Research Council, they were responsible for making certain that an independent examination of this report was carried out in accordance with institutional procedures and that all review comments were carefully considered. Responsibility for the final content of this report rests entirely with the authoring committee and the institution.

Institutional oversight for this project was provided by:

BOARD ON ATMOSPHERIC SCIENCES AND CLIMATE

ANTONIO J. BUSALACCHI, JR. (*Chair*), University of Maryland, College Park
ROSINA M. BIERBAUM, University of Michigan, Ann Arbor
RICHARD CARBONE, National Center for Atmospheric Research, Boulder, Colorado
WALTER F. DABBERDT, Vaisala, Inc., Boulder, Colorado
KIRSTIN DOW, University of South Carolina, Columbia
GREG S. FORBES, The Weather Channel, Inc., Atlanta, Georgia
ISAAC HELD, National Oceanic and Atmospheric Administration, Princeton, New Jersey
ARTHUR LEE, Chevron Corporation, San Ramon, California
RAYMOND T. PIERREHUMBERT, University of Chicago, Illinois
KIMBERLY PRATHER, Scripps Institution of Oceanography, La Jolla, California
KIRK R. SMITH, University of California, Berkeley
JOHN T. SNOW, University of Oklahoma, Norman
THOMAS H. VONDER HAAR, Colorado State University/CIRA, Fort Collins
XUBIN ZENG, University of Arizona, Tucson

Ex Officio Members

GERALD A. MEEHL, National Center for Atmospheric Research, Boulder, Colorado

NRC Staff

CHRIS ELFRING, Director
LAURIE GELLER, Senior Program Officer
IAN KRAUCUNAS, Senior Program Officer
MARTHA MCCONNELL, Program Officer
MAGGIE WALSER, Associate Program Officer
TOBY WARDEN, Associate Program Officer
JOSEPH CASOLA, Postdoctoral Fellow

RITA GASKINS, Administrative Coordinator
KATIE WELLER, Research Associate
LAUREN M. BROWN, Research Assistant
ROB GREENWAY, Program Associate
SHELLY FREELAND, Senior Program Assistant
AMANDA PURCELL, Senior Program Assistant
JANEISE STURDIVANT, Program Assistant
RICARDO PAYNE, Program Assistant
SHUBHA BANSKOTA, Financial Associate

Contents

Summary

Science has made enormous inroads in understanding climate change and its causes, and is beginning to help develop a strong understanding of current and potential impacts that will affect people today and in coming decades. This understanding is crucial because it allows decision makers to place climate change in the context of other large challenges facing the nation and the world. There are still some uncertainties, and there always will be in understanding a complex system like Earth's climate. Nevertheless, there is a strong, credible body of evidence, based on multiple lines of research, documenting that climate is changing and that these changes are in large part caused by human activities. While much remains to be learned, the core phenomenon, scientific questions, and hypotheses have been examined thoroughly and have stood firm in the face of serious scientific debate and careful evaluation of alternative explanations.

As a result of the growing recognition that climate change is under way and poses serious risks for both human societies and natural systems, the question that decision makers are asking has expanded from "What is happening?" to "What is happening and what can we do about it?". Scientific research can help answer both of these important questions. In addition to the extensive body of research on the causes and consequences of climate change, there is a growing body of knowledge about technologies and policies that can be used to limit the magnitude of future climate change, a smaller but expanding understanding of the steps that can be taken to adapt to climate change, and a growing recognition that climate change will need to be considered in actions and decisions across a wide range of sectors and interests. Advice on prudent short-term actions and long-term strategies in these three areas can be found in the companion reports *Limiting the Magnitude of Future Climate Change* (NRC, 2010c), *Adapting to the Impacts of Climate Change* (NRC, 2010a), and *Informing an Effective Response to Climate Change* (NRC, 2010b).

This report, *Advancing the Science of Climate Change* (Box S.1), reviews the current scientific evidence regarding climate change and examines the status of the nation's scientific research efforts. It also describes the critical role that climate change science, broadly defined, can play in developing knowledge and tools to assist decision makers as they act to respond to climate change. The report explores seven crosscutting research themes that should be included in the nation's climate change research enterprise and recommends a number of actions to advance the science of climate

BOX S.1
Statement of Task and Report Overview

The Panel on Advancing the Science of Climate Change, one of five groups convened under the *America's Climate Choices* suite of activities (see Foreword), was charged to address the following question: "What can be done to better understand climate change and its interactions with human and ecological systems?" The panel was asked to provide a concise overview of past, present, and future climate change, including its causes and its impacts, then to recommend steps to advance our current understanding, including new observations, research programs, next-generation models, and the physical and human assets needed to support these and other activities. The panel was instructed to consider both the natural climate system and the human activities responsible for driving climate change and altering the vulnerability of different regions, sectors, and populations as a single system, and to consider the scientific advances needed to better understand the effectiveness of actions taken to limit the magnitude of future climate change and to adapt to the impacts of climate change. (The full statement of task of the Panel on Advancing the Science of Climate Change can be found in Appendix B, and its membership can be found in Appendix A; full biographical sketches of the panel members can be found in Appendix C.)

In response to this charge, the panel first assessed what science has learned about climate change and its impacts across a variety of sectors, as well as what is known about options for responding to climate change in those sectors. An overview of this analysis is provided in Chapter 2, and the details can be found in the technical chapters (Chapters 6-17) that compose Part II of the report. The panel also identified scientific advances that could improve our present understanding of climate change or the effectiveness of actions taken to limit its magnitude or adapt to its impacts. Seven crosscutting research themes, presented in Chapter 4, were identified based on this analysis. Finally, the panel evaluated actions that could be taken to achieve these scientific advances, including the physical and human assets required. Chapter 5 includes the panel's recommendations on these important topics.

change—a science that includes and increasingly integrates across the physical, biological, social, health, and engineering sciences. Overall, the report concludes that

1. Climate change is occurring, is caused largely by human activities, and poses significant risks for a broad range of human and natural systems; and
2. The nation needs a comprehensive and integrated climate change science enterprise, one that not only contributes to our fundamental understanding of climate change but also informs and expands America's climate choices.

WHAT WE KNOW ABOUT CLIMATE CHANGE

Conclusion 1: Climate change is occurring, is caused largely by human activities, and poses significant risks for—and in many cases is already affecting—a broad range of human and natural systems.

This conclusion is based on a substantial array of scientific evidence, including recent work, and is consistent with the conclusions of recent assessments by the U.S. Global Change Research Program (e.g., USGCRP, 2009a), the Intergovernmental Panel on Climate Change's Fourth Assessment Report (IPCC, 2007a-d), and other assessments of the state of scientific knowledge on climate change. Both our assessment—the details of which can be found in Chapter 2 and Part II (Chapters 6-17) of this report—and these previous assessments place high or very high confidence[1] in the following findings:

- Earth is warming. Detailed observations of surface temperature assembled and analyzed by several different research groups show that the planet's average surface temperature was 1.4°F (0.8°C) warmer during the first decade of the 21st century than during the first decade of the 20th century, with the most pronounced warming over the past three decades. These data are corroborated by a variety of independent observations that indicate warming in other parts of the Earth system, including the cryosphere (snow- and ice-covered regions), the lower atmosphere, and the oceans.
- Most of the warming over the last several decades can be attributed to human activities that release carbon dioxide (CO_2) and other heat-trapping greenhouse gases (GHGs) into the atmosphere. The burning of fossil fuels—coal, oil, and natural gas—for energy is the single largest human driver of climate change, but agriculture, forest clearing, and certain industrial activities also make significant contributions.
- Natural climate variability leads to year-to-year and decade-to-decade fluctuations in temperature and other climate variables, as well as substantial regional differences, but cannot explain or offset the long-term warming trend.
- Global warming is closely associated with a broad spectrum of other changes, such as increases in the frequency of intense rainfall, decreases in Northern Hemisphere snow cover and Arctice sea ice, warmer and more frequent hot days and nights, rising sea levels, and widespread ocean acidification.

[1] As discussed in Appendix D, high confidence indicates an estimated 8 out of 10 or better chance of a statement being correct, while very high confidence (or a statement than an ourcome is "very likely") indicates a 9 out of 10 or better chance.

- Human-induced climate change and its impacts will continue for many decades, and in some cases for many centuries. Individually and collectively, these changes pose risks for a wide range of human and environmental systems, including freshwater resources, the coastal environment, ecosystems, agriculture, fisheries, human health, and national security, among others.
- The ultimate magnitude of climate change and the severity of its impacts depend strongly on the actions that human societies take to respond to these risks.

Despite an international agreement to stabilize GHG concentrations "at levels that would avoid dangerous anthropogenic interference with the climate system" (UNFCCC, 1992), global emissions of CO_2 and several other GHGs continue to increase. Projections of future climate change, which are based on computer models of how the climate system would respond to different scenarios of future human activities, anticipate an additional warming of 2.0°F to 11.5°F (1.1°C to 6.4°C) over the 21st century. A separate National Research Council (NRC) report, *Climate Stabilization Targets: Emissions, Concentrations, and Impacts over Decades to Millennia* (NRC, 2010i), provides an analysis of expected impacts at different magnitudes of future warming.

In general, it is reasonable to expect that the magnitude of future climate change and the severity of its impacts will be larger if actions are not taken to reduce GHG emissions and adapt to its impacts. However, as with all projections of the future, there will always be some uncertainty regarding the details of future climate change. Several factors contribute to this uncertainty:

- Projections of future climate change depend strongly on how human societies decide to produce and use energy and other resources in the decades ahead.
- Human-caused changes in climate overlap with natural climate variability, especially at regional scales.
- Certain Earth system processes—including the carbon cycle, ice sheet dynamics, and cloud and aerosol processes—are not yet completely understood or fully represented in climate models but could potentially have a strong influence on future climate changes.
- Climate change impacts typically play out at local to regional scales, but processes at these scales are not as well represented by models as continental- to global-scale changes.
- The impacts of climate change depend on how climate change interacts with other global and regional environmental changes, including changes in land use, management of natural resources, and emissions of other pollutants.
- The impacts of climate change also depend critically on the vulnerability and

adaptive capacity of human and natural systems, which can vary widely in space and time and generally are not as well understood as changes in the physical climate system.

Climate change also poses challenges that set it apart from other risks with which people normally deal. For example, many climate change processes have considerable inertia and long time lags, so it is mainly future generations that will have to deal with the consequences (both positive and negative) of decisions made today. Also, rather than smooth and gradual climate shifts, there is the potential that the Earth system could cross tipping points or thresholds that result in abrupt changes. Some of the greatest risks posed by climate change are associated with these abrupt changes and other climate "surprises" (unexpected changes or impacts), yet the likelihood of such events is not well known. Moreover, there has been comparatively little research on the impacts that might be associated with "extreme" climate change—for example, the impacts that could be expected if global temperatures rise by 10°F (6°C) or more over the next century. Thus, while it seems clear that the Earth's future climate will be unlike the climate that ecosystems and human societies have become accustomed to during the last 10,000 years, the exact magnitude of future climate change and the nature of its impacts will always remain somewhat uncertain.

Decision makers of all types, including businesses, governments, and individual citizens, are beginning to take actions to reduce the risks posed by climate change—including actions to limit its magnitude and actions to adapt to its impacts. Effective management of climate risks will require decision makers to take actions that are flexible and robust, to learn from new knowledge and experience, and to adjust future actions accordingly. The long time lags associated with climate change and the presence of differential vulnerabilities and capacities to respond to climate change likewise represent formidable management challenges. These challenges also have significant implications for the nation's climate science enterprise.

A NEW ERA OF CLIMATE CHANGE RESEARCH

Conclusion 2: The nation needs a comprehensive and integrative climate change science enterprise, one that not only contributes to our fundamental understanding of climate change but also informs and expands America's climate choices.

Research efforts over the past several decades have provided a wealth of information to decision makers about the known and potential risks posed by climate change.

Experts from a diverse range of disciplines have also identified and developed a variety of actions that could be taken to limit the magnitude of future climate change or adapt to its impacts. However, much remains to be learned. Continued investments in scientific research can be expected to improve our understanding of the causes and consequences of climate change. In addition, the nation's research enterprise could potentially play a much larger role in addressing questions of interest to decision makers as they develop, evaluate, and execute plans to respond to climate change. Because decisions always involve value judgments, science cannot prescribe the decisions that should be made. However, scientific research can play a key role by informing decisions and by expanding and improving the portfolio of available options.

Crosscutting Themes for Climate Change Research

This report identifies seven crosscutting research themes, grouped into three general categories, that collectively span the most critical research needs for understanding climate change and for informing and supporting effective responses to it.

Research to Improve Understanding of Human-Environment Systems

1. *Climate Forcings, Responses, Feedbacks and Thresholds in the Earth System.* Some examples of research needs that fall under this theme include improved understanding of climate sensitivity, ice sheet dynamics, climate-carbon interactions, crop and ecosystems responses to climate changes (in interaction with other stresses), and changes in extreme events.
2. *Climate-Related Human Behaviors and Institutions.* Some examples include improved understanding of human behavior and decision making in the climate context, institutional impediments to limiting or adaptation responses, determinants of consumption, and drivers of climate change.

Research to Support Effective Responses to Climate Change

3. *Vulnerability and Adaptation Analyses of Coupled Human-Environment Systems.* Some examples include developing methods and indicators for assessing vulnerability[2] and developing and assessing integrative management ap-

[2] Vulnerability is the degree to which a system is susceptible to, or unable to cope with, adverse effects of climate change, including changes in climate variability and extremes. Vulnerability is a function of the character, magnitude, and rate of climate variation to which a system is exposed, its sensitivity, and its adaptive capacity (NRC, 2010a).

proaches to respond effectively to the impacts of climate change on coasts, freshwater resources, food production systems, human health, and other sectors.

4. *Research to Support Strategies for Limiting Climate Change.* Some examples include developing new and improved technologies for reducing GHG emissions (such as enhanced energy efficiency technologies and wind, solar, geothermal-based, and other energy sources that emit few or no GHGs), assessing alternative methods to limit the magnitude of future climate change (such as modifying land use practices to increase carbon storage or geoengineering[3] approaches), and developing improved analytical frameworks and participatory approaches to evaluate trade-offs and synergies among actions taken to limit climate change.

5. *Effective Information and Decision-Support Systems.* Some examples include research on risk communication and risk-management processes; improved understanding of individual, societal, and institutional factors that facilitate or impede decision making; analysis of information needs and existing decision-support activities, and research to improve decision-support products, processes, and systems.

Tools and Approaches to Improve Both Understanding and Responses

6. *Integrated Climate Observing Systems.* Some examples include efforts to ensure continuity of existing observations; develop new observational capacity for critical physical, ecological, and social variables; ensure that current and planned observations are sufficient both to continue building scientific understanding of and support more effective responses to climate change (including monitoring to assess the effectiveness of responses); and ensure adequate emphasis and support for data assimilation, analysis, and management.

7. *Improved Projections, Analyses, and Assessments.* Some examples include advanced models for analysis and projections of climate forcing, responses, and impacts, especially at regional scales; and integrated assessment models and approaches—both quantitative and nonquantitative—for evaluating the

[3] The term "geoengineering" refers to deliberate, large-scale manipulations of Earth's environment designed to offset some of the harmful consequences of GHG-induced climate change. Geoengineering encompasses two very different classes of approaches: CO_2 removal and solar radiation management. See Chapter 15 for details.

advantages and disadvantages of, and the trade-offs and co-benefits[4] among, various options for responding to climate change.

These seven themes and the range of research questions within them are explored in Chapter 4, and additional discussion of specific research needs can be found in Chapters 6-17. Because progress in any one of these themes is related to progress in others, all seven will need to be pursued simultaneously or at least iteratively. The nation currently has the capabilities and capacity to make incremental progress in some of these key research areas, but making more dramatic improvements in our understanding of and ability to respond to climate change will require several fundamental alterations in the support for and organization and conduct of climate change research.

RECOMMENDATIONS

Recommendation 1: The nation's climate change research enterprise should include and integrate disciplinary and interdisciplinary research across the physical, social, biological, health, and engineering sciences; focus on fundamental, use-inspired research that contributes to both improved understanding and more effective decision making; and be flexible in identifying and pursuing emerging research challenges.

Climate change research needs to be integrative and interdisciplinary. Climate change involves many aspects of the Earth system, as well a wide range of human activities, and both climate change and actions taken to respond to climate change interact in complex ways with other global and regional environmental changes. Understanding climate change, its impacts, and potential responses thus inherently requires integration of knowledge bases from many different scientific disciplines, including the physical, social, biological, health, and engineering sciences, and across different spatial scales of analysis, from local to global. Developing the science to support choices about climate change also requires engagement of decision makers and other stakeholders, as discussed below.

Climate change research should focus on fundamental, use-inspired research. This report recognizes the need for scientific research to both improve understanding of climate changes and assist in decision making related to climate change. In categorizing these types of scientific research, we found that terms such as "pure," "basic," "applied," and "curiosity driven" have different definitions across communities, are as likely to cause

[4] A co-benefit refers to an additional benefit resulting from an action undertaken to achieve a particular purpose, but that is not directly related to that purpose.

confusion as to advance consensus, and are of limited value in discussing climate change. More compelling, however, is the categorization offered by Stokes (1997), who argues that two questions should be asked of a research topic: Does it contribute to fundamental understanding? Can it be expected to be useful? Research that can answer yes to both of these questions, or "fundamental, use-inspired research," warrants special priority in the realm of climate change research.

Climate change research should support decision making at local, regional, national, and international levels. Many choices about how to respond to climate change fundamentally involve values and ethics and, thus, cannot be based on science alone. However, scientific research can inform and guide climate-related decisions in a variety of ways. Continued research on the causes, mechanisms, and consequences of climate change will help clarify the risks that climate changes pose to human and natural systems. Science can help identify new options and strategies for limiting the magnitude of climate change or adapting to its impacts, as well as help improve existing options. Science also plays the key role of evaluating the advantages and disadvantages associated with different responses to climate change, including unintended consequences, trade-offs, and co-benefits among different sets of actions. Finally, scientific research on new, more effective information-sharing and decision-making processes and tools can assist decision making.

Climate change research needs to be a flexible enterprise, able to respond to changing knowledge needs and support adaptive risk management and iterative decision making. Many resource and infrastructure decisions, from storm sewer planning to crop planting dates, will be made in the context of continuously evolving climate conditions as well as ongoing changes in other environmental and human systems. Decision makers would thus be well advised to employ iterative and adaptive risk-management[5] strategies as they make climate-related decisions. The nation's scientific enterprise will be increasingly called upon to provide the up-to-date, decision-specific information that such strategies require. Furthermore, as actions to limit and adapt to climate change—many of them never tried before—are taken, decision makers will need to understand and take the consequences of these actions into account. This will place increased demands on scientific monitoring, modeling, and analysis activities. To meet these evolving needs, the nation's climate research enterprise will itself need to be flexible and adaptive, and to practice "learning by doing" as it provides decision makers with the information they need to make effective decisions.

[5] Adaptive (or iterative) risk management refers to an ongoing decision-making process that takes known and potential risks and uncertainties into account and periodically updates and improves plans and strategies as new information becomes available.

Recommendation 2: Research priorities for the federal climate change research program should be set within each of the seven crosscutting research themes outlined above. Priorities should be set using the following three criteria:

1. **Contribution to improved understanding;**
2. **Contribution to improved decision making; and**
3. **Feasibility of implementation, including scientific readiness and cost.**

Progress in the seven crosscutting research themes would advance the science of climate change in ways that are responsive to the nation's needs for information. Progress in all seven themes is needed, but priorities will ultimately need to be set within them. The development of more comprehensive, exhaustive, and prioritized lists of specific research needs within each theme should involve members of the relevant research communities, taking into account that it is far more challenging to identify and evaluate key uncertainties and information needs in understudied areas than in established research fields. It is critical that priority setting also include the perspective of societal need, which necessitates input from decision makers and other stakeholders. Finally, feasibility of implementation, including scientific readiness, cost, and other practical, institutional, and managerial concerns, need to be considered to ensure effectiveness. Chapter 5 provides additional details on priority setting.

Recommendation 3: The federal climate change research program, working in partnership with other relevant domestic and international bodies, should redouble efforts to develop, deploy, and maintain a comprehensive observing system that can support all aspects of understanding and responding to climate change.

Long-term, stable, and well-calibrated observations across a spectrum of human and environmental systems are essential for diagnosing and understanding climate change and its impacts. The suite of needed observations includes measurements of physical, biological, ecological, and socioeconomic processes, and includes both remotely sensed and in situ data across a range of scales. Observations are also critical for developing, initializing, and testing models of future human and environmental changes, and for monitoring and improving the effectiveness of actions taken to respond to climate change. However, many observing systems are in decline, putting our ability to monitor and understand future changes at risk. Stemming this decline should be a top priority. Responding effectively to climate change will also require new observational capabilities to monitor and evaluate progress in limiting climate change and adapting to its impacts, as well as to monitor known risks and identify new or emerging risks as climate change unfolds. All of these data need to be archived, checked for quality, and made readily accessible to a wide range of users, keeping in mind that many climate-related decisions require information of many different types and at different scales.

Hence, there is a critical need to develop, deploy, and maintain a robust infrastructure for collecting and archiving a wide range of climate and climate-related data, integrating data collected on different systems, and ensuring that the data are reliable, accurate, and easily accessible. The federal climate research program is the obvious entity for leading the development of such a coordinated, comprehensive, and integrated climate observing system, and ensuring that the system facilitates both improved understanding and more effective decision making. However, other relevant partners, including the domestic and international research communities and action-oriented programs at all spatial scales, also need to be engaged in system design, deployment, and maintenance. Critical steps include reviewing current and planned observational assets, identifying critical climate monitoring and measurement needs, and developing a comprehensive strategy to meet these needs, including data management and stewardship activities. The climate observing system should be coordinated with other environmental and social data collection efforts to take advantage of synergies and ensure interoperability. Finally, careful balancing is needed to ensure that resources are used effectively, that investments in one kind of observation do not impede the ability to invest in others, and that the full spectrum of most critical observations are collected and made available for diverse uses.

Recommendation 4: The federal climate change research program should work with the international research community and other relevant partners to support and develop advanced models and other analytical tools to improve understanding and assist in decision making related to climate change.

Enhanced modeling capabilities, including improved representations of underlying human and Earth system processes, are needed to support efforts to understand, limit, and adapt to climate change. Improvements are especially needed in integrated Earth system models to allow more thorough examination of climate-related feedbacks and the possibility of abrupt changes, regional-scale projections of climate change and its impacts, and integrated assessment activities that explicitly link coupled human-environment systems. Also critical are more informative and comprehensive scenarios of future human activities that influence or are influenced by climate, and models and analyses of the effects of different actions (and combinations of actions) taken to adapt to climate change or limit its magnitude. Information on decadal time scales is particularly relevant to many climate-related decisions. Improvements in all of these areas go hand in hand with improvements in fundamental understanding, for example of processes and mechanism of regional climate variability and change. Improvements in models and other analytical tools also support decision making by allowing more thorough and comprehensive analyses of the economic, social, and environmental consequences of climate change and of actions taken to respond.

Adequate computational resources are critical for Earth system models, regional climate models, integrated assessment models, impacts-adaptation-vulnerability models, climate forcing scenario development efforts, and other tools for projecting future changes. Near-term progress would benefit from improvements in and access to high-performance computing. As with observations, efforts are needed to ensure that the output from models, analyses, and assessments are appropriately managed, undergo continuing development, and actually inform decision-making processes at appropriate levels. The federal climate change research program should lead the development of a strategy for dramatically improving and integrating regional climate modeling, global Earth system models, and various integrated assessment, vulnerability, impact, and adaptation models. To ensure the success of this strategy, the program and its partners should take steps to increase the computational and human resources available to support a wide range of modeling efforts and ensure that these efforts are linked with both the national observing system strategy and with efforts to support effective decision making.

Recommendation 5: A single federal interagency program or other entity should be given the authority and resources to coordinate and implement an integrated research effort that supports improving both understanding of and responses to climate change. If several key modifications are made, the U.S. Global Change Research Program could serve this role.

There are several ways that climate change research at the federal level could be organized to achieve a broad, integrated, and decision-relevant research effort capable of coordinating and leading the nation's broader climate change research enterprise. After reviewing several options (see Chapter 5), the panel came to the conclusion that the Global Change Research Act of 1990, which established the USGCRP, provides the legislative authority needed to implement a strategically integrated climate change research program (Global Change Research Act, P.L. 101-606, Title 15, Chapter 56A, 1990). The USGCRP is capable of implementing the other recommendations offered in this report, provided that several key modifications are made to its current structure, goals, and practices.

The USGCRP has been highly successful on many fronts, including in elucidating the causes and some of the impacts of climate change. However, institutional issues and other factors have resulted in critical knowledge gaps, including a number of the research needs identified in this report (see also NRC, 2009k). Other persistent criticisms of the program include inadequate support for and progress in social science research, decision-support activities, and integration across disciplines. To better support improvements in our understanding of climate change and effective responses to it, the USGCRP will need to establish improved mechanisms for identifying and addressing these and other weaknesses and gaps, as well as the barriers that give rise to such gaps. The USGCRP also needs to establish more effective mechanisms to interact with decision makers and other stakeholders.

To ensure progress in the seven key research themes identified above, and implement the other recommendations offered in this report, the USGCRP will need high-level leadership. This includes effective and forward-looking leadership of the program itself as well as supportive leaders in its partner agencies. To effectively shape and govern an interagency research effort, the program also needs expanded budgeting oversight and authority to coordinate and prioritize climate change research across agencies. The importance of effective leadership, with adequate support and programmatic and budgetary authority, has been recognized in several NRC reviews of the USGCRP (see Chapter 5 and Appendix E). Support and oversight from institutions with overarching authority, such as the Office of Management and Budget, the Office of Science and Technology Policy, and relevant congressional committees, will be essential, as will a comprehensive, inclusive, and ongoing strategic planning process.

Recommendation 6: The federal climate change research program should be formally linked with action-oriented response programs focused on limiting the magnitude of future climate change, adapting to the impacts of climate change, and informing climate-related actions and decisions, and, where relevant, should develop partnerships with other research and decision-making entities working at local to international scales.

The engagement of institutions at all levels and of all sorts—academic, governmental, private-sector, and not-for-profit—will be needed to meet the challenges of climate change. By working collaboratively with action-oriented programs, both at the federal level and across the country, the federal climate change research program can help ensure that the nation's responses to climate change are as effective as possible. For example, scientific knowledge about the impacts of climate change and about the vulnerability and adaptive capacity of different human and environmental systems—which typically requires analysis focused at local to regional scales—is critical for developing and assessing adaptation measures. Likewise, research on human behavior, institutions, and decision-making processes, products, and tools can contribute to programs designed to inform decision makers and other stakeholders about climate change (including the emerging federal approach to provide "climate services"). Scientific research also underpins the development, implementation, and assessment of policies and technologies intended to limit the magnitude of climate change and, as such, is an important partner for technology development programs such as the Climate Change Technology Program. Such an "end-to-end" climate change research enterprise was also called for in the recent NRC reports *Restructuring Federal Climate Research to Meet the Challenges of Climate Change* (NRC, 2009k) and *Informing Decisions in a Changing Climate* (NRC, 2009g). Achieving this vision will require high-level coordination, ideally through formal mechanisms, between the research program and action-oriented programs at the federal level. It will also requite new and improved mechanisms for engaging with both research and action-oriented programs at state and local levels. Finally, partnerships with the international research community will be essential for maximizing the effectiveness of domestic investments in climate change research.

Recommendation 7: Congress, federal agencies, and the federal climate change research program should work with other relevant partners (including universities, state and local governments, the international research community, the business community, and other nongovernmental organizations) to expand and engage the human capital needed to carry out climate change research and response programs.

The scale, importance, and complexity of the climate challenge implies a critical need to increase the workforce performing fundamental and decision-relevant climate research, implementing responses to climate change, and working at the interface between science and decision making. Thanks to more than three decades of research on climate change, the disciplinary research community in the United States and elsewhere is strong, at least in research areas that have received significant emphasis and support. However, the more integrative and decision-relevant research program described in this report will require expanded intellectual capacity in several previously neglected fields as well as in interdisciplinary research areas. Responding effectively to climate change will also require new interdisciplinary intellectual capacity among state, local, and national government agencies, universities, and other public and private research labs, as well as among science managers coordinating efforts to advance the science of climate change. Building and mobilizing this broad research community will require a concerted and coordinated effort.

The federal climate research program, federal agencies and laboratories, universities, professional societies, and other elements of the nation's research enterprise should use a variety of mechanisms to encourage and facilitate interdisciplinary and integrative research. At the national scale, institutional changes are needed in federal research and mission agencies to increase the focus on interdisciplinary and decision-relevant research throughout government and in the nationwide research efforts the agencies support. Additional venues for presentation and publication of interdisciplinary and decision-relevant climate research are also needed, as well as professional organizations that support and reward these efforts. Finally, state and local governments, corporations, and nongovernmental organizations should be key partners in developing and engaging a workforce to implement the national climate research strategy. Further discussion of the actions needed to educate and train future generations of scientists, engineers, technicians, managers, and decision makers for responding to climate change can be found in the companion report *Informing an Effective Response to Climate Change* (NRC, 2010b).

Part I

CHAPTER ONE

Introduction: Science for Understanding and Responding to Climate Change

Humans have always been influenced by climate. Despite the wealth and technology of modern industrial societies, climate still affects human well-being in fundamental ways. Climate influences, for example, where people live, what they eat, how they earn their livings, how they move around, and what they do for recreation. Climate regulates food production and water resources and influences energy use, disease transmission, and other aspects of human health and well-being. It also influences the health of ecosystems that provide goods and services for humans and for the other species with which we share the planet.

In turn, human activities are influencing climate. As discussed in the following chapters, scientific evidence that the Earth is warming is now overwhelming. There is also a multitude of evidence that this warming results primarily from human activities, especially burning fossil fuels and other activities that release heat-trapping greenhouse gases (GHGs) into the atmosphere. Projections of future climate change indicate that Earth will continue to warm unless significant and sustained actions are taken to limit emissions of GHGs. Increasing temperatures and GHG concentrations are driving a multitude of related and interacting changes in the Earth system, including decreases in the amounts of ice stored in mountain glaciers and polar regions, increases in sea level, changes in ocean chemistry, and changes in the frequency and intensity of heat waves, precipitation events, and droughts. These changes in turn pose significant risks to both human and ecological systems. Although the details of how the future impacts of climate change will unfold are not as well understood as the basic causes and mechanisms of climate change, we can reasonably expect that the consequences of climate change will be more severe if actions are not taken to limit its magnitude and adapt to its impacts.

Scientific research will never completely eliminate uncertainties about climate change and its risks to human health and well-being, but it can provide information that can be helpful to decision makers who must make choices in the face of risks. In 2008, the U.S. Congress asked the National Academy of Sciences to "investigate and study the serious and sweeping issues relating to global climate change and make recommendations regarding what steps must be taken and what strategies must be adopted in response . . . including the science and technology challenges thereof." This report is part of the resulting study, called *America's Climate Choices* (see Foreword). In the chapters that follow, this report reviews what science has learned about climate change and its causes and consequences across a variety of sectors. The report also identifies scientific advances that could improve understanding of climate change and the effectiveness of actions taken to limit the magnitude of future climate change or adapt to its impacts. Finally, the report identifies the activities and tools needed to make these scientific advances and the physical and human assets needed to support these activities (see Appendix B for the detailed statement of task). Companion reports provide information and advice on *Limiting the Magnitude of Future Climate Change* (NRC, 2010c), *Adapting to the Impacts of Climate Change* (NRC, 2010a), and *Informing an Effective Response to Climate Change* (NRC, 2010b).

SCIENTIFIC LEARNING ABOUT CLIMATE CHANGE

Climate science, like all science, is a process of collective learning that proceeds through the accumulation of data; the formulation, testing, and refinement of hypotheses; the construction of theories and models to synthesize understanding and generate new predictions; and the testing of hypotheses, theories, and models through experiments or other observations. Scientific knowledge builds over time as theories are refined and expanded and as new observations and data confirm or refute the predictions of current theories and models. Confidence in a theory grows if it survives this rigorous testing process, if multiple lines of evidence lead to the same conclusion, or if competing explanations can be ruled out.

In the case of climate science, this process of learning extends back more than 150 years, to mid-19th-century attempts to explain what caused the ice ages, which had only recently been discovered. Several hypotheses were proposed to explain how thick blankets of ice could have once covered much of the Northern Hemisphere, including changes in solar radiation, atmospheric composition, the placement of mountain ranges, and volcanic activity. These and other ideas were tested and debated by the scientific community, eventually leading to an understanding (discussed in detail in Chapter 6) that ice ages are initiated by small recurring variations in Earth's orbit around the Sun. This early scientific interest in climate eventually led scientists working in the late 19th century to recognize that carbon dioxide (CO_2) and other GHGs have a profound effect on the Earth's temperature. A Swedish scientist named Svante Arrhenius was the first to hypothesize that the burning of fossil fuels, which releases CO_2, would eventually lead to global warming. This was the beginning of a more than

100-year history of ever more careful measurements and calculations to pin down exactly how GHG emissions and other factors influence Earth's climate (Weart, 2008).

Progress in scientific understanding, of course, does not proceed in a simple straight line. For example, calculations performed during the first decades of the 20th century, before the behavior of GHGs in the atmosphere was understood in detail, suggested that the amount of warming from elevated CO_2 levels would be small. More precise experiments and observations in the mid-20th century showed that this was not the case, and that increases in CO_2 or other GHGs could indeed cause significant warming. Similarly, a scientific debate in the 1970s briefly considered the possibility that human emissions of aerosols—small particles that reflect sunlight back to space—might lead to a long-term cooling of the Earth's surface. Although prominently reported in a few news magazines at the time, this speculation did not gain widespread scientific acceptance and was soon overtaken by new evidence and refined calculations showing that warming from emissions of CO_2 and other GHGs represented a larger long-term effect on climate.

Thus, scientists have understood for a long time that the basic principles of chemistry and physics predict that burning fossil fuels will lead to increases in the Earth's average surface temperature. Decades of observations and research have tested, refined, and extended that understanding, for example, by identifying other factors that influence climate, such as changes in land use, and by identifying modes of natural variability that modulate the long-term warming trend. Detailed process studies and models of the climate system have also allowed scientists to project future climate changes. These projections are based on scenarios of future GHG emissions from energy use and other human activities, each of which represents a different set of choices that societies around the world might make. Finally, research across a broad range of scientific disciplines has improved our understanding of how the climate system interacts with other environmental systems and with human systems, including water resources, agricultural systems, ecosystems, and built environments.

Uncertainty in Scientific Knowledge

From a philosophical perspective, science never *proves* anything—in the manner that mathematics or other formal logical systems prove things—because science is fundamentally based on observations. Any scientific theory is thus, in principle, subject to being refined or overturned by new observations. In practical terms, however, scientific uncertainties are not all the same. Some scientific conclusions or theories have been so thoroughly examined and tested, and supported by so many independent observations and results, that their likelihood of subsequently being found to be wrong is vanishingly small. Such conclusions and theories are then regarded as settled facts. This is the case for the conclusions that the Earth system is warming and that much of this warming is very likely due to human activities. In other cases, particularly for matters that are at the leading edge of active research, uncertainties may be substantial and important. In these cases, care must be taken not to draw stronger conclusions than warranted by the available evidence.

The characterization of uncertainty is thus an important part of the scientific enterprise. In some areas of inquiry, uncertainties can be quantified through a long sequence of repeated observations, trials, or model runs. For other areas, including many aspects of climate change research, precise quantification of uncertainty is not always possible due to the complexity or uniqueness of the system being studied. In these cases, researchers adopt various approaches to subjectively but rigorously assess their degree of confidence in particular results or theories, given available observations, analyses, and model results. These approaches include estimated uncertainty ranges (or error bars) for measured quantities and the estimated likelihood of a particular result having arisen by chance rather than as a result of the theory or phenomenon being tested. These scientific characterizations of uncertainty can be misunderstood, however, because for many people "uncertainty" means that little or nothing is known, whereas in scientific parlance uncertainty is a way of describing how precisely or how confidently something is known. To reduce such misunderstandings, scientists have developed explicit techniques for conveying the precision in a particular result or the confidence in a particular theory or conclusion to policy makers (see Box 1.1).

A NEW ERA OF CLIMATE CHANGE SCIENCE: RESEARCH FOR UNDERSTANDING AND RESPONDING TO CLIMATE CHANGE

In the process of scientific learning about climate change, it has become evident that climate change holds significant risks for people and the natural resources and ecosystems on which they depend. In some ways, climate change risks are different from many other risks with which people normally deal. For example, as discussed in Chapters 2 and 3, climate change processes have considerable inertia and long time lags. The actions of today, therefore, will be reflected in climate system changes several decades to centuries from now. Future generations will be exposed to risks, some potentially severe, because of today's actions, and in some cases these changes will be irreversible. Likewise, climate changes can be abrupt—they have the potential to cross tipping points or thresholds that result in large changes or impacts. The likelihood of such abrupt changes is not well known, however, which makes it difficult to quantify

BOX 1.1
Uncertainty Terminology

In assessing and reporting the state of knowledge about climate change, scientists have devoted serious debate and discussion to appropriate ways of expressing uncertainty to policy makers (Moss and Schneider, 2000). Recent climate change assessment reports have adopted specific procedures and terminology to describe the degree of confidence in specific conclusions or the estimated likelihood of a certain outcome (see, e.g., Manning et al., 2004). For example, a statement that something is "very likely" in the assessments by the Intergovernmental Panel on Climate Change indicates an estimated 9 out of 10 or better chance that a certain outcome will occur (see Appendix D).

In estimating confidence, scientific assessment teams draw on information about "the strength and consistency of the observed evidence, the range and consistency of model projections, the reliability of particular models as tested by various methods, and, most importantly, the body of work addressed in earlier synthesis and assessment reports" (USGCRP, 2009a). Teams are also encouraged to provide "traceable accounts" of how these estimates were constructed, including important lines of evidence used, standards of evidence applied, approaches taken to combining and reconciling multiple lines of evidence, explicit explanations of any statistical or other methods used, and identification of critical uncertainties. In general, statements about the future are more uncertain than statements about observed changes or current trends, and it is easier to employ precise uncertainty language in situations where conclusions are based on extensive quantitative data or models than in areas where data are less extensive, important research is qualitative, or models are in an earlier stage of development.

In this report, *Advancing the Science of Climate Change*, when we draw directly on the statements of the formal national and international assessments, we adopt their terminology to describe uncertainty. However, because of the more concise nature and intent of this report, we do not attempt to quantify confidence and certainty about every statement of the science.

the risks posed by such changes. Climate change also interacts in complex ways with other ongoing changes in human and environmental systems. Society's decisions about land use and food production, for example, both affect and are affected by climate change.

On the basis of decades of scientific progress in understanding changes in the physical climate system and the growing evidence of the risks posed by climate change, many decision makers—including individuals, businesses, and governments at all levels—are either taking actions to respond to climate change or asking what actions they might take to respond effectively. Many of these questions center on what specific actions might to be taken to limit climate change by reducing emissions of

GHGs: what gases, from what sources, when and where, through what specific technology investments or changes in management practices, motivated and coordinated by what policies, with what co-benefits[1] or unintended consequences, and monitored and verified through what means? Other questions focus on the specific impacts that are expected and the actions that can be taken to prepare for and adapt to them, such as reducing vulnerabilities or improving society's coping and adaptive capacity.

This report explores what these emerging questions and decision needs imply for future scientific learning about climate change and for the scientific research enterprise. As the need for science expands to include both *improving understanding* and *informing and supporting decision making*, the production, synthesis, and translation of scientific knowledge into forms that are useful to decision makers becomes increasingly important. It may also imply a need to change scientific practices, with scientists working more closely with decision makers to improve the scientific decision support that researchers can offer. However, even with this decision focus, scientific knowledge cannot by itself specify or determine any choice. It cannot tell decision makers what they *should* do; their responsibilities, preferences, and values also influence their decisions. Science can inform decisions by describing the potential consequences of different choices, and it can contribute by improving or expanding available options, but it cannot say what actions are required or preferred.

REPORT OVERVIEW

This report describes what has been learned about climate change. It then identifies the most critical current research needs, including research needed to improve our understanding of climate change and its impacts and research related to informing decision makers and allowing them to respond more effectively to the challenges of climate change. As directed by the charge to the panel (see Appendix B), this report covers the broad scientific territory of understanding climate change and its interactions with humans and ecosystems, including responses to climate change. Thus, it spans the breadth of "climate change science," which in this report is defined to include research in the physical, social, ecological, environmental, health, and engineering sciences, as well as research that integrates these and other disciplines.

The following chapters, which are broken into two parts, discuss the contributions that climate change science has made and can make in advancing our understanding of climate change *and* in supporting climate-related decisions. The five chapters in Part

[1] A co-benefit refers to an additional benefit resulting from an action undertaken to achieve a particular purpose, but which is not directly related to that purpose.

I include the panel's conclusions, recommendations, and supporting analysis. Chapter 2 provides an overview of available scientific knowledge about climate change. This overview is drawn from the 12 technical chapters in Part II of the report, which provide more detailed and extensively referenced information on what science has learned about climate change and its interactions with key human and environmental systems. Chapter 3 examines some of the complexities and risks associated with climate change that emerge from what has been learned and discusses the role that scientific research can play in helping decision makers manage those risks. Chapter 4 describes seven crosscutting and integrative research themes that emerge from the panel's review of key scientific research needs (the details of which can be found in the final section of each of the chapters in Part II). Chapter 5, the final chapter in Part I, provides the panel's recommendations for advancing the science of climate change, including priority-setting, infrastructural, and organizational issues.

Broadly speaking, the report concludes that the causes and many of the consequences of climate change are becoming increasingly clear, and that additional research is needed both to continue to improve understanding of climate change and to support effective responses to it. This expanded research enterprise needs to be more integrative and interdisciplinary, will demand improved infrastructural support and intellectual capacity, and will need to be tightly linked to efforts to limit and adapt to climate change at all scales. In short, the report concludes that we are entering a new era of climate change research, one in which research is needed to understand not just where the world is headed, but also how the risks posed by climate change can be managed effectively.

CHAPTER TWO

What We Know About Climate Change and Its Interactions with People and Ecosystems

Over the past several decades, the international and national research communities have developed a progressively clearer picture of how and why Earth's climate is changing and of the impacts of climate change on a wide range of human and environmental systems. Research has also evaluated actions that could be taken—and in some cases are already being taken—to limit the magnitude of future climate change and adapt to its impacts. In the United States, a series of reports by the U.S. Global Change Research Program (USGCRP, also known as the Climate Change Science Program from 2001 to 2008) have synthesized the information specific to the nation, culminating in the report *Global Climate Change Impacts in the United States* (USGCRP, 2009a). Internationally, scientific information about climate change is periodically assessed by the Intergovernmental Panel on Climate Change (IPCC), most recently in 2007. Much has been learned, and this knowledge base is continuously being updated and expanded with new research results.

Our assessment of the current state of knowledge about global climate change, which is summarized in this chapter and described in detail in Part II of the report, leads to the following conclusion.

Conclusion 1: Climate change is occurring, is caused largely by human activities, and poses significant risks for—and in many cases is already affecting—a broad range of human and natural systems.

This conclusion is based on a substantial array of scientific evidence, including recent work, and is consistent with the conclusions of the IPCC's Fourth Assessment Report (IPCC, 2007a-d), recent assessments by the USGCRP (e.g., USGRP, 2009a), and other recent assessments of the state of scientific knowledge on climate change. Both our assessment and these previous assessments place high or very high confidence[1] in the following findings:

---

[1] As discussed in Appendix D, high confidence indicates an estimated 8 out of 10 or better chance of a statement being correct, while very high confidence (or a statement than an ourcome is "very likely") indicates a 9 out of 10 or better chance.

- Earth is warming. Detailed observations of surface temperature assembled and analyzed by several different research groups show that the planet's average surface temperature was 1.4°F (0.8°C) warmer during the first decade of the 21st century than during the first decade of the 20th century, with the most pronounced warming over the past three decades. These data are corroborated by a variety of independent observations that indicate warming in other parts of the Earth system, including the cryosphere (the frozen portions of Earth's surface), the lower atmosphere, and the oceans.
- Most of the warming over the last several decades can be attributed to human activities that release carbon dioxide (CO_2) and other heat-trapping greenhouse gases (GHGs) into the atmosphere. The burning of fossil fuels—coal, oil, and natural gas—for energy is the single largest human driver of climate change, but agriculture, forest clearing, and certain industrial activities also make significant contributions.
- Natural climate variability leads to year-to-year and decade-to-decade fluctuations in temperature and other climate variables, as well as substantial regional differences, but cannot explain or offset the long-term warming trend.
- Global warming is closely associated with a broad spectrum of other changes, such as increases in the frequency of intense rainfall, decreases in Northern Hemisphere snow cover and Arctic sea ice, warmer and more frequent hot days and nights, rising sea levels, and widespread ocean acidification.
- Human-induced climate change and its impacts will continue for many decades, and in some cases for many centuries. Individually and collectively, and in combination with the effects of other human activities, these changes pose risks for a wide range of human and environmental systems, including freshwater resources, the coastal environment, ecosystems, agriculture, fisheries, human health, and national security, among others.
- The ultimate magnitude of climate change and the severity of its impacts depend strongly on the actions that human societies take to respond to these risks.

The following sections elaborate on these statements and provide a concise, high-level overview of the current state of scientific knowledge about climate change in 12 critical areas of interest to a broad range of stakeholders:

- Changes in the climate system;
- Sea level rise and risk in the coastal environment;
- Freshwater resources;
- Ecosystems, ecosystem services, and biodiversity;

- Agriculture, fisheries, and food production;
- Public health;
- Cities and the built environment;
- Transportation systems;
- Energy systems;
- Solar radiation management;
- National and human security; and
- Designing, implementing, and evaluating climate policies.

The research progress in each of these topics is explored in additional detail in Part II of the report, but even those chapters are too brief to provide a comprehensive review of the very large body of research on these issues. Likewise, this report does not cover all scientific topics of interest in climate change research, only those of most immediate interest to decision makers. Readers interested in additional information should consult the extensive assessment reports completed by the USGCRP,[2] the IPCC,[3] the National Research Council (NRC),[4] and other groups, as well as the numerous scientific papers that have been published since their completion.

CHANGES IN THE CLIMATE SYSTEM[5]

Earth's physical climate system, which includes the atmosphere, oceans, cryosphere, and land surface, is complex and constantly evolving. Nevertheless, the laws of physics and chemistry ultimately govern the system, and can be used to understand how and why climate varies from place to place and over time.

The Greenhouse Effect is a Natural Phenomenon That Is Critical for Life as We Know It

GHGs—which include water vapor, carbon dioxide (CO_2), methane (CH_4), nitrous oxide (N_2O), and several others—are present in relatively low concentrations in the atmosphere, but, because of their ability to absorb and re-radiate infrared energy, they trap heat near the Earth's surface, keeping it much warmer than it would otherwise be (Figure 2.1). The atmospheric concentrations of GHGs have increased over the past two centuries as a result of human activities, especially the burning of the fossil

[2] *http://www.globalchange.gov/publications/reports/scientific-assessments/us-impacts*

[3] *http://www.ipcc.ch/publications_and_data/publications_and_data_reports.htm*

[4] *http://national-academies.org/climatechange/*

[5] For additional discussion and references, see Chapter 6 in Part II of the report.

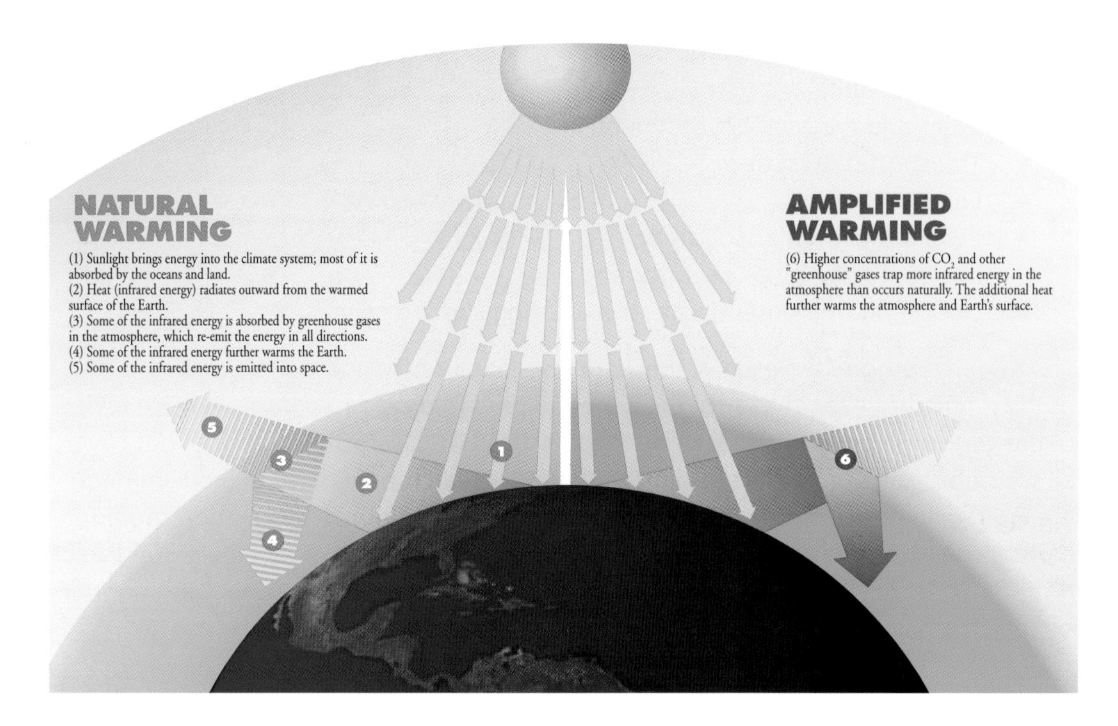

FIGURE 2.1 The greenhouse effect. SOURCE: Marian Koshland Science Museum of the National Academy of Sciences.

fuels—coal, oil, and natural gas—for energy. The increasing concentrations of GHGs are amplifying the natural greenhouse effect, causing Earth's surface temperature to rise. Human activities have also increased the number of aerosols (small liquid droplets or particles suspended in the atmosphere). Aerosols have a wide range of environmental effects, but on average they increase the amount of sunlight that is reflected back to space, a cooling effect that offsets some, but not all, of the warming induced by increasing GHG concentrations.

Earth Is Warming

There are many indications—both direct and indirect—that the climate system is warming. The most fundamental of these are thermometer measurements, enough of which have been collected over both land and sea to estimate changes in global average surface temperature since the mid- to late 19th century. A number of inde-

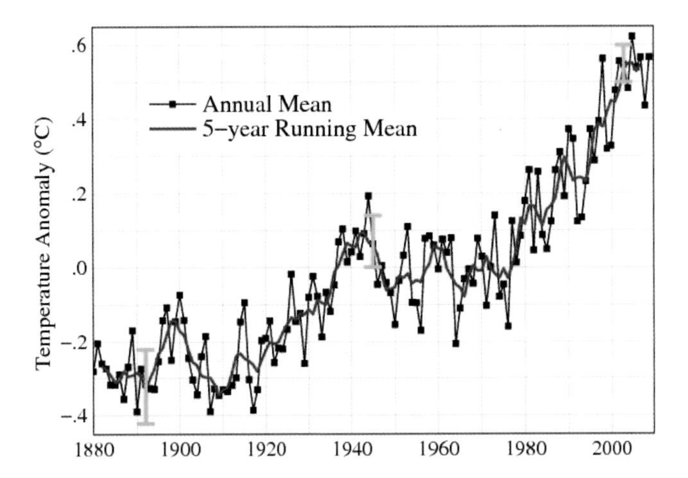

FIGURE 2.2 Global surface temperature change from 1880 to 2009 in degrees Celsius. The black curve shows annual average temperatures, the red curve shows a 5-year running average, and the green bars indicate the estimated uncertainty in the data during different periods of the record. For further details see Figure 6.13. SOURCE: NASA GISS (2010; based on Hansen et al., 2006, updated through 2009 at *http://data.giss.nasa.gov/gistemp/graphs/).*

pendent research teams collect, analyze, and correct for errors and biases in these data (for example, accounting for the "urban heat island" effect and changes in the instruments and methods used to measure ocean surface temperatures). Each group uses slightly different analysis techniques and data sources, yet the temperature estimates published by these groups are highly consistent with one another.

Surface thermometer measurements show the first decade of the 21st century was 1.4°F (0.8°C) warmer than the first decade of the 20th century (Figure 2.2). This warming has not been uniform, but rather it is superimposed on natural year-to-year and even decade-to-decade variations. Because of this natural variability, it is important to focus on trends over several decades or longer when assessing changes in the heat balance of the Earth. Physical factors also give rise to substantial spatial variations in the pattern of observed warming, with much stronger warming over the Arctic than over tropical latitudes and over land areas than over the ocean.

Other measurements of global temperature changes come from satellites, weather balloons, and ships, buoys, and floats in the ocean. Like surface thermometer measurements, these data have been analyzed by a number of different research teams around the world, corrected to remove errors and biases, and calibrated using independent observations. Ocean heat content measurements, which are taken from the top several hundred meters of the world's oceans, show a warming trend over the past several decades that is similar to the atmospheric warming trend in Figure 2.2.

Up until a few years ago, scientists were puzzled by the fact that the satellite-based record of atmospheric temperature trends seemed to disagree slightly with the data obtained from weather balloon-based measurements, and both seemed to be slightly inconsistent with surface temperature observations. Recently, researchers identified several small errors in both the satellite and weather balloon-based data sets, including errors caused by instrument replacements, changes in satellite orbits, and the effect of sunlight on the instruments carried by weather balloons. After correcting these errors, temperature records based on satellite, weather balloon, and ground-based measurements now agree within the estimated range of uncertainty associated with each type of observation.

The long-term trends in many other types of observations also provide evidence that Earth is warming. For example:

- Hot days and nights have become warmer and more frequent;
- Cold snaps have become milder and less frequent;
- Northern Hemisphere snow cover is decreasing;
- Northern Hemisphere sea ice is declining in both extent and average thickness;
- Rivers and lakes are freezing later and thawing earlier;
- Glaciers and ice caps are melting in many parts of the world (as described in more detail below); and
- Precipitation, ecosystems, and other environmental systems are changing in ways that are consistent with global warming (many of these changes are also described below).

Based on this diverse, carefully examined, and well-understood body of evidence, scientists are virtually certain that the climate system is warming. In addition, scientists have collected a wide array of "proxy" evidence that indicates how temperatures and other climate properties varied before direct measurements were available. These proxy data come from ice cores, tree rings, corals, lake sediments, boreholes, and even historical documents and paintings. A recent assessment of these data and the techniques used to analyze them concluded that the past few decades have been warmer than any other comparable period for at least the last 400 years, and possibly for the last 1,000 years or longer (NRC, 2006b).

The Climate System Exhibits Substantial Natural Variability

Earth's climate varies naturally on a wide range of timescales, from seasonal variations (such as a particularly wet spring, hot summer, or snowy winter) to geological timescales of millions or even billions of years. Careful statistical analyses have demonstrated that it is very unlikely[6] that natural variations in the climate system could have given rise to the observed global warming, especially over the last several decades. However, natural processes produce substantial seasonal, year-to-year, and even decade-to-decade variations that are superimposed on the long-term warming trend, as well as substantial regional differences. Improving understanding of natural variability patterns, and determining how they might change with increasing GHG emissions and global temperatures, is an important area of active research (see the end of this section and Chapter 6).

Natural climate variations can also be influenced by volcanic eruptions, changes in the output from the Sun, and changes in Earth's orbit around the Sun. Large volcanic eruptions, such as the eruption of Mount Pinatubo in 1991, can spew copious amounts of aerosols into the upper atmosphere. If the eruption is large enough, these aerosols can reflect enough sunlight back to space to cool the surface of the planet by a few tenths of a degree for several years.

The Sun's output has been measured precisely by satellites since 1979, and these measurements do not show any overall trend in solar output over this period. Prior to the satellite era, solar output was estimated by several methods, including methods based on long-term records of the number of sunspots observed each year, which is an indirect indicator of solar activity. These indirect methods suggest that there was a slight increase in solar energy received by the Earth during the first few decades of the 20th century, which may have contributed to the global temperature increase during that period (see Figure 2.2).

Perhaps the most dramatic example of natural climate variability is the ice age cycle. Detailed analyses of ocean sediments, ice cores, geologic landforms, and other data show that for at least the past 800,000 years, and probably the past several million years, the Earth has gone through long periods when temperatures were much colder than today and thick blankets of ice covered much of the Northern Hemisphere (including the areas currently occupied by the cities of Chicago, New York, and Seattle). These very long cold spells were punctuated by shorter, warm "interglacial" periods, including the last 10,000 years. Through a convergence of theory, observations, and

[6] As discussed in Appendix D, *very unlikely* indicates a less than 1 in 10 chance of a statement being incorrect.

modeling, scientists have deduced that the ice ages were initiated by small recurring variations in the Earth's orbit around the Sun.

GHG Emissions and Concentrations Are Increasing

Human activities have increased the concentration of CO_2 and certain other GHGs in the atmosphere. Detailed worldwide records of fossil fuel consumption indicate that fossil fuel burning currently releases over 30 billion tons of CO_2 into the atmosphere every year (Figure 2.3, blue curve). Tropical deforestation and other land use changes release an additional 3 to 5 billion tons every year.

Precise measurements of atmospheric composition at many sites around the world indicate that CO_2 levels are increasing, currently at a pace of almost 2 parts per million (ppm) per year. We know that this increase is largely the result of human activities because the chemical signature of the excess CO_2 in the atmosphere can be linked to the composition of the CO_2 in emissions from fossil fuel burning. Moreover, analyses of bubbles trapped in ice cores from Greenland and Antarctica reveal that atmospheric CO_2 levels have been rising steadily since the start of the Industrial Revolution (usually taken as 1750; see Figure 2.3, red curve). The current CO_2 level (388 ppm as of the end of 2009) is higher than it has been in at least 800,000 years.

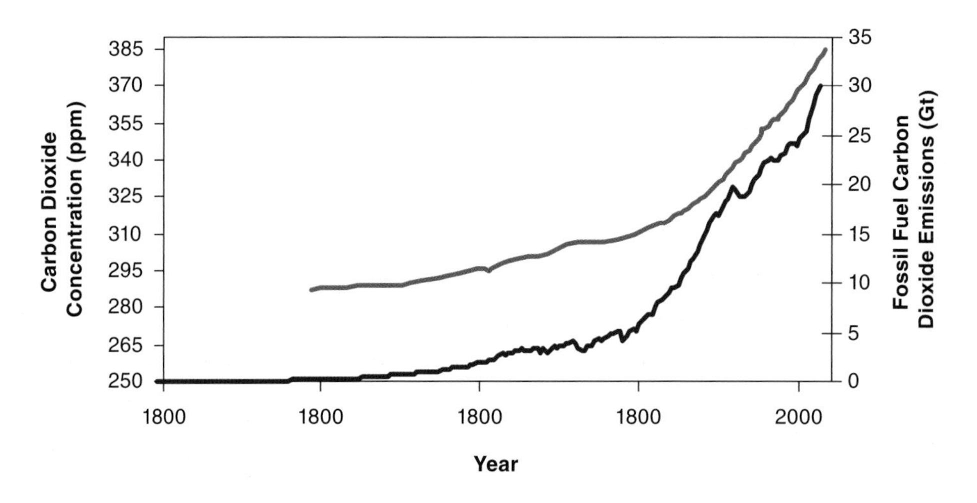

FIGURE 2.3 CO_2 emissions due to fossil fuel burning (blue line and right axis) from 1800 to 2006 and atmospheric CO_2 concentrations (red line and left axis) from 1847 to 2008. For further details see Figures 6.2, 6.3, and 6.4. Based on data from Boden et al. (2009), Keeling et al. (2009), and Neftel et al. (1994).

Only 45 percent of the CO_2 emitted by human activities remains in the atmosphere; the remainder is absorbed by the oceans and land surface. Current estimates, which are based on a combination of direct measurements and models that simulate ecosystem processes and biogeochemical cycles, indicate that roughly twice as much CO_2 is taken up annually by ecosystems on the land surface as is released by deforestation; thus, the land surface is a net "carbon sink." The oceans are also a net carbon sink, but only some of the CO_2 absorbed by the oceans is taken up and used by marine plants; most of it combines with water to form carbonic acid, which (as described below) is harmful to many kinds of ocean life. The combined impacts of rising CO_2 levels, temperature change, and other climate changes on natural ecosystems and on agriculture are described later in this chapter and in further detail in Part II of the report.

Human Activities Are Associated with a Net Warming Effect on Climate

Human activities have led to higher concentrations of a number of GHGs as well as other *climate forcing* agents. For example, the human-caused increase in CO_2 since the beginning of the Industrial Revolution is associated with a warming effect equivalent to approximately 1.6 Watts of energy per square meter of the Earth's surface (Figure 2.4). Although this may seem like a small amount of energy, when multiplied by the surface area of the Earth it is 50 times larger than the total power consumed by all human activities.

In addition to CO_2, the concentrations of methane (CH_4), nitrous oxide (N_2O), ozone (O_3), and over a dozen chlorofluorocarbons and related gases have increased as a result of human activities. Collectively, the total warming associated with GHGs is estimated to be 3.0 Watts per square meter, or almost double the forcing associated with CO_2 alone. While CO_2 and N_2O levels continue to rise (due mainly to fossil fuel burning and agricultural processes, respectively), concentrations of several of the halogenated gases are now declining as a result of action taken to protect the ozone layer, and the concentration of CH_4 also appears to have leveled off (see Chapter 6 for details).

Human activities have also increased the number of aerosols, or particles, in the atmosphere. While the effects of these particles are not as well measured or understood as the effects of GHGs, recent estimates indicate that they produce a net cooling effect that offsets some, but not all, of the warming associated with GHG increases (see Figure 2.4). Humans have also modified Earth's land surface, for example by replacing forests with cropland. Averaged over the globe, it is estimated that these land use and land cover changes have increased the amount of sunlight that is reflected back to

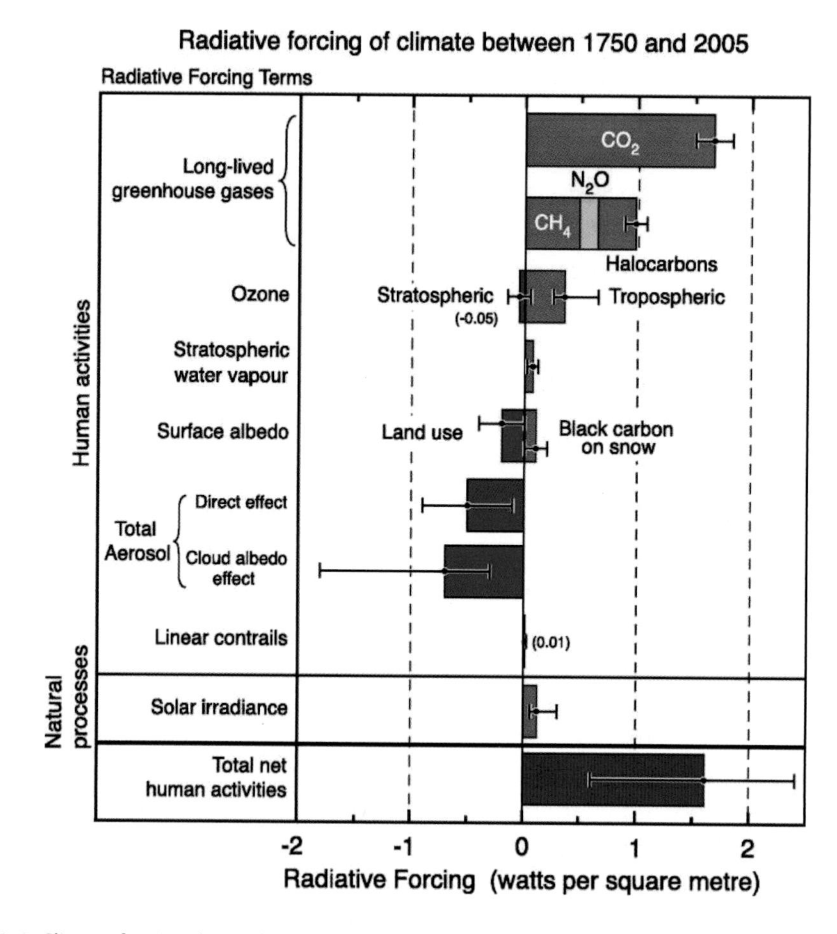

FIGURE 2.4 Climate forcing due to both human activities and natural processes, expressed in Watts per square meter (energy per unit area). Positive forcing corresponds to a warming effect. See Chapter 6 for further details. SOURCE: Forster et al. (2007).

space, producing a small net cooling effect. Other human activities can influence local and regional climate but have only a minor influence on global climate.

Feedback Processes Determine How the Climate System Responds to Forcing

The response of the climate system to GHG increases and other climate forcing agents is strongly influenced by the effects of positive and negative *feedback processes* in the climate system. One example of a positive feedback is the water vapor feedback. Water vapor is the most important GHG in terms of its contribution to the *natural* greenhouse effect (see Figure 6.1), but changes in water vapor are not considered a climate forcing because its concentration in the lower atmosphere is controlled mainly by the (natural) processes of evaporation and precipitation, rather than by human activities. Because the rate of evaporation and the ability of air to hold water vapor both increase as the climate system warms, a small initial warming will increase the amount of water vapor in the air, reinforcing the initial warming—a positive feedback loop. If, on the other hand, an initial warming were to cause an increase in the amount of low-lying clouds, which tend to cool the Earth by reflecting solar radiation back to space (especially when they occur over ocean areas), this would tend to offset some of the initial warming—a negative feedback. Other important feedbacks involve changes in other kinds of clouds, land surface properties, biogeochemical cycles, the vertical profile of temperature in the atmosphere, and the circulation of the atmosphere and oceans—all of which operate on different time scales and interact with one another in addition to responding directly to changes in temperature.

The collective effect of all feedback processes determines the *sensitivity* of the climate system, or how much the system will warm or cool in response to a certain amount of forcing. A variety of methods have been used to estimate climate sensitivity, which is typically expressed as the temperature change expected if atmospheric CO_2 levels reach twice their preindustrial values and then remain there until the climate system reaches equilibrium, with all other climate forcings neglected. Most of these estimates indicate that the expected warming due to a doubling of CO_2 is between 3.6°F and 8.1°F (2.0°C and 4.5°C), with a best estimate of 5.4°F (3.0°C). Unfortunately, the diversity and complexity of processes operating in the climate system means that, even with continued progress in understanding climate feedbacks, the exact sensitivity of the climate system will remain somewhat uncertain. Nevertheless, estimates of climate sensitivity are a useful metric for evaluating the causes of observed climate change and estimating how much Earth will ultimately warm in response to human activities.

Global Warming Can Be Attributed to Human Activities

Many lines of evidence support the conclusion that most of the observed warming since the start of the 20th century, and especially over the last several decades, can be attributed to human activities, including the following:

1. Earth's surface temperature has clearly risen over the past 100 years, at the same time that human activities have resulted in sharp increases in CO_2 and other GHGs.
2. Both the basic physics of the greenhouse effect and more detailed calcula-

tions dictate that increases in atmospheric GHGs should lead to warming of Earth's surface and lower atmosphere.

3. The vertical pattern of observed warming—with warming in the bottom-most layer of the atmosphere and cooling immediately above—is consistent with warming caused by GHG increases and inconsistent with other possible causes (see below).
4. Detailed simulations with state-of-the-art computer-based models of the climate system are only able to reproduce the observed warming trend and patterns when human-induced GHG emissions are included.

In addition, other possible causes of the observed warming have been rigorously evaluated:

5. As described above, the climate system varies naturally on a wide range of time scales, but a rigorous statistical evaluation of observed climate trends, supported by analyses with climate models, indicates that the observed warming, especially the warming since the late 1970s, cannot be attributed to natural variations.
6. Satellite measurements conclusively show that solar output has not increased over the past 30 years, so an increase in energy from the Sun cannot be responsible for recent warming. There is evidence that some of the warming observed during the first few decades of the 20th century may have been caused by a slight uptick in solar output, although this conclusion is much less certain.
7. Direct measurements likewise show that the number of cosmic rays, which some scientists have posited might influence cloud formation and hence climate, have neither declined nor increased during the last 30 years. Moreover, a plausible mechanism by which cosmic rays might influence climate has not been demonstrated.

Based on these and other lines of evidence, the Panel on Advancing the Science of Climate Change—along with an overwhelming majority of climate scientists (Rosenberg et al., 2010)—conclude that much of the observed warming since the start of the 20th century, and most of the warming over the last several decades, can be attributed to human activities.

Models and Scenarios Are Used to Estimate Future Climate Change

In order to project future changes in the climate system, scientists must first estimate how GHG emissions and other climate forcings (such as aerosols and land use) will change over time. Since the future cannot be known with certainty, a large number of different *scenarios* are developed, each using different assumptions about future economic, social, technological, and environmental conditions. These scenarios have increased in complexity over time, and the most recent scenario development efforts include sophisticated models of energy production and use, economic activity, and the possible influence of different climate policy actions on future emissions. Future climate change, like past climate change, is also subject to natural climate variations that modulate the expected warming trend.

After future forcing scenarios are developed, *climate models* are used to simulate how these changes in GHG emissions and other climate forcing agents will translate into changes in the climate system. Climate models are computer-based representations of the atmosphere, oceans, cryosphere, land surface, and other components of the climate system. All climate models are fundamentally based on the laws of physics and chemistry that govern the motion and composition of the atmosphere and oceans. The most sophisticated versions of these models—referred to as *Earth system models*—include representations of a wide range of additional physical, chemical, and biological processes such as atmospheric chemistry and ecosystems on land and in the oceans. The resolution of climate models has also steadily increased, although global models are still not able to resolve features as small as individual clouds, so these small-scale processes must be approximated in global models.

After decades of development by research teams in the United States and around the world, and careful testing against observations of climate over the past century and further into the past, scientists are confident that climate models are able to capture many important aspects of the climate system. Scientists are also confident that climate models give a reasonable projection of future changes in climate that can be expected based on a particular scenario of future GHG emissions, at least at large (continental to global) scales. A variety of *downscaling* techniques have been developed to project future climate changes at regional and local scales. These techniques are not as well established and tested as global climate models, and their results reflect uncertainties in both the underlying global projections and regional climate processes. Hence, predictions of regional and local climate change are generally much more uncertain than large-scale changes. Other key sources of uncertainty in projections of future climate change include (1) uncertainty in future climate forcing, especially how human societies will produce and use energy in the decades ahead; (2) processes that are not included or well represented in models, such as changes in ice sheets, and certain land use and ecosystem processes; and (3) the possibility that abrupt changes or other climate "surprises" (see below) may occur.

Projections of Future Climate Change Indicate Continued Warming

The most recent comprehensive modeling effort to date included more than 20 different state-of-the-art climate models from around the world. Each of these climate models projected future climate change based on a range of different scenarios of future GHG emissions and other changes in climate forcing. Continued warming is projected by all models, but the trajectory and total amount of warming varies from model to model and between different scenarios of future climate forcing. Based on these results, the IPCC estimates that global average surface temperatures will rise an additional 2.0°F to11.5°F (1.1°C to 6.4°C), relative to the 1980-1999 average, by the end of the 21st century. The wide spread in these numbers comes from uncertainty not only in exactly how much the climate system will warm in response to continued GHG emissions, but also uncertainty in how future GHG emissions will evolve.[7] Hence, the choices that human societies make over the next several decades will have an enormous influence on the magnitude of future climate change.

As with observed climate change to date, there are wide geographic variations in the magnitude of future warming, with much stronger projected warming over high latitudes and over land areas (see Figure 2.5). In the United States, temperatures are projected to warm substantially over the 21st century under all projections of future climate change (USGCRP, 2009a). Temperature increases over the next few decades primarily reflect past emissions and are thus similar across different scenarios of future GHG emissions. However, by midcentury and especially at the end of the century, higher emissions scenarios (e.g., scenarios with continued growth in global GHG emissions) lead to much warmer temperatures than lower emissions scenarios.

A Multitude of Additional Climate and Climate-Related Changes Are Projected

In addition to increasing global average temperatures, a host of other climate variables are projected to experience significant changes over the 21st century, just as they have during the past century. For example, it is very likely[8] that

- Heat waves will become more intense, more frequent, and longer lasting, while the frequency of cold extremes will continue to decrease;
- Snow and ice extent will continue to decrease;
- The intensity of precipitation events will continue to increase;

[7] As discussed in Chapter 6, none of the scenarios considered in this modeling effort attempted to represent how climate policy interventions might influence future GHG emissions.

[8] Estimated greater than 9 out of 10 chance (see Appendix D).

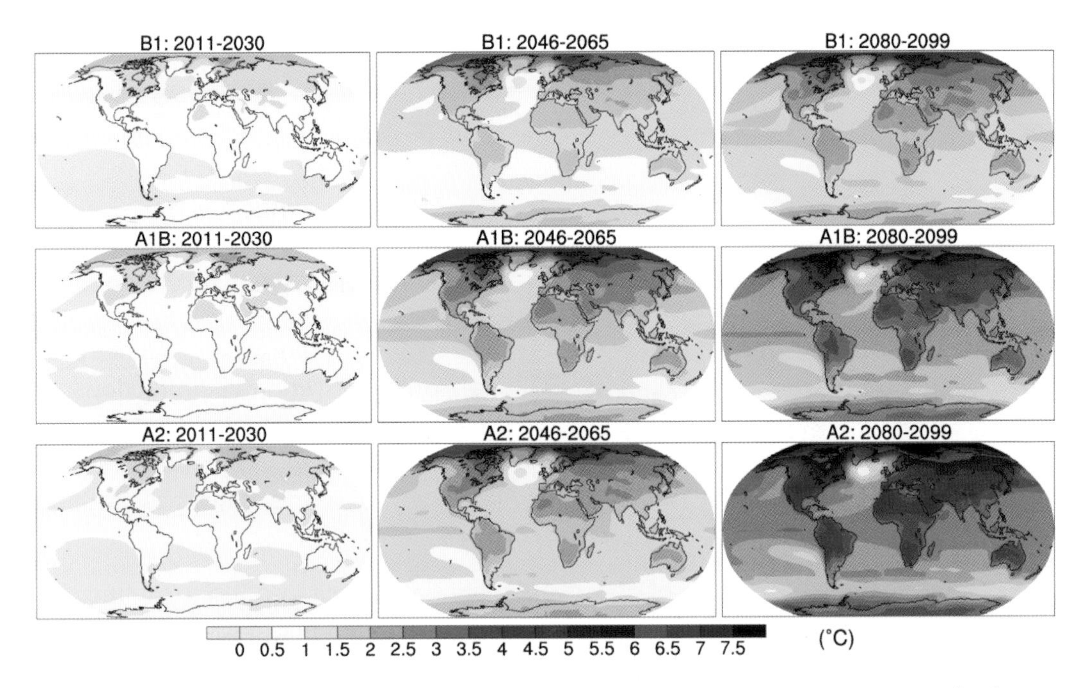

FIGURE 2.5 Worldwide projected changes in temperatures, relative to 1961-1990 averages, under three different emissions scenarios (rows) for three different time periods (columns). For further details see Figure 6.21. SOURCE: Meehl et al. (2007a).

- Glaciers and ice sheets will continue to melt; and
- Global sea level will continue to rise.

Many of these changes are discussed below and described in detail in Part II of the report.

Abrupt Changes May Occur

Confounding all projections of future climate is the possibility of abrupt changes in the climate system, other environmental systems, or human systems. Paleoclimate records indicate that the climate system can experience abrupt changes in as little as a decade. The Earth's temperature is now demonstrably higher than it has been for several hundred years, and GHG concentrations are now higher than they have been in at least 800,000 years. These sharp departures from historical climate regimes raise the possibility that "tipping points" or thresholds for stability might be crossed as the climate system warms, leading to rapid or abrupt changes in climate. Climate change may also lead to abrupt changes in human or ecological systems, especially systems that are also experiencing other environmental stresses. However, in general we have only a limited understanding of where the tipping points in the climate system, other environmental systems, or human systems might be, when they might be crossed, or what the consequences might be.

Research Needs for Advancing Climate System Science

Additional research, supported by expanded observational and modeling capacity, is needed to improve understanding of key climate processes, improve projections of future climate change (especially at regional scales), and evaluate the potential for abrupt changes in the climate system. The following are some of the most critical research needs for continued improvements in our ability to understand, observe, and project the behavior of the climate system:

- Improve understanding of how the climate system will respond to forcing.
- Refine the ability to project interannual, decadal, and multidecadal climate change, including extreme events, at regional scales.
- Advance understanding of feedbacks and thresholds that may be crossed that lead to irreversible or abrupt changes.
- Expand and maintain comprehensive and sustained climate observations to provide real-time information about climate variability and change.

For a longer discussion of these and other climate system research needs, see Chapter 6.

SEA LEVEL RISE AND THE COASTAL ENVIRONMENT[9]

The coastlines of the United States and the world are major centers of economic and cultural development, where populations and associated structural development continue to grow. The coasts are also home to critical ecological and environmental resources. Coastal areas have always experienced various risks and hazards, such as flooding from coastal storms. However, coastal managers and property owners are concerned about how these risks are being and will be intensified by sea level rise and other climate changes.

[9] For additional discussion and references, see Chapter 7 in Part II of the report.

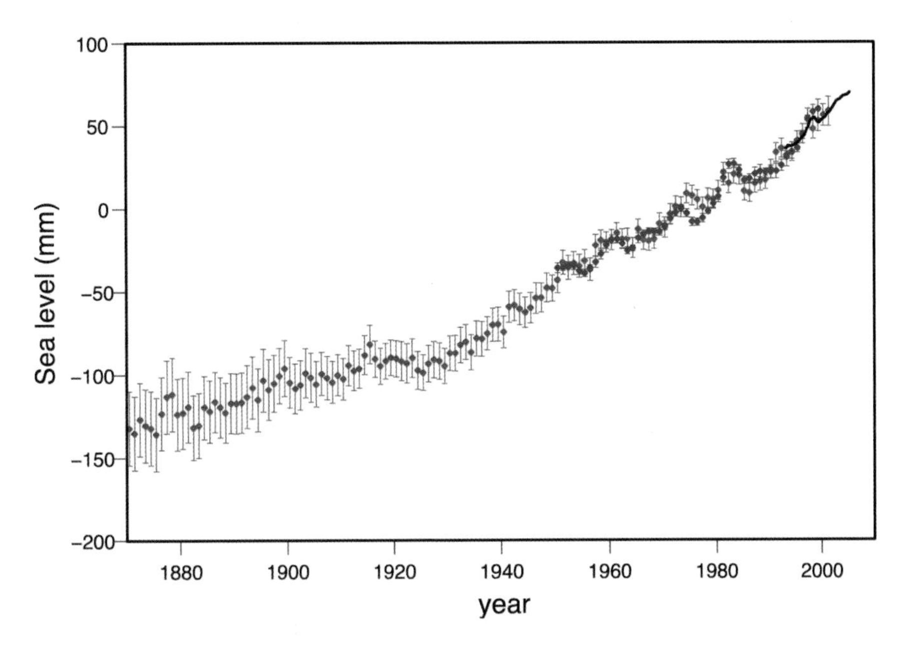

FIGURE 2.6 Annual, global mean sea level as determined by records of tide gauges (red and blue curves) and satellite altimetry (black curve). For further details see Figure 7.2. SOURCE: Bindoff et al. (2007).

Observations of Sea Level Rise

Sea level has been systematically measured by tide gauges for more than 100 years. Other direct and indirect observations have allowed oceanographers to estimate (with lower precision) past sea levels going back many thousands of years. We know that sea level has risen more than 400 feet (120 meters) since the peak of the last ice age 26,000 years ago, with periods of rapid rise predating a relatively steady level over the past 6,000 years. During the past few decades, tide gauge records augmented by satellite measurements have been used to produce precise sea level maps across the entire globe. These modern records indicate that the rate of sea level rise has accelerated since the mid-19th century, with possibly greater acceleration over the past two decades (Figure 2.6). The exact amount of sea level change experienced in different locations varies because of different rates of settling or uplift of land and because of differences in ocean circulation.

Causes of Sea Level Rise

Past, present, and future changes in global sea level are mainly caused by two fundamental processes: (1) the thermal expansion of the existing water in the world's ocean basins as it absorbs heat and (2) the addition of water from land-based sources—mainly the shrinking of ice sheets and glaciers.

Because of the huge capacity of the oceans to absorb heat, 80 to 90 percent of the heating associated with human GHG emissions over the past 50 years has gone into raising the temperature of the oceans. The subsequent thermal expansion of the oceans is responsible for an estimated 50 percent of the observed sea level rise since the late 19th century. Even if GHG concentrations are stabilized, ocean warming and the accompanying sea level rise will continue until the oceans reach a new thermal equilibrium with the atmosphere. Ice in the world's glaciers and ice sheets contributes directly to sea level rise through melt or the flow of ice into the sea. The major ice sheets of Greenland and Antarctica contain the equivalent of 23 and 197 feet (7 and 60 meters) of sea level, respectively.

Projections of Sea Level Rise

Projections of future sea level have been the subject of active discussion in the recent literature on climate change impacts. The 2007 Assessment Report by the IPCC estimated that sea level would likely rise by an additional 0.6 to 1.9 feet (0.18 to 0.59 meters) by 2100. This projection was based largely on the observed rates of change in ice sheets and projected future thermal expansion over the past several decades and did not include the possibility of changes in ice sheet dynamics. Scientists are working to improve how ice dynamics can be resolved in models. Recent research, including investigations of how sea level responded to temperature variations during the ice age cycles, suggests that sea levels could potentially rise another 2.5 to 6.5 feet (0.8 to 2 meters) by 2100, which is several times larger than the IPCC estimates. However, sea level rise estimates are rather uncertain, due mainly to limits in scientific understanding of glacier and ice sheet dynamics. For instance, recent findings of a warming ocean around Greenland suggest an explanation for the accelerated calving of outlet glaciers into the sea, but the limited data and lack of insight into the mechanisms involved prevent a quantitative estimate of the rate of ice loss at this time. Nevertheless, it is clear that global sea level rise will continue throughout the 21st century due to the GHGs that have already been emitted, that the rate and ultimate amount of sea level rise will be higher if GHG concentrations continue to increase, and that there is a risk of much larger and more rapid increases in sea level. While this risk cannot be quantified at present, the consequences of extreme and rapid sea level rise could be economically and socially devastating for highly built-up and densely populated coastal areas around the world, especially low-lying deltas and estuaries.

Ice Sheet Processes Could Potentially Lead to Abrupt Changes

In addition to rapid accelerations in the rate of sea level rise, a collapse or rapid wastage of major ice sheets could lead to other abrupt changes. For example, if the Greenland ice sheet were to shrink substantially over several decades, a large amount of freshwater would be delivered to key regions of the North Atlantic. This influx of freshwater could alter the ocean structure and influence ocean circulation, with implications for regional and global weather patterns. Compelling evidence has been assembled indicating that rapid freshwater discharges at the end of the last ice age caused abrupt ocean circulation changes, which in turn led to significant impacts on regional climate. The recent ice melting on Greenland and other areas in the Arctic, combined with increased river discharges in the Arctic region (see discussion of precipitation and runoff changes below), may have already led to changes in ocean circulation patterns. However, much work remains to develop confident projections of future ocean circulation changes—and the influence of these changes on regional climate patterns—resulting from ongoing freshwater discharges in the North Atlantic.

Sea Level Rise Is Associated with a Range of Impacts on Coastal Environments

Coastal areas are among the most densely populated and fastest-growing regions of the United States, as well as the rest of the world. Such population concentration and growth are accompanied by a high degree of development and use of coastal resources for economic purposes, including industrial activities, transportation, trade, resource extraction, fisheries, tourism, and recreation. Sea level rise can potentially affect all of these activities and their accompanying infrastructure, and it could also magnify other climate changes, such as an increase in the frequency or intensity of storms (see below). Even if the frequency or intensity of coastal storms does not change, increases in average sea level will magnify the impacts of extreme events on coastal landscapes.

The economic impacts of climate change and sea level rise on coastal areas in the United States have been an important focus for research. While economic impact assessments have become increasingly sophisticated, they remain incomplete and are subject to the well-recognized challenges of cost-benefit analyses (see Chapter 17). In addition, while studies of economic impacts may be useful at a regional level, general conclusions regarding the total magnitude of economic impacts in the United States cannot be drawn from existing studies; this is because the metrics, modeling approaches, sea level rise projections, inclusions of coastal storms, and assumptions about human responses (e.g., the type and level of protection) vary considerably across the studies.

Coastal ecosystems such as dunes, wetlands, estuaries, seagrass beds, and mangroves provide numerous ecosystem goods and services, ranging from nursery habitat for certain fish and shellfish to habitat for bird, mammal, and reptilian species, including some endangered ones; protective or buffering services for coastal infrastructure against the onslaught of storms; water filtering and flood retention; carbon storage; and the aesthetic, cultural, and economic value of beaches and coastal environments for recreation, tourism, and simple enjoyment. The impact of sea level rise on these and other nonmarket values is often omitted from economic impact assessments of coastal areas because of difficulties in assigning values.

Science for Responding to Sea Level Rise

Scientific understanding of people's vulnerability and ability to adapt to sea level rise (and other impacts of climate change on coastal systems) has increased in recent years. Developing countries are expected to face greater challenges in dealing with the impacts of rising sea levels because of lower adaptive capacity—which is largely a function of economic, technological, and knowledge resources; social capital; and well-functioning institutions. In developed countries like the United States, adaptive capacity may be higher, but this has not been thoroughly examined to date and there are a large number of assets and people at risk. Moreover, significant gaps remain in our empirical understanding of and ability to identify place-based vulnerabilities to the impacts of sea level rise along the U.S. coastline. Considerable challenges also remain in translating whatever adaptive capacity exists into real adaptation actions on the ground.

Virtually all adaptive responses to sea level rise have costs as well as social and ecological consequences, and most are complicated by having effects that extend far into the future and beyond the immediately affected coastal regions. Engineering options such as seawalls and levees are not feasible in all locations, and in many they could have negative effects on coastal ecosystems, beach recreation, tourism, aesthetics, and other socially valued aspects of coastal environments. A wide range of barriers and constraints make "soft" solutions—such as changes in land use planning and, ultimately, retreat from the shoreline—equally challenging. Such constraints and limits on adaptation are increasingly recognized, but little is currently known about how to determine the most appropriate, cost-effective, least ecologically damaging, and most socially acceptable adaptation options for different places and regions. As discussed below and in further detail in Chapter 4, continued and expanded scientific research can help to address these gaps in understanding.

Research Needs for Advancing Science on Sea Level Rise and Associated Risks in the Coastal Environment

While global sea level rise is certain to continue, the physical science of sea level rise and related climate changes remains incomplete, making future projections uncertain. Moreover, social and ecological understanding of place-based vulnerability and adaptation options in coastal regions of the United States is lacking. Thus, research is needed to improve our understanding and projections of future sea level rise, the impacts of this rise on affected human and natural systems, and the feasibility of adaptation options in the near and longer term. Specific research needs, which are explained in more detail in Chapter 7, include the following:

- Reduce the scientific uncertainties associated with changes in glaciers and ice sheets.
- Improve understanding of ocean dynamics and regional rates of sea level rise.
- Develop tools and approaches for understanding and predicting the impacts of sea level rise on coastal ecosystems and coastal infrastructure.
- Expand the ability to identify and assess vulnerable coastal regions and populations and to develop and assess adaptation strategies that can reduce vulnerability or build adaptive capacity.
- Develop decision-support capabilities for all levels of governance in response to the challenges associated with sea level rise.

FRESHWATER RESOURCES[10]

The availability of water for human and ecosystem use depends on two main factors: (1) the climate-driven global water cycle and (2) society's ability to manage, store, and conserve water resources. Each of these factors is complex, as is their interaction. Water cycling—which includes evaporation and transpiration, precipitation, and both surface runoff and groundwater movement—determines how freshwater flows and how it interacts with other processes. Precipitation amounts, intensity, geographic distribu-

[10] For additional discussion and references, see Chapter 8 in Part II of the report.

tion, and other characteristics matter for water management, and all are affected by both short-term climate variability and long-term climate change. Likewise, soils, topography, land cover, precipitation intensity, and other variables influence how much precipitation can be stored for use. Other variables such as level of consumption, pollution, conservation, pricing, distribution, and land use changes are also important for water management decisions. These complex processes prevent any easy conclusions about regional water supplies based solely on climate model-driven projections. Nonetheless, historical and current changes in some variables are becoming clear.

Global Precipitation and Extreme Rainfall Events Are Increasing

In general, changes in precipitation are harder to measure and predict than changes in temperature. Nevertheless, some conclusions and projections are robust. For example, based on the fundamental properties and dynamics of the climate system, it is expected that the intensity of the global water cycle and of precipitation extremes (droughts and extremely heavy precipitation events) should both increase as the planet warms. Increases in worldwide precipitation and in the fraction of total precipitation falling in the form of heavy precipitation have already been observed; for example, the fraction of total rainfall falling in the heaviest 1 percent of rain events increased by about 20 percent over the past century in the United States. Climate models project that these trends, which create challenges for flood control and storm and sewer management, are very likely to continue. However, models also indicate a strong seasonality in projected precipitation changes in the United States, with drier summers across much of the Midwest, the Pacific Northwest, and California, for example.

Snow Cover Is Decreasing

Another robust projection of climate change is that snow and ice cover should decrease as temperatures rise. Worldwide, snow cover is decreasing, although substantial regional variability exists. In the United States, changes in snowpack in the West currently represent the best-documented hydrological manifestation of climate change. The largest losses in snowpack are occurring in the lower elevations of mountains in the Northwest and California, as higher temperatures cause more precipitation to fall as rain rather than snow. Moreover, snowpack is melting as much as 20 days earlier in many areas of the West. Snow is expected to melt even earlier under projections of future climate change, resulting in streams that have reduced flow and higher temperatures in late summer. Such changes have major implications for ecosystems, hydropower, urban and agricultural water supplies, and other uses.

Total Runoff Is Increasing but Shows Substantial Regional Variability

Average flows of streams and rivers in the United States have increased in most areas over the past several decades, which is consistent with observed and expected trends in precipitation. There are regional differences, however, with decreased stream flow in the Columbia and Colorado Rivers, for example. Observed changes in stream flow reflect both natural variability in hydrology as well as the aggregate effects of many human influences, of which climate change is only one. In some areas, changes in climate are exacerbating decreases in river and stream flows that are already declining due to agricultural, residential, and other human uses.

Droughts and Floods Are Likely to Increase

Given the observed increases in heavy precipitation events and the expectation that this intensification will continue, the risk from floods is projected to increase in the future. Local water, land use, and flood risk-management decisions, however, can modify the actual flood vulnerability of communities.

Drought is a complex environmental impact. It is strongly affected not only by the balance between precipitation and evapotranspiration (the sum of evaporation of water from the surface and transpiration of water though the leaves of plants) and the resulting effect on soil moisture, but also by other human influences such as urbanization, deforestation, and changes in agriculture. Additionally, historical data on drought frequencies and intensities are limited, making it difficult to unambiguously attribute severe droughts to climate change. Climate model projections indicate that the area affected by drought and the number of annual dry days are likely to increase in the decades ahead. In areas where water is stored for part of the year in snowpack, reductions in snowpack and earlier snowmelt are expected to increase the risk of water limitations and drought.

Storm Patterns and Intensities May Change

How storm patterns may change in the future is of obvious importance to water managers, but considerably less is known in this area than in the hydrologic changes discussed above. Changes in the intensity of hurricanes have been documented and attributed to changes in sea surface temperatures, but the link between these changes and climate change remains uncertain and the subject of considerable research and scientific debate. The most recent climate model projections indicate that climate change may lead to increases in the intensity of the strongest hurricanes, but effects on frequency of occurrence are still in an active area of research. Relatively little is known about how climate change will affect midlatitude storm patterns, in part because of the close connection between storm patterns and regional climate variability, although shifts in predominant storm tracks have been observed.

Water Quality and Groundwater Supplies May Be Affected

Some regions of the United States rely on groundwater for drinking, residential use, or agriculture. The impacts of climate change on groundwater are far from clear; in fact, little research effort has been devoted to this topic. Changes in precipitation and evaporation patterns, plant growth processes, and incursions of sea water into coastal aquifers as sea levels rise will all affect the rate of groundwater recharge, the absolute volume of groundwater available, groundwater quality, and the physical connection between surface and groundwater bodies.

Increased temperatures generally have a negative impact on water quality in lakes and rivers, typically by stimulating growth of nuisance algae. Changes in heavy precipitation, runoff, and stream flow can also be expected to have an impact on a diverse set of water quality variables. Water quality will also be affected by saltwater intrusion into coastal aquifers as sea levels rise. In general, the water quality implications of climate change are even less understood than impacts on water supply.

Climate Change May Increase Water Management Challenges

Effective management of water supplies requires fairly precise information about current and expected future water availability. However, the complex processes involved in the hydrologic cycle prevent simplistic conclusions about how to manage water supplies based on climate model projections. In many regions, the uncertainties associated with projections of rainfall and runoff coupled with uncertainties in other changes, such as changes in land use and land cover, leads to cascading uncertainties about changes in freshwater resources. These uncertainties are compounded by uncertainties in the technical capacity to store, manage, and conserve water resources, which in turn are shaped by socioeconomic, cultural, institutional, and behavioral issues that determine the use of water. Two clear messages that emerge from research on water management is that water managers will need to make decisions while facing persistent and sometimes considerable uncertainty, and that improved decision-support tools would be helpful for planning purposes.

Research Needs for Advancing Science on Freshwater Resources in the Context of Climate Change

Changes in freshwater systems are expected to create significant challenges for flood management, drought preparedness, water supplies, and many other water resource issues. Responding to these challenges will require better data and improved model projections as well as a better understanding of both the impacts of climate change and the role of water governance on future water resources. Significant gaps remain in the knowledge base that informs both projections of climate impacts on water resources and governance strategies that can build adaptive capacity of water systems to climate effects. Key research needs, which are explored in more detail in Chapter 8, include the following:

- Improve projections of changes in precipitation and other water resources at regional and seasonal time scales.
- Develop long-term observational systems for measuring and predicting hydrologic changes and planning management responses.
- Improve tools and approaches for decision making under uncertainty and complexity.
- Develop vulnerability assessments of the diverse range of water users and integrative management approaches to respond effectively to changes in water resources.
- Increase understanding of water institutions and governance, and design effective systems for the future.
- Improve water engineering and technologies.
- Evaluate effects, feedbacks, and mitigation options of water resource use on climate.

ECOSYSTEMS, ECOSYSTEM SERVICES, AND BIODIVERSITY[11]

Decades of research on terrestrial and marine ecosystems and their biodiversity have improved our understanding of their importance for society and their links to climate. Ecosystems provide food, fuel, and freshwater. They regulate climate through the global carbon cycle and the hydrologic cycle. They buffer against storms, erosion, and extreme events and provide cultural, nonmaterial benefits such as space for recreation, education, and spiritual practices. Ecosystems are thus essential components of Earth's life support system, and understanding the impacts of climate change on ecosystems is a critical part of the research enterprise.

[11] For additional discussion and references, see Chapter 9 in Part II of the report.

Climate Change Is Already Affecting Land-Based Ecosystems

Shifts in climate are changing the geographical range of many plant and animal species. A series of place-based observations and syntheses of existing data indicate that many plants and animals have experienced range shifts over the past 30 years that approach the magnitude of those inferred for the last 20,000 years (the time of the last glacial maximum). The phenology (seasonal periodicity and timing of life-cycle events) of species is also changing with the earlier onset of spring and longer growing seasons. The implications of these changes for biodiversity, the provision of ecosystems services, and feedbacks on climate are not well understood.

Large and long-duration forest fires have increased fourfold over the past 30 years in the American West. Forest fires are influenced by many factors, including climate change, but warming has increased the length of the fire season by more than two months in some locations, and the increasing size of wildfires can be attributed in part to earlier snowmelt, temperature changes, and drought. Decomposition and respiration of CO_2 back to the atmosphere also increase as temperatures warm. Finally, populations of forest pests such as the spruce beetle, pine beetle, spruce budworm, and wooly adelgid are increasing in the western United States as a result of climate change.

Future Climate Change Will Affect Land-Based Ecosystems in a Variety of Ways

Both the amount and rate of warming will influence the ability of plants and animals to adapt. In addition, temperature changes will interact with changes in CO_2, precipitation, pests, soil characteristics, and other factors. Tree species, for example, are expected to shift their ranges northward or upslope, with some current forest types such as oak-hickory expanding, others such as maple-beech contracting, and still others such as spruce-fir disappearing from the United States altogether (Figure 2.7). Experimental and modeling studies indicate that exposure to elevated CO_2 and temperatures can lead to increases in photosynthesis and growth rates in many plant species, although at higher temperatures this trend may reverse. Projections suggest that forest productivity will increase with elevated CO_2 and climate warming, especially in young forests on fertile soils where water is adequate. Where water is scarce and drought is expected to increase, or where pests increase in response to warming, however, forest productivity is projected to decrease.

Some analyses have indicated the possibility of major changes in ecosystems due to the combined effects of changes in temperature and precipitation, potentially affect-

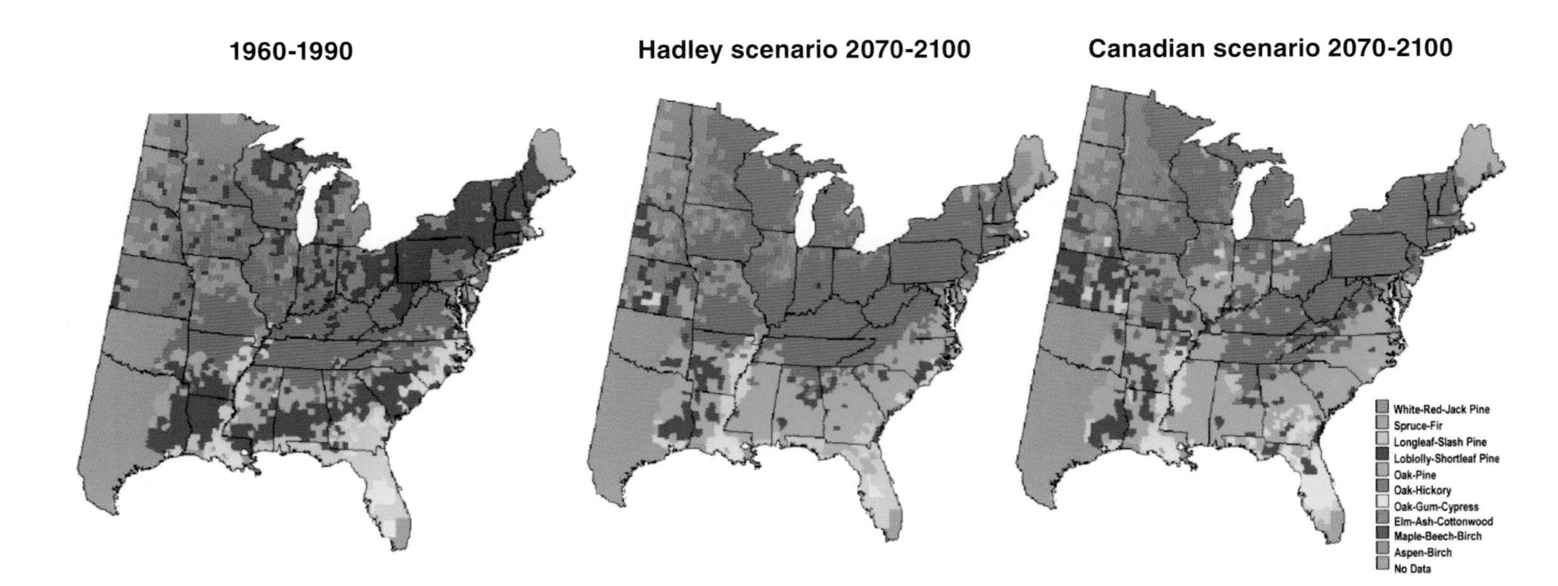

FIGURE 2.7 Potential changes in the geographic ranges of the dominant forest types in the eastern United States under projections of future climate change. Many forest types shift their ranges northward or shrink in areas, while some expand their areas. For further details see Figure 9.2. SOURCE: USGCRP (2001).

ing ecosystem productivity, carbon cycling, and the composition of plant communities. For example, drier conditions in the Amazon could potentially lead to increased susceptibility to fire, lower productivity, and shifts from forest to savanna systems in that region. The strong warming observed across the Arctic is already leading to poleward shifts of boreal forests into regions formerly covered in tundra, and these shifts are expected to continue.

Climate Change Is Also Affecting Ocean Ecosystems

Just as on land, ranges of many marine animals have shifted poleward in recent decades (Figure 2.8). The pace of these changes can be faster in the sea because of the high mobility of many marine species. Changes have also been observed in ocean productivity, which measures the photosynthetic activity of organisms at the base of the marine food web. Model projections suggest that some habitats, such as polar seas and areas with coastal upwelling, may see increases in productivity as climate change progresses. The majority of ocean areas, however, are projected to experience declines in productivity as warm, nutrient-poor surface water is increasingly isolated from the colder, nutrient-rich water below. Even in highly productive coastal upwelling systems, it is possible that even stronger upwelling could draw up deeper, low-oxygen (hy-

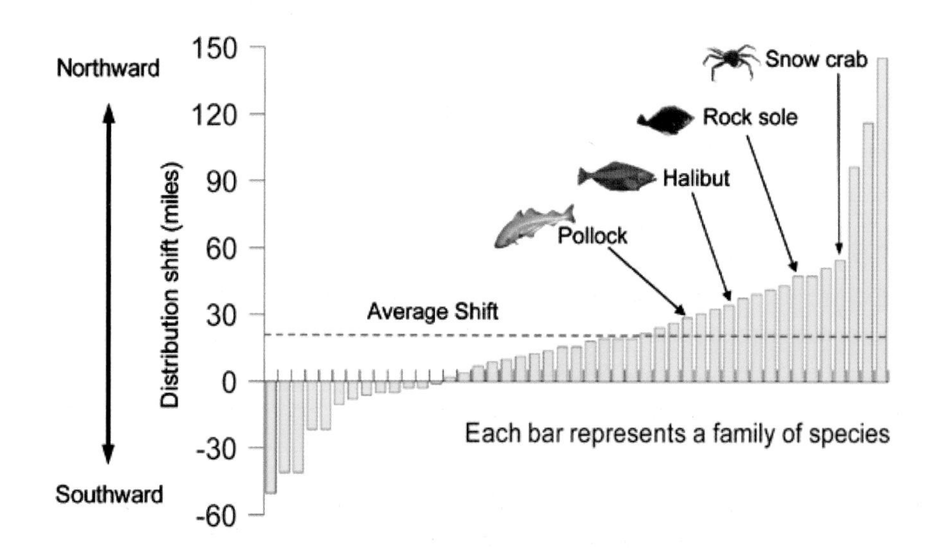

FIGURE 2.8 Observed northward shift of marine species in the Bering Sea between the years 1982 and 2006. The average shift among the species examined was approximately 19 miles north of its 1982 location (red line). For further details see Figure 9.3. SOURCES: Mueter and Litzow, (2008); USGCRP (2009a).

poxic) water, creating dead zones where few species can survive. Such hypoxic dead zones have recently appeared off the coasts of Oregon and Washington.

Continued losses of sea ice and stronger warming at higher latitudes are expected to drive major habitat alterations in Arctic ecosystems. Ice dynamics plays an important role in ocean productivity, and sea ice is a critical habitat for many species, including birds and mammals. Although the details are uncertain, polar ecosystems are at the threshold of major ecosystem changes due to climate change. Without careful management, these changes may be exacerbated by expanding human uses in polar seas as sea ice continues to decline.

In the tropics, warm temperatures pose a bleaching threat to corals. Corals are animals, but they depend on algae growing in their tissues for much of their nutrition. This tight symbiotic relationship can be disrupted by extreme temperatures, which can cause corals to eject the algae and "bleach." Mass bleaching events, which often lead to coral death, have occurred with increasing frequency over recent decades associated with severe warming events. In the most extreme case, the strong El Niño event of 1998, an estimated 16 percent of the world's coral reefs died. Models suggest that the fate of corals under future warming scenarios depends critically on the pace of warming.

The Oceans Are Becoming More Acidic, Which Poses Major Risks for Ocean Ecosystems

One of the most certain outcomes from increasing CO_2 concentrations in the atmosphere is the acidification of the world's oceans. Roughly one-quarter of the CO_2 currently released by human activities is absorbed in the sea. While some of the CO_2 is taken up by marine organisms, most if it combines with water to form carbonic acid. The result has been a roughly 30 percent increase in ocean acidity since preindustrial times. If CO_2 emissions continue to increase at present rates, ocean acidification could intensify by three to four times this amount by the end of this century. In addition, ocean acidification may reduce the ability of the ocean to take up CO_2; this represents a positive feedback on global warming because it would lead to faster CO_2 accumulation in the atmosphere.

Although the acidification of the sea is highly certain, the response of ocean ecosystems to changing ocean chemistry is highly uncertain. Acidification can disrupt many biological processes, including the rates at which marine animals can form shells. Coral reefs are particularly sensitive. If atmospheric CO_2 levels reach twice their preindustrial values, the resulting increase in acidity could mean there will be few places in the ocean that can sustain coral growth. Polar seas could also experience major changes, since many of the species at the base of the food web may be disrupted. Hence, ocean acidification poses a major threat to ocean ecosystems, but the details are only beginning to be understood. A separate report, *Ocean Acidification: A National Strategy to Meet the Challenges of a Changing Ocean* (NRC, 2010f), examines ocean acidification and its potential impacts in further detail.

Ecosystems Play a Key Role in the Global Carbon Cycle

Plants on land and in the ocean take up carbon during photosynthesis and release it through respiration. Experimental research has shown that some land ecosystems respond to higher atmospheric concentrations of CO_2 by taking up and storing more carbon in plant tissues, soils, and sediments. Based on a combination of ecosystem models and observations, it has been estimated that for the period 2000 to 2008, land ecosystems removed roughly one-third of the CO_2 emitted by human activities. However, roughly half of this carbon sink was offset by changes in land use that resulted in net CO_2 emissions back to the atmosphere (mainly through tropical deforestation).

If the balance between CO_2 absorption and emissions by ecosystems were to change in response to either future climate changes or changes in management, this could lead to a significant positive or negative feedback on atmospheric CO_2 levels. For example, the warming of ocean surface waters across much of the world may represent a positive feedback on climate change, because warming of surface waters commonly reduces the uptake of CO_2 by phytoplankton, which could lead to less ocean uptake of CO_2, faster CO_2 accumulation in the atmosphere, and accelerated greenhouse warming. However, a number of factors influence the storage of carbon in ocean- and land-based ecosystems. For instance, the availability of nutrients and water can limit uptake by land plants, and increases in temperature or large wildfires can increase GHG emissions from land-based ecosystems to the atmosphere. Other important factors modulating the carbon sink provided by terrestrial ecosystems include species redistributions and changes in growing season lengths, drought, insects, pathogens, and land use. As a result of this complex interplay of factors, projections of the future land-based carbon sink are uncertain.

Changes in terrestrial ecosystems could also potentially lead to abrupt climate changes. For example, increasing temperatures are leading to warming and thawing of permafrost (frozen soils) across the northern latitudes. These frozen soils store vast amounts of carbon. As permafrost continues to thaw, this carbon may be released to the atmosphere in large quantities in the form of the GHGs CO_2 and CH_4, which would significantly amplify global warming (and since this warming would then lead to further permafrost thawing, this represents a potential positive feedback). Other such carbon-climate feedbacks are possible, and this area of research is garnering increasing attention and concern.

Several Human Interventions Have Been Proposed to Increase Carbon Storage in Natural Ecosystems

Because productivity in the ocean is often limited by the availability of certain nutrients, it has been hypothesized that ocean fertilization could stimulate plankton blooms and thus enhance the transfer of CO_2 from the atmosphere to the oceans. For example, in some parts of the ocean, productivity is limited by the availability of iron, which suggests the potential for increasing carbon uptake via iron fertilization. Experiments to test this hypothesis have so far resulted in considerable uncertainty about its potential. While this approach could store some carbon, the maximum achievable rates might be only a small fraction of the total carbon emitted by human activities.

On land, changes in land use and land cover by human actions have been responsible, over time, for as much as 35 percent of human-induced CO_2 emissions. Today, emissions from tropical deforestation and other changes in land use account for around 17 percent of annual CO_2 emissions. Land management practices that reduce deforestation and degradation, or that enhance storage of carbon in land ecosystems, could provide potentially low-cost options to reduce GHG concentrations in the atmosphere and thus limit the magnitude of future climate change. Changes in land use can also influence temperatures by changing the reflective properties of the Earth's surface and by altering rates of transpiration of water. The overall potential to limit climate change through management of land and ocean ecosystems has not been thoroughly evaluated, however.

Research Needs for Advancing Science on Ecosystems, Ecosystems Services, and Biodiversity in the Context of Climate Change

Research is needed to better understand and project the impacts of climate change on ecosystems, ecosystem services, and biodiversity and to evaluate how land and ocean changes and management options influence the climate system. Some of the key research needs in these areas, which are described in further detail in Chapter 9, include the following:

- Improve understanding of how higher temperatures, enhanced CO_2, and other climate changes, acting in conjunction with other stresses, are influencing or may influence ecosystems, ecosystem services, and biodiversity.
- Evaluate the potential climate feedbacks associated with changes in ecosystems and biodiversity on land and in the oceans.
- Assess the potential of land and ocean ecosystems to limit or buffer the impacts of climate change through specific management actions.
- Improve assessments of the vulnerabilities of ecosystems to climate change, including methods for quantifying ecosystem benefits to society.
- Improve observations and modeling of terrestrial and marine ecosystems and their interactions with the climate system.

AGRICULTURE, FISHERIES, AND FOOD PRODUCTION[12]

Meeting the food needs of a still-growing and more affluent global human population presents a key challenge. Climate change increases the complexity of this challenge because of its multiple impacts on agricultural crops, livestock, and fisheries. Agricultural management may also provide opportunities to reduce net human GHG emissions.

Agricultural Crops Will Be Influenced in Multiple Ways by Climate Change

Temperature, length of growing season, atmospheric CO_2 levels, water availability, pests, disease, and extreme weather events can all affect crop growth and yields to varying degrees—and sometimes in conflicting ways—depending on location, agricultural system, and the degree of warming. For example, growth of some heat-loving crop plants such as melons and sweet potatoes will initially respond positively to increasing temperatures and longer growing seasons in the United States. Other crops, including grains and soybeans, respond negatively, both in vegetative growth and seed production, to even small increases in temperature. Many crop plants, such as wheat and soybeans, respond positively to the fertilization effect of increases in atmospheric CO_2, potentially offsetting some of the negative effects of warming.

In the United States, many northern states are projected to experience increases in some crop yields over the next several decades, while in the Midwest and southern Great Plains, temperature increases and possible precipitation decreases may decrease yields unless measures are taken to adapt. Likewise, global-scale studies suggest that

[12] For additional discussion and references, see Chapter 10 in Part II of the report.

moderate warming of 1.8°F to 5.4°F (1°C to 3°C), increases in CO_2, and changes in precipitation could benefit crop and pasture lands in mid- to high latitudes but decrease yields in seasonally dry and low-latitude areas. Projections also suggest that global food production is likely to decrease with increases in average temperatures of greater than 5.4°F (3°C). However, most analyses and projections of future climate change do not include critical factors such as changes in extreme events (especially intense rainfall and drought), pests and disease, and water supplies, all of which have the potential to significantly affect agricultural production.

Forestry and Livestock Will Also Be Affected by Climate Change

Commercial forestry will be affected by factors similar to those affecting crop production and natural forest ecosystems. Climate models project that global timber production will increase and shift poleward due to changes in temperature, longer growing seasons, and enhanced CO_2. However, as with projections of agricultural changes, these models typically exclude potentially important factors such as pests, diseases, and water availability, making the results somewhat uncertain.

Livestock respond to climate change directly through heat and humidity stresses and are affected indirectly by changes in forage quantity and quality, water availability, and disease. Because heat stress reduces milk production, weight gain, and reproduction in livestock, the production of pork, beef, and milk is projected to decline in the United States with warming temperatures, especially with increases above 5.4°F (3°C).

Climate Change Impacts on Fisheries and Aquaculture Are Less Well Understood

The impacts of climate change on seafood are far less well known than impacts on agriculture. Year-to-year climate variations cause large fluctuations in fish stocks, both directly and indirectly, and this has always posed a challenge for effective fisheries management. Similar sensitivities to longer time-scale variations in climate have been documented in a wide range of fish species around the globe. Shifts in fisheries distributions are expected to be most pronounced for U.S. fisheries in the North Pacific and North Atlantic, since future temperature increases are projected to be greatest at these higher latitudes and warming will be coupled to major habitat changes driven by reduced sea ice. The effects of ocean acidification (described above) may be even more important for fisheries than the effects of rising temperatures, although they are currently even more uncertain. Many fished species, including invertebrates like oysters, clams, and scallops, produce shells as adults or as larvae, and shell production could be compromised by increased acidification. Other species fished by humans rely on shelled plankton as their primary food source, and projected declines in these plankton species could have major impacts on fished species higher in the food chain. Finally, acidification can disrupt a variety of physiological processes beyond the production of shells.

Freshwater fisheries face climate challenges similar to those of marine fisheries. Complex interactions among multiple factors such as elevated temperature, reduced dissolved oxygen, increased stratification of lakes, and elevated aquatic pollutant toxicity at higher temperatures pose particular challenges to freshwater fisheries and make projections uncertain. Indirect effects such as altered streamwater flows, changing lake levels, and extreme weather events, coupled with the inability of freshwater fish to move between watersheds, will affect freshwater fisheries, but detailed projections are highly uncertain. Cold-water species such as trout and salmon appear particularly sensitive.

Aquaculture is growing rapidly in the United States and elsewhere as the availability of wild seafood declines. Impacts of climate change on aquaculture are not well studied, but ocean acidification and the difficulty of moving aquaculture infrastructure to new locations as fish habitats shift may pose significant challenges to aquaculture production.

Science for Adaptation in Agricultural Systems

The ability of farmers and the food production, processing, and distribution system to adapt to climate change will to a large extent determine the impacts of climate change on food production. Proposed short-term adaptation strategies include changes in farming locations; shifts in planting dates and crop varieties; increasing storage capacity, irrigation and chemical application; livestock management; and broader-level efforts such as investments in agricultural research (see the companion report *Adapting to the Impacts of Climate Change* [NRC, 2010a]). However, not all farmers have access to these strategies. Small farms, farmers with substantial debt, and farmers without their own land are much more likely to suffer large negative impacts on their livelihoods.

Models that incorporate possible responses of farmers and markets to climate change generally project only small impacts on the agricultural economy of the United States. However, these models do not incorporate costs of adaptation, rates of technological change, changes in pests or diseases, or extreme events like heat waves, heavy rainfall, and flooding. Further research will thus be needed to develop a comprehensive and detailed understanding of how climate change will influence U.S. agricultural production and economics. Understanding of international food supplies, distribution, trade, and food security also remains quite limited.

Food Security

Food security—which includes availability of food, access to food, safety of the food system, and resilience to income or food price shocks—is affected by climate change as well as a multitude of nonclimatic factors such as economic markets and agricultural policies. Because the global food system is interconnected, it is not possible to view U.S. food security in isolation. Food security in the developing world affects global political stability and, thereby, U.S. national security (see below). Studies that project the number of people at risk of hunger from climate change are highly uncertain but indicate that the outcome depends strongly on socioeconomic development, since affluence tends to reduce vulnerability.

Modifying Food Production Systems Could Potentially Help Limit the Magnitude of Future Climate Change

Food production systems are not only affected by climate change; they also contribute to it through GHG emissions of CO_2, CH_4 (primarily from livestock and flooded rice paddies), and N_2O (primarily from fertilizer use). Recent global assessments conclude that agriculture accounts for about 10 to 12 percent of total global human emissions of GHGs. With the intensification of agriculture that will be required to feed the world's growing and increasingly affluent population, these emissions are projected to increase. Many options are available to manage agricultural and livestock systems to reduce emissions, such as changes in feed and feeding practices, manure management, and more efficient fertilizer application. At a landscape level, management of agricultural lands presents opportunities to reduce atmospheric concentrations of CO_2 by sequestering soil carbon, shifting to crops with higher carbon storage potential, and reducing forest clearing for agricultural expansion. Neither the factors that affect the ability of farmers to adopt these types of management practices nor the incentives and institutions that would foster adaptation have been well studied.

Research Needs for Advancing Science on Agriculture, Fisheries, and Food Production in the Context of Climate Change

A broad range of research is needed to understand the impacts of climate change on food production systems and to develop strategies that assist in both limiting the magnitude of climate change through management practices and reducing vulnerability and increasing adaptive capacity in regions and populations in the United States and other parts of the world. Some critical research needs, which are explored in further detail in Chapter 10, are listed below.

- Improve understanding and models of response of agricultural crops and fisheries to climate and other environmental changes.
- Expand observing and monitoring systems.
- Assess food security and vulnerability in the context of climate change.
- Develop approaches to evaluate trade-offs and synergies in managing agricultural lands and in managing ocean resources.
- Develop and improve technologies, management strategies, and institutions to reduce GHG emissions from agriculture and fisheries and to enhance adaptation to climate change.

PUBLIC HEALTH[13]

Weather and climate influence the distribution and incidence of a variety of public health outcomes. Indeed, any health outcome that is influenced by environmental conditions may be impacted by a changing climate. However, the causal chain linking climate change to shifting patterns of health threats and outcomes is complicated by factors such as wealth, distribution of income, status of public health infrastructure, provision of preventive and acute medical care, and access to and appropriate use of health care information. Additionally, the severity of future health impacts will be strongly influenced by concurrent changes in nonclimatic factors as well as strategies to limit and adapt to climate change.

Extreme Temperatures and Thermal Stress

Heat waves are the leading causes of weather-related morbidity and mortality in the United States, and hot days and hot nights have become more frequent and more

[13] For additional discussion and references, see Chapter 11 in Part II of the report.

intense in recent decades. Their frequency, intensity, and duration are projected to increase, especially under the higher warming scenarios. Warming temperatures may also reduce exposure and health impacts associated with cold temperatures, although the extent of any reduction is highly uncertain, and analyses and projections of the impacts of temperature changes on human health are complicated by other factors. In particular, death rates depend on a range of circumstances other than temperature, including housing characteristics and personal behaviors, and these have not been extensively studied in the context of future climate projections.

Severe Weather

Deaths and physical injuries from hurricanes, tornadoes, floods, and wildfire occur annually across the United States. Direct morbidity and mortality increase with the intensity and duration of such events. As a general trend, climate change will lead to an increase in the intensity of rainfall and the frequency of heat waves, flooding, and wildfire. Uncertainties remain in projections of future storm patterns, including hurricanes. The number of deaths and injuries that result from all of these extreme events can be decreased through advanced warning and preparation. Changes in severe weather events may also lead to increases in diarrheal disease and increased incidence of respiratory symptoms, particularly in developing countries. Mental health impacts are often overlooked in the discussion of climate change and public health. Severe weather often results in increased anxiety, depression, and even posttraumatic stress disorder.

Infectious Diseases

The ranges and impacts of a number of important pathogens may change as a result of changing temperatures, precipitation, and extreme events. Increasing temperatures may increase or shift the ranges of disease vectors (and their associated pathogens), including mosquitoes (malaria, dengue fever, West Nile virus, Saint Louis encephalitis virus), ticks (Rocky Mountain spotted fever, Lyme disease, and encephalitis), and rodents (hantavirus and leptospirosis). Consequently, additional people will be exposed to infectious diseases in many parts of the world. Several pathogens that cause food- and waterborne diseases are sensitive to ambient temperature, with faster replication rates at higher temperatures. Waterborne disease outbreaks are also associated with heavy rainfall and flooding and, therefore, may also increase.

Air Quality

Poor air quality—specifically increased ground-level ozone and/or aerosol concentrations—results in increased incidence of respiratory illness. For example, acute ozone exposure is associated with increased hospital admissions for pneumonia, chronic obstructive pulmonary disease, asthma, and allergic rhinitis, and also with premature mortality. Temperature and ozone concentrations are closely connected; projected increases in temperatures in coming decades may increase the occurrence of high-ozone events and related health effects. Climate change could also affect local to regional air quality through changes in chemical reaction rates, boundary layer heights that affect vertical mixing of pollutants, and changes in airflow patterns that govern pollutant transport. In addition to air quality problems driven by pollution, preliminary evidence suggests that allergen production by species such as ragweed increases with high temperature and/or high CO_2 concentration.

The relationship between climate change, air quality, and public health is further complicated by the fact that policies designed to limit the magnitude of climate change may be at odds with improving public health outcomes. For example, reducing aerosol concentrations would reduce air pollution–related health impacts, but the resulting changes in atmospheric reflectivlity could further increase temperatures.

Vulnerable Populations

Vulnerability to the public health challenges discussed above will vary within and between populations. Overall, older adults, infants, children, and those with chronic medical conditions tend to be more sensitive to the health impacts of climate change. Susceptibility varies geographically, with the status of public health infrastructure playing a large role in determining vulnerability differences between populations.

Research Needs for Advancing Science on Climate Change and Public Health

Additional research is needed to clarify exposure-response relationships and impacts of climate change on human health, identify effective and efficient adaptation options, and quantify the trade-offs and co-benefits associated with responses to climate change in other sectors. Some key research needs, which are explored in further detail in Chapter 11, include the following:

- Systematically assess current and projected health risks associated with climate change.
- Carry out research on the feedbacks and interactions between air quality and climate change.
- Characterize the differential vulnerabilities of particular populations to climate-related impacts and the multiple stressors they already face or may encounter in the future.
- Identify effective, efficient, and fair adaptation measures to deal with health impacts of climate change.
- Develop integrated approaches to evaluate ancillary health benefits (and unintended consequences) of actions to limit or adapt to climate change.
- Develop better understanding of informing, communicating with, and educating the public and health professionals as an adaptation strategy.

CITIES AND THE BUILT ENVIRONMENT[14]

Cities now house the majority of the world's population and are expected to continue to grow more rapidly than nonurban areas. Cities and other built-up areas contribute to global climate change through their consumption—including construction materials, energy, water, and food—and their role as the focus for most industrial production. They also contribute to local climate change via the positive feedbacks on warming associated with the built environment. Given their concentration of people, industry, and infrastructure, cities and built environments are expected to face significant direct and indirect impacts from climate change. These include impacts associated with sea level rise because a large number of cities in the United States and worldwide are located in coastal zones. Just as cities help drive climate change, cities also offer opportunities for limiting the magnitude of climate change, and many cities have also started to consider options for adapting to climate change.

Cities Play a Major Role in Driving Climate Change

As the venue for the majority of the world's production and consumption, cities are the geographical loci of energy use, which is the primary source of GHG emissions. This role of cities grows even more significant when their environmental footprint is considered, including, for example, the impact of urban dwellers' emissions on local and regional air pollution and of their materials consumption on distant deforestation.

[14] For additional discussion and references, see Chapter 12 in Part II of the report.

Built-up areas also change the reflectivity of the terrestrial surface, primarily through increased dark surfaces (e.g., roads, rooftops), which contribute to the urban heat island effect.

Impacts of Climate Change on Cities and Other Human Settlements

Without effective steps to limit and adapt to climate change, cities will face a number of climate-related challenges. For example, an increase in warm temperature extremes, coupled with the heat island effect, could increase heat-related health problems, especially for vulnerable populations. Temperature increases will also increase periods of peak energy demands and, in conjunction with other climate changes, are expected to worsen urban air pollution. In many cases, this pollution could extend well beyond the boundaries of cities, potentially affecting ecosystems and crop production on regional scales. Sea level rise and more intense storm surges are of concern for the 635 million people worldwide who live less than 33 feet (10 meters) above sea level, many of them in coastal cities. Cities and settlements adjacent to fire-prone habitats are projected to confront increasing threats of fire, and desert cities, such as those in the American West, will likely confront water shortages.

Potential for Changes in Cities to Limit Future Climate Change

As the geographical focus of most production and consumption, cities offer opportunities to reduce GHG emissions in both absolute and per capita terms, while also improving air quality and urban heat island effects. Many of these opportunities are ultimately tied to the design and geometry of cities, which can foster more or less energy use and emissions per capita as well as shape urban ecosystem function and biotic diversity. Altering surface reflectivity through changes in impervious features (such as white and green roofs) is another potential action that warrants consideration in many cities (see Chapter 15).

Adapting to Climate Change in Cities

Cities face all the challenges that any other sector encounters in regard to adaptation, but research on urban adaptation has only recently begun in earnest. Attention to date has focused on infrastructure and strategies such as emergency preparedness and response. In addition, where resource stresses have already mounted, such as water shortages in the American West, local and regional entities have begun planning to address their vulnerability to climate change in the context of specific natural resources. Understanding options for adaptation and preparing to adapt in cities requires attention to differences in vulnerability among subpopulations (e.g., different economic groups, age groups) and across cities of different size, structure, and location.

Research Needs for Advancing Science on Cities and the Built Environment in the Context of Climate Change

Research on the special vulnerabilities of cities and built-up areas to climate change is needed, as is research on the response options available for cities to limit the magnitude of climate change or adapt to its impacts. Some key research needs, which are explored in further detail in Chapter 12, include the following:

- Characterize and quantify the contributions of urban areas to both local and global changes in climate.
- Assess the vulnerability of cities and their residents to climate change, including the relative vulnerability of different populations and different urban forms (e.g., design, geometry, and infrastructure) and configurations relative to other human settlements.
- Develop and test approaches for limiting and adapting to climate change in the urban context, including, for example, the efficacy of and social considerations involved in adoption and implementation of white and green roofs, landscape architecture, smart growth, and changing rural-urban socioeconomic and political linkages.
- Improve understanding of the links between air quality and climate change, including measurements, modeling, and analyses of socioeconomic benefits and trade-offs associated with different GHG emissions-reduction strategies in the context of air quality, especially strategies that may simultaneously benefit both climate and air quality.
- Improve understanding of urban governance capacity and develop effective decision-support tools and approaches for decision making under uncertainty, especially when multiple governance units may be involved.

TRANSPORTATION[15]

In the United States, the transport of goods and services is highly reliant on a single fuel—petroleum—about 60 percent of which is imported. Almost 28 percent of U.S. GHG emissions can be attributed to the transportation sector, with the overwhelm-

[15] For additional discussion and references, see Chapter 13 in Part II of the report.

ing share in the form of CO_2 emitted from combustion of petroleum-based fuels. The transportation sector also emits other pollutants that endanger human health. The transportation sector thus stands at the nexus of climate change, human health, economic growth, and national security.

Transportation Is a Major Driver of Climate Change

Between 1970 and 2007, U.S. transportation energy use and accompanying GHG emissions nearly doubled. This occurred even as the efficiency of light- and heavy-duty vehicles and aircraft increased, because increases in efficiency were offset by an even larger growth in overall transportation activity. Additionally, although the fuel efficiency of passenger vehicles improved, a large part of this improvement was offset by increases in vehicle size and weight, so the average fuel economy (miles per gallon) of new vehicles has been essentially stagnant for two decades.

Limiting Transportation-Related Emissions

Reducing the total volume of transportation activity is one way to limit GHG emissions from this sector. The most obvious target for such reductions is the transport of passengers and goods on highways, which is responsible for 75 percent of the energy used in transportation. Reducing traffic volume is difficult, however, in light of the interconnections among such factors as choices about where to live and work, the built environment (see Chapter 12), and the availability and flexibility of transportation options.

Improving the fuel economy of highway vehicles and shifting transportation activities away from highways and to modes that have the potential to be more efficient (such as rail and public transit) are also important approaches to reducing emissions. However, whether an alternative mode provides net emissions benefit depends on how it is used. For example, except in a few dense urban corridors, such as in New York City, load factors are not high enough to make public transit less energy and emissions intensive (per passenger-mile) than passenger cars, especially outside of rush hours. The "container revolution"—a shift from truck to rail (and ocean) carriers—has increased efficiency of the transportation of goods. The NRC's Transportation Research Board is currently conducting an in-depth analysis of the technical potential for reducing the energy (and hence emissions) intensity of freight movement.

Improving Efficiency

Many recent studies have pointed out opportunities to improve the efficiency of petroleum-fueled vehicles. In new vehicles, fuel consumption (and GHG emissions) per passenger-mile can be reduced by improving today's gasoline-fueled and diesel oil–fueled vehicles, by shifting to hybrid or electric vehicles, and by improving today's hybrid vehicles. The extent to which these changes result in reduced emissions will depend on consumer preferences regarding vehicle weight and power. Reductions in emissions intensity will depend crucially on consumers' willingness to opt for constant or reduced vehicle weight and power, so as not to offset efficiency improvements. *America's Energy Future* (NRC, 2009d) judged that considerable reductions in vehicle weight will be required to meet the newest U.S. fuel economy standards for light-duty vehicles. Energy efficiency improvements are also under way in commercial passenger aircraft, but they are not expected to be large enough to counter the expected growth in demand for air travel over the next several decades.

Alternative Transportation Fuels

In addition to improving the efficiency of vehicles and other transportation modes, biofuels, grid-based electricity, and hydrogen fuel cells could supplement or replace current transportation fuels. However, it is important to consider the full impact of the fuel cycle when considering such approaches. Emissions are reduced only if these alternative fuels are produced through low-emissions processes. Further details can be found in Chapter 13 and in the recent report *Liquid Transportation Fuels* (NRC, 2009g).

Climate Change Can Affect Transportation Systems in a Number of Ways

For example, increases in the number or intensity of heat waves could affect thermal expansion on bridge joints and paved surfaces, deform rail tracks, and reduce the load limits of airplanes (because warmer air provides less aerodynamic lift). Increases in Arctic temperatures are associated with thawing of permafrost and accompanying subsidence of roads, railbeds, runway foundations, and pipelines. On the other hand, higher Arctic temperatures could provide longer ocean transportation seasons and possibly make a northwest sea route available. Rising sea levels could increase flooding and erosion of transportation infrastructure in coastal areas, and changes in storm patterns could lead to disruptions in transportation services and infrastructure designed for historical climate conditions.

Adapting to Climate Change

Engineering options are already available for strengthening and protecting transportation facilities such as bridges, ports, roads, and railroads from coastal storms and flooding. The development and implementation of technologies that monitor major transportation facilities and infrastructure is an option, as is the development and reevaluation of design standards. However, relatively little attention has been given to evaluation approaches for where and when such options should be pursued, or to the potential co-benefits or unintended consequences of them.

Research Needs for Advancing Science on Climate Change and Transportation

Transportation systems contribute to GHG emissions and are affected by the resultant climate changes. Research is needed to better understand the nature of these impacts as well as ways to reduce GHG emissions from the transportation sector. Some key research needs, which are explored in further detail in Chapter 13, include the following:

- Improve understanding of what controls the volume of transportation activity and what strategies might be available to reduce volume.
- Conduct research on the most promising strategies for promoting the use of less fuel-intensive modes of transportation.
- Continue efforts to improve transportation efficiency.
- Accelerate the development and deployment of alternative propulsion systems, fuels, and supporting infrastructure.
- Advance understanding of how climate change will impact transportation systems and develop approaches for adapting to these impacts.

ENERGY SUPPLY AND USE[16]

The United States is responsible for 20 percent of worldwide energy consumption, and 86 percent of the domestic energy supply comes from combustion of fossil fuels. The CO_2 emitted by these activities constitutes a significant portion of total U.S. GHG emissions. Considerable research has focused on the role of the energy sector in emissions of GHGs and on the development of technologies and strategies that could result in increased energy efficiency as well as energy sources that release fewer or no GHGs. Another potential strategy for reducing energy-related emissions—and a key research topic—capturing CO_2 during or after combustion and sequestering it from the atmo-

[16] For additional discussion and references, see Chapter 14 in Part II of the report.

sphere. A small amount of research has also focused on the implications of deployment of energy technologies on human and environmental systems.

Energy Efficiency

Many proposed strategies to limit the magnitude of future climate change focus on increasing energy efficiency, especially in the near term. A substantial body of research backs up the technical potential for large energy efficiency improvements. For example, the recent report *Real Prospects for Energy Efficiency in the United States* (NRC, 2009c) included a comprehensive review of information on the performance, costs, and GHG emissions reducing potential of different energy efficient technologies and processes for residential and commercial buildings, industry, and transportation.

While a number of proven technologies are available, a host of economic, behavioral, and institutional factors have hampered the United States' ability to realize these efficiency improvements and associated emissions reductions. Many of these factors have been characterized in the scientific literature (see Chapter 14), and while research has shed some light on ways to overcome these barriers, more work is needed. For example, input from social science research can inform the design of policies, programs, and incentives that are more consistent with knowledge about human behavior and consumer choices.

Low-Carbon Fuels for Electricity Production

Energy systems that do not rely on fossil fuels and will ultimately be needed to limit the magnitude of future climate change. Switching from one fossil fuel to another having lower emissions (e.g., from coal to natural gas for power generation) also remains an important near-term option. Increasing the efficiency of power generation (for example, by adding combined-cycle technology to natural gas-fueled plants) can also contribute to lower carbon emissions per unit of energy produced. However, greater use of technologies with low or zero emissions would be needed to dramatically reduce emissions. These technologies include nuclear energy—which currently provides about 20 percent of U.S. electricity generation—and technologies that exploit energy from renewable resources, including solar, wind, hydropower, biomass, and geothermal energy.

Renewable sources currently account for only about 5 percent of total electricity generation, but there is potential for growth. Many will require advances in technology that optimize performance and lower cost in order to be widely adopted. Both renewable and nuclear technology have the potential to provide a large fraction of U.S. electricity supply, but there are a number of distribution, cost, risk, and public acceptance issues that remain to be addressed.

Capture and Storage of CO_2 During or After Combustion

Fossil fuels will probably remain an important part of the U.S. energy system for the near future, in part because of their abundance and the legacy of infrastructure investments. Carbon capture and storage (CCS) technology could be used to remove CO_2 from the exhaust gases of power plants fueled by fossil fuels or biomass (as well as exhaust gases from industrial facilities) and sequester it away from the atmosphere in depleted oil and gas reservoirs, coal beds, or deep saline aquifers. Research to evaluate the technical, economic, and environmental impacts, and legal aspects, of CCS is a key research need. A number of methods and strategies have also been proposed to capture and sequester CO_2 from ambient air. Some of these, such as iron fertilization of the oceans, were mentioned above. Other direct carbon capture technologies, such as air filtration, are in early phases of study.

Effects of Climate Change on Energy Systems

Climate change is expected to affect energy system operations in several ways. For example, increases in energy demands for cooling and decreases in energy demands for heating can be expected across most parts of the country. This could drive up peak electricity demand, and thus capacity needs, but could also reduce the use of heating oil and natural gas in winter. Water limitations in parts of the country, and increased demand for water for other uses, could result in less water for use in cooling at thermal electric plants. Increased water temperatures may also reduce the cooling capacity of available water resources. Water flows at hydropower sites may increase in some areas and decrease in others. Changes in extreme weather events—including hurricanes, floods, and droughts—may disrupt a wide range of energy system operations, including transmission lines, oil and gas platforms, ports, refineries, wind farms, and solar installations.

Research on Adapting to Climate Change in the Energy Sector

Actions to help the energy sector adapt to the effects of climate change include increasing regional electric power generating capacity; accounting for changes in patterns of demand; hardening infrastructure to withstand extreme events; developing electric power generation strategies that use less water; instituting contingency planning for reduced hydropower generation; increasing resilience of fuel and electricity delivery systems; and increasing energy storage capacity. Research is needed to develop and improve analytical frameworks and metrics for identifying the most vulnerable infrastructure and most effective response options.

Research Needs for Advancing Science in the Energy Supply and Consumption Sector

Because energy is a dominant component of human GHG emissions, major investments are needed in both the public and private sectors to accelerate research, development, and deployment of climate-friendly energy technologies. Research is also needed on behavioral and institutional barriers to adoption of new energy technologies. It is critical that energy research not be conducted in an isolated manner, but rather using integrated approaches and analyses that investigate energy supply and use within the greater context of efforts to achieve sustainable development goals and other societal concerns. Some specific research needs, discussed in further detail in Chapter 14, include the following:

- Develop new energy technologies along with effective implementation strategies.
- Develop improved understanding of behavioral impediments at both the individual and institutional level to reducing energy demand and adopting energy efficient technologies.
- Develop analytical frameworks to evaluate trade-offs and synergies between efforts to limit the magnitude and adapt to climate change.

SOLAR RADIATION MANAGEMENT[17]

The term *geoengineering* refers to deliberate, large-scale manipulations of Earth's environment designed to offset some of the harmful consequences of GHG-induced climate change. Geoengineering encompasses two different classes of approaches: carbon dioxide removal (CDR) and solar radiation management (SRM) (see Figure 2.9). CDR approaches (also referred to as postemission GHG management, atmospheric remediation, or carbon sequestration methods), several of which were discussed in the sections above, involve removal and long-term sequestration of atmospheric CO_2 (or other GHGs) in forests, agricultural systems, or through direct air capture and geologic

[17] For additional discussion and references, see Chapter 15 in Part II of the report.

storage. Additional details about these techniques and their implications can be found in the companion report *Limiting the Magnitude of Future Climate Change* (NRC, 2010c).

SRM approaches, the focus of this section, are those designed to increase the reflectivity of Earth's atmosphere or surface in an attempt to offset some of the effects of GHG-induced climate change. SRM approaches seek to either reduce the amount of sunlight reaching Earth's surface or reflect additional sunlight back into space. There is a limited body of research on this topic. While some SRM approaches may be technologically and economically feasible (only considering direct deployment costs), they all involve considerable risk and potential for unintended (albeit currently understudied) side effects. It is unclear at the present time, therefore, whether SRM could actually reduce the overall risk associated with climate change and whether it could realistically be employed as quickly as is technically possible, especially in light of the full range of environmental and sociopolitical complexities involved.

Although few, if any, voices are promoting SRM as a near-term alternative to GHG emissions-reduction strategies, the concept has recently been gaining more serious attention as a possible "backstop" measure, because strategies attempted to date have failed to yield significant emissions reductions, and climate trends may become significantly disruptive or dangerous. Further research is necessary to better understand the physical science of the impacts and feasibility of SRM as well as issues related to governance, ethics, social acceptability, and political feasibility of planetary-scale, intentional manipulation of the climate system.

Proposed Solar Radiation Management Approaches

The SRM approaches proposed to date can be divided into four broad categories: space, stratosphere, cloud, and surface based. Space-based proposals involve placing satellites with reflective surfaces in space. However, to counteract GHG-induced warming, 10 square miles of reflective surface would need to be put into orbit each day for as long as CO_2 emissions continue to increase at current rates. The most widely discussed option for stratosphere-based SRM is the injection of sulfate aerosols, which would reflect some amount of incoming solar radiation back to space, offsetting some of the warming associated with GHGs. Another SRM option is to "whiten" clouds, or make them more reflective, by increasing the number of water droplets in the clouds. This could potentially be achieved over remote parts of the ocean by distributing a fine seawater spray in the air. Surface-based options include whitening roofs in the built environment, and planting more reflective crops. While these proposals merit

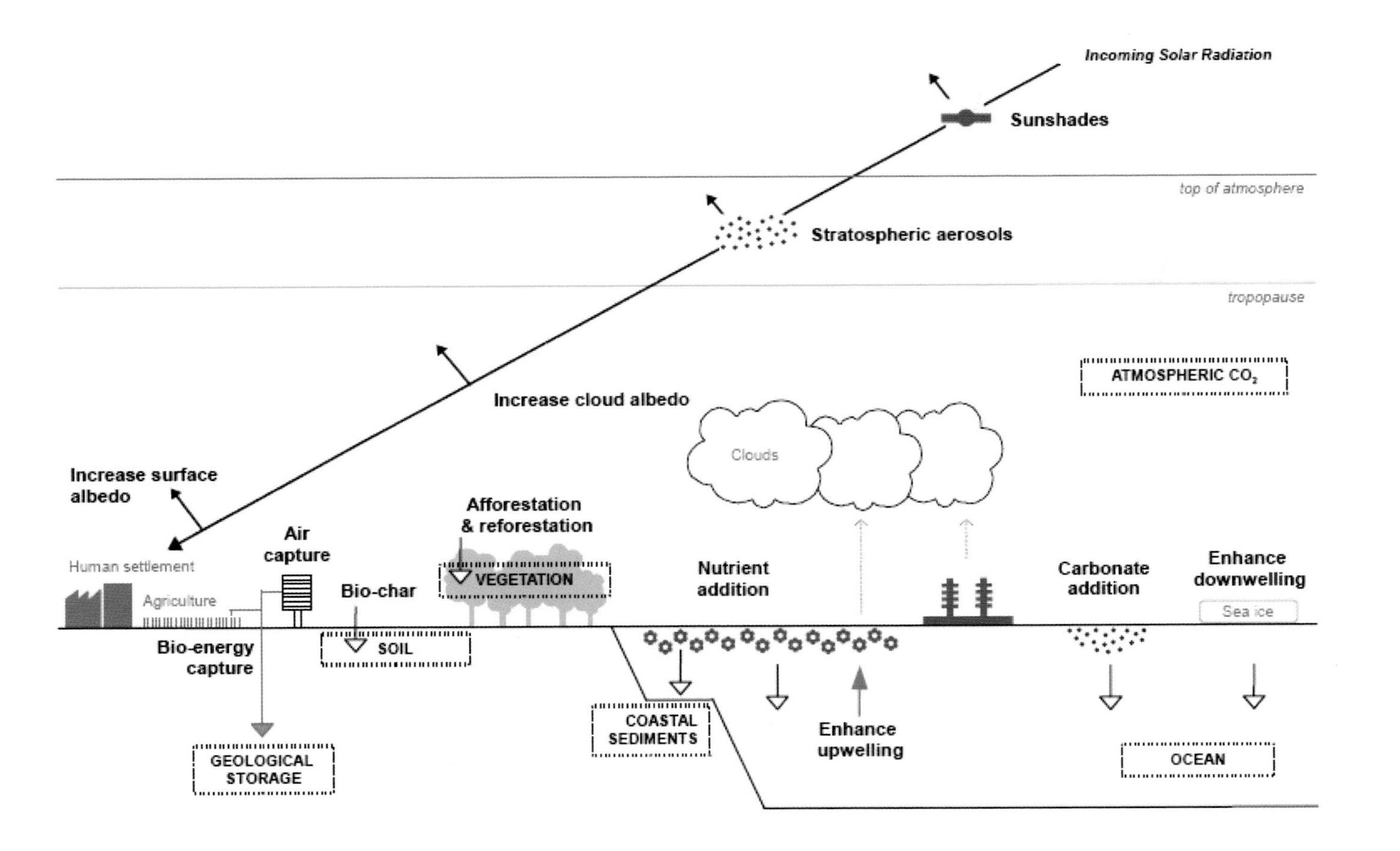

FIGURE 2.9 Various geoengineering options, including both solar radiation management and carbon dioxide removal. For further details see Figure 15.1. SOURCE: Lenton and Vaughn (2009).

further research, their efficacy and environmental consequences are not currently well understood.

Potential Drawbacks and Unintended Consequences

The overall environmental impacts of SRM approaches are not well characterized, and all proposals have the potential for unintended negative consequences. For example, approaches that are intended to offset globally averaged warming may still lead to local- or regional-scale imbalances in climate forcing that could produce large regional changes. Several analyses also suggest that a sudden increase in stratospheric sulfate aerosol could potentially enhance losses of stratospheric ozone for several decades, especially in the Arctic. Additionally, since aerosols remain in the atmosphere for a much shorter time than GHGs, abandonment of aerosol injection could cause warming at a rate far greater than what is estimated in the absence of SRM. These and other issues, including the impact of SRM on precipitation and the hydrologic cycle, are not well understood. Finally, it should be noted that a major shortcoming of SRM approaches is that, while they have the potential to offset GHG-induced warming of the atmosphere, they would not offset ocean acidification or other impacts of elevated CO_2.

Governance Issues

Due to the global nature of SRM, and especially considering the drawbacks and potential negative impacts, an international framework is needed to govern SRM research, development, and possible deployment. Important components of such a framework include a clear definition of "climate emergency" that would trigger deployment and criteria for whether, when, and how SRM approaches should be tested and/or deployed. Unilateral SRM testing or deployment could lead to international tension, distrust, or even conflict. Public involvement in SRM-related decision making, including research activities, is likewise important since public acceptance is a key issue in informing governance decisions.

Ethical Issues

Intentional climate alteration, including SRM, raises significant issues with respect to ethics and responsibility. A key consideration in the deployment of SRM, as with other responses to climate change, is the distribution of risks among population groups in the present generation, as well as future generations. Some have suggested that SRM research efforts may also pose a "moral hazard" by detracting from efforts to reduce GHG emissions or to adapt to the impacts of climate change. SRM and other geoengineering approaches also raise deep questions about humans' relationship with nature, many of which are beyond the scope of this report.

Research Needs for Advancing Solar Radiation Management

It is beyond the scope of this report to design a research program on SRM, or even to determine the scope, scale, priorities, or goals of such a program. However, the various SRM proposals and their consequences need to be examined, as long as such research does not replace or reduce research on fundamental understanding of climate change or other approaches to limiting climate change or adapting to its impacts. Some key SRM-related research needs, discussed in Chapter 15, include the following:

- Improve understanding of the physical potential and technical feasibility of SRM and other geoengineering approaches.
- Evaluate the potential consequences of SRM approaches on other aspects of the Earth system, including ecosystems on land and in the oceans.
- Develop and evaluate systems of governance that would provide a model for how to decide whether, when, and how to intentionally intervene in the climate system.
- Measure and evaluate public attitudes and develop approaches that effectively inform and engage the public in decisions regarding SRM.

NATIONAL AND HUMAN SECURITY[18]

Climate change will influence human and natural systems that are linked throughout the globe, creating important implications for bilateral and multilateral relations and for national, international, and human security. Changes in temperature, sea level, precipitation patterns, and other elements of the physical climate system can add substantial stresses to infrastructure and especially to the food, water, energy, and ecosystem resources that societies use. Key concerns regarding the interactions between climate change and security include direct impacts on military operations; potential impacts to regional strategic priorities; causal links between environmental scarcity and conflict; the role of environmental conservation and collaboration in promoting peace; and relationships between environmental quality, resource abundance, and

[18] For additional discussion and references, see Chapter 16 in Part II of the report.

human security. In general, these areas are much less well understood than the causes and more direct consequences of climate change.

Military Operations

Climate change may affect military assets and operations directly: through physical stresses on military systems and personnel, severe weather constraints on operations due to increased frequency and intensity of storms and floods, increased uncertainty about the effects of Arctic ice and ice floes on navigation safety both on and below the ocean surface, or risks to coastal infrastructure due to sea level rise. Climate change is expected to increase heavy rainfalls and floods, droughts, and fires in many parts of the world and could lead to changing storm patterns. This may generate a change in military missions because the U.S. military has substantial logistical, engineering, and medical capabilities that have been used to respond to emergencies in the United States and abroad. Finally, the U.S. military is a major consumer of fossil fuels and could potentially play a major role in reducing U.S. GHG emissions.

International Relations and National Security

Climate change has the potential to disrupt international relations and raise security challenges through impacts on specific assets and resources. For example, loss of Arctic sea ice will increase the value of Arctic navigation routes. The legal status of the Northwest Passage in particular has long been contested, but the prospect of it becoming more widely usable raises the stakes substantially. Another possible disruption to international relations is the prospect of substantial mineral reserves under the Arctic Ocean. Climate change will also affect shorelines and in some cases "exclusive economic zones" and baselines used for projecting national boundaries seaward. Boundaries that could be affected include those in the South China Sea and between the United States and Cuba. Climate-related changes in precipitation and the hydrologic cycle will likely result in changes in flow regimes in international river systems, and this raises the possibility of challenges to interstate relationships, even conflict, over shared water resources. Finally, climate-related changes in food supply and sea level rise-related land losses could potentially result in intra- and interstate migration and refugee-related conflicts.

Treaty Verification

The prospect of binding international agreements on GHG emissions will have important implications for treaty verification and compliance. In particular, measurements of GHG concentrations and emissions are needed to inform national and international policy aimed at regulating emissions, to verify compliance with emissions-reduction policies, and to ascertain their effectiveness. Measurements of GHGs for treaty verification or for financial transactions (carbon trading) will require a higher level of scrutiny than that used in the research domain. Key concerns in such a regime are data security, authentication, reliability, and transparency.

Human Security

The impact of climate change may increase the probability of conflict, and this has become a prominent argument for considering climate change in security analyses. The concept of human security, however, goes far beyond the traditional concerns of national security and conflict and instead includes considerations of access to sufficient food, water, and health care infrastructure, as well as freedom from repression and freedom and capacity to determine one's life path. Analysts have moved toward a more integrative conception of security and threats, one that reflects the lived realities that individuals and communities face. Nevertheless, there are still multiple ways of thinking about human security and no agreement on a policy agenda. Research efforts in this area to date have focused on issues of equity, fairness, vulnerability, and human dignity, and have identified conditions that are critical to maintaining or restoring human security: effective governance systems, healthy and resilient ecosystems, comprehensive and sustained disaster risk-management efforts, empowerment of individuals and local institutions, and supportive values.

Research Needs for Advancing Science on National and Human Security Implications of Climate Change

Scientific understanding of the national and human security implications of climate change are considerably less well understood than many of the other impacts of climate change. As a result, there are a wide variety of research needs for improving understanding of the relationship between climate change and security, including the following:

- Develop improved observations, models, and vulnerability assessments for regions of importance in terms of military infrastructure.
- Build understanding of observations and monitoring requirements for treaty verification.
- Identify areas of potential human insecurity and vulnerability in response to climate change impacts interacting with other social and environmental changes.

DESIGNING, IMPLEMENTING, AND EVALUATING CLIMATE POLICIES[19]

Analyzing different policy options that might be used to limit the magnitude of climate change or promote successful adaptation is a key area of scientific research. Indeed, the ability to comprehensively assess the potential consequences of various climate policies—including the costs, benefits, trade-offs, co-benefits, and uncertainties associated with their implementation—is paramount to informing public- and private-sector decision making on climate change. Despite a broad range of research focusing on policy making and evaluation in general, policy-oriented research focused specifically on climate change and its interaction with natural and social systems has been relatively limited. Because climate change is becoming an increasingly important public policy concern in the United States and many other countries, additional research to support climate policy design and implementation is needed.

International Policies for Limiting the Magnitude and Adapting to the Impacts of Climate Change

At the international level, examples of climate policies include the United Nations Framework Convention on Climate Change, the Kyoto Protocol, and the Copenhagen Accord. Policy options available at the national, regional, and local levels include direct regulation, taxes, cap-and-trade systems for emissions permits, incentive structures and subsidies for voluntary action, technical aid and incentives for the creation and implementation of new technology portfolios, and adaptation options and planning. Research in this area finds that direct regulation, when enforced, can effectively reduce emissions. It also finds that while taxes are cost-effective, they do not guarantee specific emissions-reduction levels and may be hard to adjust, and that the efficacy of tradable permits depends on the structure of the policy. Voluntary agreements can play a role in accelerating technology adoption, but they are less effective in reducing

[19] For additional discussion and references, see Chapter 17 in Part II of the report.

emissions. Finally, whereas incentives and subsidies to develop cleaner technologies maybe be slow and costly, they can complement other emissions-reduction policies.

Monitoring Compliance with Emissions-Reduction Policies

International agreements and policies, to be effective, need to be enforced, verified, and monitored. Standards and certification mechanisms for reducing GHG emissions also need to be created and implemented. Constraints to monitoring compliance with and the effectiveness of such policies include lack of adequate and reliable methods for measuring GHG emissions, lack of mechanisms for accurately accounting for GHG emissions and for offsets, and lack of technical capacity to monitor and enforce policies nationally and across international borders.

Assessing Benefits and Costs of Climate Action

Benefit-cost analyses seek to translate climate change impacts, including lost or gained ecosystem services, into a monetary metric so that they can be compared to estimates of the costs and benefits associated with policies to limit the magnitude or adapt to the impacts of climate change. Alternatively, cost-effectiveness analysis is often used when the costs and benefits of action differ greatly in character, or when the benefits are subject to greater uncertainty or controversy. Cost-effectiveness analysis allows analytically based comparisons of decisions without requiring that all impacts—in this case, damages from climate change and costs of emissions reduction—be reduced to a single metric. Both approaches can be powerful tools for informing decisions, but disagreements about (1) how to value ecosystem services or other resources for which market prices do not exist; (2) how to handle low-probability-high-consequence events, discount rates, and risk aversion; (3) prospects for technological innovation; and (4) how to incorporate distributional and intergenerational equity concerns lead to wide ranges in estimates of the social value of climate actions.

Dealing with Complex and Interacting Policies, Multilevel Governance, and Equity

Effective climate policy making requires analyses that consider the complexity of real policies, how institutions interact across levels of government from global to local, and equity issues. Climate policies are not made in a vacuum. They interact with other climate and nonclimate policies and are often nested across different scales from local to global. In the United States, rapidly emerging local and state climate policy agendas interact with federal policy. It is not yet clear how these interactions will play out and what the net effect will be. The multilevel and hybrid character of climate policy (both for limiting and adapting to climate change) presents opportunities (such as for synergistic outcomes) and challenges (such as one level of decision making constraining or negating the other). One of the most critical challenges is dealing fairly with the distributional effects of climate change impacts. Three main sources of equity concerns shape climate policy debates: historical responsibility for the problem of climate change, who will bear the brunt of its negative impacts, and who will be responsible for solving it. Scientific research cannot answer these questions, but it can provide relevant information to policy makers as they attempt to do so.

Research Needs Related to Climate Policy Development and Implementation

Research needs in this area, explored in further detail in Chapter 17, include the following:

- Continue to improve understanding of what leads to the adoption and implementation of international agreements on climate and other environmental issues and what mechanisms are most effective at achieving their goals.
- Develop and evaluate protocols, institutions, and technologies for monitoring and verifying compliance with international agreements.
- Continue to improve methods for estimating costs, benefits, and cost-effectiveness.
- Develop methods for analyzing complex, hybrid policies.
- Develop further understanding of how institutions interact in the context of multilevel governance and adaptive risk management.
- Develop analyses that examine climate policy from a sustainability perspective, taking account of the full range of effects of climate policy on human well-being, including unintended consequences and equity effects.

CHAPTER THREE

A New Era of Climate Change Research

As the preceding chapter makes clear, scientific research has steadily increased our understanding of climate change, as well as our appreciation of the complexity of the climate system and related, interacting human and environmental systems. The research summarized in Chapter 2, and described in further detail in Part II of the report, has also identified some of the challenges and risks associated with climate change, including some special characteristics and complexities that distinguish it from many other problems faced by society. In this chapter, we summarize some of these characteristics and discuss their implications in terms of the risks and choices faced by decision makers both in the United States and around the world. The chapter also briefly describes some of the actions that decision makers are taking to respond to climate risks, including actions to limit the magnitude of climate change and adapt to its impacts. These emerging responses call for a new era of climate change research, one that not only continues to improve understanding of climate change and the risks associated with it, but that also supports, facilitates, and improves actions taken to respond.

COMPLEXITIES OF CLIMATE CHANGE

Future climate will be unlike the climate of the recent past. Based on available records of atmospheric composition, sea level, and other sources (see Chapter 6), Earth's climate appears to have been relatively stable for roughly the past 10,000 years. Exceptional years, decades, and even centuries have occurred, of course, occasionally creating havoc for civilizations in some regions of the world (see, e.g., Diamond, 2005; Zhang et al., 2007b). However, human societies have generally been well served by assuming that the climate fluctuates around a relatively constant average state, with no long-term trends toward warmer or cooler temperatures, more or less precipitation, or more or fewer extreme events. This is changing, as Earth's climate system—from greenhouse gas (GHG) concentrations to ice cover, precipitation, and a host of other interrelated changes—moves outside the range within which it has fluctuated throughout the 10,000 years of recorded human history. As a result, many of our conventional practices for including climate and climate-related uncertainty in decision making—such as using historical records to plan for the "100-year flood" or the "100-year drought"—will need to be revisited, and new ways of thinking about preparing and adapting to change will need to emerge. Conventional practices may even heighten risks by encouraging us to continue planting vulnerable crop varieties, harvesting threatened resources at unsustainable levels, or building homes and communities in areas at growing risk from fires, floods, or rising sea levels.

Projections of future climate depend strongly on current and future human actions. The magnitude of future climate change and the severity of its impacts are strongly dependent on how human societies produce and use energy, manage natural resources, and take other actions to respond to climate risks and vulnerabilities in the decades ahead. Not only is it impossible to anticipate all of the actions that humans might take, but the consequences of these actions for both the climate system and related human and environmental systems are subject to a large number of uncertainties and unknowns.

Climate change processes have considerable inertia and long time lags. Until GHG emissions are brought below the rate of their removal from the atmosphere, atmospheric concentrations will continue to rise. The most important GHGs remain in the atmosphere for years to centuries and continue to affect Earth's heat balance throughout their atmospheric lifetimes. Other climate change processes also exhibit considerable inertia, which results in delays between GHG emissions and the impacts of climate change. The oceans, for example, warm much more slowly than the atmosphere in response to the buildup of heat-trapping gases. Additionally, many of the sources of GHG emissions, such as power plants and automobiles, have lifetimes of years to decades. Thus, decisions made now will shape the world for generations to come. Research has shown that individuals and organizations have trouble perceiving risks and taking action on problems with such long lead-times.

The sensitivity of the climate system is somewhat uncertain. As discussed in Chapters 2 and 6, scientists have learned a great deal about the response of the climate system to GHGs and other climate forcing agents through a combination of direct observations of recent climate change, indirect evidence of historical climate variations, and climate modeling studies. However, Earth's climate sensitivity—which dictates how much warming would be expected if future emissions were known exactly—remains somewhat uncertain. Thus, it is possible that future temperature changes will lie below the range of current climate model projections. However, it is also possible that realized temperature changes will lie above the projected range. Additionally, climate models cannot currently simulate certain feedback processes in the Earth system, such as those related to changes in ecosystems on land or in the oceans, that could potentially amplify (or reduce) the response to a given climate forcing. Uncertainty in the sensitivity of Earth's climate system also makes it difficult to precisely quantify the effectiveness of actions or strategies that might be taken to limit the magnitude of future climate change.

There may be tipping points or thresholds that, once crossed, lead to irreversible events. Some of the physical and biological feedbacks triggered by climate change can become irreversible when they pass a certain threshold or tipping point. For example, there is general scientific consensus that the Arctic, which is systematically losing summer sea ice thickness and extent on an annual basis, is expected to become permanently ice-free during summers by the middle of the 21st century, regardless of how future emissions change. This change to an ice-free summer Arctic is expected, in part, because of the positive feedback between warming and sea ice melting (see Chapter 6). A number of other possible tipping points and irreversible changes have been identified in the Earth system, and human systems can also experience tipping points, such as the collapse of an economy or political system. Because of the possibility of crossing such thresholds, simple extrapolations of recent trends may underestimate future climate change impacts. Given the complexity of coupled human-environment systems, it is difficult to forecast when a tipping point might be approaching, but the probability of crossing one increases as the climate system moves outside the range of natural variability.

Analyses of impacts resulting from higher levels of climate change are limited. Most scientific analyses of climate change have focused on the impacts associated with a global temperature change of between 3.6°F to 5.4°F (2°C to 3°C) by the end of the 21st century, relative to preindustrial conditions. Yet model-based projections of future global temperature change range from 2°F to more than 11°F, and even larger changes are possible. For comparison, the higher end of the expected range of future temperature change is comparable to the estimated temperature difference between the present climate and the climate at the height of the last ice age, when glaciers covered the sites presently occupied by New York, Chicago, and Seattle and ecosystems around the world were radically different. Although there have been some recent efforts to estimate the impacts that might be associated with global temperature changes of greater than 9°F or 10°F (5°C or 6°C) over the next century (see, for example, University of Oxford, Tyndall Centre, and Hadley Centre Met Office, 2009), relatively little scientific information is available regarding the potential risks posed by such extreme changes in global climate.

Climate change does not act in isolation. As noted in several parts of Chapter 2 and in many of the chapters in Part II, climate change is just one of many stressors affecting human and environmental systems. For example, estuaries and coral reefs are being affected by warming ocean temperatures, ocean acidification, sea level rise, and changes in runoff from precipitation, and these climate-driven impacts interact with other ongoing threats such as pollution, invasive species, coastal development, and overfishing. The impacts of these multiple stresses and interacting environmental changes on food production, water management, energy production, and other critical human activities are associated with important risks in terms of meeting human needs. The prevalence of multiple stresses and the interconnected nature of many climate-related processes also raise significant scientific and management challenges.

Vulnerabilities to climate change vary across regions, societies, and groups. Climate change will unfold in different ways across the United States and across the globe, and different sectors, populations, and regions will be differentially exposed and sensitive to the impacts of these changes. Different groups will also differ in their ability to cope with and adapt to environmental changes. In general, research suggests that the impacts of climate change will more harshly affect poorer nations and communities. Actions taken to limit future climate change and adapt to its impacts also have the potential to cause differential benefits or harm. For example, different communities or regions may experience differential exposure to the unintended side effects of alternative energy production strategies. However, our understanding and ability to predict vulnerability, adaptive capacity, and the side effects of different response strategies are less well established than our understanding of the basic causes and mechanisms of climate change.

Individually and collectively, the complexities described in the paragraphs above make it challenging to analyze the risks posed by climate change. Nonetheless, as described in Chapter 2 and discussed in detail in Part II of the report, the scientific community has high confidence in projections of a number of future climate changes and impacts. For example, (1) water availability will decrease in many areas that are already drought-prone and in areas where freshwater systems are fed by glaciers and snowpack; (2) a higher fraction of rainfall will fall in the form of heavy precipitation events, increasing the risk of flooding; (3) people and ecosystems in coastal zones will be exposed to higher storm surges, saltwater intrusion, and other risks as sea levels rise; and (4) coral reefs will experience widespread bleaching and mortality as a result of increasing temperatures, rising sea levels, and ocean acidification. There is less certainty in other projections, such as the combined impact of CO_2 increases, temperature increases, precipitation changes, and other climate and climate-related changes on agricultural crops and natural ecosystems in different regions of the world, although negative impacts are expected to increase with higher temperatures. Projections of the future—in any sphere—always entail some amount of uncertainty. Nevertheless, as described in the next two sections, decision makers are already starting to take actions to respond to these and other risks associated with climate change, and scientific research can help in a number of ways.

RESPONDING TO CLIMATE RISKS

Based on current scientific understanding of ongoing and projected future changes in climate, and the risks associated with these changes, many decisions makers are now taking or planning actions to limit the magnitude of climate change, to adapt to ongoing and anticipated changes, and to include climate considerations in their decision-making processes. These actions are detailed in the three companion reports to this study (NRC, 2010a-c). For example, in *Informing an Effective Response to Climate Change* (NRC, 2010b), it is noted that 34 states have created climate change action plans, 20 have established emissions-reduction targets, and 15 have developed adaptation plans. Many U.S. cities and counties have also begun to respond to the challenges of climate change, and there is substantial activity at the federal level as well—for example, at least 10 of the 15 cabinet-level agencies and departments have made climate-related decisions. Many private firms are also taking action. At least 475 major companies have provided information on emissions to the Carbon Disclosure Project, over 60 major companies have set emissions-reduction targets, and climate change concerns are affecting the investment decisions of many. Finally, individuals in the United States and around the world are supporting government actions or taking actions themselves. For example, in 2009 one in three Americans rewarded companies that are taking steps to reduce GHG emissions by buying their products, while more than one in four avoided buying products from companies they perceived to be recalcitrant on the issue (Leiserowitz, 2010).

As these decision makers continue to take actions in response to climate change, many issues emerge that science can address. For example, scientific research can

- Project the beneficial and adverse effects of climate change, and their likelihood;
- Identify and evaluate the likely or possible consequences—including unintended consequences—of different decisions and actions taken (or not taken) to respond to climate change;
- Monitor and evaluate the effectiveness and consequences of actions as they are taken;
- Improve the effectiveness of actions before or while they are taken;
- Expand the portfolio of possible actions that might be taken in the future; and

BOX 3.1
Adaptive Risk Management:
Iterative and Inclusive Management of Climate Risks

Because individuals and groups often have trouble making sound decisions in the face of uncertainty, many tools have been developed to enhance our ability to make decisions in the face of risk (Bernstein, 1998; Jaeger et al., 2001). This suite of tools and the logic of their application are often referred to as adaptive risk management or sometimes iterative risk management or risk governance (Arvai et al., 2006; Renn, 2005, 2008). Components of adaptive (or iterative) risk management are discussed in the following paragraphs and are developed in more detail in the companion report *Informing an Effective Response to Climate Change* (NRC, 2010b).

Risk identification, assessment, and evaluation. Risks need to be evaluated by a range of affected stakeholders (who typically have different values and preferences) and by considering a range of factors. These include the impacts of allowing risks to go unmitigated, the costs of different risk-management strategies, and public perceptions and acceptability of risks and/or responses to those risks, as well as broader societal values that tend to favor certain general approaches to managing risk over others (e.g., a precautionary approach versus a cost-benefit or risk-benefit approach).

Iterative decision making and deliberate learning. Because many climate-related decisions will have to be made with incomplete information, and new information can be expected to become available over time (including information about the effectiveness of actions taken), decisions should be revisited, reassessed, and improved over time. This will require deliberate planning and processes for "learning by doing," as well as ongoing monitoring and assessment to evaluate both evolving risks and the effectiveness of responses.

Maximizing flexibility. Whenever decisions with long-term implications can be made incrementally (i.e., in small steps rather than all at once), the risk of making the "wrong" decision now can be reduced by keeping as many future options open as possible.

- Develop, through research on human behavior and decision making itself, improved decision-making processes.

The discussion in the preceding section suggests that the climate is not a system that can be turned quickly and that responses to climate change may be necessary even as more information on risks is collected. Fundamentally, dealing with climate change requires making decisions without complete certainty. Under such conditions, adaptive risk management (Box 3.1) is a useful—and advisable—strategy for responding to climate-related risks as conditions change and we learn more about them.

Maximizing robustness. When decisions have to be made all at once (for example, whether to build a piece of infrastructure), the risk of making the "wrong" decision can be reduced by selecting robust options—options that maximize the probability of meeting identified goals and desirable outcomes while minimizing the probability of undesirable outcomes under a wide range of plausible future conditions.

Ensuring durability. Many climate-related policies will need to remain in place, albeit in modified form, for many decades in order to achieve their intended goals. This requires mechanisms that can ensure the long-term durability of policies and provide stability for investors and society, while allowing for adaptive adjustments over time to take advantage of new information—a significant challenge for policy and institutional design.

A portfolio of approaches. In the face of complex problems, where surprises are expected and much is at stake, it would be unwise to rely on only one or a small number of actions to "solve" the problem without major side effects. A more robust approach would be to employ a portfolio of actions to increase the chance that at least one will succeed in reducing risk and to provide more options for future decision makers.

Effective communication. An essential component of effective risk management is the communication of risks, including the risks associated with different responses, to all involved stakeholders, including public-, private-, and civic-sector stakeholders as well as expert and lay individuals familiar with, or potentially affected by, the risks at hand.

Inclusive process. Since climate-related risks affect different regions, communities, and stakeholders in different ways and to different degrees, stakeholders should be included in significant roles throughout the process of identifying risks and response options, determining and evaluating what risks and responses are "acceptable" and "unacceptable," as well as in the communication and management of the risks themselves (NRC, 2008h).

IMPLICATIONS FOR THE NATION'S CLIMATE RESEARCH ENTERPRISE

The past several decades of research have yielded a great deal of knowledge about climate change, but there is much still to be learned about both ongoing and future changes and the risks associated with them. Moreover, as decision makers respond to the risks posed by climate change, additional knowledge will be needed to assist them in making well-informed choices. For example, decision makers could use additional information about how the Earth system will respond to future GHG emissions, the range of impacts that could be encountered and the probabilities associated with them, the quantifiable and nonquantifiable risks posed by these changes, the options that can be taken to limit climate change and to reduce vulnerability and increase adaptive capacity of both human and environmental systems, and methods for making choices and managing risk in an environment that continues to change.

Because decisions always involve values, science cannot prescribe the decisions to be made, but scientific research can inform decisions and help to ensure and improve their effectiveness. As we enter a time when decision makers are responding to climate change, the nation's climate research enterprise can assist by including both science for understanding and science for supporting responses to climate change. The diverse and complex set of scientific issues to be addressed in this new era of climate change research span the physical, social, biological, health, and engineering sciences and require integration across them. In the next chapter, we discuss the research needs and themes for the nation's climate change science enterprise in this new era.

CHAPTER FOUR

Integrative Themes for Climate Change Research

One of the main tasks assigned to the Panel on Advancing the Science of Climate Change was to identify the additional science needed to improve our understanding of climate change and its interactions with human and environmental systems, including the scientific advances needed to improve the effectiveness of actions taken to respond to climate change. An examination of the research needs identified in the technical chapters of Part II of the report reveals that there is indeed still much to learn. However, our analysis suggests that the most crucial research needs of the coming decades can be captured in seven crosscutting research themes, whether one is interested in sea level rise, agriculture, human health, national security, or other topics of concern. For example, nearly every chapter in Part II calls for improved understanding of human behaviors and institutions, more detailed information about projected future changes in climate, and improved methods for assessing the economic, social, and environmental costs, benefits, co-benefits, and unintended consequences of actions taken in response to climate change.

Box 4.1 lists the seven crosscutting research themes that the panel has identified, grouped into three general categories: research for improving understanding of coupled human-environment systems, research for improving and supporting more effective responses to climate change, and tools and approaches needed for both of these types of research. These seven crosscutting themes are not intended to represent a comprehensive or exclusive list of research needs, nor do the numbers indicate priority order. Rather, they represent a way of categorizing and, potentially, organizing some of the nation's most critical climate change research activities. Most of these themes are integrative—they require collaboration across different fields of study, including some fields that are not typically part of the climate change science enterprise. Moreover, there are important synergies among the seven themes, and they are not completely independent. For example, research focused on improving responses to climate change will clearly benefit from increased understanding of both human systems and the Earth system, and advances in observations, models, and scientific understanding often go hand in hand. Finally, because most of the themes include research that contributes both to fundamental scientific understanding and to more informed decision making, research under all seven themes would benefit from

BOX 4.1
Crosscutting Themes for the New Era of Climate Change Research

Research to Improve Understanding of Human-Environment Systems
1. Climate Forcings, Feedbacks, Responses, and Thresholds in the Earth System
2. Climate-related Human Behaviors and Institutions

Research to Support Effective Responses to Climate Change
3. Vulnerability and Adaptation Analyses of Coupled Human-Environment Systems
4. Research to Support Strategies for Limiting Climate Change
5. Effective Information and Decision-Support Systems

Research Tools and Approaches to Improve Both Understanding and Responses
6. Integrated Climate Observing Systems
7. Improved Projections, Analyses, and Assessments

increased dialogue with decision makers across a wide range of sectors and scales. As discussed in Chapter 5, these characteristics all point to the need for an expanded and enhanced climate change science enterprise—an enterprise that is comprehensive, integrative, interdisciplinary, and better supports decision making both in the United States and around the world.

In the following sections, the seven integrative, crosscutting research themes identified by the panel are discussed in detail. Our intent is to describe some of the more important scientific issues that could be addressed within each theme, to show how they collectively span the most critical areas of climate change research, and to demonstrate the vital importance of research progress in all of these areas to the health and well-being of citizens of the United States as well as people and natural systems around the world. Issues related to the implementation of these themes are explored in the next chapter.

THEME 1: CLIMATE FORCINGS, FEEDBACKS, RESPONSES, AND THRESHOLDS IN THE EARTH SYSTEM

Scientific understanding of climate change and its interactions with other environmental changes is underpinned by empirical and theoretical understanding of the Earth system, which includes the atmosphere, land surface, cryosphere, and oceans, as well as interactions among these components. Numerous decisions about climate change, including setting emissions targets and developing and implementing adaptation plans, rest on understanding how the Earth system will respond to greenhouse gas (GHG) emissions and other climate forcings. While this understanding has improved markedly over the past several decades, a number of key uncertainties remain. These include the strength of certain forcings and feedbacks, the possibility of abrupt changes, and the details of how climate change will play out at local and regional scales over decadal and centennial time scales. While research on these topics cannot be expected to eliminate all of the uncertainties associated with Earth system processes (and uncertainties in future human actions will always remain), efforts to improve projections of climate and other Earth system changes can be expected to yield more robust and more relevant information for decision making, as well as a better characterization of remaining uncertainties.

Research on forcing, feedbacks, thresholds, and other aspects of the Earth system has been ongoing for many years under the auspices of the U.S. Global Change Research Program (USGCRP) and its predecessors (see Appendix E). Our analysis—the details of which can be found in Part II of the report—indicates that additional research, supported by expanded observational and modeling capacity, is needed to better understand climate forcings, feedbacks, responses, and thresholds in the Earth system. A list of some of the specific research needs within this crosscutting theme is included in Table 4.1, and the subsections below and the chapters of Part II include additional discussion of these topics. Many of these needs have also been articulated, often in greater detail, in a range of recent reports by the USGCRP, the National Research Council, federal agencies, and other groups.

Climate Variability and Abrupt Climate Change

Great strides have been made in improving our understanding of the natural variability in the climate system (see, e.g., Chapter 6 of this report and USGCRP, 2009b). These improvements have translated directly into advances in detecting and attributing human-induced climate change, simulating past and future climate in models, and understanding the links between the climate system and other environmental and human systems. For example, the ability to realistically simulate natural climate variations, such as the El Niño-Southern Oscillation, has been a critical driver for, and test of, the development of climate models (see Theme 7). Improved understanding of natural variability modes is also critical for improving regional climate projections, especially on decadal time scales. Research on the impacts of natural climate variations can also provide insight into the possible impacts of human-

TABLE 4.1 Examples of Research Needs Related to Improving Fundamental Understanding of Climate Forcings, Feedbacks, Responses, and Thresholds in the Earth System

- Extend understanding of natural climate variability on a wide range of space and time scales, including events in the distant past.
- Improve understanding of transient climate change and its dependence on ocean circulation, heat transport, mixing processes, and other factors, especially in the context of decadal-scale climate change.
- Improve estimates of climate sensitivity, including theoretical, modeling, and observationally based approaches.
- Expand observations and understanding of aerosols, especially their radiative forcing effects and implications for strategies that might be taken to limit the magnitude of future climate change;
- Improve understanding of cloud processes, and cloud-aerosol interactions, especially in the context of radiative forcing, climate feedbacks, and precipitation processes.
- Improve understanding of ice sheets, including the mechanisms, causes, dynamics, and relative likelihood of ice sheet collapse versus ice sheet melting.
- Advance understanding of thresholds and abrupt changes in the Earth system.
- Expand understanding of carbon cycle processes in the context of climate change and develop Earth system models that include improved representations of carbon cycle processes and feedbacks.
- Improve understanding of ocean dynamics and regional rates of sea level rise.
- Improve understanding of the hydrologic cycle, especially changes in the frequency and intensity of precipitation and feedbacks of human water use on climate.
- Improve understanding and models of how agricultural crops, fisheries, and natural and managed ecosystems respond to changes in temperature, precipitation, CO_2 levels and other environmental and management changes.
- Improve understanding of ocean acidification and its effects on marine ecosystems and fisheries.

SOURCE: These research needs (and those included in each of the other six themes in this chapter) are compiled from the detailed lists of key research needs identified in the technical chapters of Part II of this report.

induced climate change. Continued research on the mechanisms and manifestations of natural climate variability in the atmosphere and oceans on a wide range of space and time scales, including events in the distant past, can be expected to yield additional progress.

Some of the largest risks associated with climate change are associated with the potential for abrupt changes or other climate "surprises" (see Chapters 3 and 6). The paleoclimate record indicates that such abrupt changes have occurred in the past, but our ability to predict future abrupt changes is constrained by our limited understanding of thresholds and other nonlinear processes in the Earth system. An improved understanding of the likelihood and potential consequences of these changes will be important for setting GHG emissions-reduction targets and for developing adaptation strategies that are robust in the face of uncertainty. Sustained observations will be critical for identifying abrupt changes and other climate surprises if and when they occur, and for supporting the development of improved abrupt change simulations in climate models. Finally, since some abrupt changes or other climate surprises may result from complex interactions within or among different components of coupled human-environment systems, improved understanding is needed on multiple stresses and their potential role in future climate shifts (NRC, 2002a).

Improved understanding of forcings, feedbacks, and natural variability on regional scales is also needed. Many decisions related to climate change impacts, vulnerability, and adaptation could benefit from improvements in regional-scale information, especially over the next several decades. As discussed in Theme 7, these improvements require advances in understanding regional climate dynamics, including atmospheric circulation in complex terrain as well as modes of natural variability on all time scales. It is especially important to understand how regional variability patterns may change under different scenarios of global climate change and the feedbacks that regional changes may in turn have on continental- and global-scale processes. Regional climate models, which are discussed later in this chapter, are a key tool in this area of research.

The Atmosphere

Many research needs related to factors that influence the atmosphere and other components of the physical climate system are discussed in the chapters of Part II, and many of these needs have also been summarized in other recent reports. For example, many of the conclusions and research recommendations in *Understanding Climate Change Feedbacks* (NRC, 2003b) and *Radiative Forcing of Climate Change* (NRC, 2005d), such as those highlighted in the following two paragraphs, remain highly relevant today:

> The physical and chemical processing of aerosols and trace gases in the atmosphere, the dependence of these processes on climate, and the influence of climate-chemical interactions on the optical properties of aerosols must be elucidated. A more complete understanding of the emissions, atmospheric burden, final sinks, and interactions of carbonaceous and other aerosols with clouds and the hydrologic cycle needs to be developed. Intensive regional measurement campaigns (ground-based, airborne, satellite) should be con-

> ducted that are designed from the start with guidance from global aerosol models so that the improved knowledge of the processes can be directly applied in the predictive models that are used to assess future climate change scenarios.
>
> The key processes that control the abundance of tropospheric ozone and its interactions with climate change also need to be better understood, including but not limited to stratospheric influx; natural and anthropogenic emissions of precursor species such as NO_x, CO, and volatile organic carbon; the net export of ozone produced in biomass burning and urban plumes; the loss of ozone at the surface, and the dependence of all these processes on climate change. The chemical feedbacks that can lead to changes in the atmospheric lifetime of CH_4 also need to be identified and quantified. (NRC, 2003b)

Two particularly important—and closely linked—research topics related to forcing and feedback processes in the physical climate system are clouds and aerosols. Aerosols and aerosol-induced changes in cloud properties play an important role in offsetting some of the warming associated with GHG emissions and may have important implications for several proposed strategies for limiting the magnitude of climate change (see Theme 4). Cloud processes modulate future changes in temperature and in the hydrologic cycle and thus represent a key feedback. As noted later in this chapter, the representation of cloud and aerosol processes in climate models has been a challenge for many years, in part because some of the most important cloud and aerosol processes play out at spatial scales that are finer than global climate models are currently able to routinely resolve, and in part because of the complexity and limited understanding of the processes themselves. Continued and improved observations, field campaigns, process studies, and experiments with smaller-domain, high-resolution models are needed to improve scientific understanding of cloud and aerosol processes, and improved parameterizations will be needed to incorporate this improved understanding into global climate models.

The Cryosphere

Changes in the cryosphere, especially the major ice sheets on Greenland and Antarctica, represent another key research area in the physical climate system. Comprehensive, simultaneous, and sustained measurements of ice sheet mass and volume changes and ice velocities are needed, along with measurements of ice thickness and bed conditions, both to quantify the current contributions of ice sheets to sea level rise (discussed below) and to constrain and inform ice sheet model development. These measurements, which include satellite, aircraft, and in situ observations, need to overlap for several decades in order to enable the unambiguous isolation of ice melt, ice dynamics, snow accumulation, and thermal expansion. Equally important are investments in improving ice sheet process models that capture ice dynamics as well as ice-ocean and ice-bed interactions. Efforts are already underway to improve modeling capabilities in these critical areas, but fully coupled ice-ocean-land models will ultimately be needed to reliably assess ice sheet stability, and considerable work remains to develop and validate such models. Glaciers and ice caps outside Greenland and Antarctica are also expected to remain significant contributors to sea level rise in the near term, so observations and analysis of these systems remain critical for understanding decadal and century-scale sea level rise. Finally, additional paleoclimate data from ice cores, corals, and ocean sediments would be valuable for testing models and improving our understanding of the impacts of sea level rise.

The Oceans

A variety of ocean processes are important for controlling the timing and characteristics of climate change. For a given climate forcing scenario, the timing of atmospheric warming is strongly dependent on the north-south transport of heat by ocean currents and mixing of heat into the ocean interior. Changes in the large-scale meridional overturning circulation could also have a significant impact on regional and global climate and could potentially lead to abrupt changes (Alley et al., 2003; NRC, 2002a). The relative scarcity of ocean observations, especially in the ocean interior, makes these factors among the more uncertain aspects of future climate projections. Changes in ocean circulations and heat transport are also connected to the rapid disappearance of summer sea ice in the Arctic Ocean. A better understanding of the dependence of ocean heat uptake on vertical mixing and the abrupt changes in polar reflectivity that follow the loss of summer sea ice in the Arctic are some of the most critical improvements needed in ocean and Earth system models.

Ice dynamics and thermal expansion are the main drivers of rising sea levels on a global basis, but ocean dynamics and coastal processes lead to substantial spatial variability in local and regional rates of sea level rise (see Chapters 2 and 7). Direct, long-term monitoring of sea level and related oceanographic properties via tide gauges, ocean altimetry measurements from satellites, and an expanded network of in situ measurements of temperature and salinity through the full depth of the ocean water column are needed to quantify the rate and spatial variability of sea level change and to understand the ocean dynamics that control global and local rates of sea level rise. In addition, oceanographic, geodetic, and coastal models are needed to predict the rate and spatial dynamics of ocean thermal expansion, sea level rise, and coastal inundation. The need for regionally specific information creates additional challenges. For example, coastal inundation models require better bathymetric data, better data on precipitation rates and stream flows, ways of dealing with storm-driven sediment transport, and the ability to include the effects of built structures on coastal wind stress patterns (see Chapter 7). Such improvements in projections of sea level changes are critical for many different decision needs.

The Hydrosphere

There is already clear evidence that changes in the hydrologic cycle are occurring in response to climate change (see, e.g., Trenberth et al., 2007; USGCRP, 2009a). Improved regional projections of changes in precipitation, soil moisture, runoff, and groundwater availability on seasonal to multidecadal time scales are needed to inform water management and planning decisions, especially decisions related to long-term infrastructure investments. Likewise, projections of changes in the frequency and intensity of severe storms, storm paths, floods, and droughts are critical both for water management planning and for many adaption decisions. Developing improved understanding and projections of hydrological and water resource changes will require new multiscale modeling approaches, such as nesting cloud-resolving climate models into regional weather models and then coupling these models to land surface models that are capable of simulating the hydrologic cycle, vegetation, multiple soil layers, groundwater, and stream flow. Improved data collection, data analysis, and linkages with water managers are also critical. See Chapter 8 for additional details.

Ecosystems on Land

Climate change interacts with ecosystem processes in a variety of ways, including direct and indirect influences on biodiversity, range and seasonality shifts in both plants and animals, and changes in productivity and element cycling processes, among others (NRC, 2008b). Research is needed to understand how rapidly species and ecosystems can or cannot adjust in response to climate-related changes and to understand the implications of such adjustments for ecosystem services. In addition, improved analyses of the interactions of climate-related variables—especially temperature, moisture, and CO_2—with each other and in combination with other natural and human-caused changes (e.g., land use change, water diversions, and landscape-scale management choices) are needed, as such interactions are more relevant than any individual change acting alone. Climate change-related changes in fire, pest, and other disturbance regimes have also not been well assessed, especially at regional scales.

Research is needed to identify the ecosystems, ecosystem services, species, and people reliant on them that are most vulnerable. See Chapter 9 for additional details.

The Carbon Cycle

Changes in the carbon cycle and other biogeochemical cycles play a key role in modulating atmospheric and oceanic concentrations of CO_2 and other GHGs. Scientists have learned a great deal over the past 50 years about the exchange of carbon between the atmosphere, ocean, and biosphere and the effects of these changes on temperature and other climate change (CCSP, 2007a). However, key uncertainties remain. For example, we have an incomplete understanding of how interacting changes in temperature, precipitation, CO_2, and nutrient availability will change the processing of carbon by land ecosystems and, thus, the amount of CO_2 emitted or taken up by ecosystems in the decades ahead (see Chapter 9). As noted in Chapters 2 and 6, some of these feedbacks have the potential to dramatically accelerate global warming (e.g., the possibility that the current warming of permafrost in high-latitude regions will lead to melting of frozen soils and release huge amounts of CO_2 and CH_4 into the atmosphere). Changes in biogeochemical processes and biodiversity (including changes in reflectance characteristics due to land use changes) also have the potential to feed back on the climate system on a variety of time scales. Models and experiments that integrate knowledge about ecosystem processes, plant physiology, vegetation dynamics, and disturbances such as fire are needed, and such models should be linked with climate models.

As the ocean warms and ocean circulation patterns change, future changes in the ocean carbon cycle are also uncertain. For example, it is unclear whether the natural "biological pump," which transports enormous amounts of carbon from the surface to the deep ocean, will be enhanced (Riebesell et al., 2007) or diminished (Mari, 2008) by ocean acidification and by changes in ocean circulation. Recent observational and modeling results suggest that the rate of ocean uptake of CO_2 may in fact be declining (Khatiwala et al., 2009). Because the oceans currently absorb over 25 percent of human-caused CO_2 emissions (see Chapter 6), changes in ocean CO_2 uptake could have profound climate implications. Results from the first generation of coupled carbon-climate models suggest that the capacity of the oceans and land surface to store carbon will decrease with global warming, which would represent a positive feedback on warming (Friedlingstein et al., 2006). Improved understanding and representation of the carbon cycle in Earth system models is thus a critical research need.

Interactions with Managed Systems and the Built Environment

Feedbacks and thresholds within human systems and human-managed systems, and between the climate system and human systems, are a closely related research need that spans both this research theme and several of the other research themes described in this chapter. For example, crops respond to multiple and interacting changes in temperature, moisture, CO_2, ozone, and other factors, such as pests, diseases, and weeds. Experimental studies that evaluate the interactions of multiple factors are needed, especially in ecosystem-scale experiments and in environments where temperature is already close to optimal for crops. Of particular concern are water resources for agriculture, which are influenced at regional scales by competition from other uses as well as by changing frequency and intensity of rainfall. Assessments that evaluate crop response to climate-related variables should explicitly include interactions with other resources that are also affected by climate change. Designing effective agricultural strategies for limiting and adapting to climate change will require models and analyses that reflect these complicated interactions and that also incorporate the response of farmers and markets not only to production and prices but to policies and institutions (see Themes 3, 4, and 7 below).

In fisheries, sustainable yields require matching catch limits with the growth of the fishery. Climate variability already makes forecasting the growth of fish populations difficult, and future climate change will increase this uncertainty. There is considerable uncertainty about—and considerable risk associated with—the sensitivity of fish species to ocean acidification. Further studies of connections between climate and marine population dynamics are needed to enhance model frameworks for effective fisheries management. Most fisheries are also subject to other stressors, such as increasing levels of pollution, and the interactions of these other stresses should be analyzed and incorporated into models. Finally, all of these efforts should be linked to the analysis of effective institutions and policies for managing fisheries. (See Chapter 9 for additional details of links between climate change and agriculture and fisheries.)

The role of large built environments (including the transportation and energy systems associated with them) in shaping GHG emissions, aerosol levels, ground-level air pollution, and surface reflectivity need to be examined in a systematic and comparative way to develop a better understanding of their role in climate forcing. This should include attention to the extended effect of urban areas on other areas (such as deposition of urban emissions on ocean and rural land surfaces) as well as interactions between urban and regional heat islands and urban vegetation-evapotranspiration feedbacks to climate. Examination of both local and supralocal institutions, markets, and policies will be required to understand the various ways urban centers drive climate change and to identify leverage points for intervention. (See Chapter 10 and Theme 4 later in this chapter for additional details.)

Finally, the identification and evaluation of unintended consequences of proposed or already-initiated strategies to limit the magnitude of climate change or adapt to its impacts will need to be evaluated as part of the overall evaluation of the efficacy of such approaches. This topic is explored in more detail later in the chapter, but it depends on a robust Earth system research enterprise.

THEME 2: CLIMATE-RELATED HUMAN BEHAVIORS AND INSTITUTIONS

Knowledge gained from research involving physical, chemical, and ecological processes has been critical for establishing that climate change poses sufficiently serious risks to justify careful consideration and evaluation of alternative responses. Emerging concerns about how best to respond to climate change also bring to the fore questions about human interactions with the climate system: how human activities drive climate change; how people understand, decide, and act in the climate context; how people are affected by climate change; and how human and social systems might respond. Thus, not surprisingly, many of the research needs that emerge from the detailed analyses in Part II focus on human interactions with climate change (see Table 4.2).

Human and social systems play a key role in both causing and responding to climate change. Therefore, in the context of climate change, a better understanding of human behavior and of the role of institutions and organizations is as fundamental to effective decision making as a better understanding of the climate system. Such knowledge underlies the ability to solve focused problems of climate response, such as deciding how to prioritize investments in protecting coastal communities from sea level rise, choosing policies to meet federal or state targets for reducing GHG emissions, and developing better ways to help citizens understand what science can and cannot tell them about potential climate-driven water supply changes. Such fundamental understanding provides the scientific base for making informed choices about climate responses in much the same way that a fundamental understanding of the physical climate system provides the scientific base for projecting the consequences of climate change.

Research investments in the behavioral and social sciences would expand this knowledge base, but such investments have been lacking in the past (e.g., NRC, 1990a, 1999a, 2003a, 2004b, 2005a, 2007f, 2009k). Barriers and institutional factors, both in research funding agencies and in academia more broadly, have also constrained progress in

TABLE 4.2 Examples of Research Needs on Human Behavior, Institutions, and Interactions with the Climate System (from Part II)

- Improve understanding of water-related institutions and governance.
- Improve understanding of human behaviors and institutional and behavioral impediments to reducing energy demand and adopting energy-efficient technologies.
- Improve understanding of what leads to the adoption and implementation of international agreements on climate and other environmental issues and what forms of such agreements most effectively achieve their goals.
- Improve understanding of how institutions interact in the context of multilevel governance and adaptive management.
- Improve understanding of the behaviors, infrastructure, and technologies that influence human activities in the transportation, urban, agricultural, fisheries, and other sectors.
- Improve understanding of the relationship between climate change and institutional responses that affect national security, food security, health, and other aspects of social well-being.

these areas (NRC, 1992a). This section outlines some of the key areas of fundamental research on human behavior and institutions that need to be developed to support better understanding of human interactions with the climate system and provide a scientific basis for informing more effective responses to climate change. It draws on several past analyses and assessments of research gaps and needs (NRC, 1992a, 1997a, 2001, 2002b, 2005a, 2009g, 2009k).

How People Understand Climate Change and Climate Risks

Climate change represents a special challenge for human comprehension (Fischhoff, 2007; Marx and Weber, 2009). To make decisions about climate change, a basic understanding of the processes of climate change and of how to evaluate the associated risks and potential benefits would be helpful for most audiences. However, despite several decades of exposure to information about climate change, such understanding is still widely lacking. A number of recent scientific analyses (Leiserowitz, 2007; Maibach et al., 2010; Moser and Tribbia, 2006, 2007; Wilson, 2002; see also NRC, 2010b) have identified some of the comprehension challenges people—including both the general public and trained professional in some fields—face in making decisions about how to respond to climate change.

First, because of the inherent uncertainties, projections of future climate change are often presented in terms of probabilities. Cognitive studies have established that humans have difficulty in processing probabilistic information, relying instead on cognitive shortcuts that may deviate substantially from what would result from a careful analysis (e.g., Gigerenzer, 2008; Nichols, 1999).

Second, the time scale of climate change makes it difficult for most people to observe these changes in their daily lives. Climate change impacts are not yet dramatically noticeable in the most populated regions of the United States, and even rapid climate change takes place over decades, making it difficult for people to notice unless they look at historical records (Bostrom and Lashof, 2007; Moser, 2010). Scientists are only beginning to understand how recent and longer-term trends in weather influence perceptions of climate change (Hamilton and Keim, 2009; Joireman et al., in press). It is also difficult to unambiguously attribute individual weather events to climate change, and climate change is easily displaced by events people perceive as exceptional or simply as more important at any one time (Fischhoff, 2007; Marx and Weber, 2009; Marx et al., 2007; Weber, 2006).

Third, people commonly use analogies, associations, or simplified mental models to communicate or comprehend climate change, and these simplifications can result in significant misunderstandings. For example, climate change is sometimes confused with other types of pollution or with other global atmospheric problems (especially the stratospheric ozone "hole," which some people erroneously think leads to global warming by allowing more solar radiation to enter the atmosphere) (Bostrom et al., 1994; Brechin, 2003; Kempton, 1991). Likewise, confusing the atmospheric lifetimes of GHGs with those of conventional air pollutants sometimes leads people to the erroneous inference that if emissions stop, the climate change problem will rapidly go away (Bostrom and Lashof, 2007; Morgan et al., 2001; Sterman, 2008; Sterman and Booth Sweeney, 2007).

Fourth, individual information processing is influenced by social processes, including the "frames" people apply when deciding how to assess new information, the trust they have in sources providing new information, and the views of those to whom they are connected in social networks (Durfee, 2006; Morgan et al., 2001; Moser and Dilling, 2007; Nisbet and Mooney, 2007; NRC, 2010b; Pidgeon et al., 2008). Information that is consistent with, rather than incongruent with, existing beliefs and values is more likely to be accepted, as is information from trusted sources (Bishr and Mantelas, 2008; Cash et al., 2003; Critchley, 2008; Cvetkovich and Loefstedt, 1999).

These challenges demonstrate the importance of understanding how people—acting as consumers, citizens, or members of organizations and social networks—comprehend climate change, and how these cognitive processes influence climate-relevant decisions and behaviors. Fundamental knowledge of risk perception provides a basis for this understanding (e.g., NRC, 1996; Pidgeon et al., 2003; Renn, 2008; Slovic,

2000), but this knowledge needs to be extended and elaborated (e.g., Lorenzoni et al., 2005; Lowe, 2006; O'Neill and Nicholson-Cole, 2009). A wide range of relevant theories and concepts have been advanced in various branches of psychology, sociology, and anthropology, as well as the political, pedagogic, and decision sciences (among others), but these have yet to be more fully synthesized and applied to climate change (Moser, 2010). Improved knowledge of how individuals, groups, networks, and organizations understand climate change and make decisions for responding to environmental changes can inform the design and evaluation of tools that better support decision making (NRC, 2009g).

Institutions, Organizations, and Networks

Individual decisions about climate change, important as they are, are not the only human decisions that shape the trajectory of climate change. Some of the most consequential climate-relevant decisions and actions are shaped by institutions—such as markets, government policies, and international treaties—and by public and private organizations.

Institutions shape incentives and the flow of information. They can also either encourage or help us avoid situations where individual actions lead to outcomes that are undesirable for both the individual and the group (sometimes called "the tragedy of the commons"). The problem of decision making for the collective good has been extensively studied around localized resources such as forests or fisheries (Chhatre and Agrawal, 2008; Dietz and Henry, 2008; McCay and Jentoft, 2009; Moran and Ostrom, 2005; NRC, 2002b; Ostrom, 2007, 2010; Ostrom and Nagendra, 2006). This body of research can provide important guidance for shaping effective responses to climate change at local and regional levels. It can also inform the design and implementation of national and international climate policies (see Chapter 17). However, improving our understanding of the flexibility and efficacy of current institutions and integrating this body of knowledge with existing work on international treaties, national policies, and other governance regimes remains a significant research challenge.

Many environmentally significant decisions are made by organizations, including governments, publicly traded companies, and private businesses. Research on environmental decision making by businesses covers a broad range of issues. These include responses to consumer and investor demand, management of supply chains and production networks, standard setting within sectors, decisions about technology and process, how environmental performance is assessed and reported, and the interplay between government policy and private-sector decision making (NRC, 2005a). Responses to climate change in the private sector have not been studied as extensively, but such research efforts might yield important insights.

A number of state and local governments have also been proactive in developing policies to adapt to climate change and reduce GHG emissions. To learn from these experiences, which is a key aspect of adaptive risk management, research is needed on both the effectiveness of these policies and the various factors that influenced their adoption (Brody et al., 2008; Teodoro, 2009; Zahran et al., 2008). In the United States, local policies are almost always embedded in state policies, which in turn are embedded in national policies, raising issues of multilevel governance—another emerging research area (see Chapter 17).

Finally, it is clear that public policy is shaped not only by the formal organizations of government, but also by policy networks that include government, the private sector, and the public. An emerging challenge is to understand how these networks influence policy and how they transmit and learn from new information (Bulkeley, 2005; Henry, 2009).

Environmentally Significant Consumption

Decisions about consumption at the individual, household, community, business, and national levels have a profound effect on GHG emissions. For example, voluntary consumer choices to increase the efficiency of household energy use could reduce total U.S. GHG emissions by over 7 percent if supportive policies were in place (Dietz et al., 2009b). Consumer choices also influence important aspects of vulnerability and adaptation; for example, increasing demand for meat in human diets places stresses on the global food system as well as on the environment (Fiala, 2008; Stehfest et al., 2009), and demand for beachfront homes increases vulnerability and shapes adaptation options related to sea level rise, storm surges, and other coastal impacts.

Considerable research on consumption decision making has been carried out in economics, psychology, sociology, anthropology, and geography (NRC, 1997a, 2005a), but much of this research has been conducted in isolation. For example, economic analyses often take preferences as given. Studies in psychology, sociology, and anthropology, on the other hand, focus on the social influences on preferences but often fail to account for economic processes. Decisions based on knowledge from multiple disciplines are thus much more likely to be effective than decisions that rely on the perspective of a single discipline, and advances in the understanding of climate and related environmental decision making are likely to require collaboration across multiple social science disciplines (NRC, 1997a, 2002b). This is an area of research where theories and methodologies are in place but progress has been slowed by a lack of support for experiments and large-scale data collection efforts that integrate across disciplines.

Human Drivers of Climate Change

Ultimately, it is desirable to understand how choices, and the factors that shape them, lead to specific environmental outcomes (Dietz et al., 2009c; Vayda, 1988). A variety of hypotheses have been offered and tested about the key societal factors that shape environmental change—what are often called the drivers of change (NRC, 1992a). Growth in population and consumption, technological change, land and resource use, and the social, institutional, and cultural factors shaping the behavior of individuals and organizations have all been proposed as critical drivers, and some empirical work has elucidated the influence of each of them (NRC, 1997b, 1999c, 2005a, 2008b). However, much of this research has focused on only one or a few factors at a time and has used highly aggregated data (Dietz et al., 2009a). To understand the many human drivers of climate change as a basis for better-informed decision making, it will be necessary to develop and test integrative models that examine multiple driving forces together, examine how they interact with each other at different scales of human activity and over time, and consider how their effects vary across different contexts.

To evaluate the effectiveness of policies or other actions designed to limit the magnitude of climate change, increased understanding is needed about both the *elasticity* of climate drivers—the extent to which changes in drivers produce changes in climate impacts—and the *plasticity* of drivers, or the ease with which the driver can be changed by policy interventions (York et al., 2002). For example, analyses of the effects of population growth on GHG emissions suggest an elasticity of about 1 to 1.5; that is, for every 1 percent increase in human population, there is roughly a 1 to 1.5 percent increase in environmental impact (Clark et al., 2010; Dietz et al., 2007; Jorgenson, 2007, 2009; Shi, 2003; York et al., 2003). On the other hand, recent research suggests that environmental impact is more directly related to the number of households than to the number of people (Cole and Neumayer, 2004; Liu et al., 2003). Thus, a shift to smaller average household size could offset or even overwhelm the reduction in climate drivers resulting from reduced population growth. Similarly, it has been argued that increasing affluence leads at first to increased environmental impact but, once a threshold level of affluence has been reached, environmental impact declines (Grossman and Krueger, 1995; Selden and Song, 1994). In the case of GHG emissions, however, emissions apparently continue to increase with increasing affluence (Carson,

2010; Cavlovic et al., 2000; Dasgupta et al., 2002; Dietz et al., 2007; Stern, 2004), suggesting that economic growth alone will not reduce emissions.

Processes that Induce or Constrain Innovation

The adoption of new technology is yet another area in which institutions, organizations, and networks have an important influence on decision making. New and improved technologies will be needed to meet the challenges of limiting climate change and adapting to its impacts (NRC, 2010a,c). However, the mere existence of a new technology with desirable properties is not sufficient to ensure its use. For example, individuals and organizations are currently far less energy efficient than is technologically feasible or economically optimal (Jaffe and Stavins, 1994; Weber, 2009). There are also many examples of differential use of or opposition to new technologies among individuals, communities, and even nations. Although adoption of and resistance to innovation, especially in new technologies, have been extensively studied (e.g., Stern et al., 2009), much of this research has been technology specific. A validated theoretical framework has not yet been developed for analyses of adoption issues related to new technologies to reduce GHG emissions or enhance resilience of particular systems, or of proposals to intentionally modify the climate system (see Chapter 15). One lesson from the existing literature is worth highlighting—the earlier in the process of technological development that social acceptance is considered, the more likely it is that technologies will be developed that will actually be used (Rosa and Clark, 1999). Another is that, beyond the character of the innovation itself, it is essential to understand the role of the decision and institutional environment in fostering or constraining its adoption (Lemos, 2008; Rayner et al., 2005). Many of these concepts and research needs also emerge from the next two themes in this chapter.

THEME 3: VULNERABILITY AND ADAPTATION ANALYSES OF COUPLED HUMAN-ENVIRONMENT SYSTEMS

Not all people, activities, environments, and places are equally vulnerable[1] or resilient to the impacts of climate change. Identification of differences in vulnerability across space and time is both a pivotal research issue and a critical way in which scientific research can provide input to decision makers as they make plans to adapt to climate

[1] Vulnerability is the degree to which a system is susceptible to, or unable to cope with, adverse effects of climate change, including changes in climate variability and extremes. Vulnerability is a function of the character, magnitude, and rate of climate variation to which a system is exposed, its sensitivity, and its adaptive capacity (NRC, 2010a).

BOX 4.2
Vulnerability and Adaptation Challenges in Coastal Regions

Coastal regions house most of the world's people, cities, and economic activities. For example, in 2000, the coastal counties of California were home to 77 percent of the state's residents, 81 percent of jobs, and 86 percent of the state's gross product—which represents nearly 19 percent of the total U.S. economy (Kildow and Colgan, 2005). A number of climate and climate-related processes have the potential to damage human and environmental systems in the coastal zone, including sea level rise; saltwater intrusion; storm surge and damages from flooding, inundation, and erosion; changes in the number and strength of coastal storms; and overall changes in precipitation amounts and intensity. Under virtually all scenarios of projected future climate change, coastal areas face increased risks to their transportation and port systems, real estate, fishing, tourism, small businesses, power generating and supply systems, other critical infrastructure (such as hospitals, schools, and police and fire stations), and countless managed and natural ecosystems.

Coastal regions are not homogenous, however, and climate change impacts will play out in different ways in different places. Some areas of the coast and some industries and populations are more vulnerable, and thus more likely to suffer harm, than others. Thus, managers and decision makers in the coastal zone—including land use planners, conservation area managers, fisheries councils, transportation planners, water supply engineers, hazard and emergency response personnel, and others—will face a wide range of challenges, many of them place specific, regarding how to respond to the risks posed by climate change. What does a coastal land use planner need to know about climate change impacts in order to make decisions about land use in a particular region? How can a research program provide information that will assist decision makers in such regions?

Knowledge and predictions about just how much sea level will rise in certain regions over time is a fundamental question. However, as noted in Chapter 7, precise projections are not easy to provide. Moreover, sea level rise projections are, by themselves, not sufficient to meet coastal managers' information needs. Managers also need to know how changes in sea level translate into erosion rates, flooding

change. Indeed, the companion report *Adapting to the Impacts of Climate Change* (NRC, 2010a) identifies vulnerability assessments as a key first step in many if not all adaptation-related decisions and actions. An example of the use of vulnerability assessments in the context of climate-related decision making in the coastal zone can be found in Box 4.2.

In addition to merely identifying and characterizing vulnerabilities, scientific research can help identify and assess actions that could be taken to reduce vulnerability and increase resilience and adaptive capacity in human and environmental systems. Combined vulnerability and adaptation analyses can, for example, identify "no-regrets" actions that could be taken at little or no cost and would be beneficial regardless of frequencies, storm surge levels, risks associated with different development setback limits, numbers of endangered species in exposed coastal ecosystems, habitat changes, and changes in water supply and quality parameters. In addition to these climate and other environmental changes, coastal managers need to consider the numbers of hospitals, schools, and senior citizens in potentially affected areas; property tax dollars generated in the coastal zone; trends in tourism; and many other factors.

Vulnerability assessments of human, social, physical, and biological resources in the coastal zone can help decision makers identify the places and people that are most vulnerable to climate change and help them to identify effective steps that can be taken to reduce vulnerability or increase adaptive capacity. To help coastal managers and other decision makers assess risks, evaluate trade-offs, and make adaptation decisions, they need a scientific research program that improves understanding and projections of sea level rise and other climate change impacts at regional scales, integrates this understanding with improved understanding or nonclimatic changes relevant to decision making, identifies and evaluates the advantages and disadvantages of different adaptation options, and facilitates ongoing assessment and monitoring. Such a program would require the engagement of many different kinds of researchers, including those focusing on resource and land use institutions; social dynamics; economic resilience; developing or evaluating regional climate models; sea level and ocean dynamics; coastal ocean circulation; spatial geomorphologic, geologic, and geographical characteristics; and aquatic and terrestrial ecosystem dynamics, goods, and services. In addition to interdisciplinary interactions, research teams would benefit from interactions with decision makers to improve knowledge and understanding of the specific challenges they face (Cash et al., 2003; NRC, 2008h, 2009k). The knowledge gained by these researchers needs to be integrated and synthesized in decision-support frameworks that actively involve and are accessible to decision makers (e.g., Kates et al., 2006; Moser and Luers, 2008). Finally, a research enterprise that includes the development, testing, and implementation of improved risk assessment approaches and decision-support systems will enhance the capacity of decision makers in the coastal zone—as well as other sectors—to respond effectively to climate change.

how climate change unfolds. They can also help to identify sectors, regions, resources, and populations that are particularly vulnerable to changes in climate considered in the context of changes in related human and environmental systems. Finally, scientific research can assist adaptation planning by helping to develop, assess, and improve actions that are taken to adapt, and by identifying barriers to adaptation and options to overcome those barriers. Indeed, many of the chapters in Part II of the report identified vulnerability and adaptation analyses, developing the scientific capacity to perform such analyses, and developing and improving adaptation options as key research needs. Table 4.3 lists some of these needs.

TABLE 4.3 Examples of Research Needs Related to Vulnerability and Adaptation (from Part II)

- Expand the ability to identify and assess vulnerable coastal regions and populations and to develop and assess adaptation strategies, including barriers to their implementation.
- Assess food security and vulnerability of food production and distribution systems to climate change impacts, and develop adaptation approaches.
- Develop and improve technologies, management strategies, and institutions to enhance adaptation to climate change in agriculture and fisheries.
- Develop vulnerability assessments and integrative management approaches and technologies to respond effectively to changes in water resources.
- Assess vulnerabilities of ecosystems and ecosystem services to climate change.
- Assess current and projected health risks associated with climate change and develop effective, efficient, and fair adaptation measures.
- Assess the vulnerability of cities and other parts of the built environment to climate change, and develop methods for adapting.
- Advance understanding of how climate change will affect transportation systems and how to reduce vulnerability to these impacts.
- Develop improved vulnerability assessments for regions of importance in terms of military operations and infrastructure.

Characteristics of Vulnerability and Adaptation Analyses

Vulnerability and adaptation analyses can be performed in many contexts and have a wide range of uses. In general, vulnerability analyses assess exposure to and impacts from a disturbance, as well as sensitivity to these impacts and the capacity to reduce or adapt to the negative consequences of the disturbance. These analyses can then be used by decision makers to help decide where, how much, and in what ways to intervene in human or environmental systems to reduce vulnerability, enhance resilience, or improve efficient resource management (Eakin et al., 2009; Turner, 2009). In the context of climate change, vulnerability analyses seek to evaluate and estimate the harm to populations, ecosystems, and resources that might result from changes in climate, and to provide useful information for decision makers seeking to deal with these changes (Füssel and Klein, 2006; Kates et al., 2001; Kelly and Adger, 2000).

A major lesson learned from conventional vulnerability analyses is that they often miss the mark if they focus on a single system or set of interactions—for example, a certain population or ecosystem in isolation—rather than considering the larger system in which people and ecosystems are embedded (O'Brien and Leichenko, 2000; Turner et al., 2003a). The Hurricane Katrina disaster (Box 4.3) illustrates the importance of interactions among the human and environmental components in influencing vulnerability: land and water management decisions interacted with environmental, social, and economic dynamics to make the people and ecosystems of New Orleans and surrounding areas particularly vulnerable to storm surges, with tragic results.

As recognition has grown that vulnerability should be assessed in a wider context, attention has increasingly turned to integrated approaches focused on coupled human-environment systems. Such analyses consider both the natural characteristics and the human and social characteristics of a system, evaluate the consequences of climate change and other stresses acting on the integrated system, and explore the potential actions that could be taken to reduce the negative impacts of these consequences, including the trade-offs among efforts to reduce vulnerability, enhance resilience, or improve adaptive capacity (see Figure 4.1) (Eakin and Luers, 2006; Kasperson et al., 2009; Turner et al., 2003a). Integrated approaches that allow the evaluation of the causal structure of vulnerabilities (i.e., the long-term drivers and more immediate causes of differential exposure, sensitivity, and adaptive capacity) can help identify the resources and barriers that can aid or constrain implementation of adaptation options, including

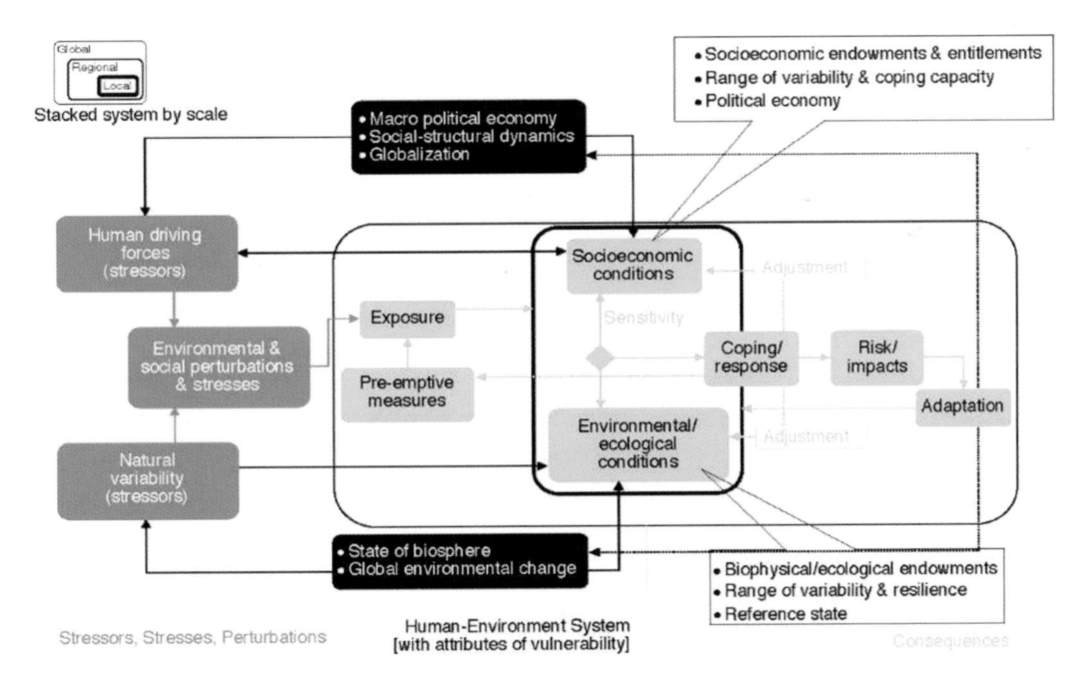

FIGURE 4.1 A framework for analyzing vulnerabilities, focusing on a coupled human-environment system in which vulnerability and response depend on both socioeconomic and human capital as well as natural resources and changes in the environment. From left to right, the figure includes the stresses on the coupled system, the degree to which those stresses are felt by the system, and the conditions that shape the ability of the system to adapt. SOURCE: Kasperson et al. (2009), adapted from Turner et al. (2003a).

BOX 4.3
Vulnerability of New Orleans to Hurricane Katrina

The Mississippi River, especially in and around New Orleans, has been intensively engineered to control flooding and provide improved access for ships to the port of New Orleans. These hydraulic works significantly reduce the river's delivery of sediments to the delta between the city and the Gulf of Mexico, and thus the land-building processes that would otherwise offset the gradual subsidence and erosion of the delta. In addition, the construction of channels and levees and other changes in the lower delta have affected vegetation, especially the health of cypress swamps. Together, these changes in elevation and vegetation have weakened the capacity of the lower delta to serve as a buffer to storm surges from the Gulf of Mexico.

Various assessments of the condition of the lower Mississippi Delta—which together form a quasi-integrated vulnerability study—revealed that in the event of a direct hurricane strike, the vegetation and land areas south of New Orleans were insufficient to protect the city from large storm surges, and also that various hydraulic works would serve to funnel flood waters to parts of the city (Costanza et al., 2006; Day et al., 2007). Despite this knowledge, little was done to reduce the region's vulnerabilities prior to 2005. When Hurricane Katrina struck in late August of that year, the human-induced changes in the region's hydrology, vegetation, and land-building processes, together with the failure to maintain adequate protective structures around New Orleans, resulted in extensive flooding of the city and surrounding area over the following week (see figure below). This, combined with a lack of institutional preparedness and other social factors, led to a well-documented human disaster, especially for the poorest sections of the city (Costanza et al., 2006; Day et al., 2007; Kates et al., 2006).

While climate change may or may not have contributed to the Katrina disaster (see Chapter 8 for a discussion of how climate change might influence the frequency or intensity of hurricanes and other storms), it does illustrate how scientific analysis can help identify vulnerabilities. The Katrina disaster also illustrates how scientific analyses alone are not sufficient to ensure an effective response.

ecological, cognitive, social, cultural, political, economic, legal, institutional, and infrastructural hurdles (e.g., Adger et al., 2009a,b). Integrated vulnerability analyses also allow improved understanding and identification of areas in which climate change works in combination with other disturbances or decisions (e.g., land-management practices) to increase or decrease vulnerability (Cutter et al., 2000; Luers et al., 2003; Turner et al., 2003b).

Challenges of Analyzing Vulnerability

Because of the complexity of interactions within and the variance among coupled human-environment systems, integrated vulnerability and adaptation analyses often rely

New Orleans, Louisiana, in the aftermath of Hurricane Katrina, showing Interstate 10 at West End Boulevard, looking toward Lake Pontchartrain. This photo is from the U.S. Coast Guard's initial Hurricane Katrina damage assessment overflights of New Orleans. SOURCE: U.S. Coast Guard, Petty Officer 2nd Class Kyle Niemi.

on place-based (local and regional) assessments for decision making (e.g., Cutter et al., 2000; O'Brien et al., 2004; Turner et al., 2003b; Watson et al., 1997). However, with few notable exceptions (e.g., Clark et al., 1998; Cutter et al., 2000), there is little empirical research on the vulnerability of places, communities, economies, and ecological systems in the United States to climate change, nor is there much empirically grounded understanding of the range of adaptation options and associated constraints (Moser, 2009a; NRC, 2010a).

The development of common metrics and frameworks for vulnerability and adaptation assessments is needed to assist cross-sectoral and interregional comparison and learning. While some research has focused on useful outputs for decision making and adaptation planning (Luers et al., 2003; Moss et al., 2002; Polsky et al., 2007;

Schmidtlein et al., 2008), developing comparative metrics has been challenging due to a lack of baseline data and insufficient monitoring; difficulty in measuring critical and dynamic social, cultural, and environmental variables across scales and regions; limitations in accounting for the indirect impacts of adaptation measures; and uncertainties regarding changes in climate variability, especially changes in the frequency or severity of extreme events, which often dominate vulnerability (Eakin and Luers, 2006; NRC, 2010a; O'Brien et al., 2004).

Assessing adaptive capacity has also been difficult because of its latent character; that is, although capacity can be characterized, it can only be "measured" after it has been realized or mobilized. Hence, adaptive capacity can often only be assessed based on assumptions about different factors that might facilitate or constrain response and action (Eakin and Luers, 2006; Engle and Lemos, 2010) or through the use of model projections. Progress here will rely on improved understanding of human behavior relevant to adaptation; institutional barriers to adaptation; political and social acceptability of adaptation options; their economic implications; and technological, infrastructure, and policy challenges involved in making certain adaptations.

THEME 4: RESEARCH TO SUPPORT STRATEGIES FOR LIMITING CLIMATE CHANGE

Decisions about how to limit the magnitude of climate change, by how much, and by when demand input from research activities that span the physical, biological, and social science disciplines as well as engineering and public health. In addition to assessing the feasibility, costs, and potential consequences of different options and objectives, research is critical for developing new and improving existing technologies, policies, goals, and strategies for reducing GHG emissions. Scientific research, monitoring, and assessment activities can also assist in the ongoing evaluation of the effectiveness and unintended consequences of different actions or set of actions as they are taken—which is critical for supporting adaptive risk management and iterative decision making (see Box 3.1). This section highlights some pressing research needs related to efforts to limit the magnitude of future climate change.

Commonly discussed strategies for limiting climate change (see Figure 4.2) include reducing energy consumption, for instance by improving energy efficiency or by reducing demand for energy-intensive goods and services; reducing emissions of GHGs from energy production and use, industrial processes, agriculture, or other human activities; capturing CO_2 from power plants and industrial processes, or directly from the atmosphere, and sequestering it in geological formations; and increasing CO_2

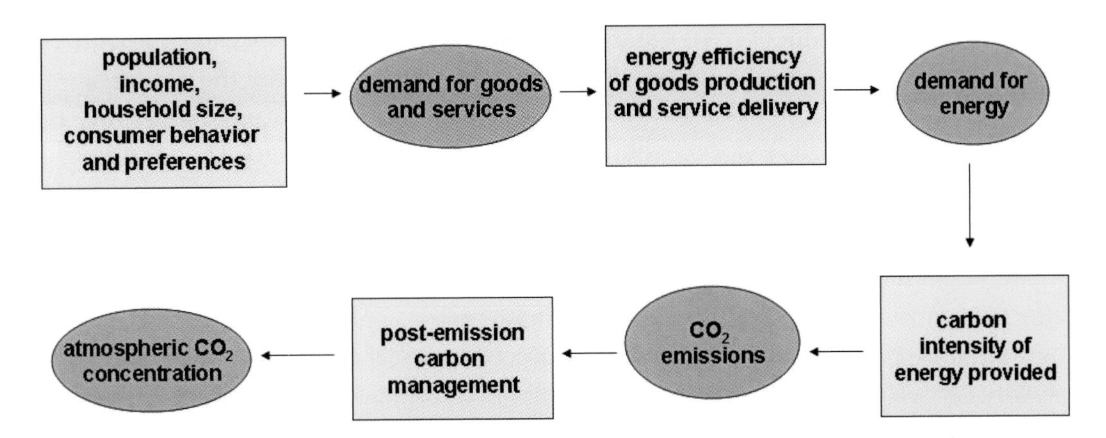

FIGURE 4.2 The chain of factors that determine how much CO_2 accumulates in the atmosphere. The blue boxes represent factors that can potentially be influenced to affect the outcomes in the purple circles. SOURCE: NRC (2010c).

uptake by the oceans and land surface. There is also increasing interest in solar radiation management and other geoengineering approaches (see Chapters 9, 14, and 15). While much is known about some of these strategies, others are not well understood, and there are many scientific research needs related to the development, improvement, implementation, and evaluation of virtually all technologies, policies, and other approaches for limiting climate change.

Setting goals for limiting the magnitude of climate change involves ethical and value questions that cannot be answered by scientific analysis. However, scientific research can help inform such efforts by providing information about the feasibility and potential implications of specific goals. The companion report *Limiting the Magnitude of Future Climate Change* (NRC, 2010c) suggests that the U.S. goal be framed in terms of a cumulative budget for GHG emissions over a set time period. The report does not recommend a specific budget goal, but it examines a "representative" budget in the range of 170 to 200 Gt CO_2-eq[2] for the period 2012 to 2050.[3] As the *Limiting* report notes, reaching a goal in this range will be easier and less costly overall if actions to limit GHG emissions are undertaken sooner rather than later. It will also require pursuing multiple emissions-reduction strategies across a range of sectors, as well as continued research and development aimed at creating new emissions-reduction opportunities. Finally, to support adaptive risk management and iterative decision making with re-

[2] Gt CO_2-eq indicates gigatons (or billion tons) of CO_2 equivalent emissions; this metric converts emissions of other GHGs to an equivalent concentration of CO_2.

[3] This range was derived from recent integrated assessment modeling exercises carried out by the Energy Modeling Forum (*http://emf.stanford.edu*).

spect to emissions reductions or other climate goals, scientific research will be needed to monitor and improve implementation approaches and to evaluate the potential trade-offs, co-benefits, and unintended consequences of different strategies, as well as the interaction of multiple approaches working in concert. These and other examples of research needs for supporting actions to limit climate change are listed in Table 4.4.

The challenge of limiting climate change also engages many of the other research themes identified in this chapter. For example, understanding and comparing the full effects of various energy technologies or climate policies (including their comparative benefits, costs, risks, and distributional effects) typically requires an integration of climate models with energy and economic models (Theme 7), which in turn are based on fundamental understanding of the climate system (Theme 1) and human systems

TABLE 4.4 Examples of Research Needs Related to Limiting the Magnitude of Climate Change (from Part II)

- Advance the development, deployment, and adoption of energy and transportation technologies that reduce GHG emissions.
- Develop and evaluate strategies for promoting the use of less-emission-intensive modes of transportation.
- Characterize and quantify the contributions of urban areas to both local and global changes in climate, and develop and test approaches for limiting these contributions.
- Continue to support efforts to improve energy efficiency in all sectors and develop a better understanding of the obstacles to improved efficiency.
- Improve understanding of behavioral and sociological factors related to the adoption of new technologies, policies, and practices.
- Develop and improve integrated approaches for evaluating energy services in a systems context that accounts for a broad range of societal and environmental concerns, including climate change.
- Develop and improve technologies, management strategies, and institutions to reduce net GHG emissions from agriculture, while maintaining or enhancing food production potential.
- Assess the potential of land, freshwater, and ocean ecosystems to increase net uptake of CO_2 (and other GHGs) and develop approaches that could take advantage of this potential without major adverse consequences.
- Improve understanding of links between air quality and climate change and develop strategies that can limit the magnitude of climate change while improving air quality.
- Improve understanding of the potential efficacy and unintended consequences of solar radiation management approaches and direct air capture of CO_2, provided that this research does not detract from other important research areas.
- Establish and maintain monitoring systems capable of supporting evaluations of actions and strategies taken to limit the magnitude of future climate change, including systems that can verify compliance with international GHG emissions-reduction agreements.

(Theme 2), as well as the observations (Theme 6) that underpin such understanding. Similarly, setting and evaluating goals and policies for limiting the magnitude of future climate change involves decision-making processes at a variety of scales that would benefit from decision-support tools that aid in handling uncertainty and understanding complex value trade-offs (Theme 5). These decisions would similarly benefit from integrated analyses or linked "end-to-end" models (Theme 7) of how policies and other actions influence emissions, how the climate system and related environmental systems will respond to these changes in emissions, and how human and natural systems will be affected by all of these changes—all of which again depend critically on observations (Theme 6). Thus, while the following subsections describe a number of key research needs related to limiting the magnitude of future climate change, progress in many other research areas will also be needed.

Developing New Technologies

Efforts to reduce transportation- and energy-related GHG emissions focus on reducing total energy demand (through, for example, conservation or changes in consumption patterns); improving energy efficiency; reducing the GHG intensity of the energy supply (by using energy sources that emit fewer or no GHGs); and direct capture and sequestration of CO_2 during or after the combustion of fossil fuels (see Figure 4.2 and Chapters 13 and 14). The strategy of reducing demand is discussed earlier (under Theme 2: Human Behavior and Institutions). Technology development is directed primarily toward the other three strategies: efficiency, lower carbon intensity, and carbon capture and storage.

Numerous scientific and engineering disciplines contribute to the development and implementation of energy technology options: the physical, biological, and engineering sciences, for example, are all critical for the development of new technologies, while the social sciences play a key role in both technology development and technology deployment and adoption. In many cases, these diverse disciplines need to work together to evaluate, improve, and expand energy technology options. A coordinated strategy for promoting and integrating energy-related research is needed to ensure the most efficient use of investments among these disciplines and activities.

A number of reports (e.g., *Technology and Transformation* [NRC, 2009d] and the Strategic Plan of the U.S. Climate Change Technology Program [DOE, 2009c]) have suggested that priority areas for strategic investment in the energy sector should include energy end use and infrastructure, sustainable energy supply, carbon sequestration, and reduction of non-CO_2 GHG emissions. These are discussed in Chapter 14. In the transportation sector, key research and development topics include vehicle efficiency, vehicles that run on electricity or non-petroleum-based transportation fuels, and technologies and policies that could reduce travel demand (including, for example, communication technologies like video conferencing). Chapter 13 includes additional discussion on these topics.

Technology developments in the energy and transportation sector are interrelated. For example, widespread adoption of batteries and fuel cells would switch the main source of transportation energy from petroleum to electricity, but this switch will only result in significant GHG emissions reductions if the electricity sector can provide low- and no-GHG electricity on a large scale. This and other codependencies between the energy and transportation sectors underscore the need for an integrated, holistic national approach to limit the magnitude of future climate change as well as related research investments. Widespread adoption of new transportation or energy technologies would also demand significant restructuring of the nation's existing transportation and energy infrastructure, and scientific and engineering research will play an important role in optimizing that design.

As described in Chapter 12, urban design presents additional opportunities for limiting climate change. The design of urban developments can, for example, reduce the GHG "footprints" of buildings and the level of demand they create for motorized travel. However, the success of new urban and building designs will depend on better understanding of how technology design, social and economic considerations, and attractiveness to potential occupants can be brought together in the cultural contexts where the developments will occur. Research is also needed to consider the implication of new designs for human vulnerability to climate change as well as other environmental changes.

Finally, as discussed in Chapter 10, there are a number of potential options for reducing GHG emissions from the agricultural, fisheries, and aquaculture sectors through new technologies or management strategies. Development of new fertilizers and fertilizer management strategies that reduce emissions of N_2O is one area of interest—one that may also yield benefits in terms of agricultural contributions to other forms of pollution. Reducing CH_4 emissions through changes in rice technologies or ruminant feed technologies are two additional areas of active research. Further research is needed in these and other areas, and also on the effectiveness, costs and benefits, and perceptions of farmers, fish stock managers, and consumers when considering implementation of new technologies in these sectors.

Facilitating Adoption of Technologies

There are a number of barriers to the adoption of technologies that could potentially reduce GHG emissions. For example, the Environmental Protection Agency (EPA) recently suspended Energy Star certification for programmable thermostats because it was unable to show that they save energy in actual use (EPA, 2009a). Similar difficulties could be in store for "smart meters," which are promoted as devices that will allow households to manage energy use to save money and reduce emissions, but which are often designed mainly for the information needs of utility companies rather than consumers. Research on improved designs of these and other types of monitoring and control equipment could help reduce energy use by helping users operate homes, motor vehicles, and commercial and industrial facilities more efficiently.

There are similar opportunities for improved energy efficiency through behavioral change. For example, U.S. households could significantly reduce their GHG emissions (and save money) by adopting more energy-efficient driving behaviors and by properly maintaining automobiles and home heating and cooling systems (Dietz et al., 2009b). Research on behavioral change suggests that a good portion of this potential could actually be achieved, but further analysis is needed to develop and assess specific strategies, approaches, and incentives.

In general, barriers to technology adoption have received only limited research attention (e.g., Gardner and Stern, 1996; NRC, 2005a; Pidgeon et al., 2003). Such research can identify barriers to faster adoption of technologies and develop and test ways to overcome these barriers through, for example, better technological design, policies for facilitating adoption, and practices for addressing public concerns. This research can also develop more realistic estimates of technology penetration rates given existing barriers and assess the perceived social and environmental consequences of technology use, some of which constitute important barriers to or justifications for adoption. Finally, the gap between technological potential and what is typically accomplished might be reduced by integrating knowledge from focused, problem-solving research on adoption of new technologies and practices (e.g., Stern et al., 2009, in press).

Institutions and Decision Making

The 20th century saw immense social and cultural changes, many of which—such as changes in living patterns and automobile use—have had major implications for climate change. Many societal and cultural changes can be traced to the confluence of individual and organizational decision making, which is shaped by institutions that reward some actions and sanction others, and by technologies. New institutions, such as GHG emissions trading systems, voluntary certification systems for energy-efficient building design, bilateral international agreements for emissions reduction, agreements on emissions monitoring, and carbon offset markets, are critical components of most of the plans that have been proposed to limit human GHG emissions during the next few decades (see Theme 2 above and also the companion reports *Limiting the Magnitude of Future Climate Change* [NRC, 2010c] and *Informing an Effective Response to Climate Change* [NRC, 2010b]). Many such mechanisms are already in operation, and these constitute natural experiments, but the scientific base for evaluating these experiments and designing effective institutions is limited (see, e.g., Ostrom, 2010; Prakash and Potoski, 2006; Tietenberg, 2002). Institutional design would likely be enhanced by more systematic research to evaluate past and current efforts, compare different institutional approaches for reaching the same goals, and develop and pilot-test new institutional options.

A large number of individual, community, and organizational decisions have a substantial effect on GHG emissions and land use change as well as on vulnerability to climate change. Many of these decisions are not currently made with much or any consideration of climate change. For example, individual and household food choices, the layout of communities, and the design of supply chains all have effects on climate. Understanding social and cultural changes is important for projecting future climate change, and, in some cases, these changes may provide substantial leverage points for reducing climate change. Thus, enhanced understanding of the complex interplay of social, cultural, and technological change is critical to any strategy for limiting future climate change.

Geoengineering Approaches

Available evidence suggests that avoiding serious consequences from climate change poses major technological and policy challenges. If new technologies and institutions are insufficient to achieve critical emissions-reduction targets, or if a "climate emergency" emerges, decision makers may consider proposals to manage Earth's climate directly. Such efforts, often referred to as geoengineering approaches, encompass two very different categories of approaches: carbon dioxide removal (CDR) from the atmosphere, and solar radiation management (SRM). Two proposals for CDR—iron fertilization in the ocean and direct air capture—are discussed briefly in Chapters 9 and 14, respectively. As noted in Chapter 2 and discussed in greater detail in Chapter 15, little is currently known about the efficacy or potential unintended consequences of SRM approaches, particularly how to approach difficult ethical and governance questions. Therefore, research is needed to better understand the feasibility of different geoengineering approaches; the potential consequences (intended and unintended) of such approaches on different human and environmental systems; and the related physical, ecological, technical, social, and ethical issues, including research that could inform societal debates about what would constitute a "climate emergency" and on governance systems that could facilitate whether, when, and how to intentionally intervene in the climate system. It is important that such research not distract or take away from other important research areas, including research on understanding the climate system and research on "conventional" strategies for limiting the magnitude of climate change and adapting to its impacts.

THEME 5: EFFECTIVE INFORMATION AND DECISION-SUPPORT SYSTEMS

Global climate changes are taking place within a larger context of vast and ongoing social and environmental changes. These include the globalization of markets and communication, continued growth in human population, land use change, resource degradation, and biodiversity loss, as well as persistent armed conflict, poverty, and hunger. There are also ongoing changes in cultural, governance, and economic conditions, as well as in technologies, all of which have substantial implications for human well-being. Thus, decision makers in the United States and around the world need to balance climate-related choices and goals with other social, economic, and environmental objectives (Burger et al., 2009; Lindseth, 2004; Schreurs, 2008), as well as issues of fairness and justice (Page, 2008; Roberts and Parks, 2007; Vanderheiden, 2008) and questions of risk (Bulkeley, 2001; Jacques, 2006; Lorenzoni and Pidgeon, 2006; Lubell et al., 2007; Vogler and Bretherton, 2006), all while taking account of a changing context for those decisions. Accordingly, in addition to climate and climate-related information, decision makers need information about the current state of human systems and their environment, as well as an appreciation of the plausible future outcomes and net effects that may result from their policy decisions. They also need to consider how their decisions and actions could interact with other environmental and economic policy goals, both in and outside their areas of responsibility.

The research needs highlighted in this report are intended to both improve fundamental understanding of and support effective decision making about climate change. As explored in the companion report *Informing an Effective Response to Climate Change* (NRC, 2010b), there is still much to be learned about the best ways of deploying science to support decision making. Indeed, available research suggests that, all too often, scientists' efforts to provide information are of limited practical value because effective decision-support systems are lacking (NRC, 2009g). Scientific research on decision-support models, processes, and tools can help improve these systems.

TABLE 4.5 Examples of Scientific Research Needs Pertaining to Decision Support in the Context of Climate Change (from Part II)

- Develop a more comprehensive and integrative understanding of factors that influence decision making.
- Improve knowledge and decision-support capabilities for all levels of governance in response to the challenges associated with sea level rise.
- Develop effective decision-support tools and approaches for decision making under uncertainty, especially when multiple governance units may be involved, for water resource management, food and fiber production issues, urban and human health issues, and other key sectors.
- Develop protocols, institutions, and technologies for monitoring and verifying compliance with international climate agreements.
- Measure and evaluate public attitudes and test communication approaches that most effectively inform and engage the public in climate-related decision making.

Effective decision support also requires interactive processes involving both scientists and decision makers. Such processes can inform decision makers about anticipated changes in climate, help scientists understand key decision-making needs, and work to build mutual understanding, trust, and cooperation—for example, in the design of decision tools and processes that make sense both scientifically and in the actual decision-making context. Table 4.5 provides a list of the related scientific research needs that emerge from the chapters in Part II of the report.

Decision Processes

Even when viable technologies or actions that could be effective in limiting the magnitude or adapting to the impacts of climate change exist, and appropriate institutions and policies to facilitate their implementation or adoption are in place (see Themes 2, 3, and 4), success can depend strongly on decision-making processes in populations or organizations (NRC, 2005a, 2008h). One of the major contributions the social sciences can make to advancing the science of climate change is in the understanding, development, assessment, and improvement of these decision-making processes. Scientific research can, for example, help identify the information that decision makers need, devise effective and broadly acceptable decision-making processes and decision-support mechanisms, and enhance learning from experience. Specific research agendas for the science of decision support are available in a number of other reports (NRC, 2009g, 2010b), and other sections of this chapter describe some of the tools that have been or could be developed to inform or assist decision makers in their deliberations (for example, vulnerability and adaptation analyses of coupled human environmental systems, which are described in Theme 3).

One of the most important and well-studied approaches to decision making is *deliberation with analysis* (also called analytic deliberation or linked analysis and deliberation). Deliberation with analysis is an iterative process that begins with the many participants in a decision working together to define a decision problem and then to identify (1) options to consider and (2) outcomes and criteria that are relevant for evaluating those options. Typically, participants work with experts to generate and interpret decision-relevant information and then revisit the objectives and choices based on that information. This model was developed in the broad context of environmental risks (NRC, 1996) and has been elaborated in the context of climate-related decision making (NRC, 1999b, 2009g)

The deliberation with analysis approach allows repeated structured interactions among the public, decision makers, and scientists that can help the scientific community understand the information needs of and uses by decision makers, and appreciate the opportunities and constraints of the institutional, material, and organizational contexts under which stakeholders make decisions (Lemos, 2008; Rayner et al., 2005; Tribbia and Moser, 2008). It also helps decision makers and other stakeholders better understand and trust the science being produced. While research on deliberation with analysis has provided a general framework that has proven effective in local and regional issues concerning ecosystem, watershed, and natural resource management, more research is needed to determine how this approach might be employed for national policy decisions or international decision making around climate change (NRC, 1996, 2005a, 2007a, 2008h).

Effective Decision-Support Systems

A decision-support system includes the individuals, organizations, networks, and institutions that develop decision-relevant knowledge, as well as the mechanisms to share and disseminate that knowledge and related products and services (NRC, 2009g). Agricultural or marine extension services, with all their strengths and weaknesses, are an important historical example of a decision-support system that has helped make scientific knowledge relevant to and available for practical decision making in the context of specific goals. The recent report *Informing Decisions in a Changing Climate* (NRC, 2009g) identified a set of basic principles of effective decision support that are applicable to the climate change arena: "(1) begin with users' needs; (2) give priority to process over products; (3) link information producers and users; (4) build connections across disciplines and organizations; (5) seek institutional stability; and (6) design processes for learning."

Effective decision-support systems work to both guide research toward decision relevance and link scientific information with potential users. Such systems will thus play an important role in improving the linkages between climate science and decision making called for both in this report and in many previous ones (e.g., Cash et al., 2003; NRC, 1990a, 1999b, 2009g). Research on the use of seasonal climate forecasts exemplifies current understanding of decision-support systems (see Box 4.4).

The basic principles of effective decision support are reasonably well known (see, e.g.,

BOX 4.4
Seasonal Climate Forecasts

For the past 20 years, the application of seasonal climate forecasts for agricultural, disaster relief, and water management decision making has yielded important lessons regarding the creation of climate knowledge systems for action in different parts of the world at different scales (Beller-Sims et al., 2008; Gilles and Valdivia, 2009; NRC, 1999b; Pagano et al., 2002; Vogel and O'Brien, 2006). Successful application of seasonal climate forecasting tends to follow a systems approach where forecasts are contextualized to the decision situation and embedded within an array of other information relevant for risk management. For example, in Australia, users and producers of seasonal climate forecasts have created knowledge systems for action in which the forecasts are part of a broader range of knowledge that informs farmers' decision making (Cash and Buizer, 2005; Lemos and Dilling, 2007). In the U.S. Southwest, potential flooding from the intense 1997-1998 El Niño was averted in part because the 3- to 9-month advance forecasts were tailored to the needs of water managers and integrated into water supply outlooks (Pagano et al., 2002).

The application of seasonal climate forecasts is not always perfect. Seasonal forecasts have proven useful in certain U.S. regions directly affected by El Niño events but may have limited predictive skill outside those regions and outside the extremes of the El Niño-Southern Oscillation cycle (see Chapter 6). There is evidence that too much investment in climate forecasting may crowd out more sustainable alternatives to manage risk or even harm some stakeholders (Lemos and Dilling, 2007). For example, even under high uncertainty, a forecast of El Niño and the prospect of a weak fishing season give companies in Peru an incentive to accelerate seasonal layoffs of workers (Broad et al., 2002). More recent efforts to apply the lessons from seasonal climate forecasting to inform climate adaptation policy argue for the integration of climate predictions within broader decision contexts (Johnston et al., 2004; Klopper et al., 2006; Meinke et al., 2009). In such cases, rather than "perfect" forecasts, the best strategy for supporting decision making is to use integrated assessments and participatory approaches to link climate information to information on other stressors (Vogel et al., 2007).

NRC, 2009g). However, they need to be applied differently in different places, with different decision makers, and in different decision contexts. Determining how to apply these basic principles is at the core of the science of decision support—the science needed for designing information products, knowledge networks, and institutions that can turn good information into good decision support (NRC, 2009g). The base in fundamental science for designing more effective decision-support systems lies in the decision sciences and related fields of scholarship, including cognitive science, communications research, and the full array of traditional social and behavioral science disciplines.

Expanded research on decision support would enhance virtually all the other research called for in this report by improving the design and function of systems that seek to make climate science findings useful in adaptive management of the risks of climate change. The main research needs in this area are discussed in *Informing Decisions in a Changing Climate* (NRC, 2009g), *Informing an Effective Response to Climate Change* (NRC, 2010b), and several other studies (e.g., NRC, 2005a, 2008g). A recent review of research needs for improved environmental decision making (NRC, 2005a) emphasized the need for research to identify the kinds of decision-support activities and products that are most effective for various purposes and audiences. The report recommended studies focused on assessing decision quality, exploring decision makers' evaluations of decision processes and outcomes, and improving formal tools for decision support.

The key research needs for the science of decision support fall into the following five areas (NRC, 2009g):

- *Information needs.* Research is needed to identify the kinds of information that would add greatest value for climate-related decision making and to understand information needs as seen by decision makers.
- *Communicating risk and uncertainty.* People commonly have difficulty making good sense and use of information that is probabilistic and uncertain. Research on how people understand uncertain information about risks and on better ways to provide it can improve decision-support processes and products.
- *Decision-support processes.* Research is needed on processes for providing decision support, including the operation of networks and intermediaries between the producers and users of information for decision support. This research should include attention to the most effective channels and organizational structures to use for delivering information for decision support; the ways such information can be made to fit into individual, organizational, and institutional decision routines; the factors that determine whether potentially useful information is actually used; and ways to overcome barriers to the use of decision-relevant information.

- *Decision-support products.* Research is needed to design and apply decision tools, data analysis platforms, reports, and other products that convey user-relevant information in ways that enhance users' understanding and decision quality. These products may include models and simulations, mapping and visualization products, websites, and applications of techniques for structuring decisions, such as cost-benefit analysis, multiattribute decision analysis, and scenario analysis.
- *Decision-support "experiments."* Efforts to provide decision support for various decisions and decision makers are already under way in many cities, counties, and regions. These efforts can be treated as a massive national experiment that can, if data are carefully collected, be analyzed to learn which strategies are attractive, which ones work, why they work, and under what conditions. Research on these experiments can build knowledge about how information of various kinds, delivered in various formats, is used in real-world settings; how knowledge is transferred across communities and sectors; and many other aspects of decision-support processes.

THEME 6: INTEGRATED CLIMATE OBSERVING SYSTEMS

Nearly all of the research called for in this report either requires or would be considerably improved by a comprehensive, coordinated, and continuing set of observations—physical, biological, and social—about climate change, its impacts, and the consequences (both intended and unintended) of efforts to limit its magnitude or adapt to its impacts (Table 4.6). Extensive, robust, and well-calibrated observing systems would support the research that underpins the scientific understanding of how and why climate is changing, provide information about the efficacy of actions and strategies taken to limit or adapt to climate change, and enable the routine dissemination of climate and climate-related information and products to decision makers. Unfortunately, many of the needed observational assets are either underdeveloped or in decline. In addition, a variety of institutional factors—such as distributed responsibility across many different entities—complicate the development of a robust and integrated climate observing system.

The breadth of information needed to support climate-related decision making implies an observational strategy that includes both remotely sensed and in situ observations and that provides information about changes across a broad range of natural and human systems. To be useful, these observations must be

- Sustained for decades to separate long-term trends from short-term variability;
- Well calibrated and consistent through time to ensure that observed changes are real;

- Spatially extensive to account for variability across scales and to ensure that assessments of change are not overly influenced by local phenomena;
- Supported by a robust data management infrastructure that supports effective data archiving, accesses, and stewardship; and
- Sustained by defined roles and responsibilities across the federal government as well as state and local governments, the research community, private businesses, and the international community.

Space-Based Platforms

Our understanding of the climate system and other important human and environmental systems has benefitted significantly through the use of satellite observations over the past 30 years (NRC, 2008c). For example, data from the Earth Observing System (EOS) series of satellites deployed in the late 1990s and early 2000s provide critical input into process and climate models that have provided key insights into Artic sea ice decline, sea level rise, changes in freshwater systems, ozone changes over Antarctica, changes in solar activity, ocean ecoystem change, and changes in land use, to name just a few. Box 4.5 provides an example of a key satellite-based measurement that has promoted enhanced understanding of the physical climate system and how it is changing over time.

TABLE 4.6 Examples of Science Needs Related to Observations and Observing Systems (see Part II for additional details)

- Extend and expand long-term observations of atmosphere and ocean temperatures; sea level; ice extent, mass, and volume; and other critical physical climate system variables.
- Extend and expand long-term observations of hydrologic changes and related changes relevant for water management decision making.
- Expand observing and monitoring systems for ecosystems, agriculture and fisheries, air and water quality, and other critical impact areas.
- Improve observations that allow analysis of multiple stressors, including changes in climate, land use changes, pollutant deposition, invasions of nonnative species, and other human-caused changes.
- Develop improved observations and monitoring capabilities to support vulnerability assessments of coupled human-environment systems at the scale of cities, states, nations, and regions, and for tracking and analyzing human health and well-being.
- Develop improved observations for vulnerability assessments related to military operations and infrastructure.
- Establish long-term monitoring systems that are capable of monitoring and assessing the effectiveness of actions taken to limit or adapt to climate change.
- Develop observations, protocols, and technologies for monitoring and verifying compliance with international emissions-reduction agreements.

BOX 4.5
Ocean Altimetry[a]

Ocean altimetry measurements provide an illustrative example of how satellites have advanced scientific understanding of climate and climate change. Sea level changes are a fundamental indicator of changes in global climate and have profound socioeconomic implications (see Chapter 7). Variations in sea level also provide insight into natural climate processes such as the El Niño-Southern Oscillation cycle (see Chapter 6) and have the potential to inform a broad array of other climate science disciplines including ocean science, cryospheric science, hydrology, and climate modeling applications (see, e.g., Rahmstorf et al., 2007).

Prior to the satellite era, tide gauge measurements were the primary means of monitoring sea level change. However, their limited spatial distribution and ambiguous nature (e.g., vertical land motion can cause erroneous signals that mimic the effects of climate change at some sites) limited their use for climate research. With the launch of TOPEX/Poseidon in 1992, satellite altimeter measurements with sufficient accuracy and orbital characteristics to monitor small (on the order of millimeters per year) sea level changes became available (Cazenave and Nerem, 2004). Jason-1, launched at the end of 2001, continued the TOPEX/Poseidon measurements in the same orbit, including a critical 6-month overlap that allowed intercalibration to ensure the continuity of records. It is important to note that tide gauges remain a critical component of the sea level observing system, providing an independent source of data in coastal areas that can be used to calibrate and interpret satellite data records. The integration of tide gauge and satellite data provides an excellent example of how satellite and surface-based observations are essential complements to one another within an integrated observing system.

Together, the TOPEX/Poseidon and Jason-1 missions have produced a continuous 15-year time series of precisely calibrated measurements of global sea level. These measurements show that sea level rose at an average rate of ~3.5 mm/year (0.14 inches/year) during the TOPEX/Jason-1 period, nearly double the rate inferred from tide gauges over the 20th century (Beckley et al., 2007; Leuliette et al., 2004). Since sea level rise is driven by a combination of ocean warming and shrinking glaciers and ice sheets (see Chapter 7), these altimetry results are also important for refining and constraining estimates of ocean heat content and ice loss. Another powerful aspect of satellite altimetry is that it provides maps of the spatial variability of the sea level–rise signal (see figure on facing page), which is valuable for the identification of sea level "fingerprints" associated with climate change (see also Mitrovica et al., 2001). Sea level measurements are also used extensively in ocean reanalysis efforts and short-term climate predictions.

Jason-2,[b] which carries similar but improved instrumentation, was launched in June 2008. By design, Jason-2 overlaps with the Jason-1 mission, thus providing the requisite intercalibration period and securing the continuity of high-accuracy satellite altimetry observations. Funds have been requested in the President's 2011 budget to support a 2013 launch of Jason-3, a joint effort between NOAA and EUMETSAT (the European meteorological satellite program), as part of a transition of satellite altimetry from research to "operational" status. Researchers hope to avoid a gap in the satellite record because measurements from tide gauges and other satellite measurements would not be sufficient to accurately determine the bias between the two time series on either side of the gap. It should also be emphasized that ocean altimetry, despite the challenges of ensuring overlap and continuity, is on a much better trajectory than many other important climate observations, as described in the text.

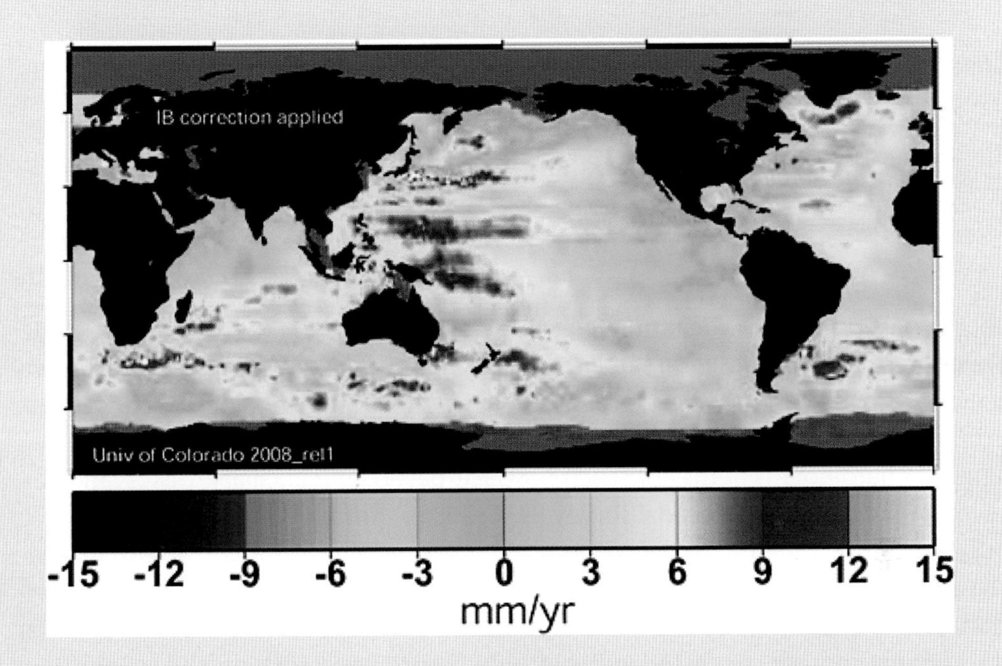

Trends (mm/year) in sea level change over 1993-2007 from TOPEX/Poseidon and Jason-1 altimeter measurements. SOURCE: Courtesy of Colorado Center for Astrodynamics Research, University of Colorado at Boulder (http://sea-level.colorado.edu).

---

[a] Material in this box is adapted from *Ensuring the Climate Record from the NPOESS and GOES-R Spacecraft: Elements of a Strategy to Recover Measurement Capabilities Lost in Program Restructuring* (NRC, 2008d).

[b] Also called the Ocean Surface Topography Mission/Jason-2.

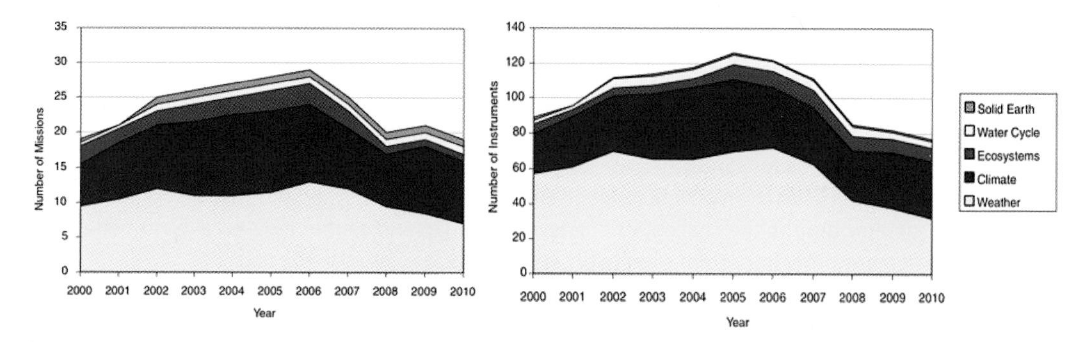

FIGURE 4.3 Number of U.S. space-based Earth observation missions (left) and instruments (right) in the current decade. An emphasis on climate and weather is evident, as is a decline in the number of missions near the end of the decade. For the period from 2007 to 2010, missions were generally assumed to operate for 4 years past their nominal lifetimes. SOURCE: NRC (2007c), based on information from NASA and NOAA websites for mission durations.

Over the past decade, a wide range of problems have plagued the maintenance and development of environmental satellites. In response to a request from several federal agencies, the NRC conducted a "decadal survey" in 2004-2006 to generate consensus recommendations from the Earth and environmental science and applications communities regarding a systems approach to space-based (and ancillary) observations. The interim report of the decadal survey (NRC, 2005b) described the national system of environmental satellites as being "at risk of collapse." That judgment was based on a sharp decline in funding for Earth observation missions and the consequent cancellation, descoping, and delay of a number of critical satellite missions and instruments. An additional concern expressed in the interim report was attracting and training scientists and engineers and providing opportunities for them to exploit new technology and apply new theoretical understanding in the pursuit of both discovery science and high-priority societal applications.

These concerns only increased in the 2 years following the publication of the interim report as additional missions and sensors were cancelled. The final decadal survey report (NRC, 2007c) presented near- and longer-term recommendations to address these troubling trends. The report outlined near-term actions meant to stem the tide of capability deterioration and continue critical data records, as well as forward-looking recommendations to establish a balanced Earth observation program designed to directly address the most urgent societal challenges (see Figure 4.3). The final report also noted the lack of clear agency responsibility for sustained research programs and for transitioning proof-of-concept measurements into sustained measurement systems (see Box 4.6).

The National Polar-orbiting Operational Environmental Satellite System (NPOESS) was created in 1994 to merge various military and civil meteorological and environmental monitoring programs. Unfortunately, by 2005, cost overruns triggered a mandatory

BOX 4.6
The Development of a Long-Term, Space-Based Earth Observation System[a]

"There is a crisis not only with respect to climate change . . . but also [with respect to] the absence of a coherent, coordinated federal environmental policy to address the challenges. In the nearest term possible, aging space- and ground-based environmental sensors must be replaced with technologically improved instruments. Beyond replacing aging instruments, there is a need to enhance continuity in the observations, so that policy makers, informed by science, will have the necessary tools to detect trends in important Earth indicators and craft wise and effective long-term policies. However, continuity, or sustained long-term observations, is not an explicitly stated requirement for either the 'operational' or 'research' space systems that are typically associated with [NOAA] and [NASA] programs, respectively.

The present federal agency paradigm of 'research *to* operations' with respect to NASA and NOAA is obsolete and nearly dysfunctional, in spite of best efforts by both agencies. This paradigm currently has NASA developing and demonstrating new observational techniques and measurements deemed useful for prediction or other applications. These are then transitioned to NOAA (or sometimes DOD) and used on a sustained, multi-decadal basis. However, this paradigm is not working for a number of reasons. The two agencies have responsibilities that are in many cases mismatched with their authorities and resources: institutional mandates are inconsistent with agency charters; budgets are not well matched to the needs; agency responsibilities are not clearly defined, and shared responsibilities are supported inconsistently by ad hoc mechanisms for cooperation. . . . A new paradigm of 'research *and* operations' is urgently needed to meet the challenge of vigilant monitoring of all aspects of climate change. . . .

Our ability as a nation to sustain climate observations has been complicated by the fact that no single agency has both the mandate and requisite budget for providing ongoing climate observations, prediction, and services. While interagency collaborations are sometimes valuable, a robust, effective program of Earth observations from space requires specific responsibilities to be clearly assigned to each agency and adequate resources provided to meet these responsibilities."

[a] Excerpted from testimony by Richard A. Anthes, President of the University Corporation for Atmospheric Research, Past President, American Meteorological Society, and Co-Chair, Committee on Earth Science and Applications from Space (2003-2007), before the Subcommittee on Commerce, Justice, Science, and Related Agencies, Committee on Appropriations, U.S. House of Representatives, March 19, 2009.

review of the NPOESS program, resulting in reductions in the number of planned satellite acquisitions as well as reductions in the instruments carried on each platform—with climate-related sensors suffering the majority of the cuts, in part because of conflicting agency priorities. More recently, there have been several efforts to restore some of the lost sensor capabilities. However, these short-term, stop-gap measures are only designed to preserve the most critical long-term records and do not represent a long-term, comprehensive strategy to observe critical climate and climate-related processes and trends from space (NRC, 2008d). The President's 2011 budget seeks to restructure the NPOESS program, but details were not available in time to inform the development of this report. An additional blow to the nation's Earth observing program was the July 2009 launch failure of NASA's Orbiting Carbon Observatory (OCO), which was expected to provide high-resolution satellite-based measurements of CO_2 and other GHGs (NRC, 2009h). The President's 2011 budget request for NASA includes $170 million for a reflight of the OCO mission, which will be called OCO-2.

Given the global scope of satellite observations and the expense of designing, launching, and operating satellites, the decadal survey (NRC, 2007c) and other reviews call for international coordination as a key component of the nation's satellite observation strategy. Collaborations with other nations not only save scarce resources for all partners, they also promote scientific collaboration and sharing of ideas among the international scientific community. However, international collaborations come at a cost. Any time partners are involved, control must be shared, and the success of the mission depends critically on the performance of all partners. A successful collaboration also requires assurance that data will be shared and that U.S. scientists are full partners on teams that ensure adequate prelaunch instrument characterization and postlaunch instrument calibration and validation.

Finally, there is a wealth of classified data that have been and continue to be collected by the intelligence community that could potentially provide useful information on understanding the nature and impacts of climate change. Declassified data from the 1960s have already been used for this purpose with great success (Csatho et al., 1999; Joughin et al., 2002; Stokes et al., 2006). More recently, a large amount of sea ice imagery was released for scientific study (NRC, 2009l). Given the importance of the climate change challenge, and the recent struggles of the civilian satellite program, the climate science community should take advantage of such data sets to the extent that they can be made available for scientific purposes.

Ground-Based and In Situ Observations of the Earth System

Ground-based in situ measurements—ranging from thermometer measurements to ecosystem surveys—are the oldest and most diverse type of environmental observations, and they remain a fundamental component of an integrated climate observing system. Over the past 60 years, direct ground-based measurements have been supplemented by airborne in situ measurements, from both aircraft and balloons, and by ground-based, remotely sensed data, such as weather radars and vertical profilers of atmospheric composition. Collectively, these observations span a broad range of instruments and types of information, from instruments initially deployed as part of research experiments to operational networks at the local, state, regional, national, and international levels deployed by a range of public and private institutions. In addition to directly supporting research on the Earth system and specific decision-making needs, these observations are critical for calibrating and validating satellite measurements and for developing and testing climate and Earth system model parameterizations.

There have been significant advances in in situ and ground-based monitoring networks over the past several decades. Examples include the Arctic observing network, the Tropical-Atmosphere Ocean (TAO) array constructed primarily to monitor temperature profiles in the upper equatorial Pacific ocean and support predictions of the El Niño-Southern Oscillation, "Argo" floats that provide dispersed observations of temperature and salinity of the upper ocean, the FLUXNET network of ecosystem carbon exchange with the atmosphere, the Aerosol Robotic Network (AERONET) that provides observations of atmospheric optical properties, and the Atmosphere Radiation Measurement (ARM) program. In addition, there is a wealth of observations from a broad range of public and private systems designed primarily for other purposes—such as wind monitoring for port safety—that could potentially be tapped to supplement existing climate observations and yield new and valuable insights. These systems will have to be integrated and maintained for decades to realize their full potential as components of a climate observing system.

The recent study *Observing Weather and Climate from the Ground Up: A Nationwide Network of Networks* (NRC, 2009j) discusses the value and challenges of coordinating the wide range of ground-based weather, climate, and climate-related observing systems to create a more integrated system that could be greater than the sum of its individual parts. The report calls for improved coordination across existing public and private networks of in situ observations. However, the number and diversity of entities involved make this a major organizational and governance challenge. If properly developed, an integrated, nationwide network of weather, climate, and related observations would undoubtedly be a tremendous asset for supporting improved understanding of climate change as well as climate-related decision making.

In addition to maintaining and enhancing observational capacity, research on new methods of observation, such as the miniaturization of instruments for in situ data collection, could both enhance data collection capabilities and lower the often substantial costs associated with data collection systems. To become effective components of an integrated climate observing system, these observational capacities, whether they represent the continuation of existing capabilities or the development of new ones, should be developed with a view toward providing meaningful, accurate, well-calibrated, integrated, and sustained data across a range of climate and climate-related variables.

Observations of Human Systems

Other sections of this chapter highlight the importance of social science research in understanding the causes, consequences, and opportunities to respond to climate change. As with research on the physical and biological components of the climate system, this research depends on the availability of high-quality, long-term, and readily accessible observations of human systems, not only in the United States but also in areas of the world with relevant U.S. interests. Census data, economic productivity and consumption data, data on health and disease patterns, insurance coverage, crop yields, hazards exposure, and public perceptions and preferences are just some of the types of information that can be relevant for developing an improved understanding of human interactions with the climate system and for answering various decision-relevant questions related to the human dimensions of climate change. Socioeconomic data are also critical for linking environmental observations with assessments of climate-related risk, vulnerability, resilience, and adaptive capacity in human systems. As with other types of observations, long time series are needed to monitor changes in the drivers of climate change and trends in resilience and vulnerability. Such observational data are most useful when geocoded (linked to specific locations) and matched (aggregated or downscaled) to scales of interest to researchers and decision makers, and when human and environmental data are collected and archived in ways that facilitate linkages between these data.

Studies conducted in the 1970s and 1980s demonstrate the feasibility of data collection efforts that integrate across the engineering and social sciences to better understand and model energy consumption (Black et al., 1985; Cramer et al., 1984; Harris and Blumstein, 1984; Socolow, 1978). Linkage of data on land-cover change and its social and economic drivers has also been productive (NRC, 2005c, 2007i). Large-scale social science data collection efforts, ranging from the census to federally funded surveys such as the National Longitudinal Study of Adolescent Health, the Panel Study of Income Dynamics, the General Social Survey, and the National Election Studies show the feasibility and value of long-term efforts to collect high-quality social data. However, to date there has been no sustained support to collect comparable data at the individual or organizational level on environmentally significant behaviors, such as energy use and GHG emissions. As states and other entities adopt policies to limit GHG emissions, sustained and integrated efforts to collect data on environmentally significant consumption will be extremely helpful for monitoring progress and honing programs and policies.

Likewise, data on the impacts of climate change on human systems and on vulnerability and adaptation of human systems to global environmental changes are critically needed (NRC, 2009g,k). Examples include morbidity and mortality data associated with air and water quality, expanded data sets focusing on household risk-pooling strategies and adaptation options, and data on urban infrastructure vulnerabilities to extreme weather and climate events. Methods that allow aggregation of data from across a range of regions to develop national-scale understanding will sometimes be necessary, but local and regional vulnerability assessments will also be needed, and these depend on both local and appropriately downscaled information (Braden et al., 2009). The potential exists for greater use of remote sensing to develop indicators of vulnerability to various climate-related hazards and of the socioeconomic drivers of climate change. If validated against in situ measurements, such measures can allow for monitoring of human-climate interactions at much finer spatial and temporal scales than is currently feasible with surveys or other in situ measures of human variables.

There is also great potential in the use of mobile communications technology, such as cell and smart phones, as a vehicle for social science research that has fine temporal and spatial scales (Eagle et al., 2009; Raento et al., 2009; Zuwallack, 2009). Many data collection efforts previously undertaken for governmental administrative purposes, business purposes, or social science research not related to climate change could potentially support the research needed for understanding the human aspects of climate change and climate-related decision making, but only if they are geocoded and linked to other data sets. International, longitudinal databases such as the International Forestry and Institutions database (e.g., Chhatre and Agrawal, 2008) also have great potential to serve as a bridge between local, regional, national, and global processes, as well as for assessing the dynamics of change across time and space.

Finally, because most major social and economic databases have been developed for purposes unrelated to climate change, these data have significant gaps from the perspective of climate science. However, all climate-relevant socioeconomic and other human systems data need not necessarily be held in a single common observing system. They simply need to be inventoried, archived, and made broadly accessible to enable the kinds of integrative analyses that are necessary for the new climate change research. A major effort is needed both to develop appropriate local data collection efforts and to coordinate them into national and global systems. Initial progress can be made by coordination across specific domains and sectors (e.g., coastal vulnerabilities, health vulnerabilities) and across scales so that locally useful information also contributes to larger-scale indicators and vice versa. Data integration is also a critical need. Some of these issues are explored in the next subsection.

Data Assimilation, Analysis, and Management

Data assimilation refers to the combination of disparate observations to provide a comprehensive and internally consistent data set that describes how a system is changing over time. Improvements in data assimilation systems have led directly to substantial improvements in numerical weather prediction over the past several decades by improving the realism of the initial conditions used to run weather forecast models. Improved data assimilation techniques have also led to improved data sets for analyses of climate change.

Climate data records (see NRC, 2004a) are generated by a systematic and ongoing process of climate data integration and reprocessing. Often referred to as reanalysis, the fundamental idea behind such efforts (see, e.g., Kalnay et al., 1996) is to use data assimilation methods to capitalize on the wealth of disparate historical observations and integrate them with newer observations, such as space-based data. Data assimilation, analysis, and reanalysis are also becoming increasingly important for areas other than regional and global atmospheric models, such as ocean models, land models, marine ecosystems, cryosphere models, and atmospheric chemistry models.

Improvements have occurred in all components of data assimilation and reanalysis, including data assimilation models, the quality and quantity of the observations, and methods for statistical interpolation (see, e.g., Daley, 1991; Kalnay, 2002). However, additional advances are needed. For example, data for the ocean, atmosphere, and land are typically assimilated separately in different models and frameworks. Given that these systems are intrinsically coupled on climate time scales, for instance through exchanges of water and energy, coupled data assimilation methodologies are needed to take into account their interactions. Next-generation data assimilation and reanalysis systems should aim to fully incorporate all aspects of the Earth system (and, eventually, human systems) to support integrated understanding and facilitate analyses of coupled human-environment systems.

Finally, and critically, all observing systems and data analysis activities depend on effective data management—including data archiving, stewardship, and access systems. Historically, support for data management has often lagged behind support for initial data collection (NRC, 2007d). As the demand for sustained climate observations is realized and actions are taken to improve, extend, and coordinate observations, there will be an increase in the demands on both technology and human capacity to ensure that the resulting data are securely archived, quality controlled, and made available to a wide range of users (Baker et al., 2007; NRC, 2004a, 2005e, 2007d). Likewise, as data volume and diversity expand new computational approaches as well as greater computing power will be needed to process and integrate the different data sets on a schedule useful for planning responses to climate change. Finally, because some data have the potential for violating personal privacy norms and legal guarantees, proper safeguards must be in place to protect confidentiality.

Toward Integrated Observations and Earth System Analysis

An integrated climate observing system and improved data analysis and data management systems will be needed to support all of the other themes described in this chapter. Regular observations of the Earth system, for example, are needed to improve climate models, monitor climate and climate-related changes, assess the vulnerability of different human and environmental systems to these change, monitor the effectiveness of actions taken to limit the magnitude of climate change, warn about impending tipping points, and inform decision making. However, creating such systems and making the information available in usable formats to a broad range of researchers and decision makers involves a number of formidable challenges, such as improving linkages between human and environmental data, ensuring adequate support for data archiving and management activities, and creating improved tools for data access and dissemination.

An *integrated Earth system analysis capability,* or the ability to create an accurate, internally consistent, synthesized description of the evolving Earth system, is a key research need identified both in this report and in many previous reports (NRC, 2009k). Perhaps the single greatest roadblock to achieving this capability is the lack of comprehensive, robust, and unbiased long-term global observations of the climate system and other related human and environmental systems. Other scientific and technical challenges include identifying the criteria for optimizing assimilation techniques for different purposes, estimating uncertainties, and meeting user demands for higher spatial resolution.

The NRC report *Informing Decisions in a Changing Climate* (NRC, 2009g) recommends that the federal government "expand and maintain national observation systems to provide information needed for climate decision support. These systems should link existing data on physical, ecological, social, economic, and health variables to each other and develop new data and key indicators as needed" for estimating climate change vulnerabilities and informing responses intended to limit and adapt to climate change. It also notes the need for geocoding existing social and environmental databases; developing methods for aggregating, disaggregating, and integrating such data sets with each other and with climate and other Earth system data; creating new databases to fill critical gaps; supporting modeling and process studies to improve methods for making the data useful; and engaging decision makers in the identification of critical data needs. That study's recommendations set appropriate strategic directions for an integrated data system. Ultimately, the collection and archiving of data for such a system would need to be evaluated on the basis of potential and actual use in research and decision making.

The recommendations in Chapter 5 provide advice on some steps that can be taken to address these challenges.

THEME 7: IMPROVED PROJECTIONS, ANALYSES, AND ASSESSMENTS

Nearly every scientific challenge associated with understanding and responding to climate change requires an assessment of the interactions among different components of the coupled human-environment system. A wide range of models, tools, and approaches, from quantitative numerical models and analytic techniques to frameworks and processes that engage interdisciplinary research teams and stakeholders, are needed to simulate and assess these interactions. While decisions are ultimately the outcome of individual, group, and political decision-making processes, scientific tools and approaches can aid decision making by systematically incorporating complex information, projecting the consequences of different choices, accounting for uncertainties, and facilitating disciplined evaluation of trade-offs as the nation turns its attention to responding to climate change. Table 4.7 lists some of the specific research needs identified in Part II of the report that are related to the development of models, tools, and approaches for improving projections, analyses, and assessments of climate change.

TABLE 4.7 Examples of Science Needs Related to Improving Projections, Analyses, and Assessments of Climate Change (from Part II)

- Continue to develop and use scenarios as a tool for framing uncertainty and risk, understanding human drivers of climate change, forcing climate models, and projecting changes in adaptive capacity and vulnerability.
- Improve model projections of future climate change, especially at regional scales.
- Improve end-to-end models through coordination and linkages among models that connect emissions, changes in the climate system, and impacts on specific sectors.
- Develop tools and approaches for understanding and predicting the impacts of sea level rise on coastal ecosystems and infrastructure.
- Improve models of the response of agricultural crops, fisheries, transportation systems, and other human systems to climate and other environmental changes.
- Develop integrated approaches and analytical frameworks to evaluate the effectiveness and potential unintended consequences of actions taken to respond to climate change, including trade-offs and synergies among various options.
- Explore cross-sector interactions between impacts of and responses to climate change.
- Continue to improve methods for estimating costs, benefits, and cost effectiveness of climate mitigation and adaptation policies, including complex or hybrid policies.
- Develop analyses that examine climate policy from a sustainability perspective, taking account of the full range of effects of climate policy on human and environmental systems, including unintended consequences and equity effects.

The boundaries between various tools and approaches for integrated analysis of climate impacts, vulnerabilities, and response options are not rigid; often, a combination of several tools or approaches is needed for improved understanding and to support decision making. This section highlights a few of the integrated tools and approaches that can be used, including

- Scenarios of future GHG emissions and other human activities;
- Climate and Earth system models;
- Process models of ecological functions and ecosystem services;
- Integrated assessment approaches, which couple human and environmental systems;
- Policy-oriented heuristic models and exercises; and
- Process-based decision tools.

This discussion is not intended to be an exhaustive treatment of these approaches—more detailed discussions can be found in Part II of the report and in other reports (e.g., NRC, 2009g)—nor is it intended as a complete list of important tools and approaches for integrated analysis. Rather, it provides examples of the kinds of approaches that need to be developed, improved, and used more extensively to improve scientific understanding of climate change and make this scientific knowledge more useful for decision making.

Scenario Development

Scenarios help improve understanding of the key processes and uncertainties associated with projections of future climate change. Scenarios are critical for helping decision makers establish targets or budgets for future GHG emissions and devise plans to adapt to the projected impacts of climate change in the context of changes in other human and environmental systems. Scenario development is an inherently interdisciplinary and integrative activity requiring contributions from many different scientific fields as well as processes that link scientific analysis with decision making. Chapter 6 describes some recent scenario development efforts as well as several key outstanding research needs.

Climate Models

Climate models simulate how the atmosphere, oceans, and land surface respond to increasing concentrations of GHGs and other climate drivers that vary over time (see Chapter 6). These models are based on numerical representations of fundamental Earth system processes, such as the exchange of energy, moisture, and materials between the atmosphere and the underlying ocean or land surface. Climate models have been critically important for understanding past and current climate change and remain an essential tool for projecting future changes. They have also been steadily increasing in detail, sophistication, and complexity, most notably by improving spatial resolution and incorporating representations of atmospheric chemistry, biogeochemical cycling, and other Earth system processes. These improvements represent an important integrative tool because they allow for the evaluation of feedbacks between the climate system and other aspects of the Earth system.

As discussed in Chapter 6, there are a number of practical limitations, gaps in understanding, and institutional constraints that limit the ability of climate models to inform climate-related decision making, including the following

- The ability to explicitly simulate all relevant climate processes (for example, individual clouds) on appropriate space and time scales;
- Constraints on computing resources;

- Uncertainties and complexities associated with data assimilation and parameterization;
- Lack of a well-developed framework for regional downscaling;
- Representing regional modes of variability;
- Projecting changes in storm patterns and extreme weather events;
- Inclusion of additional Earth system processes, such as ice sheet dynamics and fully interactive ecosystem dynamics;
- Ability to simulate certain nonlinear processes, including thresholds, tipping points, and abrupt changes; and
- Representing all of the processes that determine the vulnerability, resilience, and adaptability of both natural and human systems.

As discussed in Chapter 6, climate modelers in the United States and around the world have begun to devise strategies, such as decadal-scale climate predictions, for improving the utility of climate model experiments. These experimental strategies may indeed yield more decision-relevant information, but, given the importance of local- and regional-scale information for planning responses to climate change, continued and expanded investments in regional climate modeling remain a particularly pressing priority. Expanded computing resources and human capital are also needed.

Progress in both regional and global climate modeling cannot occur in isolation. Expanded observations are needed to initialize models and validate results, to develop improved representations of physical processes, and to support downscaling techniques. For example, local- and regional-scale observations are needed to verify regional models or downscaled estimates of precipitation, and expanded ocean observations are needed to support decadal predictions. Certain human actions and activities, including agricultural practices, fire suppression, deforestation, water management, and urban development, can also interact strongly with climate change. Without models that account for such interactions and feedbacks among all important aspects of the Earth system and related human systems, it is difficult to fully evaluate the costs, benefits, trade-offs and co-benefits associated with different courses of action that might be taken to respond to climate change (the next subsection describes modeling approaches that address some of these considerations). An advanced generation of climate models with explicit and improved representations of terrestrial and marine ecosystems, the cryosphere, and other important systems and processes, and with improved representations and linkages to models of human systems and actions, will be as important as improving model resolution for increasing the value and utility of climate and Earth system models for decision making.

Models and Approaches for Integrated Assessments

Integrated assessments combine information and insights from the physical and biological sciences with information and insights from the social sciences (including economics, geography, psychology, and sociology) to provide comprehensive analyses that are sometimes more applicable to decision making than analyses of human or environmental systems in isolation. Integrated assessments—which are done through either formal modeling or through informal linkages among relevant disciplines—have been used to develop insights into the possible effectiveness and repercussions of specific environmental policy choices (including, but not limited to, climate change policy) and to evaluate the impacts, vulnerability, and adaptive capacity of both human and natural systems to a variety of environmental stresses. Several different kinds of integrated assessment approaches are discussed in the paragraphs below.

Integrated Assessment Models

In the context of climate change, integrated assessment models typically incorporate a climate model of moderate or intermediate complexity with models of the economic system (especially the industrial and energy sectors), land use, agriculture, ecosystems, or other systems or sectors germane to the question being addressed. Rather than focusing on precise projections of key system variables, integrated assessment models are typically used to compare the relative effectiveness and implications of different policy measures (see Chapter 17). Integrated assessment models have been used, for instance, to understand how policies designed to boost production of biofuels may actually increase tropical deforestation and lead to food shortages (e.g., Gurgel et al., 2007) and how policies that limit CO_2 from land use and energy use together lead to very different costs and consequences than policies that address energy use alone (e.g., Wise et al., 2009a). Another common use of integrated assessments and integrated assessment models is for "impacts, adaptation, and vulnerability" or IAV assessments, which evaluate the impacts of climate change on specific systems or sectors (e.g., agriculture), including their vulnerability and adaptive capacity, and explore the effectiveness of various response options. IAV assessments can aid in vulnerability and adaptation assessments of the sort described in Theme 3 above.

An additional and valuable role of integrated assessment activities is to help decision makers deal with uncertainty. Three basic approaches to uncertainty analysis have been employed by the integrated assessment community: sensitivity analysis, stochastic simulation, and sequential decision making under uncertainty (DOE, 2009b; Weyant, 2009). The aim of these approaches is not to overcome or reduce uncertainty, but rather to characterize it and help decision makers make informed and robust decisions in the face of uncertainty (Schneider and Kuntz-Duriseti, 2002), for instance by adopting an adaptive risk-management approach to decision making (see Box 3.1). Analytic characterizations of uncertainty can also help to determine the factors or processes that dominate the total uncertainty associated with a specific decision and thus potentially help identify research priorities. For example, while uncertainties in climate sensitivity and future human energy production and consumption are widely appreciated, improved methods for characterizing the uncertainty in other socioeconomic drivers of environmental change are needed. In addition, a set of fully integrated models capable of analyzing policies that unfold sequentially, while taking account of uncertainty, could inform policy design and processes of societal and political judgment, including judgments of acceptable risk.

Enhanced integrated assessment capability, including improved representation of diverse elements of the coupled human-environment system in integrated assessment models, promises benefits across a wide range of scientific fields as well as for supporting decision making. A long-range goal of integrated assessment models is to seamlessly connect models of human activity, GHG emissions, and Earth system processes, including the impacts of climate change on human and natural systems and the feedbacks of changes in these systems on climate change. In addition to improved computational resources and improved understanding of human and environmental systems, integrated assessment modeling would also benefit from model intercomparison and assessment techniques similar to those employed in models that focus on Earth system processes.

Life-Cycle Assessment Methods[4]

The impacts of a product (or process) on the environment come not only when the product is being used for its intended purpose, but also as the product is manufactured and as it is disposed of at the end of its useful life. Efforts to account for the full set of environmental impacts of a product, from production of raw materials through manufacture and use to its eventual disposition, are referred to as life-cycle analysis (LCA). LCA is an important tool for identifying opportunities for reducing GHG emissions and also for examining trade-offs between GHG emissions and other environmental impacts. LCA has been used to examine the GHG emissions and land use requirements of renewable energy technologies (e.g., NRC, 2009) and other technolo-

[4] This subsection was inadvertently left out of prepublication copies of the report.

gies that might reduce GHG emissions (e.g., Jaramillo et al., 2009, Kubiszewski et al., 2010, Lenzen, 2008, Samaras and Meisterling, 2008).

LCA of corn-based ethanol and other liquid fuels derived from plant materials (e.g., Davis et al., 2009; Kim et al., 2009; Robertson et al., 2008; Tilman et al., 2009) illustrate both the value of the method and some of the complexities in applying it. Because corn ethanol is produced from sugars created by photosynthesis, which removes CO_2 from ambient air, it might be assumed that substituting corn ethanol for gasoline produced from petroleum would substantially reduce net GHG emissions. However, LCA shows that these emissions reductions are much smaller (and in some cases may even result in higher GHG emissions) when the emissions associated with growing the corn, processing it into ethanol, and transporting it are accounted for. A substantial shift to corn-based ethanol (or other biofuels) could also lead to significant land use changes and changes in food prices. LCA also points out the importance of farming practices in shaping agricultural GHG emissions and to the potential for alternative plant inputs, such as cellulose, as a feedstock for liquid fuels.

The utility and potential applications of LCA have been recognized by government agencies in the United States and around the world (EPA, 2010a; European Commission Joint Research Centre, 2010) and by the private sector. For example, Walmart is emphasizing LCA in the sustainability assessment it is requiring of all its suppliers.[5] Useful as it is, LCA, like any policy analysis tool, has limitations. For example, the boundaries for the analysis must be defined, materials used for multiple purposes must be allocated appropriately, and the databases typically consulted to estimate emissions at each step of the analysis may have uncertainties. There is currently little standardization of these databases or of methods for drawing boundaries and allocating impacts. LCA may also identify multiple environmental impacts. For example, nuclear reactors or hydroelectric systems produce relatively few GHG emissions but have other environmental impacts (see, e.g., NRC, 2009d; NRC, 2009f), and it is not clear how to weight trade-offs across different types of impacts (but see Huijbregts et al., 2008). Finally, LCA is not familiar to most consumers and policy makers so its ultimate contribution to better decision making will depend on processes that encourage its use. These and other scientific challenges are starting to be addressed by the research community (see, e.g., Finnveden et al., 2009; Horne et al., 2009; Ramaswami et al., 2008); additional research on LCA would allow its application to an expanding range of problems and improve its use as a decision tool in adaptive risk-management strategies.

[5] See *http://walmartstores.com/Sustainability/9292.aspx.*

Environmental Benefit-Cost and Cost-Effectiveness Analyses

Integrated assessment models are intended to help decision makers understand the implications of taking different courses of action, but when there are many outcomes of concern, the problem of how to make trade-offs remains. Benefit-cost analysis is a common method for making trade-offs across outcomes and thus linking modeling to the decision-support systems (see Chapter 17). Benefit-cost analysis defines each outcome as either a benefit or a cost, assigns a value to each of the projected outcomes, weights them by the degree of certainty associated with the projection of outcomes, and discounts outcomes that occur in the future. Then, by comparing the ratio of benefits to costs (or using a similar metric), benefit-cost analysis allows for comparisons across alternative decisions, including across different policy options.

As discussed in Chapter 17, the current limits of benefit-cost analysis applied to global climate change decision making are substantial. A research program focused on improvements to benefit-cost analysis and other valuation approaches, especially for ecosystem services (see below), could yield major contributions to improved decision making. Equity and distributional weighting issues, including issues related to weighting the interests of present versus future generations, are areas of particular interest. In all, five major research needs are identified in Chapter 17: (1) estimating the social value of outcomes for which there is no market value, such as for many ecosystem services; (2) handling low-probability/high-consequence events; (3) developing better methods for comparing near-term outcomes to those that occur many years hence; (4) incorporating technological change into the assessment of outcomes; and (5) including equity consideration in the analysis.

In contrast to benefit-cost analysis, cost-effectiveness analysis compares costs of actions to predefined objectives, without assigning a monetary value to those objectives. Cost-effectiveness analysis, which is also discussed in Chapter 17, can be especially useful when there is only one policy objective, such as comparing alternative policies for pricing GHG emissions to reach a specific emissions budget or concentration target. Cost-effectiveness analysis avoids some of the difficulties of benefit-cost analysis. However, when more than one outcome matters to decision makers, cost-effectiveness analysis requires a technique for making trade-offs. Again, additional research can help to extend and improve such analyses.

Ecosystem Function and Ecosystem Services Models

Dynamic models of ecosystem processes and services translate what is known about biophysical functions of ecosystems and landscapes or water systems into information about the provision of goods and services that are important to society (Daily and Matson, 2008). Such models are critical in allowing particular land, freshwater, or ocean use decisions to be evaluated in terms of resulting values to decision makers and society; for evaluating the effects of specific policies on the provision of goods and services; or for assessing trade-offs and side benefits of particular choices of land or water use. For example, Nelson et al. (2009) used ecosystem models to determine the potential for policies aimed at increasing carbon sequestration to also aid in species conservation. Such analyses can yield maps and other methods for conveying complex information in ways that can effectively engage decision makers and allow them to compare alternative decisions and their impacts on the ecosystem services of interest to them (MEA, 2005; Tallis and Kareiva, 2006).

Ecosystem process models and other methods for assessing the effects of policies on ecosystem goods and services (MEA, 2005; Turner et al., 1998; Wilson and Howarth, 2002) also provide critical information about the impacts and trade-offs associated with both climate-related and other choices, including impacts that might not otherwise be considered by decision makers (Daily et al., 2009). If and when such information is available, various market-based schemes and "payments for ecosystem services" approaches have been developed to provide a mechanism for compensating resource managers for the ecosystem services provided to other individuals and communities. The design and evaluation of such mechanisms requires collaboration across disciplines (including, for example, ecology and economics) and improvements in the ability to link incentives with trade-offs and synergies among multiple services (Jack et al., 2008). Valuation of goods and services that typically fall outside the realm of economic analysis remains a significant research challenge, although a number of approaches have been developed and applied (Farber et al., 2002).

Policy-Oriented Heuristic Models

Policy-oriented simulation methods can be a useful tool for informing policy makers about the basic characteristics of climate policy choices. These simulation methods can either involve informal linkages between policy choices, climate trajectories, and economic information, or be implemented in a formal integrated modeling framework. For example, the C-ROADS model[6] divides the countries of the world into blocs

[6] See *http://www.climateinteractive.org/simulations/C-ROADS.*

with common situations or common interests (such as the developed nations), takes as input the commitments to GHG emissions reductions each bloc might be willing to make, and generates projected emissions, atmospheric CO_2 concentrations, temperature, and sea level rise over the next 100 years. The underlying model is simple enough to be used in real time by policy makers to ask "what if" questions that can inform negotiations. It can also be used in combination with gaming simulations in which individuals or teams take on the roles of blocs of countries and negotiate with each other to simulate not only the climate system but also the international negotiation process. When such simplified models are used, however, it is important to ensure that the simplified representations of complex processes are backed up, supported, and verified by more comprehensive models that can simulate the full range of critical processes in both the Earth system and human systems.

Heuristic models and exercises have also been developed that engage decision makers, scientists, and others in planning exercises and gaming to explore futures. Such tools are particularly well developed for military and business applications but have also been applied to climate change, including in processes that engage citizens (Poumadère et al., 2008; Toth and Hizsnyik, 2008). Though not predictive, such models and exercises can provide unexpected insights into future possibilities, especially those that involve human interactions. They can also be powerful tools for helping decision makers understand and develop strategies to cope with uncertainty, especially if coupled with improved visualization techniques (Sheppard, 2005; Sheppard and Meitner, 2005).

Metrics and Indicators

Metrics and indicators are critical tools for monitoring climate change, understanding vulnerability and adaptive capacity, and evaluating the effectiveness of actions taken to respond to climate change. While research on indicators has been a focus of attention for several decades (Dietz et al., 2009c; Orians and Policansky, 2009; Parris and Kates, 2003; York, 2009), progress is needed to improve integration of physical indicators with emerging indicators of ecosystem health and human well-being (NRC, 2005c). Developing reliable and valid approaches for measuring and monitoring sustainable well-being (that is, approaches that account for multiple dimensions of human well-being, the social and environmental factors that contribute to it, and the relative efficiency with which nations, regions, and communities produce it) would greatly aid adaptive risk management (see Box 3.1) by providing guidance on the overall effectiveness of actions taken (or not taken) in response to climate change and other risks.

Development of and improvements in metrics or indicators that span and integrate all relevant physical, chemical, biological, and socioeconomic domains are needed to help guide various actions taken to respond to climate change. Such metrics should focus on the "vitals" of the Earth system, such as freshwater and food availability, ecosystem health, and human well-being, but should also be flexible and, to the extent allowed by present understanding, attempt to identify possible indicators of tipping points or abrupt changes in both the climate system and related human and environmental systems. Many candidate metrics and indicators exist, but additional research will be needed to test, refine, and extend these measures.

One key element in this research area is the development of more refined metrics and indicators of social change. For example, gross domestic product (GDP) is a well-developed measure of economic transactions that is often interpreted as a measure of overall human well-being, but GDP was not designed for this use and may not be a good indicator of either collective or average well-being (Hecht, 2005). A variety of efforts are under way to develop alternative indicators of both human well-being and of human impact on the environment that may help monitor social and environmental change and the link between them (Frey, 2008; Hecht, 2005; Krueger, 2009; Parris and Kates, 2003; Wackernagel et al., 2002; World Bank, 2006).

Certification Systems and Standards

A number of certification systems have emerged in recent decades to identify products or services with certain environmental or social attributes, assist in auditing compliance with environmental or resource management standards, and to inform consumers about different aspects of the products they consume (Dilling and Farhar, 2007; NRC, 2010d). In the context of climate change, certification systems and standards are sets of rules and procedures that are intended to ensure that sellers of credits are following steps that ensure that CO_2 emissions are actually being reduced (see Chapter 17). Certification systems typically span a product's entire supply chain, from source materials or activities to end consumer. Performance standards are frequently set and monitored by third-party certifiers, and the "label" is typically the indicator of compliance with the standards of the system.

Natural resource certification schemes, many of which originated in the forestry sector, have inspired use in fisheries, tourism, some crop production, and park management (Auld et al., 2008; Conroy, 2006). Variants are also used in the health and building sectors and in even more complicated supply chains associated with other markets. Certification schemes are increasingly being used to address climate change issues, especially issues related to energy use, land use, and green infrastructure, as well as broader sustainability issues (Auld et al., 2008; Vine et al., 2001). With such a diversification and proliferation of certification systems and standards, credibility, equitability, usability, and unintended consequences have become important challenges. These can all be evaluated through scientific research efforts (NRC, 2010d; Oldenburg et al., 2009). For example, research will be needed to improve understanding and analysis of the credibility and effectiveness of specific approaches, including positive and negative unintended consequences. Analysis in this domain, as with many of the others discussed in this chapter, will require integrative and interdisciplinary approaches that span a range of scientific disciplines and also require input from decision makers.

CHAPTER CONCLUSION

Climate change has the potential to intersect with virtually every aspect of human activity, with significant repercussions for things that people care about. The risks associated with climate change have motivated many decision makers to begin to take or plan actions to limit climate change or adapt to its impacts. These actions and plans, in turn, place new demands on climate change research. While scientific research alone cannot determine what actions should be taken in response to climate change, it can inform, assist, and support those who must make these important decisions.

The seven integrative, crosscutting research themes described in this chapter are critical elements of a climate research endeavor that seeks to both improve understanding and to provide input to and support for climate-related actions and decisions, and these themes would form a powerful foundation for an expanded climate change research enterprise. Such an enterprise would continue to improve our understanding of the causes, consequences, and complexities of climate change from an integrated perspective that considers both human systems and the Earth system. It would also inform, evaluate, and improve society's responses to climate change, including actions that are or could be taken to limit the magnitude of climate change, adapt to its impacts, or support more effective climate-related decisions.

Several of the themes in this chapter represent new or understudied elements of climate change science, while others represent established research programs. Progress in all seven themes is needed (either iteratively or concurrently) because they are synergistic. Meeting this expanded set of research requirements will require changes in the way climate change research is supported, organized, and conducted. Chapter 5 discusses how this broader, more integrated climate change research enterprise might be formulated, organized, and conducted, and provides recommendations for the new era of climate change research.

CHAPTER FIVE

Recommendations for Meeting the Challenge of Climate Change Research

Meeting the diverse information needs of decision makers as they seek to understand and address climate change is a formidable challenge. The research needs and cross-cutting themes discussed in Chapter 4 (and listed in Box 4.1) argue for a new kind of climate change science enterprise, one that builds on the strengths of existing activities and

- Focuses not only on improving understanding, but helps to inform solutions for problems at local, regional, national, and global levels;
- Integrates diverse kinds of knowledge and explicitly engages the social, ecological, physical, health, and engineering sciences;
- Emphasizes coupled human-environment systems rather than individual human or environmental systems in isolation;
- Evaluates the implications of particular choices across sectors and scales so as to maximize co-benefits, avoid unintended consequences, and understand net effects across different areas of decision making;
- Develops and employs decision-support resources and tools that make scientific knowledge useful and accessible to decision makers;
- Focuses, where appropriate, on place-based analyses to support decision making in specific locations or regions, because the dynamics of both human and environmental systems play out in different ways in different places and decisions must be context-specific; and
- Supports adaptive decision making and risk management in the face of inevitable uncertainty by remaining flexible and adaptive and regularly assessing and updating research priorities.

These points, and the discussion in the preceding chapters, lead to the following conclusion.

Conclusion 2: The nation needs a comprehensive and integrative climate change science enterprise, one that not only contributes to our fundamental understanding of climate change but also informs and expands America's climate choices.

This comprehensive, integrative program of science will need to continue current research but also engage in new research themes and directions, including research in the physical, social, ecological, environmental, health, and engineering sciences, as well as research that integrates these and other disciplines. Creating and implementing this more integrated and decision-relevant scientific enterprise will require fundamental changes in the way that research efforts are organized, the way research priorities are set, the way research is linked with decision making across a broad range of scales, the way the federal scientific program interfaces and partners with other entities, and the way that infrastructural assets and human capital are developed and maintained. This chapter examines some of the steps that will be needed to implement this new era of climate change research.

AN INTEGRATIVE, INTERDISCIPLINARY, AND DECISION-RELEVANT RESEARCH PROGRAM

Climate change research efforts that address the seven crosscutting themes described in Chapter 4 have several important distinguishing characteristics.

Climate Change Research Needs to Be Integrative and Interdisciplinary

Climate change affects a wide range of human, ecological, and physical properties and processes, and it interacts in complex ways with other global and regional environmental changes. The response of human and environmental systems to this spectrum of changes is likewise complex. Given this complexity, understanding climate change, its impacts, and potential responses inherently requires integration of knowledge bases from different areas of the physical, biological, social, health, and engineering sciences. Science that supports effective responses to climate change also will require integration of information across spatial and temporal scales. For example, global- or regional-scale information about changes in the climate system often needs to be analyzed in the context of local data on economic activity, vulnerable assets and resources, human well-being, and other place-specific information. Climate change science in the coming decades will need to be more multi- and interdisciplinary and integrative than in the past.

In some ways, the call for cross-disciplinary and cross-scale integration is a step, albeit a large one, in a progression that has been under way in national and international climate science for quite some time. As described later in the chapter, a number of domestic and international scientific programs have organized the research community to focus on climate and other regional and global environmental changes. These programs have played a critical role in establishing our present understanding. However, in general they have not been as successful in bridging the gaps between those who study the physical climate system; those who study the impacts of and responses to climate change in human, ecological, and coupled human-environment systems; and those who study the technical, economic, political, behavioral, and other aspects of various responses to climate change (ICSU-IGFA, 2009; NRC, 2009k). Moreover, a concerted effort is needed to increase the engagement of some disciplines, such as the social, behavioral, economic, decision, cognitive, and communication sciences.

Achieving better integration will require significant increases in interdisciplinary science capacity among scientists, managers, and decision makers. It will require changes in cultures within and actions across a range of institutions, including universities, government, the private sector, research institutes, professional societies, and other nongovernmental organizations, including the National Research Council. It will also require the creation of new institutions to facilitate the needed research at the appropriate scales and in appropriate contact with decision makers.

Climate Change Research Efforts Should Focus on Fundamental, Use-Inspired Research

This report recognizes the need for research to both understand climate changes and assist in decision making related to climate change. In categorizing types of scientific research, we have found that terms such as "pure," "basic," "applied," and "curiosity driven" have different definitions across communities, are as likely to cause confusion as to advance consensus, and are of limited value in discussing climate change. A more compelling categorization is offered by Stokes (1997), who argues that two questions should be asked of a research topic: Does it contribute to fundamental understanding? Can it be expected to be useful? Research that can answer yes to both of these questions, or "fundamental, use-inspired research," warrants special priority in a climate science enterprise that seeks to both increase understanding and assist in decision making. Research that addresses one or the other of Stoke's questions, which describes the full range of scientific inquiry, is also valuable. Priority setting is discussed in further detail in the next section.

Climate Change Research Should Support Decision Making at Local, Regional, National, and International Levels

Although making choices about how to respond to climate change fundamentally involves values, ethics, and trade-offs, science can inform and guide such decisions.

In particular, science can help identify possible courses of action, evaluate the advantages and disadvantages associated with different choices (including trade-offs, unintended consequences, and co-benefits among different sets of actions), develop new options, and improve the options that are available. It can also assist in the development of new, more effective decision-making processes and tools. These goals require interactive processes that engage both scientists and decision makers to identify research topics and improve methods for linking scientific analysis with decision making. Active dialogue with stakeholders at local, regional, national, and international levels can also enhance the utility and credibility of, and support for, scientific research. Strategies, tools, and approaches for improving linkages between science and decision making are described in Chapter 4 and discussed in detail in the companion volume *Informing an Effective Response to Climate Change* (NRC, 2010b).

Climate Change Research Needs to Be a Flexible Enterprise, Able to Respond to Changing Knowledge Needs and Support Adaptive Risk Management and Iterative Decision Making

As climate change progresses, past climate conditions and human experiences will serve as less and less reliable guides for decision makers (see Chapter 3 and also NRC, 2009g). Even with continued advances in scientific understanding, projections of the future will always include some uncertainties. Moreover, because climate changes interact with so many resource and infrastructure decisions, from power plant design to crop planting dates, responses to climate change will need to be developed and implemented in the context of continuously evolving conditions. Furthermore, as actions are taken to limit the magnitude of future climate change and adapt to its impacts, decision makers will need to understand and take the effectiveness and unintended consequences of these actions into account.

As a direct result of these complexities and uncertainties, all responses to climate change, including the next generation of scientific research, will require deliberate "learning by doing." Actions and strategies will need to be periodically evaluated and revised to take advantage of new information and knowledge, not only about climate and climate-related changes but also about the effectiveness of responses to date and about other changes in human and environmental systems. The nation's scientific enterprise should support adaptive risk management (i.e., an ongoing decision-making process that takes known and potential risks and uncertainties into account and periodically updates and improves plans and strategies as new information becomes available—see Box 3.1) by monitoring climate change indicators, providing timely information about the effectiveness of actions taken to respond to climate risks, improving the effectiveness of our responses over time, developing new responses, and continuing to build our understanding of climate change and its impacts. These tasks require flexible mechanisms for identifying and addressing new scientific challenges as they emerge and also ongoing interactions with decision makers as their needs change over time. Continued progress will also be needed in monitoring, projecting, and assessing climate change, especially abrupt changes and other "surprises." Individually and collectively, these demands will require significant changes in the way research is funded, conducted, evaluated, and rewarded.

Recommendation 1: The nation's climate change research enterprise should include and integrate disciplinary and interdisciplinary research across the physical, social, biological, health, and engineering sciences; focus on fundamental, use-inspired research that contributes to both improved understanding and more effective decision making; and be flexible in identifying and pursuing emerging research challenges.

SETTING PRIORITIES

Recommendation 1 calls for a broad, integrative research program to assist the nation in understanding climate change and in supporting well-crafted and coordinated opportunities to adapt to and limit the magnitude of climate change. In Chapter 4, seven crosscutting, integrative research themes were identified that would provide effective focal points for such a program:

1. Climate Forcings, Feedbacks, Responses, and Thresholds in the Earth System
2. Climate-Related Human Behaviors and Institutions
3. Vulnerability and Adaptation Analyses of Coupled Human-Environment Systems
4. Research to Support Strategies for Limiting Climate Change
5. Effective Information and Decision-Support Systems
6. Integrated Climate Observing Systems
7. Improved Projections, Analyses, and Assessments

Progress in these areas would advance the science of climate change in ways that are responsive to the nation's needs for information, and progress in all seven themes is needed (either iteratively or concurrently) because they are synergistic. However, due to limits in capacity—for example, many of the key research needs are in fields that have not yet been fully incorporated into or developed within the nation's climate change science enterprise—and in financial resources, priorities will ultimately need to be set within these themes, and perhaps also across them.

Setting priorities has been and will continue to be a critical part of the scientific process. Priority setting can be accomplished via community-based long-range planning mechanisms, national and international assessments and advisory reports, federal agency and interagency advisory and strategy planning processes, and federal budget development processes. Indeed, the U.S. federal government has already developed and established legislation, policies, and practices for developing climate and global change research budgets and priorities (for example, see Appendix E for a description of some of the USGCRP's past and current priority-setting practices).

Given these detailed, well-established processes, this panel can contribute to priority setting only at a comparably coarse level—for example, by suggesting the high-level research themes discussed in Chapter 4. The development of more comprehensive, exhaustive, and prioritized lists of specific research needs within each theme will need to involve members of the relevant research communities. It is critical, however, that priority setting also include the perspective of societal need, which necessitates input from decision makers and other stakeholders. Implementation of such priority-setting activities will further require the establishment of agreed-upon priority-setting criteria, strong leadership of and support for the research program, and new mechanisms for stakeholder input.

Priority-Setting Criteria

The establishment of criteria by which prospective priorities should be evaluated is critical for effective priority setting. There have been a number of efforts to establish priority-setting criteria for climate-related research (see, e.g., NRC, 2005a, 2009k). Drawing on these analyses, we identify the following three main criteria for setting research priorities for the nation's climate change research enterprise, including (but not limited to) the entity or program responsible for coordinating and implementing research at the federal level (see Recommendation 5 later in this chapter). The numbering of these criteria do not imply relative importance; rather, it is important to consider all three criteria. Bulleted points after each criterion are ways of thinking about priorities in the context of that criterion, not separate criteria.

1. Contribution to fundamental understanding

- Addresses key theoretical, observational, process, or modeling uncertainties;
- Adds new information to important scientific debates; and/or
- Extends research to understudied areas and questions.

2. Contribution to improved decision making

- Addresses topics that have been identified as decision-maker needs or that are key to the nation's economic vitality, its security, or the well-being of its citizens;
- Provides scientific foundations for new solutions or options, especially those that have co-benefits for other environmental or socioeconomic challenges;
- Contributes useful results that can be communicated effectively to decision makers and affected parties or have the potential to establish ongoing dialogue between researchers and users of scientific information; and/or
- Supports risk assessment and management by improving projections or predictions, providing information on probabilities, clarifying societal consequences of key outcomes, or creating decision-support resources.

3. Feasibility of implementation (practical, institutional, and managerial concerns)

- Is ready for implementation (infrastructure, personnel, and facilities are available or could be available to execute the research);
- Will provide usable results on time scales relevant for decision making or improved understanding;
- Contributes to more than one application or scientific discipline; and/or
- Is cost effective (anticipated outcomes or value of information generated by the activity is sufficient to justify both financial and opportunity costs).

The climate change research program envisioned by the Panel on Advancing the Science of Climate Change and encapsulated by these criteria focuses on fundamental, use-inspired research that increases understanding and supports decision making. To develop research that is both fundamental and useful, assessments of research priorities will need to engage both the scientific community and those who will make use of new scientific understanding in decision making, ideally through interactive and ongoing dialogues. A multidirectional flow of information between the decision-making and research communities helps decision makers understand the uses and limits of scientific information and helps the scientific community understand what information and innovations would be most useful to decision makers. This should not be a process in which decision makers have undue influence on the conduct of science or scientific conclusions. Rather, our vision is one of ongoing dialogues that lead to better understanding and improved collaboration. Interactions between decision makers and scientists have the additional benefits of enhancing the trust decision makers place in the scientific process and ensuring that researchers use actual input from decision makers, rather than educated guesswork, to help identify and prioritize research topics.

The research program envisioned in this report involves a broad range of scientific disciplines, including multi- and interdisciplinary science. Identifying and setting research priorities across such a broad and diverse range of scientific activities is much more challenging than priority setting within individual disciplines, which usually share common practices, understandings, and language. Working across areas of research where no unified community has yet been assembled represents an additional challenge, one that requires both careful sampling of views across communities and time to develop mutual understanding.

Because the costs associated with the different climate change research themes described in Chapter 3 are likely to vary by several orders of magnitude, appropriate ranking requires an understanding of the budget constraints agencies will face as well as the benefits that could potentially be realized. As discussed in the preceding recommendation, climate change research should be a flexible and adaptive enterprise, so priorities, and priority-setting criteria and processes, need to be revisited regularly. In addition to changing knowledge needs, advances in methodology or research technology can also motivate a reassessment of priorities in the context of evolving environmental conditions, changing budgets, and other variables that inform research agendas. Given that both climate change and responses to it are ongoing, and that they interact with each other as well as with other changes, such reassessments will be a key element of a healthy research program.

Recommendation 2: Research priorities for the federal climate change research program should be set within each of the seven crosscutting research themes outlined above. Priorities should be set using the following three criteria:

1. **Contribution to improved understanding;**
2. **Contribution to improved decision making; and**
3. **Feasibility of implementation, including scientific readiness and cost.**

INFRASTRUCTURAL ELEMENTS OF THE RESEARCH PROGRAM

Scientific progress in measuring climate change, attributing it to human activities, projecting future changes, and informing decisions about how to respond has and will continue to rely on significant investments in a wide range of global observational programs and modeling efforts. As noted in Chapter 4, these efforts are limited in part by infrastructure, especially the lack of a comprehensive, integrated climate observing system and of reliable, detailed projections of climate and climate-related changes at regional and local scales. Because these infrastructural areas underpin progress in virtually all other areas of climate change science, we have identified observations and modeling as critical themes in climate science research, and below we offer specific recommendations related to these key themes. Many previous reviews of climate science needs (e.g., NRC, 2009k) have also highlighted observations and models as key research needs.

Observing Systems

As discussed in Chapter 4, long-term, stable, and well-calibrated observations across a range of scales and a spectrum of human and environmental systems are essential for diagnosing, understanding, and responding to climate change and its impacts. Observations provide ongoing information about the health of the planet and clues about which components of the Earth system are at risk due to climate change and other environmental stressors. Observations are also critical for developing, initializing, and testing models of future human and environmental changes, and for monitoring and improving the effectiveness of actions taken to respond to climate change. Unfortunately, many of the critical observational assets needed to support climate research and climate change responses are either in decline or seriously underdeveloped, and the data that are being collected are not always managed as effectively or used as widely as they could be. A number of specific steps are needed to rectify this situation and develop a coordinated, comprehensive, and integrated climate observing system.

A Careful, Comprehensive Review Should Be Undertaken to Identify Current and Planned Observational Assets and Identify Critical Climate Monitoring and Measurement Needs

An observing system strategy for the new era of climate change research will need to consider not only existing and planned assets, which have largely been developed by the scientific community without much input from decision makers, but also the observations needed to support effective responses to climate change. In considering available resources and data sources, federal programs should work with international partners to identify opportunities for collaboration, leveraging, and synergy with observational systems in other countries. A special effort should be made to evaluate observations and databases from areas that have historically been neglected. Where possible, the review should consider assets in the intelligence community that could serve scientific purposes without compromising national security interests. Finally, the climate observing system should be coordinated with other environmental and social data collection efforts to take advantage of synergies and ensure interoperability.

The federal climate change research program (see Recommendation 5) is the entity best suited to lead a comprehensive assessment of current and planned observational assets and needs in support of climate research. However, the research community will need to work closely with a broad range of responsible entities and stakeholders, including programs for adapting to, limiting the magnitude of, and supporting effective decisions related to climate change, to ensure that the scope and structure of the observing system can support both fundamental research on and responses to climate change. Such partnerships are critical in light of the costs of creating and maintaining a comprehensive and long-term observing system. As the recent problems with NPOESS have demonstrated (NRC, 2008d), planning for climate observations will require clearly defined roles and responsibilities of the partners and a systems approach to the design of the overall architecture of the observing system.

A Comprehensive and Integrated Climate Observing System Should Be Developed, Built, and Maintained by the Federal Program and Relevant National and International Partners

The climate observing system should be able to monitor a broad spectrum of changes, including changes in the physical climate system (such as sea level rise, sea ice declines, and soil moisture changes); changes in related biological systems (such as species shifts and changes in crop yield or the amount of carbon stored in forests); the impacts of these changes on human systems (including human health and economic impacts); trends in human systems (such as human population and consumption changes and GHG emission trends); indicators of climate vulnerability and adaptive capacity across a range of sectors and spatial scales; and indicators of the effectiveness of actions taken to limit the magnitude of climate change and adapt to its impacts. In addition to a robust and flexible network of remote and in situ assets to monitor physical, chemical, and biological changes, observations and data sets from a wide range of human systems are needed. Observations of emission trends and the effectiveness of various climate policies and action plans are particularly important for informing actions taken to limit the magnitude of future climate change, while observations of climate change impacts at regional to local scales are particularly important for informing adaptation decisions.

The observing system, like other research activities and responses to climate change, should be integrated and flexible, and it should support adaptive risk management and decision making. For example, although observational assets with long-term and global coverage will play a critical role in monitoring climate change, its impacts, and our responses to it, we may also need easily deployable short-term observational technologies to monitor potential abrupt changes or important regional trends. The observing system should also be designed both to take advantage of advances in technology and to explicitly support adaptive risk management and decision making. External advisory boards, user councils, and other formal and informal stakeholder groups (see Recommendation 5) can play an important role in ensuring that the observing system is supplying the information required by stakeholders.

Adequate Climate Data Access, Management, and Stewardship Are Needed

Linking, integrating, and providing access to data of dramatically different types and scales will call for new and improved approaches and standards for climate and climate-related data management, including data collection, storage, and stewardship. To ensure a stable, long-term record of climate and climate-related changes, funding for data-generating activities should always include resources for long-term data management (NRC, 2007d). An equally important activity, described in further detail in Chapter 4, is the integration of data from different sources through data assimilation, analysis, and reanalysis. Finally, the system should allow ready access to data by a wide range of users, including decision makers. This will require the federal climate change program to work closely with programs involved in informing and supporting effective responses to climate change.

Recommendation 3: The federal climate change research program, working in partnership with other relevant domestic and international bodies, should redouble efforts to design, deploy, and maintain a comprehensive observing system that can support all aspects of understanding and responding to climate change.

Enhanced Modeling Capabilities and Other Analytical Tools

Improved predictions and scenarios of future climate change, its impacts, and related changes in ecosystems and human systems are critical for understanding and guiding plans to respond to climate, many of which require local- or regional-scale information at decadal time scales. As discussed in Chapters 4 and 6, great strides have been made in improving the spatial resolution, comprehensiveness, and fidelity of global climate and Earth system models. However, improvements are still needed in the ability of climate models to represent key climate feedback processes (such as the carbon cycle and changes in ice sheets) and to resolve and simulate the physical processes, interactions, and feedbacks that govern climate change at regional scales. Another emerging research need is integrated assessment models that can connect emissions projections, GHG concentrations, climate trends, and the social, economic, and environmental impacts of these trends on human and environmental systems. Other kinds of assessment tools and models, including those that allow integrated analysis of trade-offs and unintended consequences among combinations of actions or across different sectors, would also be valuable both to improve understanding and to support climate-related decision making. Chapter 4 includes a more extended description of these and other research needs related to improved projections, analyses, and assessments.

As noted in Chapter 4, and in many previous reports (e.g., NRC, 2009k), a national strategy is needed to improve (and to coordinate existing efforts to improve) regional climate modeling, global Earth system modeling, and various integrated assessment, vulnerability, impact, and adaptation modeling efforts. Developing improved models and analytical tools is strongly dependent on the availability of high-performance computing capacity as well as the infrastructure and human resources needed to develop, manage, analyze, and improve modeling approaches. The output from such models needs to be made readily available to a wide range of decision makers in formats that allow them to incorporate model analyses and projections into their decision-making processes. As with the integrated climate observing system—and perhaps more so, given the highly technical and interdisciplinary nature of many model development activities—the federal climate change research program is the logical entity for coordinating and integrating these development efforts. Input and buy-in will be needed from its partner agencies, action-oriented response programs, and other stakeholders. Likewise, international coordination and leveraging will be vital.

Recommendation 4: The federal climate change research program should work with the international research community and other relevant partners to support and develop advanced models and other analytical tools to improve understanding and assist in decision making related to climate change.

ORGANIZING THE RESEARCH

A research effort that can improve understanding of and support effective responses to climate change across a broad range of scales will require the engagement of universities, professional societies, nongovernmental organizations, corporations, and governments at many levels, including international partners. To date, the federal government, under the auspices of the USGCRP, has played a leadership role in the nation's climate change research enterprise. This section summarizes the history, structure, and current goals of the USGCRP, evaluates its capacity to carry out the seven research themes identified in Chapter 4, and recommends elements that would be needed for the USGCRP, or another federal entity, to lead and coordinate the nation's climate change research efforts. Additional background on the history and organization of the USGCRP can be found in Appendix E.

Evaluation of the U.S. Global Change Research Program

Congress established the USGCRP with the U.S. Global Change Research Act of 1990 (P.L. 101-606, Title 15, Chapter 56A). The act set the objective of "assist[ing] the Nation and the world to understand, assess, predict, and respond to human-induced and natural processes of global change systems." With federal support ranging from $2.2 billion in 1990 (in 2008 dollars) to $1.8 billion in 2008, the USGCRP[1] has made enormous contributions to the understanding of climate change over the past two decades, including a considerable fraction of the advances summarized in Chapter 2 (see, e.g., USGCRP, 2009b).

There have been a wide range of assessments, observations, and reviews of the USGCRP, including mandated formal reviews by the NRC (1999a, 2003a, 2004b, 2005e, 2007f, 2009k) and assessments and observations by other groups (e.g., the Congressional Research Service). Most of these reviews have praised the USGCRP for its support and facilitation of major advances in our understanding of the natural science aspects of global change, including the physical climate system, atmospheric chemistry, hydrology, and ecosystems, and also for supporting national and international scientific assessments. Earlier reviews of the program also noted significant progress in establishing a comprehensive and inclusive national climate change assessment process and in providing strategic guidance that promoted major advances in observations and modeling, although later reviews have noted a decline in the support for and effectiveness of these activities.

One persistent area of criticism has been the scope and balance of the program. In its early years, the primary research emphasis of USGCRP was on the physical climate system. The program has consistently aspired to call increasing attention to human interactions with the Earth's climate and other environmental systems, but these aspirations have for the most part fallen short (NRC, 2007f). Another persistent criticism has focused on decision support, including progress in decision-support science and whether the program has lived up to its mandate of providing useful information for decision making (NRC, 2007f, 2009k). Identified reasons for these shortcomings include a lack of consistent and adequate funding and institutional support for fundamental

[1] Known as the U.S. Climate Change Science Program from 2002 through 2008.

and applied research in the social sciences and a lack of adequate integration across scientific disciplines. Moreover, the failure to follow through with periodic, comprehensive national climate change assessments weakened the program's ability to build a consistent and sustainable relationship with stakeholders. Other troubling signs include a decline in congressional oversight of and interest in the program—measured, for example, by the number of hearings convened to review aspects of the program—and an overall decline of 18 percent (in constant 2008 dollars) in program funding from 1990 to the present (NRC, 2007f, 2009k).

Past NRC reviews have also pointed out weaknesses in the program's structure and institutional processes. For example, the program has relied on individual federal agencies to identify and engage in areas of climate change research aligned with their missions, with only a few, typically episodic and informal, mechanisms for supporting research in areas that do not map onto agency missions. One result of this "stove-piping" has been uneven progress, with some research elements receiving significant funding and making excellent progress, while other research areas—including those associated with several of the crosscutting themes identified in Chapter 4—receiving much less attention. Moreover, without strong coordination, leadership, and buy-in from the full range of federal agencies affected by climate change, the program has been limited in its ability to support the evolving needs for climate science, including research that could support more effective responses to climate change. Additionally, as discussed in the next section of the chapter, the activities of the USGCRP have not been very well coordinated with the Climate Change Technology Program (NRC, 2007f, 2009k) or with preliminary efforts to establish mechanisms to provide "climate services."

Needs for the Climate Change Research Program of the Future

The USGCRP currently involves 13 federal departments and agencies, and many of these organizations have several different agencies or groups active in climate change research. This scope of engagement is essential given the broad range of public interests that will be affected by climate change. However, this broad scope and inconsistencies between the mandates of the Global Change Research Act and the narrower missions of the participating agencies create a difficult and complex management environment. For example, as noted in the previous subsection, gaps between agency missions have led to weaknesses and gaps in certain research areas. Furthermore, progress on several key crosscutting issues, such as maintaining and improving climate-related observational programs, have suffered from a lack of leadership and coordination (e.g., NRC, 2008d). Thus, it is not clear that the USGCRP as presently constituted can adequately address the full set of research challenges posed by current demands for climate change research.

How then might the federal climate change research effort be structured to meet these new challenges? The Panel on Advancing the Science of Climate Change considered several alternatives, each with its strengths and weaknesses. One model is to create a new office or agency that aggregates all federal climate change research into one organizational structure. This harkens to the call in the 1990s for a National Institute of the Environment to be the home for all federal environmental research. An NRC report examined this proposal and saw several problems with as well as several advantages to this approach (NRC, 1993). For example, a consolidated aproach would improve cross-agency coordination and planning, while a more distributed responsibility for environmental research leads to improved linkages between researchers and decision makers. The report called for "cultural changes" in the practice of environmental research in the federal government, including greater engagement of the ecological, social, and engineering sciences. It also considered several options for organizing these changes and suggested that the best immediate step would be to retain the multiagency support of environmental research but with better coordination and attention to neglected priorities, rather than consolidate research into a single agency.

For climate change research, a consolidated agency would facilitate coordination and allow for priority setting based on a systematic analysis, such as the seven research themes identified in Chapter 4. Neglected high-priority areas could thus be allocated the resources needed to move them forward. Decision makers and the public would also have a single entity to consult on climate change. However, there are many disadvantages to consolidation, and those may outweigh its benefits. First, given the many challenges facing the federal government at present, it is unlikely that a proposal to create a new agency that would pull current research functions out of existing agencies could reach an actionable level on the federal agenda. Second, one of the strengths of the USGCRP is that it encourages engagement by multiple agencies and, thus, has been able to adapt relatively easily as concerns about and needs for scientific understanding of climate change have spread to affect the missions of more and more agencies. This trend is likely to continue, especially as more agencies are involved in efforts to respond to climate change. Openness to new partnerships, which is a strength of the current USGCRP, would likely be reduced with the creation of a new agency. In addition, a single agency would be limited in its ability to draw on the strengths of the non-USGCRP components of currently participating agencies. For example, NASA's Earth science activities benefit significantly from their integration with other complementary aspects of NASA's portfolio. These benefits could be significantly compromised if the Earth science activity were moved from NASA to another entity.

While there are surely other alternative arrangements for organizing federal climate research, the main alternative to a consolidated agency is some sort of interagency program, ideally one that retains the current and historical strengths of the USGCRP while addressing its known weaknesses. The major advantage of the continuance of the USGCRP in this coordinating role is that the program already exists and has the legal authority and mandate to engage in a cross-agency research program. In fact, a careful reading of the Global Change Research Act of 1990 indicates that the program was intended to accomplish many of the goals identified in this report. The main disadvantages of a continuation of the USGCRP are the weaknesses highlighted in the previous subsection. Hence, provided that these weaknesses can be addressed, the panel finds that a modified USGCRP could serve the role of leading and coordinating an integrated, decision-relevant, and expanded climate change research enterprise that continues to pursue an enhanced understanding of the causes and consequences of climate change while also improving understanding of and support for responses to climate change. Indeed, as of the writing of this report, the USGCRP is already engaging in a strategic planning process to address the weaknesses and pursue the opportunities identified in past reviews of the program. The next two subsections describe ways in which the current program might be reshaped to better meet the challenges of the new era of climate change research while maintaining its existing strengths.

Improving the Relevance of the USGCRP to Decision Making

A common finding among many past reports (for example, NRC, 2007a, 2008b, 2009g), and this one, is that improving the relevance and utility of scientific research and infusing scientific information into the decision-making process require increased dialogue and engagement between scientists and decision makers. Several mechanisms help connect scientific and decision-making entities in the context of the USGCRP. For example, the USGCRP could establish an *external advisory board* to provide input on research needs, to review and provide advice on research priorities, and to guide activities designed to enhance communication and interactions with the broader stakeholder community. An external advisory board would help to ensure that priorities for research are informed by and responsive to the needs of decision makers and other information users, and it could assess and improve the program's decision-support capabilities. If established, such a board should be composed of decision makers and stakeholders from a broad range of communities (e.g., leaders in state, local, and tribal governments; relevant businesses and industries; citizen groups; and other nongovernmental organizations), including communities that are currently not strongly linked with the program, as well as members from across the scientific research community.

Mechanisms should also be developed for regular interaction between users and researchers at the individual research project level. A number of federal agencies have already taken steps to increase such engagement, but more comprehensive and coordinated efforts are needed. For example, "user councils" focusing on a particular type of decision or research area could help researchers understand the questions that are most critical for decision makers and other stakeholders, help users understand the information that science can and cannot provide, and assist in the development of enhanced decision-support processes and tools. Workshops and dialogues, such as the "listening sessions" USGCRP has held at various venues across the country, are also a valuable contribution.

There are two important caveats that need to be kept in mind when designing and implementing strategies to increase interactions between the research community and its stakeholders. First, and most important, interactions between users and producers of scientific information should always preserve the integrity of the research process in reaching factual conclusions. Second, input from stakeholders needs to be considered in the context of the tractability of the proposed research and the resources required, and mechanisms are needed to ensure that the scientific enterprise is not totally dominated by near-term decision-support activities.

Next Steps for the USGCRP

A careful reading of the Global Change Research Act indicates that the legislation provides most of the necessary authority for implementing a strategically integrated climate change research program (see Appendix E). The act envisions a program that covers the full spectrum of activities from understanding climate change and its interactions with other global changes and stresses through developing and improving responses to these changes. The act also mandates research that is closely aligned with decision-making needs, including decisions related to the nation's energy, natural resources, and public policy programs.

The USGCRP has achieved many of the original goals of the act. However, as discussed above, in other areas some critical weaknesses and shortcomings have emerged. As the climate research program expands to include a greater emphasis on use-inspired and decision-relevant research, additional gaps and barriers are likely to arise unless steps are taken to address these deficiencies and help the program evolve. Some of the specific changes that are needed include

- Setting priorities more effectively and transparently using clear criteria and evaluative information on program performance;
- Promoting a closer connection between research and decision making by engaging a broader range of federal agencies whose stakeholders and mandates will be affected by climate change and by establishing mechanisms for sustained engagement of users in program decision making;
- Addressing known weaknesses in the program, including development of decision-support resources and engagement of the social and behavioral sciences;
- Fostering integration through targeted research funding opportunities, decision-relevant interdisciplinary research centers, and other means that build on established capacity in universities, national laboratories, and the private sector; and
- Strengthening budget coordination and management to ensure the research activities of participating agencies are sufficiently focused on USGCRP priorities.

As a first step toward developing a more comprehensive and integrated program, a program-level effort could be initiated to identify, recruit, and leverage new partner agencies, including some that have not participated heretofore in climate change issues, and to expand participation by current partner agencies. For example, programs within the Departments of Agriculture or Interior that are responsible for protected lands, national parks, conservation reserves, and activities such as agricultural extension or water resources management have not been very active players in the USGCRP to date, yet they are in the process of developing responses to climate change because their missions will be directly affected by it. Such agencies and programs could play important roles in improving understanding of climate impacts and vulnerabilities and in formulating, evaluating, and improving response strategies. Data collection and research efforts performed by agencies and programs not specifically focused on climate change, such as those at the U.S. Census Bureau or the Centers for Disease Control and Prevention, could likewise contribute substantially to our understanding of both climate change and its interactions with other human and environmental systems. The relative ease with which the USGCRP structure can integrate new agencies or departments is a key advantage over a single consolidated entity or agency. Likewise, some of the traditional research and mission agencies could be more actively involved in engaging with decision makers to help shape the program's scientific agenda and ensure the results are used effectively.

Flexibility is a key advantage of the current USGCRP structure, but, as noted above, this organizational structure has also led to research gaps. One problem is that while the USGCRP might be able to reach agreement about research priorities, the budgets to implement those priorities reside within the partner agencies, where climate change research needs compete with other agency priorities. To address this problem, mechanisms are needed to ensure that research priorities identified by USGCRP are given greater weight by participating agencies, and the USGCRP needs the budgetary authority to implement the priorities it identifies. Improved review and oversight mechanisms, such as coordinated reviews of participating agency budgets (as opposed to merely designating established agency activities as contributions to USGCRP), would help promote accountability and would assist in evaluations of how well the priorities identified by USGCRP are reflected in the programs and budget requests of the participating agencies. Mechanisms are also needed to ensure that critical areas of research that are currently underrepresented in federal agency activities receive appropriate attention. Finally—and perhaps most difficult—program managers need to have the authority, willingness, and capability to emphasize the interdisciplinary, decision-relevant science needed to both improve understanding and support effective responses to climate change. Changes to the Global Change Research Act or other mechanisms, such as an Executive Order or performance measures, may be appropriate means to implement these changes and strengthen the program's budget coordination and alignment with identified research priorities.

Changes in the USGCRP will require strong leadership. The importance of effective leadership, with adequate support and programmatic and budgetary authority to coordinate and prioritize across agencies, has been recognized in a number of previous NRC reviews of the USGCRP (NRC, 2004b, 2005e, 2007f, 2009k). Such leadership will be essential for setting priorities and building a more balanced and integrative program, ensuring effective interactions between federal research activities and action-oriented programs (as discussed in the next section of the chapter) and executing the other recommendations in this report. One step that could be taken to improve program leadership could be achieved by assigning higher-level leaders within the partner agencies and organizations to be liaisons to the program. The assignment of senior-level, experienced program managers to staff the USGCRP coordination office could increase buy-in from the participating agencies; experienced staff will be needed to address program gaps and help lead interagency program prioritization and coordination. Effective guidance and budget review could be provided by organizations such as the Office of Science and Technology Policy and the units within the Office of Management and Budget responsible for USGCRP partner agencies.

As noted in the first two sections of the chapter, setting research priorities needs to be an ongoing, iterative process. Such adaptive management of the federal research program would be facilitated by regular strategic planning and reviews that address both specific research areas and the program as a whole. The USGCRP is already required to conduct a strategic review and submit a new strategic plan every 3 years. While there have occasionally been delays in the process, these review and planning exercises have provided useful opportunities for the program to remain flexible and to support emerging priorities. A major focus of future reviews and other ongoing assessment activities within the program should be mapping the priorities, activities, and capabilities of participating agencies onto the goals of the overall research program to identify weaknesses and gaps. Identifying impediments and obstacles that may be contributing to these weaknesses and gaps would also help the program develop specific actions to address these shortcomings and build a more balanced and effective program.

Finally, it might be beneficial to coordinate future reviews of the nation's climate change research program with reviews of the effectiveness of the nation's overall response to climate change in terms of limiting climate change, developing adaptation approaches, and informing effective climate-related decisions. Because coordinated federal efforts to inform, limit, and adapt to climate change are still in early stages of development, it is difficult to offer suggestions as to how this coordination can be achieved, but attention to such coordination will be important (see also Recommendation 6).

Recommendation 5: A single federal entity should be given the authority and resources to coordinate and implement an integrated research effort that supports improving both understanding of and responses to climate change. If several key modifications are made, the U.S. Global Change Research Program could serve this role.

These modifications are described in the paragraphs above and include

- An expanded mission that includes both understanding climate change and supporting effective decisions and actions taken to respond to climate change;
- Establishing a wide range of activities and mechanisms to support two-way flows of information between science and decision making, including improved mechanisms for input from decision makers and other stakeholders on research priorities;
- Establishing more effective mechanisms for identifying and addressing gaps and weaknesses in climate research, as well as the barriers that give rise to such gaps;
- High-level leadership both within the program and among its partner agencies; and
- Budgeting oversight and authority.

BROADER PARTNERSHIPS

Climate change is both a global problem and a local problem, and its impacts have implications for and interact with nearly every sector of human activity, including energy and food production, water and other natural resources, human health, business and industrial activities, and, in turn, political stability and international security. Efforts to limit climate change are also inherently cross-sectoral and international in scope—national efforts to limit GHG emissions are connected by the global climate system, making it necessary for the United States to formulate and coordinate its strategies for reducing emissions in the context of international agreements and the actions of other nations. At the same time, many of the actions taken to limit or adapt to climate change ultimately play out at local and regional scales. Thus, the engagement of institutions at all levels and of all sorts—academic, governmental, private-sector, and not-for-profit—will be required to meet the challenges of climate change.

The scientific enterprise is also inherently local to global in scope—scientific contributions to understanding or responding to climate change appear in international journals, get assessed by international scientific bodies, and contribute to improved understanding and responses to climate change worldwide. The international research community has established a number of scientific programs to coordinate and facilitate international participation in global change research. Some of these programs and partnerships include the following:

- The World Climate Research Program (WCRP) was established by the United Nations World Meteorological Organization (WMO) and the (nongovernmental) International Council of Scientific Unions (ICSU) in 1980 with the aim of determining the predictability of climate and the effect of human activities on climate.
- ICSU established the International Geosphere-Biosphere Program (IGBP) in 1987 to more broadly address global environmental changes and their interactions in the biosphere and physical Earth system.
- The Intergovernmental Panel on Climate Change (IPCC) was established in 1988 by WMO and the United Nations Environment Programme (UNEP) to provide an international assessment of the science that all governments could use in negotiating an international approach to addressing climate change.
- In 1990, in partnership with the International Social Science Council, ICSU established the International Human Dimensions Program (IHDP) to address the social science components of global change research.
- In 1991, DIVERSITAS was established with the goal of developing an interna-

tional, nongovernmental umbrella program that would address the complex scientific questions posed by the loss of and change in global biodiversity. It was founded jointly by the United Nations Educational, Scientific and Cultural Organization; the Scientific Committee on Problems of the Environment; and the International Union of Biological Science.
- In 1992, the START (System for Analysis, Research and Training) Programme was formed jointly by ICSU and its four international global change science programs. START is designed to assist developing countries, through research and education, in building the expertise and knowledge they need to explore the drivers of and solutions to global and regional climate and environmental change.
- In 2002, the Earth System Science Partnership (ESSP) was established in order to provide integrated studies of the Earth system. ESSP is a joint initiative of WCRP, IGBP, IHDP and DIVERSITAS.

The United States has been a key scientific contributor to all of these programs—the U.S. policy of making satellite data freely available around the world is just one example—and has also been a beneficiary of international research efforts. Since the early 1990s, the International Group of Funding Agencies for Global Change Research (IGFA) has provided a forum through which national agencies that fund global change research identify issues of mutual interest and look for appropriate ways to coordinate. Continued participation in these international activities will be crucial to an effective climate science enterprise in the United States. In particular, as noted in this report and others (e.g., NRC, 2009k), the science needs for improved climate observing systems and improved model projections of future climate change can best be met through collaborations and partnerships at the international scale. Moreover, climate change is a global challenge; impacts on ecosystems and societies span the globe and some of these impacts will cascade from one region to another. Climate change science conducted in the United States can thus play an essential role in improving the knowledge of and scientific capacity to respond to climate challenges in the developing areas of the world, where knowledge about possible responses to climate change is much more limited.

National and international coordination are essential, but decision-relevant research is often focused at regional and local scales. Thus, there are many opportunities for states, municipalities, and other subnational entities to work with each other and with the federal government to build expertise, fund relevant research and research infrastructure, and create the kinds of networks and partnerships that enable effective collaborations among the research and decision-making communities. For example, research in many academic and nonacademic institutions is supported in part by state funds, including the system of agricultural experiment stations and targeted initiatives on water and other resources. Because so many climate change challenges play out at local to regional scales, new kinds of partnerships and programs will be needed to link federal and local research and response approaches and to make research useful to decision making at all scales. So-called "boundary organizations" that purposefully link researchers and decision makers provide one model for doing so (see, e.g., Brooke, 2008; Moser and Luers, 2008; Pohl, 2008; Tribbia and Moser, 2008). The Regional Integrated Science and Assessments (RISA) program and, until recently, the Sectoral Applications Research Program (SARP) organized by NOAA are examples of such programs (NRC, 2007h). Examples can also be found in other countries (for example, the United Kingdom Climate Impacts Program). Shared funding and governance can help ensure such programs provide both effective decision support and decision-relevant research.

Partnerships with Programs to Limit, Adapt to, and Inform Decisions Makers About Climate Change

As discussed in Chapters 3 and 4, climate change science can make a wide variety of contributions to action-oriented programs that focus on responses to climate change. Working collaboratively with action-oriented programs, both at the federal level and across the country, would help response programs take more effective actions and would help the federal climate change research program ensure that its research activities support effective decision making, in addition to improving fundamental understanding. The recent NRC reports *Restructuring Federal Climate Research to Meet the Challenges of Climate Change* (NRC, 2009k) and *Informing Decisions in a Changing Climate* (NRC, 2009g) also called for an integrated, "end-to-end" climate change research program that is closely linked with relevant action-oriented programs. Achieving this integration will require careful and deliberate coordination, perhaps through an oversight committee that coordinates all federal actions to understand and respond to climate change, or through less formal partnerships led by dedicated managers. In this panel's opinion, formal mechanisms have a greater chance of long-term success.

Limiting Climate Change

As discussed in Chapter 4 and in the companion report *Limiting the Magnitude of Future Climate Change* (NRC, 2010c), scientific research can help support actions taken to limit the magnitude of future climate change in a variety of ways. Some technology options in the energy sector are already commercially viable and could be implemented to achieve emissions reductions in the near term. However, research and development are needed to improve implementation success, lower costs, increase the effectiveness of current options, and expand the number of options available. Expanded investments will be needed in a wide range of research areas, such as energy sources that emit few or no GHGs, carbon capture and storage, energy efficiency and conservation approaches (including strategies to promote adoption and use of energy-efficient technologies), and technologies to reduce emissions from agriculture and other land uses. Technologies that remove GHGs from the air or reflect more sunlight back to space (geoengineering approaches) may also warrant attention, provided that they do not replace other important research efforts (see Chapter 15).

A variety of research programs on transportation and energy technology development and deployment already exist in the federal government (for example and most notably, the Climate Change Technology Program led by the Department of Energy), in several states (e.g., California's PIER program), in corporations, and through public-private partnerships such as corporate-funded university research efforts (NRC, 2009a,b,c,d). The climate change research enterprise envisioned in this report—including the USGCRP—would complement and build on these efforts. For example, research will be needed to evaluate the overall effectiveness of different technologies, possible unintended consequences of large-scale deployment, and possible trade-offs and co-benefits with other types of responses. New scientific knowledge about human behavior, public perception, and institutional structures can help identify potential barriers to widespread implementation of promising technologies or policies to limit climate change. Research is also needed on a wide range of technology implementation and deployment issues, such as research on cost and cost effectiveness, governance issues, barriers to technology adoption, and policies and programs designed to overcome these barriers. Finally, research can help to develop frameworks for decision making that allow these barriers, costs, benefits, co-benefits, and trade-offs to be explicitly evaluated and incorporated into strategies for reducing emissions.

As noted in Chapter 4, an effective national research effort on limiting the magnitude of climate change will require integration of knowledge across a wide range of fields and collaboration with engineers, policy makers, and others involved in developing and implementing actions to limit climate change. However, collaboration and linkages between the USGCRP and existing programs relevant to limiting climate change—most notably the Climate Change Technology Program—are currently weak (NRC, 2009k). These linkages need to be improved, and any new programs that emerge to focus on limiting the magnitude of future climate change would surely benefit from formal linkages to the USGCRP as well as other scientific research organizations, efforts, and activities.

Adapting to Climate Change

The companion report *Adapting to the Impacts of Climate Change* (NRC, 2010a) concludes that there is an urgent need to better understand and project climate change and its impacts (especially at local and regional scales), convey this information to decision makers and other stakeholders, and develop options and strategies for reducing the vulnerability and increasing the resilience and adaptive capacity of both human and natural systems in the United States and abroad. As discussed in Chapter 4, science can make major contributions in all of these areas. A national climate change research enterprise that has an expanded focus on adaptation strategies could, for example, provide region- and sector-specific information about climate change impacts and vulnerabilities in the context of multiple stressors acting on coupled human-environment systems. It could also evaluate and verify the feasibility and effectiveness of, trade-offs among, and the secondary environmental, social, and economic consequences of different adaptation options. Moreover, because it is difficult to assign a monetary value to some kinds of impacts (for example, biodiversity loss or threats to national security), the development of alternative metrics and assessment strategies is needed. Science can also support adaptation through research-based development and testing of decision-support strategies and tools designed to connect scientific information with decision making. Finally, there is a need for further research on human behavior and institutional barriers to implementation in the context of adaptation options and choices.

The companion report *Adapting to the Impacts of Climate Change* (NRC, 2010a) recommends that a national adaptation strategy be established to engage decision makers, stakeholders, and researchers at all levels in developing and implementing adaptation plans. The USGCRP and other elements of the nation's climate change science research enterprise will be essential partners in the success of these adaptation efforts. Connecting adaptation programs with scientific research is complicated, however, by the fact that many adaptation decisions are inherently local or regional in scale and can take years to implement. Federal centers established to address climate challenges may not effectively assist at these scales unless there are regional or local entities to provide integration in a place-based context and facilitate connections with local decision makers. Local, state, and regional partnerships between academic, public, and private institutions could serve the role of coupling adaptation efforts with scientific research to create end-to-end knowledge systems. Approaches for linking knowledge about adaptation responses across these scales, and to international adaptation research efforts, will also be needed.

Informing About Climate Change

To respond effectively to climate change, decision makers at all scales from local to international will need up-to-date, cogent, accessible, and usable information. The companion report *Informing an Effective Response to Climate Change* (NRC, 2010b) provides analysis and advice on how to ensure that scientific information is used, and used effectively, by decision makers. Many previous reports (e.g., NRC, 2008h, 2009g) have also analyzed the information sources, assessment tools, decision-support mechanisms, and other aspects of informing effective climate-related decision making.

There have been several recent efforts at the federal level to establish programs to provide climate-related information, such as NOAA's announcement of its intent to form a climate service (NOAA, 2010) and the Department of the Interior's announcement of a coordinated climate change research and resource management strategy (DOI, 2009), as well as an international agreement to establish a global framework for climate services (WMO, 2009b). As discussed in Chapter 4, these efforts, and those established in the future, will require the climate change science community's assistance in providing more and better decision-relevant information, as well as scientific research on improved communication and decision-support tools and structures.

Scientific assessments are another way the climate research program can work collaboratively with national or international initiatives to inform effective climate-related decisions and responses. Climate change assessment processes, if carefully and deliberately designed, can engage a broad range of stakeholders in the assessment of risks, costs, and potential responses to climate change impacts (Farrell and Jäger, 2005; NRC, 2007a, 2008h). Assessment activities represent an important opportunity to improve linkages between the scientific and decision-making communities. The recent NRC report *Restructuring Federal Climate Research* (NRC, 2009k) called for the USGCRP to begin planning a comprehensive national assessment of climate change impacts, adaptation options, and actions to reduce climate forcing, as called for in the Global Change Research Act, and it is encouraging that planning for such an activity is now under way.

Recommendation 6: The federal climate change research program should be formally linked with action-oriented response programs focused on limiting the magnitude of future climate change, adapting to the impacts of climate change, and informing climate-related actions and decisions, and, where relevant, should

develop partnerships with other research and decision-making entities working at local to international scales.

CAPACITY BUILDING

The scale, importance, and complexity of the climate challenge implies a critical need to increase the workforce performing fundamental and decision-relevant climate research, implementing responses to climate change, and working at the interface between science and decision making. Thanks to more than three decades of research on climate change, the research community in the United States and elsewhere is strong, at least in research areas that have received significant emphasis and support. However, level or declining climate research funding over the past decade (as documented, for example, in NRC [2009k]) has limited the number of young scientists and engineers entering the research workforce at just the point when an influx of young scientists and engineers is critically needed to revitalize the nation's climate research. Moreover, the more integrative and decision-relevant research program described in Chapter 4 will require expanded intellectual capacity in several previously neglected fields as well as in interdisciplinary research areas. It will also require greater intellectual capacity among state, local, and national government agencies, universities, and other public and private research labs, as well as among science managers coordinating efforts to advance the science of climate change. Building and mobilizing this broad research community will require both a concerted effort and a new approach.

Challenges Posed by the New Era of Climate Change Research

The broad, interdisciplinary, and integrated research enterprise envisioned in this report presents a number of implementation challenges. Among others, it requires scientists to work together in ways that are not well supported by many existing institutional structures, such as discipline-specific academic departments. It also requires researchers to engage with decision makers and other stakeholders to identify research topics and develop mechanisms for transferring research results, activities that are not a traditional strength or focus of scientific training. These challenges suggest that changes are needed within universities, federal laboratories, vocational training centers, and other research and educational institutions.

At the national scale, institutional changes are needed in federal research and mission agencies to increase the focus on interdisciplinary and decision-relevant research both in government laboratories and in the nationwide research efforts the agencies support. Some agencies will need to recruit or train scientists and program managers with the expertise needed to organize and manage such programs, especially expertise in the behavioral and social science fields that have not been as well represented or supported as the more "traditional" areas of climate and climate-related research.

Many universities are already experimenting with new interdisciplinary departments or schools focused on the environment, while others have developed multidepartment programs, centers, or institutes on sustainability, climate change, and other crosscutting topics. Many of these same university experiments include the training of undergraduate and graduate students through interdisciplinary academic programs, some of which are funded by special federal programs (such as the National Science Foundation's Integrative Graduate Education and Research Traineeship program). Although in great demand by students, these programs face challenges from a lack of long-term funding and commitment by faculty and administrators.

Changes are also needed in professional societies, journals, and other institutions that influence rewards and incentives for scientists, engineers, managers, and others involved in the climate research enterprise. For example, venues for presentation and publication of interdisciplinary and decision-relevant climate research, as well as professional organizations that support and reward these efforts, are needed to build networks and provide professional rewards. Likewise, organizational changes in advice-giving bodies (such as the NRC) may help by enabling them to emphasize the integrative nature of climate change science when providing advice for the government and the larger science community. Other needed investments include fellowships and early career awards that can help direct researchers toward interdisciplinary work, and "summer institutes" and other training opportunities that provide extended interaction and promote cross-disciplinary engagement.

Finally, at the international scale, interdisciplinary science efforts focused on climate and global change have started to emerge (for example, the ESSP projects under ICSU). Not only do these programs facilitate engagement and capacity building for scientists from developing countries, they provide a way for U.S. scientists to contribute to international programs that focus on integrative research in support of both basic understanding of and responses to climate change. Strengthening these programs will require improving international research funding capacity (through IGFA and other mechanisms) and developing new mechanisms to engage the U.S. research community with international partners. One obstacle that impairs both international collaboration and U.S. research capacity is the difficulty that non-U.S. scientists encounter in obtaining visas to visit or train in the United States; another is the fact that most federal programs will not fund non-U.S. citizens as researchers or students.

Challenges Posed by Linkages with Other Activities and Programs

State and local governments, corporations, and nongovernmental organizations are key partners in the nation's climate change research enterprise (see Recommendation 6). These partners will need a workforce that can engage effectively with the scientific community. There are many opportunities for sponsorship and leadership on climate-related research and decision support at the state and local levels. State, local, and tribal entities should work together with federal and nongovernmental partners to build expertise and create the kinds of networks, partnerships, and institutions that enable effective collaborations between the research community and decision makers. Progress in this direction is already being made. For example, climate advisory councils composed of experts from state universities, research institutions, nongovernmental organizations, municipalities, tribal governments, and agencies have been mandated by executive orders or state legislatures in a number of states. In other cases, science-based nongovernmental organizations have provided leadership in developing impact assessments and climate action plans (both for GHG emissions reductions and adaptation) that have proven helpful for informing policy makers.

A number of corporations have also taken a leadership role in reducing GHG emissions (NRC, 2010b) and promoting other sustainable business practices. These efforts can be expected to increase intellectual capacity and practical experience, both of which will be useful to both the research community and society at large. Partnerships between the research community and the private sector are critical for building effective science-decision maker relationships, for linking knowledge and action, and for identifying critical science workforce needs. Federal programs, such as NOAA's RISA program and the Regional Climate Centers, can aid in these efforts.

Finally, a strategy is needed for educating and training the next generation of climate change researchers as well as the personnel needed to design, build, and maintain the physical infrastructure and institutional assets needed to respond effectively to climate change. Climate researchers and research managers will also need training in decision-support and outreach activities needed to shape a decision-relevant science agenda. In addition, growing demands for climate information will require more people with skills and practice in effective communication, science-policy interaction, and activities at the interface between research and decision making. Much of the training in these areas will presumably need to take place at regional and local scales, but federal leadership and support are essential. Further discussion of the actions needed to educate and train future generations of scientists, engineers, technicians, managers, and decision makers for responding to climate change can be found in the companion report *Informing an Effective Response to Climate Change* (NRC, 2010b).

Recommendation 7: Congress, federal agencies, and the federal climate change research program should work with other relevant partners (including universities, state and local governments, the international research community, the business community, and other nongovernmental organizations) to expand and engage the human capital needed to carry out climate change research and response programs.

A NEW ERA OF CLIMATE CHANGE RESEARCH

We have entered a new era of climate change research. Although there are some uncertainties in the details of future climate change, it is clear that climate change is occurring, is largely due to human activities, and poses significant risks for people and the ecosystems on which we depend. Moreover, climate change is not just an environmental problem; it is a sustainability challenge that affects and interacts with other environmental changes and efforts to provide food, energy, water, shelter, and other fundamental needs of people today and in the future. In response to the risks posed by climate change, actions are now being taken both to limit the magnitude of future climate change and to adapt to its unavoidable impacts. These responses to climate change should be informed by the best possible scientific knowledge. Research is needed to improve understanding of the climate system and related human and environmental systems, to maximize the effectiveness of actions taken to respond to climate change, and to avoid unintended consequences for human well-being and the Earth system that sustains us. Acquiring the needed scientific knowledge, and making it useful to decision makers, will require an expanded climate change research enterprise. The challenge is tremendous, and so, too, should be our response, both in magnitude and in breadth.

Part II
Technical Chapters

CHAPTER SIX

Changes in the Climate System

Scientific understanding of the factors and processes that govern the evolution of Earth's climate has increased markedly over the past several decades, as has the ability to simulate and project future changes in the climate system. As noted in Chapter 2, this knowledge has been regularly assessed, synthesized, and summarized by the Intergovernmental Panel on Climate Change (IPCC), the U.S. Global Climate Research Program (USGCRP, referred to as the U.S. Climate Change Science Program from 2000 to 2008), and other groups to provide a thorough and detailed description of what is known about past, present, and projected future changes in climate and related human and environmental systems. This chapter provides an updated overview of the current state of knowledge about the climate system, followed by a list of some of the key scientific advances needed to further improve our understanding.

To help frame the sections that follow, it is useful to consider some questions that decision makers are asking or will be asking about changes in the climate system:

- How are temperature and other aspects of climate changing?
- How do we know that humans are responsible for these changes?
- How will temperature, precipitation, severe weather, and other aspects of climate change in my city/state/region over the next several decades?
- Will these changes be steady and gradual, or abrupt?
- Will seasonal and interannual climate variations, like El Niño events, continue the same way or will they be different?
- Why is there so much uncertainty about future changes?

This chapter attempts to answer these questions or explain what additional research would be needed to answer them. The chapters that follow focus on the impacts of climate change on a range of human and environmental systems, the role of these systems in driving climate change, and the state of scientific knowledge regarding actions that could potentially be taken to adapt to or limit the magnitude of climate change in those systems. All of the chapters in Part II follow a similar structure and are more detailed and extensively referenced than the concise overview of climate change science found in Chapter 2. However, these chapters represent only highlights of a broad and extensive collection of scientific research; readers desiring further detail are encouraged to consult other recent assessment reports and the primary literature.

FACTORS INFLUENCING EARTH'S CLIMATE

The Greenhouse Effect

The Earth's physical climate system, which includes the atmosphere, oceans, cryosphere, and land surface, is complex and constantly evolving. Nevertheless, the laws of physics, chemistry, and biology ultimately govern the system and can be used to understand how and why climate varies from place to place and over time. For example, the energy balance of the Earth as a whole is determined by the difference between incoming and outgoing energies at the top of the atmosphere. The only significant incoming energy is radiation from the sun, which is concentrated at short wavelengths (visible and ultraviolet light), while the outgoing energy includes both infrared (long-wavelength) radiation emitted by the Earth and the portion of incoming solar radiation (about 30 percent on average) that is reflected back to space by clouds, small particles in the atmosphere, and the Earth's surface. If the outgoing energy is slightly lower than the incoming energy for a period of time, then the climate system as a whole will warm until the outgoing radiation from the Earth balances the incoming radiation from the sun.

The temperature of the Earth's surface and lower atmosphere depends on a broader range of factors, but the transfer of radiation again plays an important role, as does the composition of the atmosphere itself. Nitrogen (N_2) and oxygen (O_2) make up most of the atmosphere, but these gases have almost no effect on either the incoming radiation from the sun or the outgoing radiation emitted by the Earth's surface. Certain other gases, however, absorb and reemit the infrared radiation emitted by the surface, effectively trapping heat in the lower atmosphere and keeping the Earth's surface much warmer—roughly 59°F (33°C) warmer—than it would be if greenhouse gases were not present.[1] This is called the *greenhouse effect*, and the gases that cause it—including water vapor, carbon dioxide (CO_2), methane (CH_4), and nitrous oxide (N_2O)—are called greenhouse gases (GHGs). GHGs only constitute a small fraction of the Earth's atmosphere, but even relatively small increases in the amount of these gases in the atmosphere can amplify the natural greenhouse effect, warming the Earth's surface (see Figure 2.1).

[1] This difference includes the greenhouse effect associated with clouds, which are composed of water droplets, but it assumes that the total reflectivity of the Earth—including the reflection by clouds—does not change.

Carbon Dioxide

The important role played by CO_2 in the Earth's energy balance has been appreciated since the late 19th century, when Swedish scientist Svante Arrhenius first proposed a link between CO_2 levels and temperature. At that time, humans were only beginning to burn fossil fuels—which include coal, oil, and natural gas—on a wide scale for energy. The combustion of these fuels, or any material of organic origin, yields mostly CO_2 and water vapor, but also small amounts of other by-products, such as soot, carbon monoxide, sulfur dioxide, and nitrogen oxides. All of these substances occur naturally in the atmosphere, and natural fluxes of water and CO_2 between the atmosphere, oceans, and land surface play a critical role in both the physical climate system and the Earth's biosphere. However, unlike water vapor molecules, which typically remain in the lower atmosphere for only a few days before they are returned to the surface in the form of precipitation, CO_2 molecules are only exchanged slowly with the surface. The excess CO_2 emitted by fossil fuel burning and other human activities will thus remain in the atmosphere for many centuries before it can be removed by natural processes (Solomon et al., 2009).

A number of agencies and groups around the world, including the Carbon Dioxide Information Analysis Center at Oak Ridge National Laboratory and the International Energy Agency, produce estimates of how much CO_2 is released to the atmosphere every year by human activities. The most recent available estimates indicate that, in 2008, human activities released over 36 Gt (gigatons, or billion metric tons) of CO_2 into the atmosphere—including 30.6 ± 1.7 Gt from fossil fuel burning, plus an additional 4.4 ± 2.6 Gt from land use changes and 1.3 ± 0.1 Gt from cement production (Le Quéré et al., 2009). Emissions from fossil fuels have increased sharply over the last two decades, rising 41 percent since 1990 (Figure 6.1). CO_2 emissions due to land use change—which are dominated by tropical deforestation—are estimated based on a variety of methods and data sources, and the resulting estimates are both more uncertain and more variable from year-to-year than fossil fuel emissions. Over the past decade (2000–2008), Le Quéré et al. (2009) estimate that land use changes released 5.1 ± 2.6 Gt of CO_2 each year, while fossil fuel burning and cement production together released on average 28.2 ± 1.7 Gt of CO_2 per year.

Up until the 1950s, most scientists thought the world's oceans would simply absorb most of the excess CO_2 released by human activities. Then, in a series of papers in the late 1950s (e.g., Revelle and Suess, 1957), American oceanographer Roger Revelle and several collaborators hypothesized that the world's oceans could not absorb all the excess CO_2 being released from fossil fuel burning. To test this hypothesis, Revelle's colleague C. D. Keeling began collecting canisters of air at the Mauna Loa Observatory

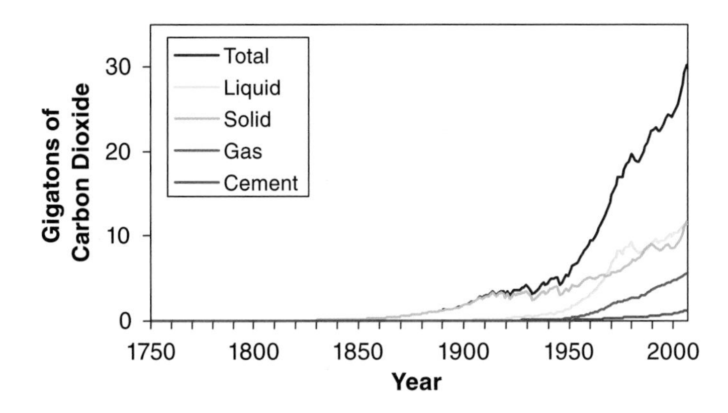

FIGURE 6.1 Estimated global CO_2 emissions from fossil fuel sources, in gigatons (or billion metric tons). Based on data from Boden et al. (2009; available at *http://cdiac.ornl.gov/trends/emis/tre_glob.html).*

in Hawaii, far away from major industrial and population centers, and analyzing the composition of these samples to determine whether CO_2 levels in the atmosphere were increasing. Similar in situ measurements continue to this day at Mauna Loa as well as at many other sites around the world. The resulting high-resolution, well-calibrated, 50-year-plus time series of highly accurate and precise atmospheric CO_2 measurements (Figure 6.2), commonly referred to as the Keeling curve, is both a major scientific achievement and a key data set for understanding climate change.

The Keeling curve shows that atmospheric CO_2 levels have risen by more than 20 percent since 1958; as of January 2010, they stood at roughly 388 ppm, rising at an average annual rate of almost 2.0 ppm per year over the past decade (Blasing, 2008; Tans, 2010). When multiplied by the mass of the Earth's atmosphere, this increase corresponds to 15.0 ± 0.1 Gt CO_2 added to the atmosphere each year, or roughly 45 percent of the excess CO_2 released by human activities over the last decade. The remaining 55 percent is absorbed by the oceans and the land surface. The size of these CO_2 "sinks" is estimated via both modeling and direct observations of CO_2 uptake in the oceans and on land. These estimates indicate that the oceans absorbed on average 8.4 ± 1.5 Gt CO_2 annually over the last decade (or 26 percent of human emissions), while the land surface took up 11.0 ± 3.3 Gt per year (29 percent), with a small residual of 0.3 Gt (Le Quéré et al., 2009).

A careful examination of the Keeling curve reveals that atmospheric CO_2 concentrations are currently increasing twice as fast as they did during the first decade of the record (compare the slope of the black line in Figure 6.2). This acceleration in the rate of CO_2 rise can be attributed in part to the increases in CO_2 emissions due to increasing

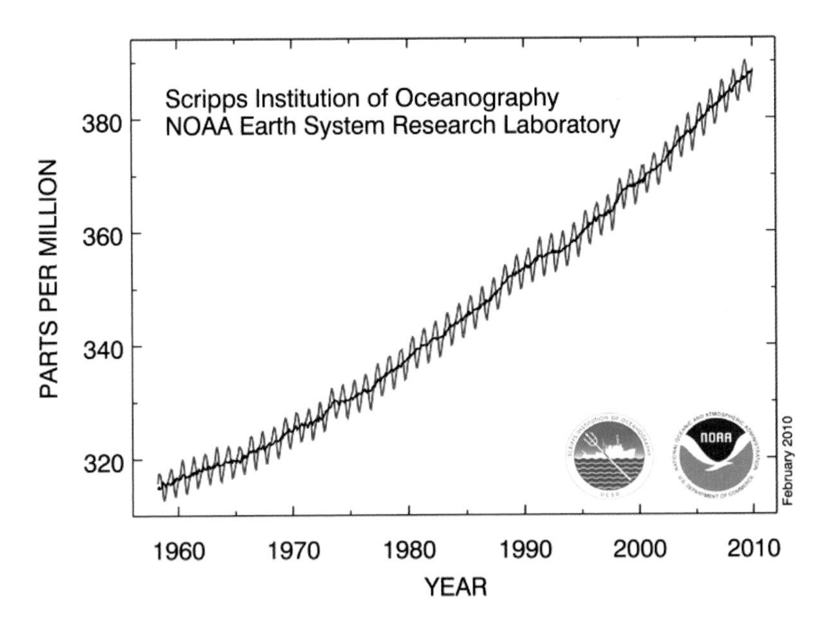

FIGURE 6.2 Atmospheric CO_2 concentrations (in parts per million [ppm]) at Mauna Loa Observatory in Hawaii. The red curve, which represents the monthly averaged data, includes a seasonal cycle associated with regular changes in the photosynthetic activity in plants, which are more widespread in the Northern Hemisphere. The black curve, which represents the monthly averaged data with the seasonal cycle removed, shows a clear upward trend. SOURCE: Tans (2010; available at *http://www.esrl.noaa.gov/gmd/ccgg/trends/*).

energy use and development worldwide (as indicated in Figure 6.1). However, recent studies suggest that the rate at which CO_2 is removed from the atmosphere by ocean and land sinks may also be declining (Canadell et al., 2007; Khatiwala et al., 2009). The reasons for this decline are not well understood, but, if it continues, atmospheric CO_2 concentrations would rise even more sharply, even if global CO_2 emissions remain the same. Improving our understanding and estimates of current and projected future fluxes of CO_2 to and from the Earth's surface, both over the oceans and on land, is a key research need (research needs are discussed at the end of the chapter).

To determine how CO_2 levels varied prior to direct atmospheric measurements, scientists have studied the composition of air bubbles trapped in ice cores extracted from the Greenland and Antarctic ice sheets. These remarkable data, though not as accurate and precise as the Keeling curve, show that CO_2 levels were relatively constant for thousands of years preceding the Industrial Revolution, varying in a narrow band between 265 and 280 ppm, before rising sharply starting in the late 19th century (Figure 6.3). The current CO_2 level of 388 ppm is thus almost 40 percent higher

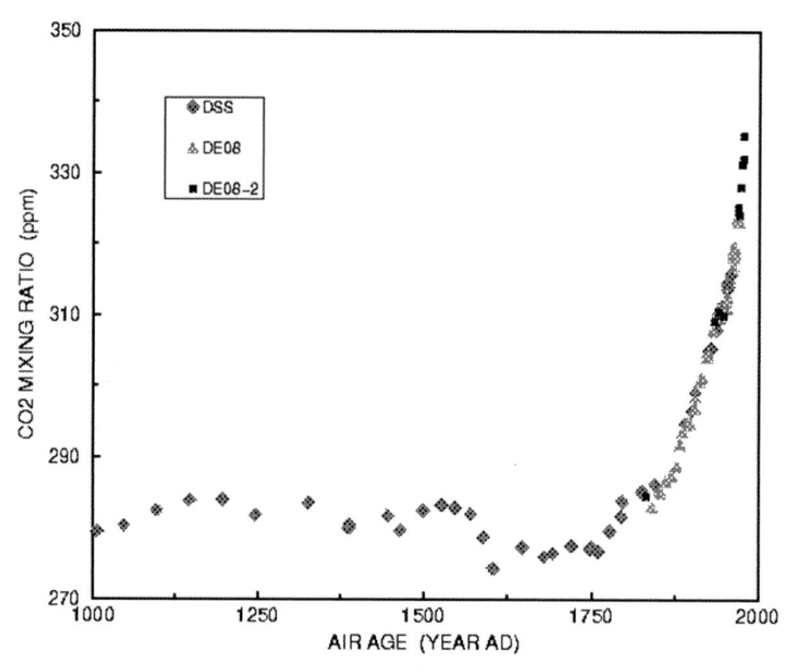

FIGURE 6.3 CO_2 variations during the last 1,000 years, in parts per million (ppm), obtained from analysis of air bubbles trapped in an ice core extracted from Law Dome in Antarctica. The data show a sharp rise in atmospheric CO_2 starting in the late 19th century, coincident with the sharp rise in CO_2 emissions illustrated in Figure 6.1. Similar data from other ice cores indicate that CO_2 levels remained between 260 and 285 ppm for the last 10,000 years. SOURCE: Etheridge et al. (1996).

than preindustrial conditions (usually taken as 280 ppm). As discussed in further detail in the next section, data from even longer ice cores extracted from the hearts of the Greenland and Antarctic ice sheets—the bottoms of which contain ice that was formed hundreds of thousands of years ago—indicate that the current CO_2 levels are higher than they have been for at least 800,000 years.

Collectively, the in situ measurements of CO_2 over the past several decades, ice core measurements showing a sharp rise in CO_2 since the Industrial Revolution, and detailed estimates of CO_2 sources and sinks provide compelling evidence that CO_2 levels are increasing as a result of human activities. There is, however, an additional piece of evidence that makes the human origin of elevated CO_2 virtually certain: measurements of the isotopic abundances of the CO_2 molecules in the atmosphere—a chemical property that varies depending on the source of the CO_2—indicate that most of the excess CO_2 in the atmosphere originated from sources that are millions of years old. The only source of such large amounts of "fossil" carbon are coal, oil, and natural gas (Keeling et al., 2005).

Climate Forcing

Changes in the radiative balance of the Earth—including the enhanced greenhouse effect associated with rising atmospheric CO_2 concentrations—are referred to as climate *forcings* (NRC, 2005d). Climate forcings are estimated by performing detailed calculations of how the presence of a forcing agent, such as excess CO_2 from human activities, affects the transfer of radiation through the Earth's atmosphere.[2] Climate forcings are typically expressed in Watts per square meter (W/m^2, or energy per unit area), with positive forcings representing warming, and are typically reported as the change in forcing since the start of the Industrial Revolution (usually taken to be the year 1750). Figure 6.4 provides a graphical depiction of the estimated globally averaged strength of the most important forcing agents for recent climate change. Each of these forcing agents are discussed below.

Well-Mixed Greenhouse Gases

Carbon dioxide (CO_2). The CO_2 emitted by human activities is the largest single climate forcing agent, accounting for more than half of the total positive forcing since 1750 (see Figure 6.4). As of the end of 2005, the forcing associated with human-induced atmospheric CO_2 increases stood at 1.66 ± 0.17 W/m^2 (Forster et al., 2007). This number may seem small relative to the total energy received by the Earth from the sun (which averages 342 W/m^2, of which 237 W/m^2 is absorbed by the Earth system, after accounting for reflection of 30 percent of the solar energy back to space). When multiplied by the surface area of the Earth, however, the CO_2 forcing is roughly 850 terawatts, which is more than 50 times the total power consumed by all human activities.

Human activities have also led to increases in the concentrations of a number of other "well-mixed" GHGs—those that are relatively evenly distributed because their molecules remain in the atmosphere for at least several years on average. Many of these gases are much more potent warming agents, on a molecule-for-molecule basis, than CO_2, so even small changes in their concentrations can have a substantial influence. Collectively, they produce an additional positive forcing (warming) of 1.0 ± 0.1 W/m^2, for a total well-mixed GHG-induced forcing (including CO_2) of 2.63 ± 0.26 W/m^2

[2] As discussed in NRC (2005e): "Radiative forcing traditionally has been defined as the instantaneous change in energy flux at the tropopause resulting from a change in a component external to the climate system. Many current applications [including the radiative forcing values discussed in this chapter] use an 'adjusted' radiative forcing in which the stratosphere is allowed to relax to thermal steady state, thus focusing on the energy imbalance in the Earth and troposphere system, which is most relevant to surface temperature change."

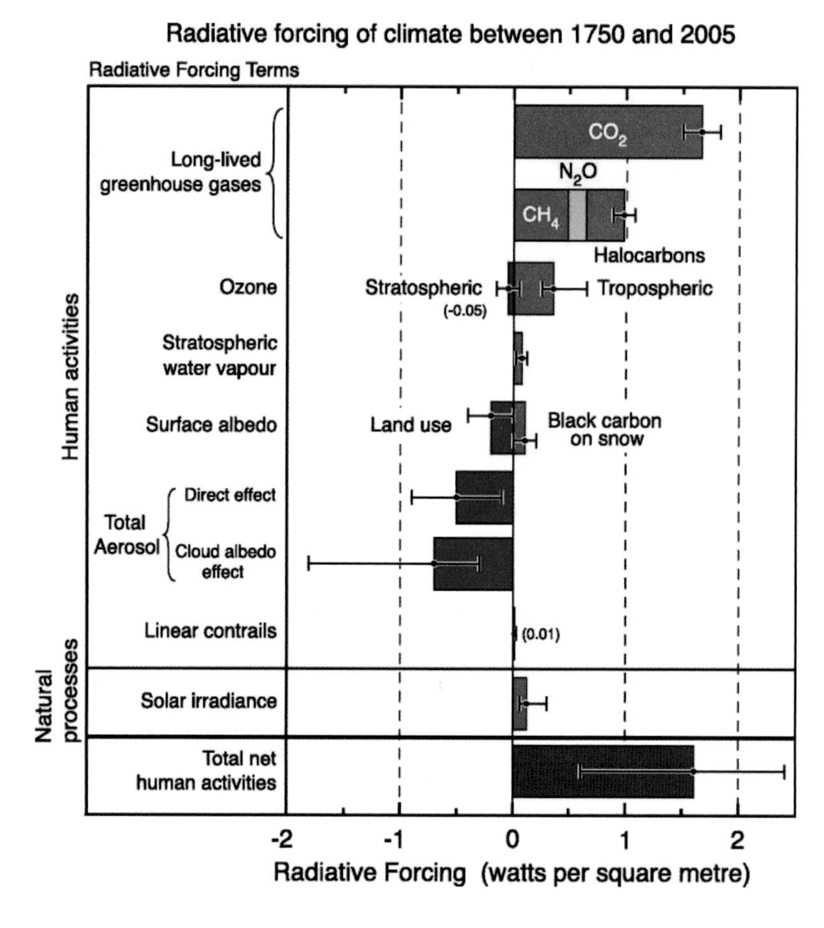

FIGURE 6.4 Radiative forcing of climate between 1750 and 2005 due to both human activities and natural processes, expressed in Watts per square meter (energy per unit area). Positive values correspond to warming. See text for details. SOURCE: Forster et al. (2007).

(Forster et al., 2007) (see Figure 6.4). Forcing estimates for all of the well-mixed GHGs are quite accurate because we have precise measurements of their concentrations, their influence on the transfer of radiation through the atmosphere is well understood, and they become relatively evenly distributed across the global atmosphere within a year or so of being emitted.

Methane (CH_4). Methane is produced from a wide range of human activities, including natural gas management, fossil fuel and biomass burning, animal husbandry, rice cultivation, and waste management (Houweling et al., 2006). Natural sources of CH_4—which are smaller than human sources—include wetlands and termites, and both of

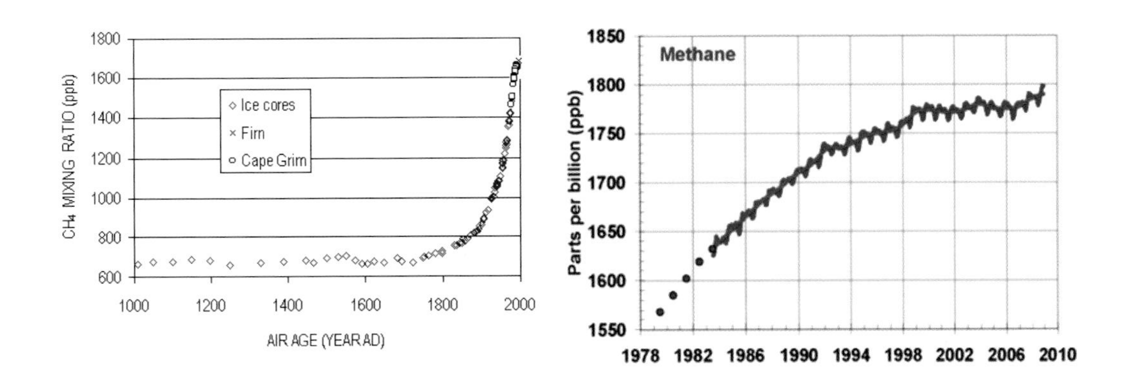

FIGURE 6.5 Atmospheric CH_4 concentrations in parts per billion (ppb), (left) during the past millennium, as measured in Antarctic ice cores, and (right) since 1979, based on direct atmospheric measurements. SOURCES: Etheridge et al. (2002) and NOAA/ESRL (2009).

these sources are actually influenced to some degree by changes in land use. Recent measurements have suggested that plants and crops may also emit trace amounts of CH_4 (Keppler et al., 2006), although the size of this source has been questioned (Dueck et al., 2007).

The atmospheric concentration of CH_4 rose sharply through the late 1970s before starting to level off, ultimately reaching a relatively steady concentration of around 1775 ppb—which is more than two-and-a-half times its average preindustrial concentration—from 1999 to 2006 (Figure 6.5). There have been several theories proposed for the apparent leveling off of CH_4 concentrations, including a decline in industrial emissions during the 1990s and a slowdown of natural wetland-related emissions (Dlugokencky et al., 2003). As discussed at the end of the chapter, there are also concerns that warming temperatures could lead to renewed rise in CH_4 levels as a result of melting permafrost across the Arctic (Schuur et al., 2009) or, less likely, the destabilization of methane hydrates[3] on the seafloor (Archer and Buffet, 2005; Overpeck and Cole, 2006). The causes of the recent uptick in concentrations in 2007 and 2008 are currently being studied (Dlugokencky et al., 2009).

Unlike CO_2, which is only removed slowly from the atmosphere by processes at the land surface, the atmospheric concentration of CH_4 is limited mainly by a chemical reaction in the atmosphere that yields CO_2 and water vapor. As a result, molecules of CH_4 spend on average less than 10 years in the atmosphere. However, CH_4 is a much

[3] Methane hydrates are crystalline structures composed of methane and water molecules that can be found in significant quantities in sediments on the ocean floor.

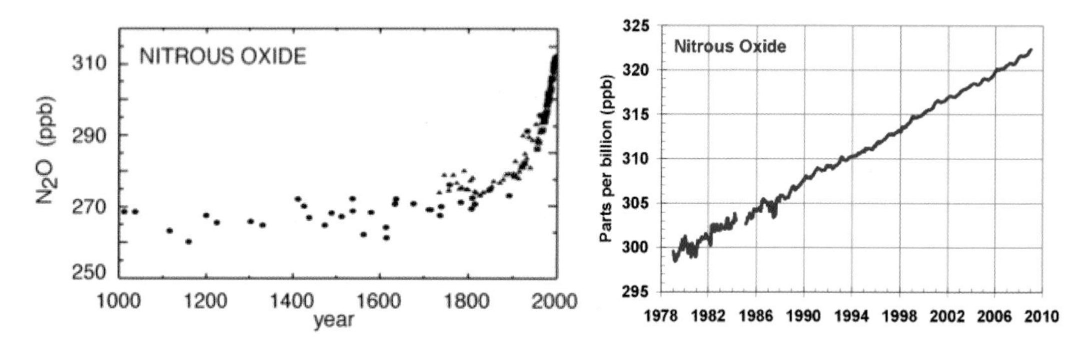

FIGURE 6.6 N_2O concentrations in the atmosphere, in parts per billion (ppb), (left) during the last millennium, and (right) since 1979. SOURCES: Etheridge et al. (1996) and NOAA/ESRL (2009).

more potent warming agent, on a molecule-for-molecule basis,[4] than CO_2, and its relative concentration in the atmosphere has risen by almost four times as much as CO_2. Hence, the increases in CH_4 since 1750 are associated with a climate forcing of roughly 0.48 ± 0.05 W/m^2 (Forster et al., 2007), or around 18 percent of the total forcing by well-mixed GHGs.

Nitrous oxide (N_2O). Concentrations of nitrous oxide in the atmosphere have increased around 15 percent since 1750, primarily as a result of agricultural activities (especially the application of chemical fertilizers) but also as a by-product of fossil fuel combustion and certain industrial process. The average atmospheric concentration of N_2O continues to grow at a steady rate of around 0.8 ppb per year and, as of the end of 2008, stood at just over 322 ppb (Figure 6.6) (see also NASA, 2008). N_2O is an extremely potent warming agent—more than 300 times as potent as CO_2 on a molecule-by-molecule basis—and its molecules remain in the atmosphere more than 100 years on average. Thus, even though N_2O concentrations have not increased nearly as much since 1750 as CH_4 or CO_2, N_2O still contributes a climate forcing of 0.16 ± 0.02 W/m^2 (Forster et al., 2007), or around 6 percent of total well-mixed GHG forcing. N_2O and its decomposition in the atmosphere also have a number of other environmental effects—for example, N_2O is now the most important stratospheric ozone-depleting substance being emitted by human activities (Ravishankara et al., 2009).

Halogenated gases. Over a dozen halogenated gases, a category that includes ozone-depleting substances such as chlorofluorocarbons (CFCs), hydrofluorocarbons, per-

[4] The relative (molecule-by-molecule) radiative forcing of a GHG over a particular time scale (usually taken as 100 years), compared to carbon dioxide, is sometimes expressed as the global warming potential of the gas. Another common comparative metric is carbon dioxide equivalent (CO_2-eq), which describes the equivalent amount of carbon dioxide that would produce the same forcing.

fluorocarbons, and sulfur hexafluoride, also contribute to the positive climate forcing associated with well-mixed GHGs. Although relatively rare—their concentrations are typically measured in parts per trillion—many of the halogenated gases have very long residence times in the atmosphere and are extremely potent forcing agents on a molecule-by-molecule basis (Ravishankara et al., 1993). Collectively they contribute an additional 0.33 ± 0.03 W/m^2 of climate forcing. Most halogenated gases do not have any natural sources (see, e.g., Frische et al., 2006) but rather arise from a variety of industrial activities. Emissions of many of these ozone-depleting compounds have declined sharply over the past 15 years because of the Montreal Protocol (see below). As a result, their atmospheric concentrations, and hence climate forcing, are now declining slightly each year as they are slowly removed from the atmosphere by natural processes (Figure 6.7) (NASA, 2008). It has been estimated that the forcing associated with halogenated gases would be 0.2 W/m^2 higher than it is today if emissions reductions due to the Montreal Protocol had not taken place (Velders et al., 2007; see also Chapter 17).

Other Greenhouse Gases

Ozone (O_3). Ozone plays a number of important roles in the atmosphere, depending on location, and its concentration varies substantially, both vertically and horizontally. The highest concentrations of ozone are found in the stratosphere—the layer of the atmosphere extending from roughly 10 to 32 miles (15 to 50 km) in height (Figure 6.8)—where it is produced naturally by the dissociation of oxygen molecules by ultraviolet light. This chemical reaction, along with the photodissociation of ozone itself, plays the beneficial role of absorbing the vast majority of incoming ultraviolet radiation, which is harmful to most forms of life, before it reaches the Earth's surface. Levels of ozone in the stratosphere have been declining over the past several decades, especially over Antarctica. Scientific research has definitively shown that CFCs, along with a few other related man-made halogenated gases (see above), are responsible for these ozone losses in the stratosphere; thus, halogenated gases contribute to both global warming and stratospheric ozone depletion. The Montreal Protocol, which was originally signed in 1987 and has now been revised several times and ratified by 196 countries, has resulted in a rapid phase-out of these gases (see Figure 6.7). Recent evidence suggests that ozone levels in the stratosphere are starting to recover as a result, although it may be several more decades before the ozone layer recovers completely (CCSP, 2008a).

Near the Earth's surface, ozone is considered a pollutant, causing damage to plants and animals, including humans, and it is one of the main components of smog (see Chapter 11). Most surface ozone is formed primarily when sunlight strikes air that

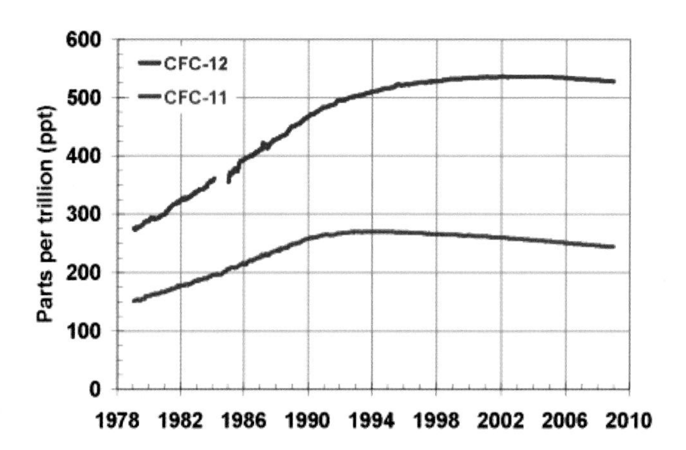

FIGURE 6.7 Atmospheric concentrations of the two halogenated gases with the largest individual climate forcings, CFC-11 and CFC-12, from 1979 to 2008. The Montreal Protocol limited the production of these and other compounds, and so their atmospheric concentrations are now slowly declining. SOURCE: NOAA/ESRL (2009).

contains nitrogen oxides (NO_x) in combination with carbon monoxide (CO) or certain volatile organic compounds (VOCs). All of these substances have natural sources, but their concentrations have increased as a result of human activities. Much of the NO_x and CO in the troposphere comes from man-made sources that involve burning, including automobile exhaust and power plants, while sources of VOCs include vegetation, automobiles, and certain industrial activities.

Ozone is also found in the upper troposphere, where its sources include local formation, horizontal and vertical mixing processes, and downward transport from the stratosphere. In general, tropospheric ozone levels show a lot of variability in both space and time, and there are only a few locations with long-term records, so it is difficult to estimate long-term ozone trends. Observational evidence to date shows increases in ozone in various parts of the world (e.g., Cooper et al., 2010). Models that include explicit representations of atmospheric chemistry and transport have also been used to estimate long-term ozone trends. These models, which are generally able to simulate observed ozone changes, indicate that tropospheric ozone levels have increased appreciably during the 20th century (Forster et al., 2007).

In addition to its role in near-surface air pollution and absorbing ultraviolet radiation in the stratosphere, ozone is a GHG, and so changes in its concentration yield a climate forcing. The losses of ozone in the stratosphere are estimated to yield a small negative forcing (cooling) of -0.05 ± 0.10 W/m^2, while increases in tropospheric ozone, which

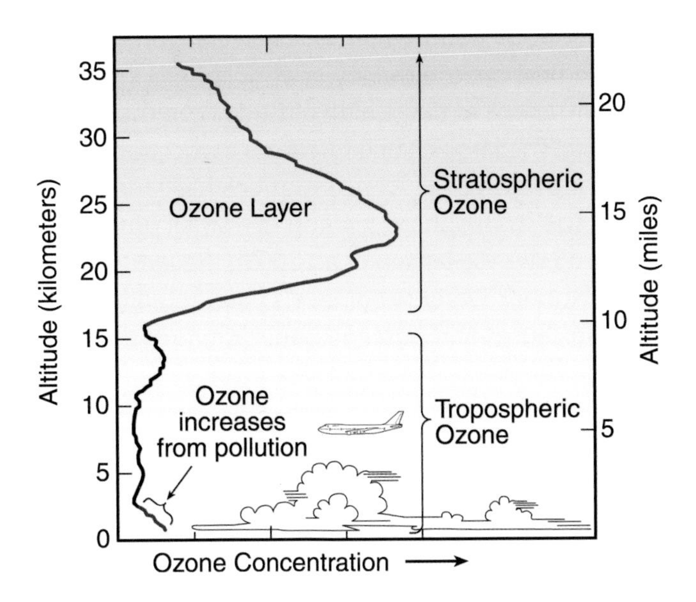

FIGURE 6.8 The vertical distribution of ozone with height, showing the protective layer of ultraviolet-absorbing ozone in the stratosphere, the harmful ozone (smog) near the Earth's surface, and the lesser—but still important—amounts of ozone in the upper troposphere. SOURCE: UNEP et al. (1994).

are comparatively larger, are estimated to yield a positive forcing of between 0.25 and 0.65 W/m^2, with a best estimate of 0.35 W/m^2 (Forster et al., 2007) (see Figure 6.4). Thus, in total, the changes in atmospheric ozone are responsible for a positive forcing that is on par with the halogenated gases and possibly as large as or slightly larger than the forcing associated with CH_4. However, the exact ozone forcing is more uncertain than for the well-mixed GHGs.

Water vapor (H_2O). Water vapor is technically the most abundant GHG and also the most important in terms of its contribution to the *natural* greenhouse effect (see Figure 2.1). A number of human activities (primarily agricultural irrigation but also through cooling towers, aircraft exhaust, and other sources) can influence local water vapor levels. However, on a global basis the concentration of water vapor in the lower atmosphere is controlled by the rate of evaporation and precipitation, which are processes that occur on a relatively fast time scale and are much more strongly influenced by changes in atmospheric temperature and circulation than by human activities directly. Thus, water vapor is usually considered to be part of the climate system—and indeed, it is involved in a number of important climate feedback processes, as described below—rather than a climate forcing agent.

In the stratosphere, on the other hand, water vapor is relatively rare and somewhat isolated from the hydrological cycle in the lower atmosphere. Processes that influence water vapor concentrations at these high altitudes can thus lead to a small but discernible climate forcing. The largest such forcing is associated with the oxidation of CH_4 into water vapor and CO_2: as CH_4 concentrations have increased, so has this source of water vapor in the stratosphere, leading to a small positive climate forcing estimated to be 0.05 ± 0.05 W/m^2 (Hansen et al., 2005).[5] Recent satellite-based observations reveal that stratospheric water vapor levels have actually declined since 2000 (Solomon et al., 2010); the causes and possible implications of this decline are still being studied.

Other Climate Forcing Agents

Aerosols. Small liquid or solid particles suspended in the atmosphere—aerosols—can be composed of many different chemicals, come from many different sources (including both natural sources and human activities), and have a wide range of effects. Fossil fuel burning, industrial activities, land use change, and other human activities have generally increased the number of aerosol particles in the atmosphere, especially over and downwind of industrialized counties. The net climate forcing associated with aerosols is estimated to be -1.2 W/m^2 (Forster et al., 2007; see also Murphy et al., 2009), which offsets roughly one-third of the total positive forcing associated with human emissions of GHGs (see Figure 6.4). However, the forcing associated with aerosols is more uncertain than the forcing associated with GHGs, in part because the global distribution and composition of aerosols are not very well known and in part because of the diversity and complexity of aerosol radiative effects.

Two separate types of effects contribute to the net cooling associated with aerosols: (1) a "direct effect," which occurs because most aerosols scatter a portion of the incoming sunlight that strikes them back to space, and (2) "indirect effects," which arise because aerosols play an important role in the formation and properties of cloud droplets, and on average the increasing number of aerosols have caused clouds to reflect more sunlight back to space. Certain kinds of aerosols, including dust particles

[5] Exhaust from jet aircraft also adds water vapor to the stratosphere, which can both directly contribute to the greenhouse effect and also form linear contrails, which tend to warm the Earth slightly. While contrails were once thought to potentially contribute a significant climate forcing, more recent estimates—including some based on measurements taken during the days following the September 11 attacks, when air travel over North America was sharply curtailed—show that aircraft exhaust has only a small effect on climate forcing, although contrails do appear to have a discernible effect on regional day-night temperature differences (Travis et al., 2002).

and black carbon (soot), absorb both incoming solar energy and the outgoing infrared energy emitted by the Earth. These aerosols tend to warm the atmosphere, offsetting some (but not all) of the cooling associated with the direct and indirect effects. Black carbon particles that settle on snow and ice surfaces can also accelerate melting; however, this positive forcing is typically included in estimates of the forcing associated with land use change, which is discussed below.

It is worth noting the sources of a few key types of aerosols to illustrate their diversity: Dust and some organic aerosols arise from natural processes, but some human activities such as land use change also lead to changes in the abundance of these species. Black carbon particles are produced from the burning of both fossil fuels and vegetation. Sulfate (SO_4) aerosols—which are a major contributor to the aerosol direct and indirect effects—have three notable sources: fossil fuel burning, marine phytoplankton, and volcanoes. The composition and size of each of these aerosol species affect how they absorb or scatter radiation, how much water vapor they absorb, how effectively they act to form cloud droplets, and how long they reside in the atmosphere—although in general most aerosols only remain in the atmosphere for a few weeks on average.

In addition to their role in global climate forcing, aerosols also have a number of other important environmental effects. The same industrial emissions that give rise to SO_4 aerosols also contribute to acid rain, which has a major detrimental effect on certain ecosystems. One of the major objectives—and successes—of the 1990 Clean Air Act (P.L. 101-549) was to reduce the amount of sulfur emissions in the United States. Similar laws in Europe have also been successful in reducing SO_4 aerosol concentrations (Saltman et al., 2005). The relationship between aerosols and cloud formation also means that changes in aerosols play an important role in modulating precipitation processes (see Chapters 8 and 15). Also, many aerosols are associated with negative impacts on public health, as discussed in further detail in Chapter 11.

Finally, aerosol emissions represent an important dilemma facing policy makers trying to limit the magnitude of future climate change: If aerosol emissions are reduced for health reasons, or as a result of actions taken to reduce GHG emissions, the net negative climate forcing associated with aerosols would decline much more rapidly than the positive forcing associated with GHGs due to the much shorter atmospheric lifetime of aerosols, and this could potentially lead to a rapid acceleration of global warming (see, e.g., Arneth et al., 2009). Understanding the many and diverse effects of aerosols is also important for helping policy makers evaluate proposals to artificially increase the amount of aerosols in the stratosphere in an attempt to offset global warming (see Chapter 15).

Changes in land cover and land use. Human modifications of the land surface can have a strong local or even regional effect on climate. One notable example is the "urban heat island" effect on temperatures, described below and in Chapter 12. Globally, land cover and land use changes are important sources of several GHGs, such as the release of CO_2 from deforestation or CH_4 from rice paddies. Land use and land cover change can also yield a global climate forcing by altering the reflectivity of the Earth's surface—for example, by replacing forests (which absorb most incident sunlight) with cropland (which is generally somewhat more reflective). Satellite measurements provide an excellent record of how changes in land cover have influenced surface reflectivity over the last few decades, although in some cases there is uncertainty as to whether observed changes are directly human-induced, part of a feedback process, or attributable to natural changes. To estimate global patterns of land use change for the last several hundred years, scientists use historical and paleoecological records combined with land use models that can simulate changes in vegetation over time in response to both climatic and nonclimatic effects.

Most recent published estimates of the global climate forcing associated with land use and land cover change are in the range of −0.1 to −0.3 W/m^2, although some estimates are as large as −0.5 W/m^2, while others indicate a small positive net forcing (Forster et al., 2007). As noted above, an additional land-surface effect is the deposition of black carbon aerosols (soot) on white snow and ice surfaces, which leads to melting and has been estimated to yield a positive forcing of up to 0.2 W/m^2, although more recent estimates have suggested a somewhat smaller warming effect (Hansen et al., 2005). Thus, the total climate forcing associated with modifications to the land surface due to human activities since 1750 could potentially be positive or negative, but the balance of evidence seems to suggest a slight cooling effect.

Changes in solar radiation. As discussed in the next section, even small variations in the amount or distribution of energy received from the sun can have a major influence on Earth's climate when they persist for many thousands of years. However, satellite measurements of solar output show no net increase in solar forcing over the last 30 years, only small periodic variations associated with the 11-year solar cycle (Figure 6.9). Changes in solar activity prior to the satellite era are estimated based on a variety of techniques including observations of sunspot numbers, which correspond roughly with solar output (Figure 6.10). The available evidence suggest that solar activity has been roughly constant (aside from the 11-year solar cycle) since the mid-20th century but that it increased slightly during the late 19th and early 20th centuries. The total solar forcing since 1750 is estimated to be less than 0.3 W/m^2 (Forster et al., 2007).

Cosmic rays. Finally, it has been proposed that cosmic rays might influence Earth's cli-

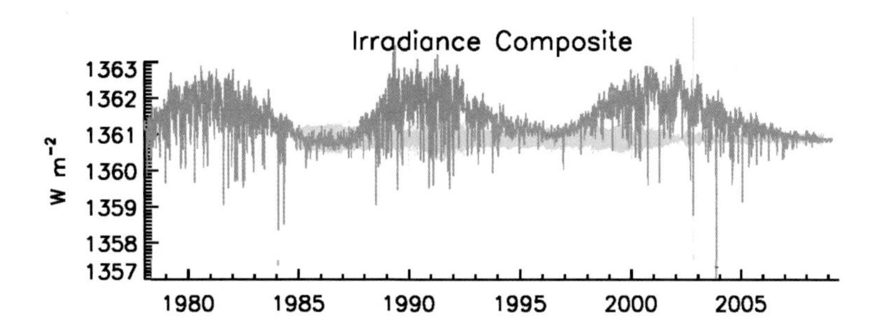

FIGURE 6.9 Solar irradiance observed at the top of the Earth's atmosphere by satellites. There is no overall trend in irradiance since 1979, but the ~11-year solar cycle produces small variations in irradiance of roughly 1.5 W/m^2. Due to the geometry of the Earth and the reflection of some of the incoming sunlight back to space, this 1.5 W/m^2 variation in irradiance corresponds to a periodic oscillation in climate "forcing" of around 0.3 W/m^2 (although climate forcing is usually defined as the overall change in forcing since 1750). SOURCE: Lean and Woods (in press).

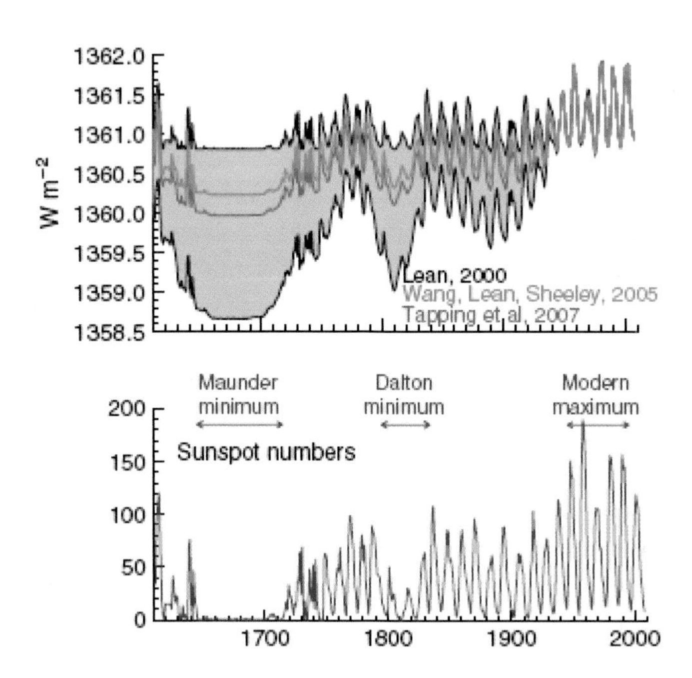

FIGURE 6.10 Estimated variations in solar irradiance at the top of the atmosphere by three different research teams during (top) the last 400 years based on (bottom) observations of sunspot numbers. All three irradiance reconstructions indicate drops in solar output during extended periods with low sunspot numbers, especially the Maunder and Dalton minimums (which are indicated in the bottom panel), and an increase in solar irradiance during the first several decades of the 20th century. The estimated total climate forcing associated with changes in solar irradiance since 1750 is 0.3 W/m^2. (As noted in the caption for Figure 6.9, the climate forcing associated with solar irradiance changes must be scaled to account for Earth's geometry and the reflection of some of the incident solar energy back to space.) SOURCE: Lean and Woods (in press).

mate by modifying cloud properties (Shaviv, 2002; Svensmark, 1998, 2006) or through a variety of other mechanisms (Gray et al., 2005). Cosmic rays are influenced by solar activity, so it is difficult to study the effect of cosmic rays in isolation. However, direct observations of cosmic ray fluxes do not show any net change over the last several decades (Benestad, 2005), and a plausible physical mechanism linking changes in cosmic rays to changes in climate has not been demonstrated. Hence, cosmic rays are not regarded as an important climate forcing (Forster et al., 2007).

Climate Feedbacks and Sensitivity

The influence of climate forcings on Earth's temperature is modulated by the effects of *feedbacks* in the climate system. One example of a positive feedback is the ice-reflectivity feedback: If a positive climate forcing leads to a slight warming that melts ice, especially (white, highly reflective) sea ice floating on the (dark, highly absorptive) ocean surface, the surface of the Earth will reflect less sunlight back to space, and the increased absorption of solar radiation reinforces the initial warming. On the other hand, if warming were to cause an increase in the amount of low-lying clouds, which tend to cool the Earth by reflecting solar radiation back to space (especially when they occur over ocean areas), this would tend to offset some of the initial warming—a negative feedback. Other important feedbacks involve changes in evaporation, other kinds of clouds, land-surface properties, the vertical profile of temperature in the atmosphere, and the circulation of the atmosphere and oceans—all of which operate on different time scales and interact with one another and with other environmental changes in addition to responding directly to changes in temperature.

The net effect of all feedback processes determines the *sensitivity* of the climate system, or the response of the system to a given set of forcings (NRC, 2003b). Climate sensitivity is typically expressed as the temperature change expected if atmospheric CO_2 levels were fixed at twice their preindustrial concentration, with all other forcings neglected (or 560 ppm of CO_2, which corresponds to a climate forcing of 3.7 W/m^2), and then remained there until the climate system reaches equilibrium. A variety of methods have been used to estimate climate sensitivity, including statistical analysis of climate forcing and observed temperature changes, analyses based on estimates of forcing and temperature variations from paleoclimatic records (see below), energy balance models, and climate models of varying complexity (e.g., Annan et al., 2005; Hegerl et al., 2006; Knutti et al., 2006; Murphy et al., 2004; Wigley et al., 2005). The IPCC's latest comprehensive assessment of climate sensitivity based on these techniques indicates that the expected warming due to a doubling of CO_2 is between 3.6°F and 8.1°F (2.0°C and 4.5°C), with a best estimate of 5.4°F (3.0°C) (Hegerl et al., 2007). Unfortunately, the diversity and complexity of processes operating in the climate system mean that, even with continued progress in understanding climate feedbacks and monitoring global climate forcing and temperature changes, the exact sensitivity of the climate system may remain uncertain (Roe and Baker, 2007).

The concept of climate sensitivity technically only applies to equilibrium climate states, that is, the total warming after the oceans, cryosphere, and biosphere have had ample time to fully adjust to the imposed forcing. In reality, the strength of climate forcings and feedbacks are continuously varying, and it takes the climate system—especially the oceans—a long time to warm up in response to a positive climate forcing. In addition, many estimates of climate sensitivity do not include climate feedbacks associated with processes that operate on decadal to centennial time scales, such as the disappearance of glaciers, changes in vegetation distribution, or changes to the carbon cycle on land and in the oceans; several recent studies that consider some of these processes have suggested that Earth's climate sensitivity may be substantially higher than the aforementioned "best estimate" (Hansen et al., 2008; Sokolov et al., 2009). Nevertheless, estimates of climate sensitivity are a useful metric for evaluating the causes of observed climate change and estimating how much the Earth will ultimately warm in response to past, present, and future human activities. Climate feedbacks and climate sensitivity also remain an important area for future research (see Research Needs at the end of this chapter).

OBSERVED CLIMATE CHANGE

Natural Climate Variability

Earth's climate varies naturally on a wide range of time scales. Many of these variations are caused by complex interactions between the fast-moving, less-dense atmosphere and the more massive, slower-to-respond oceans. For example, the El Niño-Southern Oscillation (ENSO), which is caused by ocean-atmosphere interactions in the tropical Pacific Ocean, is a source of significant year-to-year variability around the world. The "warm" or "El Niño" phase is characterized by warmer-than-normal sea surface temperatures in the eastern equatorial Pacific. El Niño years are often associated with significant, predictable regional variations in temperature and rainfall across many remote parts of the world; in the United States, for example, El Niño years typically exhibit wetter-than-normal conditions in Southern California and the southern Great Plains. Global temperatures also tend to be slightly warmer during years with strong El Niño events, such as 1998, and slightly cooler during "cool" or "La Niña" years, such as 2008.

A multitude of other patterns of natural climate variability have also been identified, and many of these are associated with strong regional climate variations. Higher-latitude oscillations, such as the Northern and Southern Annular Modes (Thompson and Wallace, 2000, 2001), the Pacific Decadal Oscillation (Guan and Nigam, 2008; Mantua et al., 1997), the North Atlantic Oscillation, and the Atlantic Multidecadal Oscillation (Guan and Nigam, 2009), have a large influence on regional climate at decadal time scales, with impacts on, for example, salmon fisheries in the Pacific Northwest (Hare et al., 1999; Mantua and Hare, 2002) and the number of hurricanes making landfall in North America (Dailey et al., 2009). The exceptionally cold and snowy winter experienced on the East Coast of the United States during 2009-2010, which was balanced by warmer-than-normal temperatures in much of northeastern Canada and the high Arctic, can be attributed in part to a strong North Atlantic Oscillation event. Natural climate oscillations on multidecadal and longer time scales could also exist (e.g., Enfield et al., 2001; Schlesinger and Ramankutty, 1994), though the instrumental record is too short and too sparse to unambiguously attribute their causal mechanisms (e.g., Zhang et al., 2007a).

Climate variations can also be forced by natural processes including volcanic eruptions, changes in the output from the sun, and changes in Earth's orbit around the sun. Large, explosive volcanic eruptions, like Tambora in 1815, Krakatoa in 1883, El Chichón in 1983, and Pinatubo in 1991, spew copious amounts of sulfate aerosols into the stratosphere, cooling the Earth for several years (Briffa et al., 1998). The Pinatubo eruption is particularly noteworthy because it occurred in an era with widespread satellite and ground-based observations that allowed for the resulting aerosol distribution and climate response to be accurately quantified. These data indicate that aerosols induced a peak climate forcing of -2.5 W/m^2 several months after the Pinatubo eruption (Harries and Futyan, 2006) and that global surface temperatures dipped approximately 0.9°F (0.5°C) 2 years later, then recovered over the next several years as aerosol levels gradually declined (Trenberth and Dai, 2007). Data from Pinatubo and other volcanic eruptions have been used to estimate the strength of climate feedbacks that operate on relatively short time scales, such as the feedback associated with the correlation between temperature and water vapor in the atmosphere, and for calibrating and validating climate model results (e.g., Soden et al., 2002).

While there has not been a net increase in the Sun's energy output over the past few decades (see Figure 6.9), the small variations in solar output associated with the 11-year solar cycle do lead to temperature and circulation change in the upper atmosphere (Shindell et al., 1999), may affect weather patterns in the tropical Pacific (Meehl et al., 2009a), and could potentially be associated with small variations in Earth's average surface temperature (Camp and Tung, 2007; Lean and Woods, in press). There is also evidence that changes in solar activity influence Earth's climate on longer time scales. For example, the "Little Ice Age" (Matthes, 1939), a period with slightly cooler temperatures between the 17th and 19th centuries, may have been caused in part by a low solar activity phase from 1645 to 1715 called the Maunder Minimum (Eddy, 1976; Shindell et al., 2001) (see Figure 6.10). Estimates of variations in solar output on even longer time scales—going back thousands of years—have also been produced by analyzing cosmogenic isotopes in tree rings and ice cores (e.g., Weber et al., 2004). However, these estimates, and hence the extent of solar influence on global climate on these time scales, are even more uncertain (Lean and Woods, in press).

Perhaps the most dramatic example of natural climate variability is the Ice Age cycle (Figure 6.11). Detailed analyses of ocean sediments, ice cores, and other data (see, e.g.,

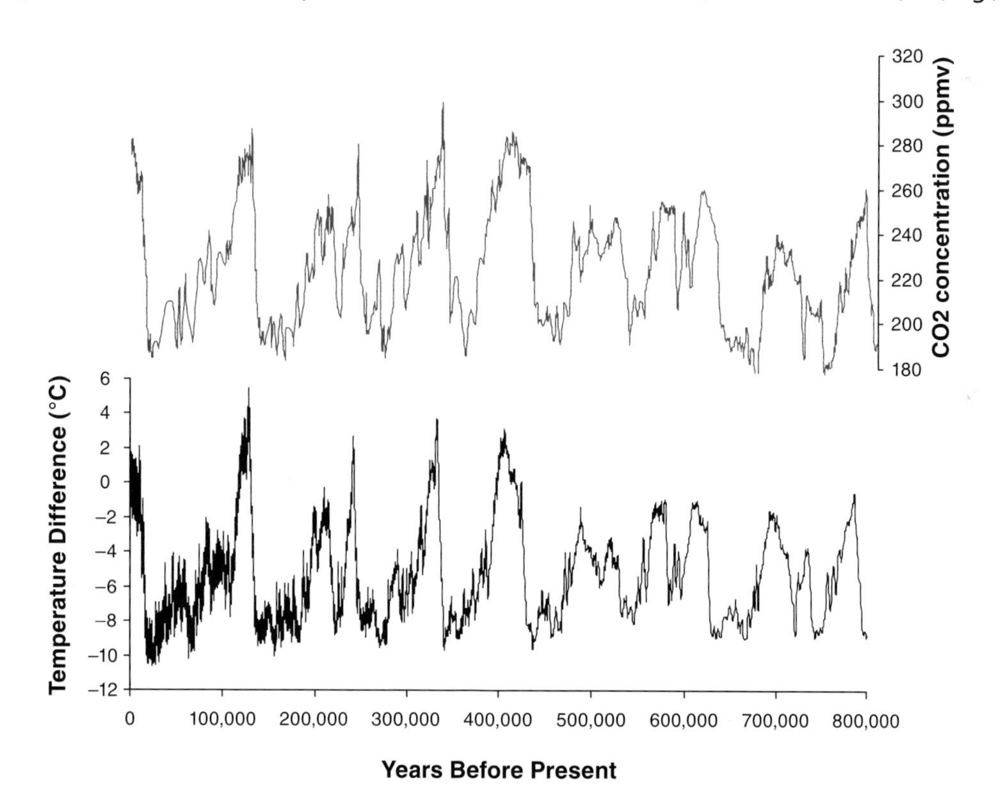

FIGURE 6.11 Analysis of ice core data extending back 800,000 years documents (top) the Earth's changing CO_2 concentration and (bottom) estimated temperatures in the Antarctic region. Until the past century, natural factors caused atmospheric CO_2 concentrations to vary within a range of about 180 to 300 ppm. Note that time progresses from right to left in this figure, and that neither temperature changes nor the rapid CO_2 rise (to 388 ppm) over the past century are shown. SOURCES: Based on data from (top) Lüthi et al. (2008) and (bottom) Jouzel et al. (2007). Data available at *http://www.ncdc.noaa.gov/paleo/icecore/antarctica/domec/domec_epica_data.html.*

Lüthi et al., 2008) show that for at least the past 800,000 years, and probably the past several million years, the Earth has gone through long periods when temperatures were much colder than today and thick blankets of ice covered much of the Northern Hemisphere (including Chicago, New York, and Seattle). These very long cold spells were punctuated by shorter "interglacial" periods—including the last 10,000 years, during which time the climate appears to have been relatively stable.

Through a convergence of theory, observations, and modeling, scientists have deduced that the ice ages were initiated by small recurring variations in Earth's orbit around the sun, which modulated the magnitude and seasonality of sunlight received at the Earth's surface in a persistent way. Over many thousands of years, these relatively small changes in solar forcing resulted in gradual changes in and feedbacks between the cryosphere and biosphere that slowly but persistently changed the abundance of GHGs in the atmosphere, reinforcing the changes in solar forcing and ultimately driving a global temperature change on the order of 9°F ± 2°F (5°C ± 1°C) between glacial and interglacial periods (EPICA Community Members, 2004; Jansen et al., 2007). Because GHGs acted as a feedback rather than as a forcing during the Ice Age cycles, temporal variations in GHGs typically lag, rather than lead, the estimated temperature changes in Figure 6.11.

Human-Induced Climate Change

Surface Temperature Measurements

Widespread thermometer measurements of sufficient accuracy to reliably estimate large-scale changes in near-surface air temperature over land areas did not become available until the mid-19th century, and routine measurements of ocean temperatures did not become available until the late 19th century. In addition to missing data and individual measurement errors, there are a variety of artificial biases present in long-term temperature records that must be removed to yield records of sufficient accuracy to evaluate climate trends. Equipment and measurement procedures have changed over time—for example, ocean temperatures have been measured by satellites, buoys, and ships, and the ship-based measurements have included readings taken by hull sensors, in water drawn in to cool the engines, and in buckets pulled up by hand from the water surface. Temperature measurements are also not evenly distributed in space or time; observing stations are common in densely populated land areas, while the southern oceans were only sparsely observed before satellite measurements became available in the late 1970s. Finally, temperature measurements can be affected by a number of local factors, such as the "urban heat island" effect (see

Chapter 12) and other changes in land use; although these changes represent real changes in local climate, they need to be quantified and corrected when evaluating large-scale changes in climate.

Several research groups around the world, including NASA's Goddard Institute for Space Studies (GISS), NOAA's National Climatic Data Center, and the Climate Research Unit at the University of East Anglia in the United Kingdom, collect and maintain databases of both historical and present-day meteorological data and use them to produce estimates of regional and global climate change. Producing these estimates requires each individual record to be quality-controlled and corrected to remove the artificial biases described above, and then additional steps are needed to convert the assemblage of individual records into representative large-scale averages. Each group uses somewhat different data sources and analysis procedures (see, e.g., Hansen et al., 1999, 2001; Karl and Williams, 1987; Menne and Williams, 2009; Menne et al., 2009). Most of these data and methods are publicly available.

Each of the research teams that produce large-scale temperature estimates has developed methods for dealing with the potential biases and sources of error such as those described in the preceding paragraph. For example, NASA GISS uses a linear interpolation procedure to "fill in" missing data and temperatures in areas between observing stations, and data from urban stations (which are identified based on either population density data or "nightlight" levels observed by satellite) are adjusted so their long-term trends match those of neighboring rural stations. The University of East Anglia instead corrects the station-level data first and then uses a simple averaging procedure to combine the data. These procedures have been developed over several decades (e.g., Hansen and Lebedev, 1987) and are constantly reevaluated to identify and correct for additional sources of error. It was recently determined, for example, that a change in the way that certain ship-based temperatures were treated introduced a spurious signature into the mid-20th-century temperature record including an abrupt drop of ~0.5°F (0.3°C) in 1945 (Thompson et al., 2008).

For the GISS data, the uncertainties associated with corrections to the raw data and with the underlying raw data themselves are estimated to yield a total uncertainty in global-average surface temperature estimates of about 0.09°F (0.05°C) during the past several decades. During the first few decades of the record, the estimated uncertainty is twice as large (0.18°F or 0.10°C), as might be expected due to the smaller number of measurements and their lower precision relative to modern instruments (Hansen et al., 2006; see also Thompson et al., 2009). Global temperature estimates produced by other research teams yield results that agree within these estimated uncertainties. Changes in temperature, or other climate variables, are typically reported as anomalies (differences) relative to a specified time period because this minimizes errors associated with calibration to absolute temperature.

Surface Temperature Changes

Global surface temperature records indicate that the Earth has warmed substantially over the past century (Figure 6.12). For example, the first decade of the 21st century (2000-2009) was 1.4°F (0.77°C) warmer than the first decade of the 20th century (1900-1909). This warming has not been uniform but rather is superimposed on substantial year-to-year and decadal-scale variability (see Box 6.1), with the most pronounced warming occurring during the last 30 years. Several hypotheses have been put for-

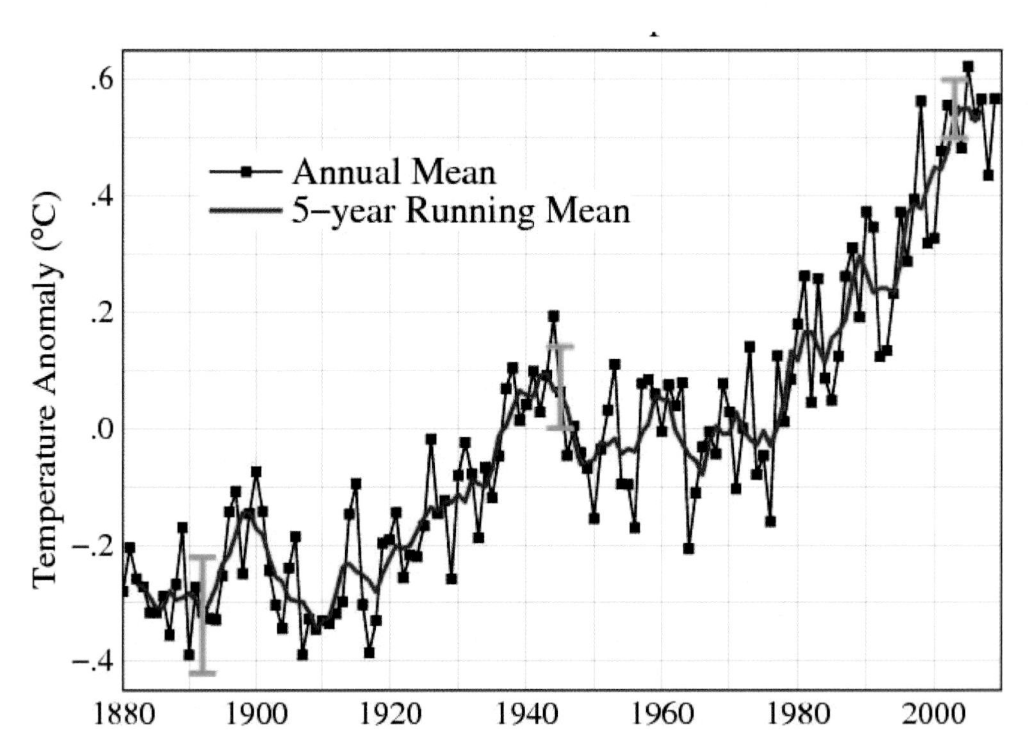

FIGURE 6.12 Global surface temperature (near-surface air temperature over land and sea surface temperatures over ocean areas) change for 1880-2009, reported as anomalies relative to a reference period of 1951-1980, as estimated by NASA GISS (estimates produced by other research teams are very similar). The black curve shows annual average temperatures, the red curve shows a 5-year running average, and the green bars indicate the estimated uncertainty in the data during different periods of the record. SOURCES: NASA GISS (2010; Hansen et al., 2006, updated through 2009; data available at *http://data.giss.nasa.gov/gistemp/graphs/*).

ward to explain the substantial decadal-scale variability in the surface temperature record, especially the period of relatively flat temperatures from the early 1940s through the late 1970s. Probably the most widely cited hypothesis, which is supported by some statistical analyses and model simulations, is that increasing levels of sulfate aerosols from fossil fuel combustion introduced a cooling effect that offset much of the positive forcing from GHGs during the "flat" part of the record (e.g., Hegerl et al., 2007). This hypothesis seems to be supported by the more pronounced "flattening" in the Northern Hemisphere, relative to the more steady increase in the Southern Hemisphere (where aerosol levels are generally much lower). However, other recent analyses (e.g., Swanson et al., 2009) suggest that natural variations in ocean circulation might also give rise to some of the decadal-scale variations in the global temperature record.

The observed warming is also unevenly distributed around the planet (Figure 6.13). In general, the largest increases in temperature worldwide have occurred over land areas and over the Arctic, which is consistent with the horizontal pattern of warming expected from a positive climate forcing. In the continental United States, on average temperatures rose by 1.5°F (0.81°C) between the first decade of the 20th century and the first decade of the 21st century, or about the same as the global temperature change over this period. There is also a rich tableau of ongoing regional, seasonal, diurnal, and local temperature changes associated with these large-scale, long-term, annual-mean surface warming trends:

- Recent analyses of temperature trends over the Midwest and northern Great Plains have revealed that winter temperatures in that region have increased by 7°F (4°C) over the past 30 years (USGCRP, 2009a).
- Late spring and early summer daytime maximum temperatures in the southeastern United States, on the other hand, declined slightly from the 1950s to the mid-1990s (Portmann et al., 2009).
- An analysis of daily temperature records reveals that during the last decade nearly twice as many extreme record high temperatures have been recorded globally than extreme record low temperatures (Meehl et al., 2009c).
- Hot days and nights have become warmer and more common, while cold days and nights have become warmer and fewer in number (IPCC, 2007a).

Many of these changes are consistent with the spatial and temporal patterns of temperature change expected to result from increasing GHG concentrations.

BOX 6.1
Short-Term Variability Versus Long-Term Trends

When conducting scientific analyses, it is important to analyze data in a manner that is consistent with the phenomenon being studied. Climate, for example, is typically defined based on 30-year averages (Burroughs, 2003; Guttman, 1989). This averaging period is chosen, in part, to minimize the influence of natural variability on shorter time scales and facilitate the analysis of long-term trends, especially trends associated with long-term changes in the Earth's radiative balance. Individual years, or even individual decades, can deviate from the long-term trend due to natural climate variability. Thus, it is not appropriate to look at only a short period of the overall record (such as changes over just the last 5 or 10 years) to infer major changes in the trajectory of global warming.

An example of a more familiar temperature trend—one associated with the seasonal cycle—illustrates the importance of analyzing trends over appropriate time scales. The figure below shows daily average temperatures for New York City for the period of January 1 through July 1, 2009. Temperatures would obviously be expected to increase on average over this 6-month period due to the seasonal cycle, but natural variability (which in this case is largely due to the passage of individual weather systems) also gives rise to significant daily, weekly, and even monthly fluctuations in these data. For example, on February 12, 2009, temperatures reached 51°F and then generally declined to 20°F on March 3 (red arrows). Similarly, temperatures reached 78°F on two days in late April before generally declining to 61°F in mid-June (green arrows). It would be incorrect to conclude that summer was not coming based on these two subsets of the data.

In a similar manner, one could potentially draw erroneous conclusions about the long-term trend in global surface temperature by focusing exclusively on a subset of the data in the figure—such as data from just the last 10 or 12 years (see also Easterling and Wehner, 2009; Fawcett, 2007; Knight et al., 2009). As discussed in the text, the climate system exhibits substantial year-to-year and even decade-to-decade variability, while global temperature increases due to rising GHG increases, and other radiative forcing factors all operate on longer time scales. Robust analyses of global climate change thus tend to focus on trends over at least several decades.[a] Scientists often average climate data over several years or decades, or use more sophisticated statistical methods, to make long-term trends more readily apparent. Statistical methods can also be used to identify other important climate patterns and trends, such as changes in extreme events or shifts in modes of natural variability.

Atmospheric Temperatures

In addition to surface-based thermometer measurements, regular and widespread measurements of the vertical profile of atmospheric temperatures are available from both satellites and weather balloons for the last several decades. Weather balloons, which are launched twice per day from over 800 sites around the world, carry instruments known as radiosondes that directly measure atmospheric conditions and radio these data back to receiving stations. Although these measurements are taken primar-

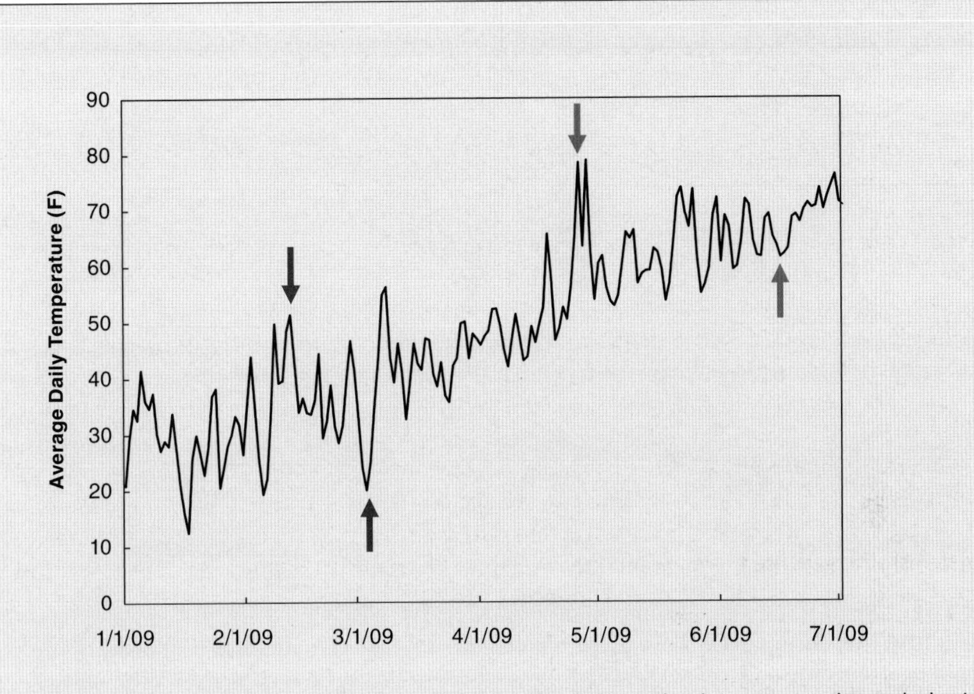

New York City daily average temperature for the first 6 months of 2009. Red and green arrows denote the beginning and end of two periods when temperatures declined on average, despite an overall warming trend due to the seasonal cycle. SOURCE: NCDC (2006).

---

[a] It should be noted, however, that there are some aspects of the climate system—such as global sea level rise due to the slow thermal expansion of the oceans (see Chapter 7)—that naturally tend to reflect longer-term changes in radiative forcing, and that short-term (e.g., decadal-scale) trends are important for identifying and studying the potential for "abrupt" climate changes, which are discussed later in the chapter.

ily to support weather prediction, researchers have developed methods for aggregating the data, removing a variety of systematic biases (including changes in instrumentation, the fact that all of the balloons are launched at the same two times each day, which means they are launched at different local times, and a recently identified bias associated with the sun heating the instruments) to yield a record of three-dimensional changes in atmospheric temperature over the last 50 years (McCarthy et al., 2008).

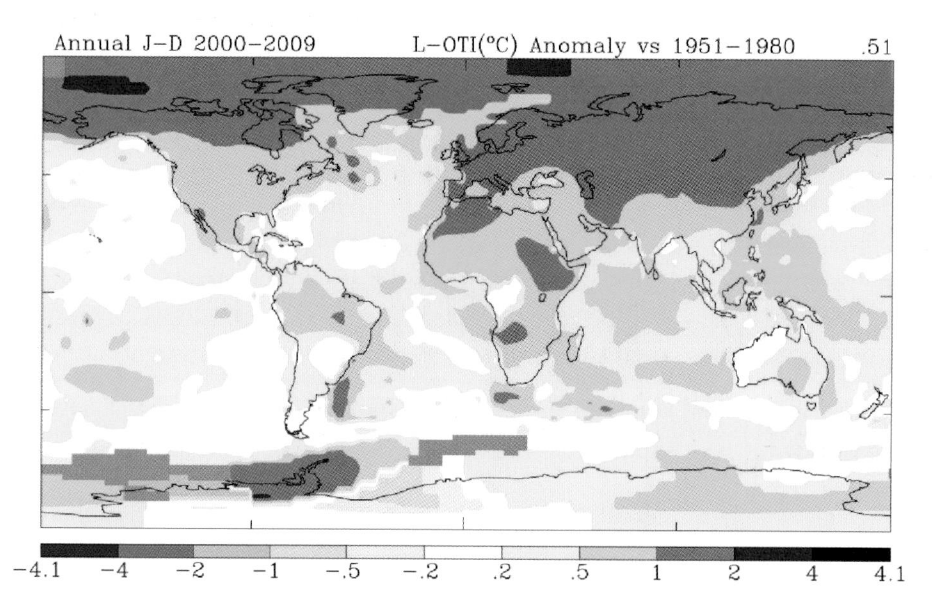

FIGURE 6.13 Average surface temperature trends (degrees per decade) for the decade 2000-2009 relative to the 1950-1979 average. Warming was more pronounced at high latitudes, especially in the Northern Hemisphere, and over land areas. SOURCES: NASA GISS (2010; Hansen et al., 2006, with 2009 update).

Regular satellite-based observations of temperature and other atmospheric properties began in the late 1970s. Rather than directly sampling atmospheric conditions, satellites measure the upwelling radiation from the Earth at specific wavelengths, and this information can be used to infer the average temperature of different layers in the atmosphere underneath. As with surface temperature records, the raw satellite data are analyzed by several different research teams, each using its own techniques and assumptions, to produce estimates of inferred temperature changes (Christy et al., 2000, 2003; Mears and Wentz, 2005). While satellite-derived data offer the advantage of excellent global coverage, they still require corrections to remove artificial biases, such as the slow decay of satellite orbits and changes in instrumentation when satellites are replaced. The fact that satellite-inferred temperatures represent layers of the atmosphere rather than specific points in space also leads to some uncertainties in the analysis and interpretation of the data—for example, it was recently demonstrated that previous estimates of lower-atmosphere warming from satellites were biased slightly downward due to the inclusion of some data from the stratosphere, which has cooled (see next paragraph; also Fu et al., 2004). As discussed in Chapter 4 and in some of the other chapters in Part II, satellite data also offer a wealth of information about other changes in the Earth system.

Radiosonde and satellite-derived data both show that the troposphere (the lowest layer of the atmosphere, extending up to roughly 10 miles [16 km] in the tropics and 6 miles [10 km] near the poles) has warmed substantially over the past several decades (Figure 6.14). The most recent analyses of satellite data from 1979 through the end of 2009 estimate a tropospheric warming of +0.23°F (+0.13°C) per decade (Christy et al., 2000, 2003) to +0.28°F (+0.15°C) per decade (Mears and Wentz, 2005; RSS, 2009), while radiosonde-derived temperature estimates yield +0.30°F (+0.17°C) per decade for the same time period and +0.29°F (+0.16°C) for the full radiosonde record starting in 1948 (HadAT2; McCarthy et al., 2008). For comparison, surface temperatures increased +0.29°F (+0.16°C) per decade since 1979 and +0.23°F (0.13°C) per decade since 1948.

Additionally, radiosondes and satellites both indicate that the stratosphere has cooled even more strongly than the troposphere has warmed (Figure 6.14, top panel). This

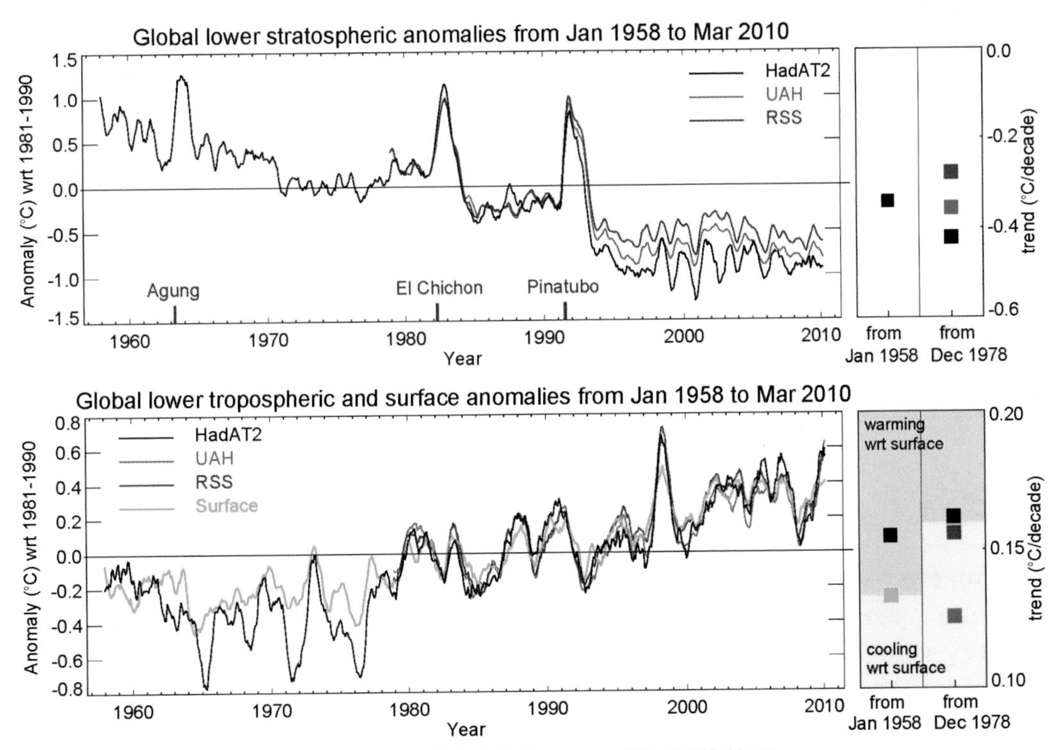

FIGURE 6.14 Radiosonde- (black) and satellite-based (blue and red) estimates of temperature anomalies for 1958-2009 in the (top) stratosphere and (bottom) troposphere. The squares on the right-hand side of the figure indicate the trends in each data series from two different start dates. SOURCE: Hadley Center (data available at *http://hadobs.metoffice.com/hadat/images.html*).

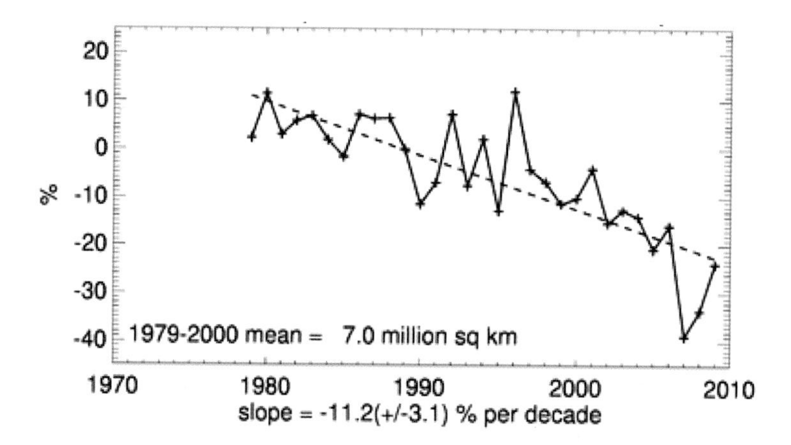

FIGURE 6.15 Satellite-based trend of September (end of summer) Arctic sea ice extent for the period 1979 to 2009, expressed as percentage difference from 1979-2000 average sea ice extent (which was 7.0 million square miles). These data show substantial year-to-year variability, but a long-term decline in sea ice extent is clearly evident, as highlighted by the dashed linear trend line. As discussed in the text, the average thickness of Arctic sea ice has also declined markedly over the last 50 years. SOURCE: NSIDC (2010).

vertical pattern of temperature change, with warming in the troposphere and cooling in the upper atmosphere, is consistent with the pattern expected due to increasing GHG concentrations (Roble and Dickinson, 1989). Current research on temperature trends focuses on, among other issues, regional, seasonal, and day-night differences in temperature trends, especially in the tropics, where climate models predict a stronger warming in the upper troposphere than has been observed to date (e.g., Fu and Johanson, 2005).

Other Indicators of Climate Change

Additional direct indicators of a warming trend over the last several decades can be found in the cryosphere and oceans. As discussed in detail in Chapter 7, the vast majority of the total heating associated with human-caused GHG emissions has actually gone into the world's oceans, which have warmed substantially over the last several decades (Levitus et al., 2009). In the cryosphere, mountain glaciers and ice-caps are melting (these changes are also discussed in detail in Chapter 7), rivers and lakes are thawing earlier and freezing later in the year (Rosenzweig et al., 2007), and winter snow cover (Trenberth et al., 2007) and summer sea ice (Figure 6.15) are both decreasing in the Northern Hemisphere. Analyses of recently declassified data from naval submarines (as well as more recent data from satellites) show that the average thickness of sea ice in the Arctic Ocean has declined substantially over the past half-century, which is yet another indicator of a long-term warming trend (Kwok and Rothrock, 2009). Warming can also be inferred from a host of ecosystem changes: flowers are blooming earlier, bird migration and nesting dates are shifting, and the ranges of many insect and plant species are expanding poleward and to higher elevations—these and other trends in biological systems are discussed in detail in Chapter 9.

Scientists have collected a wide array of indirect evidence of how temperature and other climate properties varied before instrumental measurements became available. These so-called "proxy" climate data are derived from a diverse range of sources including ice cores, tree rings, corals, lake sediments, records of glacier length, borehole temperature measurements, and even historical documents. A recent assessment of these data and the techniques used to analyze them (NRC, 2006b) concluded that, although proxy data generally become scarcer, less consistent, and more uncertain going back in time, temperatures during the past few decades were warmer than during any other comparable period for at least the last 400 years, and possibly for the last 1,000 years or longer (Figure 6.16). Proxy-based temperature and forcing estimates for the past millennium, and for longer time periods such as the Ice Age cycles described above, illustrate the natural variability of the climate system on a wide range of time scales. These estimates are also used to help constrain estimates of climate sensitivity.

While temperature and temperature-related changes are the most widely cited and typically the best-understood changes in the physical climate system, a host of concomitant and related changes have also been observed. For example, the absorption of CO_2 by the oceans is causing widespread ocean acidification, with significant implications for natural ecosystems and fisheries (as discussed in Chapters 9 and 10, respectively). There have also been significant changes in the overall amount, patterns, and timing of precipitation both globally and in the United States, and the characteristics of these precipitation changes are consistent with what would be expected for GHG-induced warming (see Chapter 9). A number of changes in atmospheric circulation patterns have also been observed (e.g., Fu et al., 2006).

Finally, it should be noted that the observed changes in the climate system to date represent only a fraction of the total expected changes associated with the GHGs currently in the atmosphere: Even if the current climate forcing were to persist indefinitely, it is estimated that the Earth would warm another 0.6°C (1.1°F) over the next several decades as the oceans slowly warm in response to the current GHG forcing, with concomitant changes in other parts of the Earth system (this so-called "commitment warming" is discussed in further detail below). In addition, since CO_2 and many other GHGs remain in the atmosphere for hundreds or even thousands of years

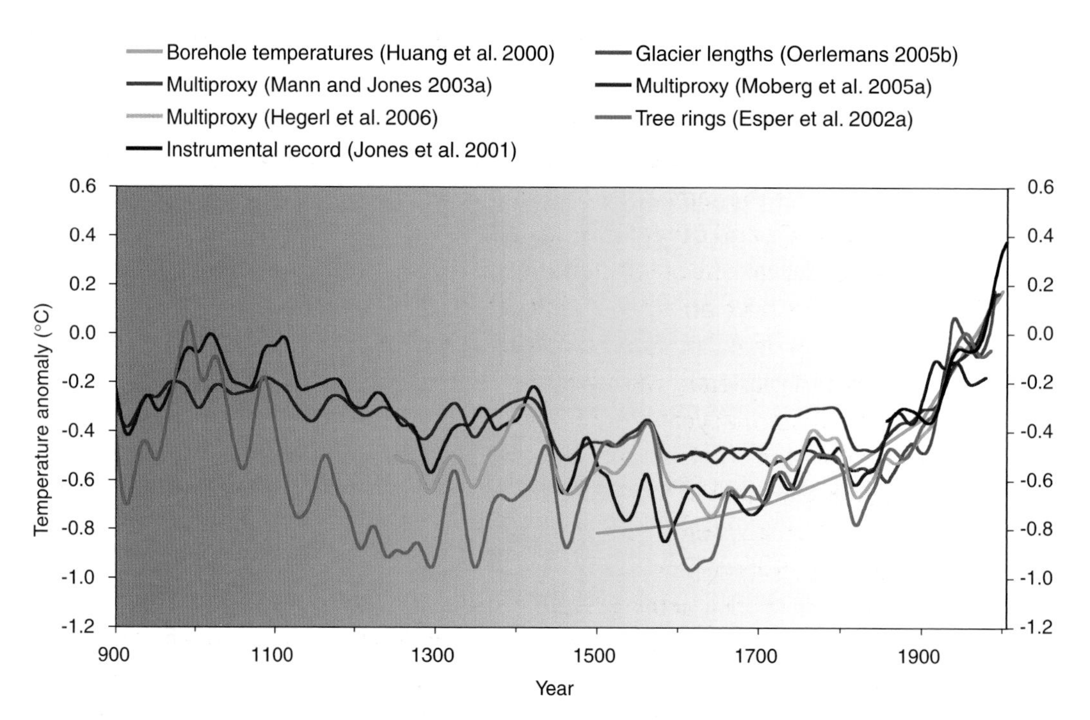

FIGURE 6.16 Estimates of surface temperature variations for the last 1,100 years derived from different combinations of proxy evidence (colored lines). Each curve portrays a somewhat different history of temperature variations and is subject to a somewhat different set of uncertainties that generally increase going backward in time (as indicated by the gray shading), but collectively these data indicate that the past few decades were warmer than any comparable period for at least the last 400 years, and possibly for the last 1,000 years. SOURCE: NRC (2006b).

(Solomon et al., 2009), an additional 2.5°F (1.4°C) of global warming is possible over the next several centuries due to ice sheet disintegration, vegetation change, and other long-term feedbacks in the climate system (Hansen et al., 2008). However, these processes are generally less well understood than the feedbacks that give rise to climate change on shorter (e.g., decadal) time scales.

Attribution of Observed Climate Change to Human Activities

Many lines of evidence support the conclusion that most of the observed warming over at least the last several decades is due to human activities:

- Both the basic physics of the greenhouse effect and more detailed calculations using sophisticated models of atmospheric radiative transfer indicate that increases in atmospheric GHGs should lead to warming of the Earth's surface and lower atmosphere (NRC, 2005d).

- Earth's surface temperature has unequivocally risen over the past 100 years, to levels not seen in at least several hundred years and possibly much longer (NRC 2006b), at the same time that human activities have resulted in sharp increases in CO_2 and other GHGs (as discussed above).
- Detailed observations of temperatures, GHG increases, and other climate forcing factors from an array of instruments, including Earth-orbiting satellites, reveal an unambiguous correspondence between human-induced GHG increases and planetary warming over at least the past three decades, in addition to substantial year-to-year natural climate variability (Hegerl et al., 2007).
- The vertical pattern of atmospheric temperature change over the past few decades, with warming in the lower atmosphere and cooling in the stratosphere (see Figure 6.15), is consistent with the pattern expected due to GHG increases and inconsistent with the pattern expected if other climate forcing agents (e.g., changes in solar activity) were responsible (Roble and Dickinson, 1989).
- Estimates of changes in temperature and forcing factors over the first seven decades of the 20th century are slightly more uncertain and also reveal significant decadal-scale variability (see Figure 6.12), but nonetheless indicate a consistent relationship between long-term temperature trends and estimated forcing by human activities.
- The horizontal pattern of observed surface temperature change over the past century, with stronger warming over land areas and at higher latitudes (Figure 6.13), is consistent with the pattern of change expected from a persistent positive climate forcing (see, e.g., Schneider and Held, 2001).
- Detailed numerical model simulations of the climate system (see the following section for a discussion of climate models) are able to reproduce the observed spatial and temporal pattern of warming when anthropogenic GHG emissions and aerosols are included in the simulation, but not when only natural climate forcing factors are included (Randall et al., 2007).
- Both climate model simulations and reconstructions of temperature variations over the past several centuries indicate that the current warming trend cannot be attributed to natural variability in the climate system (Jansen et al., 2007; NRC, 2006b).
- As discussed earlier in this chapter, estimates of climate forcing and temperature changes on a range of time scales, from the several years following volcanic eruptions to the 100,000+ year Ice Age cycles, yield estimates of climate

sensitivity that are consistent with the observed magnitudes of observed climate change and estimated climate forcing.

- Finally, there is not any compelling evidence for other possible explanations of the observed warming, such as changes in solar activity (Lean and Woods, in press), changes in cosmic ray flux (Benestad, 2005), natural climate variability (Hegerl et al., 2007), or release of heat stored in the deep ocean or other climate system components (Barnett et al., 2005a).

FUTURE CLIMATE CHANGE

Climate Forcing Scenarios

In order to project future changes in the climate system, scientists must first estimate how GHG emissions and other climate forcings will evolve over time. Since the future cannot be known with certainty, a large number of scenarios of future emissions are developed using different assumptions about future economic, social, technological, and environmental conditions. Emissions scenarios are not forecasts and do not attempt to predict "short-term" fluctuations such as business cycles or oil market price spikes. Instead, they focus on long-term (e.g., decades to centuries) trends in energy and land use that ultimately affect the radiation balance of the Earth.

For the past decade, the most widely used scenarios of 21st-century GHG emissions have been those produced for the IPCC's Special Report on Emissions Scenarios (SRES) (Nakicenovic, 2000). The SRES scenarios are quantitative realizations of qualitative storylines that sketched a range of alternative assumptions regarding 21st-century population growth and economic and technological development. The SRES scenarios were all intended to represent alternative baseline (or "business as usual") GHG emissions trajectories, with no explicit policy interventions to limit emissions. In addition, probability distributions were not estimated for either the range or individual SRES scenarios, and so there was no explicit characterization of the likelihood that actual emissions might fall outside the range of the included scenarios.

Since 2000, major scenario exercises have put less emphasis on alternative no-policy baselines and instead concentrated primarily on elaborating the socioeconomic, technological, and policy aspects of alternative GHG trajectories over the next century, with an emphasis on changes over the next few decades; on improving the realism and comprehensiveness of both individual scenarios and the suite of scenarios, for example by adding or improving representations of all important forcing agents and developing scenarios with widely spaced total radiative forcing estimates; and on developing these trajectories in a more integrated and iterative manner with climate model projections and assessments of current and future climate impacts. Recent scenario development exercises stressing these characteristics have been undertaken by the CCSP (2007c), the Energy Modeling Forum (Clarke et al., 2009), and other groups (Moss et al., 2010). These exercises have yielded a number of important insights, such as the challenges associated with reaching certain GHG emissions or temperature goals.

The aim of developing more useful climate forcing scenarios is subject to several pressures that are in tension with each other, such as providing more sophisticated and increasingly detailed representations of socioeconomic, environmental, and policy factors, while at the same time keeping the origin of the assumptions used transparent, plausible, and understandable. Additional challenges to scenario development include balancing and integrating the qualitative and quantitative elements of scenarios; developing scenarios that provide socioeconomic and environmental information (which is useful, for example, for adaptation planning) that is consistent with the corresponding emissions trajectories; and making more explicit, transparent, and defensible judgments of probabilities associated with scenario-based ranges of key variables (CCSP, 2007b; Parson, 2008).

In response to these issues, climate modelers, integrated assessment modelers, and researchers focusing on impacts, adaptation, and vulnerability collaborated to develop a new process for preparing and applying scenarios in climate research (Moss et al., 2010). In contrast to the traditional approach in which scenarios are developed and applied in a linear causal chain from socioeconomic "drivers" of emissions, to atmospheric and climate processes, to impacts, the new process starts with four scenarios of future radiative forcing called "Representative Concentration Pathways." These pathways are defined by their radiative forcing in 2100 and include (1) a high scenario of 8.5 W/m^2, and still rising; (2) an "overshoot scenario" in which radiative forcing peaks midcentury and then declines to a level of 2.6 W/m^2 (which is lower than any of the SRES scenarios) in 2100, and (3) two intermediate scenarios that stabilize in 2100 at 6 and 4.5 W/m^2. These representative concentration pathways will be used to conduct new climate model experiments and produce new climate change scenarios. In parallel, new socioeconomic and emissions scenarios will be developed to explore detailed scenarios of socioeconomic drivers, adaptation, mitigation, and other issues such as feedbacks. The process rests on the simple observation that any particular radiative forcing trajectory can be realized by many different socioeconomic, technology, and policy futures. The new process facilitates research into a number of key issues including feedbacks, the ease or difficulty of achieving overshoot scenarios (and the climate and ecosystem consequences of these trajectories, which are highly uncertain), as well as process issues discussed in the previous paragraph.

Climate Models

Climate models encapsulate scientists' best understanding of climate and related Earth system processes and are important tools for understanding past, present, and future climate change. While there are many different kinds of climate models, all are based fundamentally on the laws of physics that govern atmospheric and oceanic motions, including the conservation of mass, energy, and angular momentum and laws that govern the propagation of radiation through the atmosphere. Most modern climate models also include representations of the oceans, cryosphere, and land surface, as well as the exchanges of energy, moisture, and materials among these components. *Earth system models* additionally simulate a wide range of biophysical processes including atmospheric chemistry and the biogeochemistry of ecosystems on land and in the oceans (Figure 6.17).

Climate and Earth system models (for simplicity, referred to hereafter as climate models) use computer-based numerical techniques to solve a system of mathematical equations that embody these laws, systems, and processes, yielding a predicted evolution of the climate system over time (see, e.g., DOE, 2008b, 2009b; Donner and

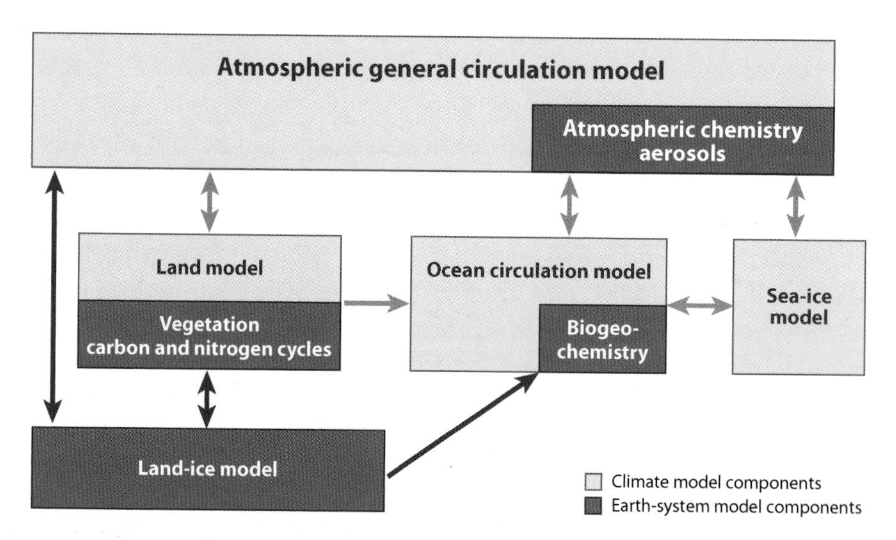

FIGURE 6.17 Schematic illustration of the components of climate and Earth system models. The components of climate models are in gray and the additional components in Earth system models are in green. The connecting arrows indicate exchanges that couple the model components. SOURCE: Donner and Large (2008).

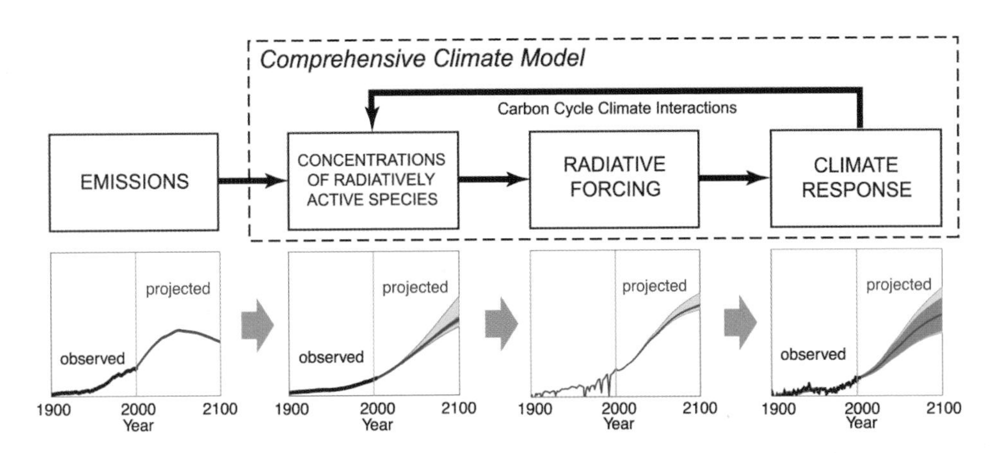

FIGURE 6.18 Schematic overview of the translation from a specified trajectory of emissions of GHGs and other climate forcing agents to trajectory to climate response. Simulated climate changes will include both the forced response and internal (natural) variability. The specific model results in the bottom row are for illustrative purposes only. SOURCE: Meehl et al. (2007a).

Large, 2008). Climate models are based on the same basic equations that are used to predict short-term weather variations. However, rather than trying to predict the exact future evolution of the atmosphere (i.e., the weather), climate models instead focus on accurately simulating the processes that govern interannual and longer-term climate trends (see Box 6.1).

Climate models are used to simulate both natural climate variability and the evolution of the climate system under specified climate forcing, including both historical data and scenarios of future forcing changes (Figure 6.18). Our confidence in the ability of climate models to reliably project certain aspects of future climate stems from the extensive development and testing processes used to design models and evaluate their performance—including simulations of 20th-century climate when the climate forcing and response are both reasonably well known (up to the limits of observations and recordkeeping) and simulations of the response to volcanic eruptions (e.g., Randall et al., 2007). Moreover, by assessing many different models, each with different emphases, strengths, and weaknesses, or many different runs of the same model (which provides an indication of natural variability), the most robust features of future projections emerge. These results are presented in the next section of the chapter.

Advances in climate modeling over the past 50 years have been driven by two main factors: (1) increases in computer power, which have allowed improved spatial resolution, the inclusion of additional Earth system components, more explicit representations of processes, and multiple model experiments to explore different assumptions and model specifications; and (2) improvements in theoretical and mechanistic understanding of the climate system and the processes being modeled, which in turn are tied to basic research and improvements in observational capabilities. Today, continued improvements in computational power, scientific understanding, and supporting observations are still the primary factors driving improvements in climate models—or stated conversely, even if the evolution of future climate forcing were known exactly, limits in computer power, observational data, and scientific understanding of the climate system would still constrain the ability of models to produce perfect predictions of future climate (Shapiro et al., in press).

For example, the typical horizontal grid spacing of a state-of-the-art global climate model is on the order of 60 miles (100 km), but climatically relevant features such as clouds, topography, and land cover often vary at a scales of a half-mile or less. These subgridscale features and processes must be *parameterized*—approximated using numerical techniques that specify the large-scale influence of small-scale processes—or *upscaled* through statistical or "nested model" approaches that extend representative small-scale simulations to larger spatial scales. As a result of these approximations and other factors (described below), global climate models generally only provide consistent and reliable simulations of temperature, precipitation, and other relevant climate variables at continental to global scales.

Regional Climate Projections

The lack of regionally specific climate information from global climate models poses a major challenge, because many climate-related decisions, especially those related to adaptation, demand information on regional to local scales. A variety of *downscaling* approaches have been developed to obtain this regional information. One widely used approach is statistical downscaling, wherein empirical relationships between past observations of local- and regional-scale climate variations are used to translate large-scale projections from global climate models to smaller space scales and shorter time scales. Alternatively, finer-scale regional models can be "nested" within coarser-resolution global models to simulate regional climate changes (e.g., Hay et al., 2002; Leung et al., 2003; UCAR, 2007). A related approach is linking models currently used to predict weather and seasonal to interannual climate variations with those that predict climate change on decadal to centennial time scales (this is sometimes called "seamless prediction").

In general, downscaling techniques are not as well developed or understood as global models, and key technical and scientific issues remain. For example, regional modeling efforts have been limited by constraints on computing resources, uncertainties and complexities associated with data assimilation and parameterization, the lack of a well-developed framework for downscaling, and the limitations of the large-scale simulations on which the downscaling is performed (Held and Soden, 2006; NRC, 2009k). An additional challenge for regional projections is representing regional modes of variability, such as ENSO and the Pacific Decadal Oscillation (described earlier in this chapter). Not only do these regional modes have a strong influence on local and regional climate change, but many also have global signatures, and they could potentially change themselves as the climate system warms. Finally, climate forcing scenarios that project human influences on local and regional climate, such as regional aerosol loading and land use change, are needed because these forcings may have a large influence on local and regional climate change (CCSP, 2008c).

Projections of 21st-Century Climate

The most comprehensive suite of climate modeling experiments performed to date were completed in 2005 as part of the World Climate Research Programme's Coupled Model Intercomparison Project phase 3 (CMIP3; Meehl et al., 2007b) in support of the IPCC's Fourth Assessment Report. CMIP3 included 23 different state-of-the-art models from groups around the world, all of which were run with a specific set of emissions scenarios (based on the SRES report described above) to facilitate comparison and synthesis of results. As described in detail by the IPCC (Meehl et al., 2007a), the CMIP3 climate models project increases in mean surface temperatures over the 21st century ranging from 2.0°F to 11.5°F (1.1°C to 6.4°C), relative to the 1980-1999 average, by the end of the century.

Figure 6.19 shows projected global temperature changes associated with three representative scenarios of high, medium-high, and low future GHG emissions. The separation between the three curves illustrates the uncertainty associated with the choice of scenario, while the uncertainties associated with differences among different models in simulating the climate system can be inferred from the shading surrounding each curve. The "commitment warming" associated with emissions through the year 2000 and, for two of the future forcing scenarios, through 2100 are also shown. These "commitment" runs, which are performed by instantaneously stabilizing atmospheric GHG concentrations, show that the climate system will continue to warm for several centuries after GHG emissions are stabilized—illustrating the inherent time lag between GHG emissions and the long-term climate response.

As with observed climate change to date, regional manifestations of future climate projections vary substantially, with stronger warming over higher latitudes and land areas (Figure 6.20). The similarity between the three panels on the left-hand side of Figure 6.20 also illustrates how temperature increases over the next few decades reflect past emissions as well as somewhat similar GHG emissions over the next few decades for the three selected SRES scenarios (none of which include explicit policy interventions). By midcentury and especially at the end of the century, however, the medium- and high-emissions scenarios clearly lead to much warmer temperatures than the lower-emissions scenario. U.S. temperatures are projected to warm substantially over the 21st century under all emissions scenarios (USGCRP, 2009a).

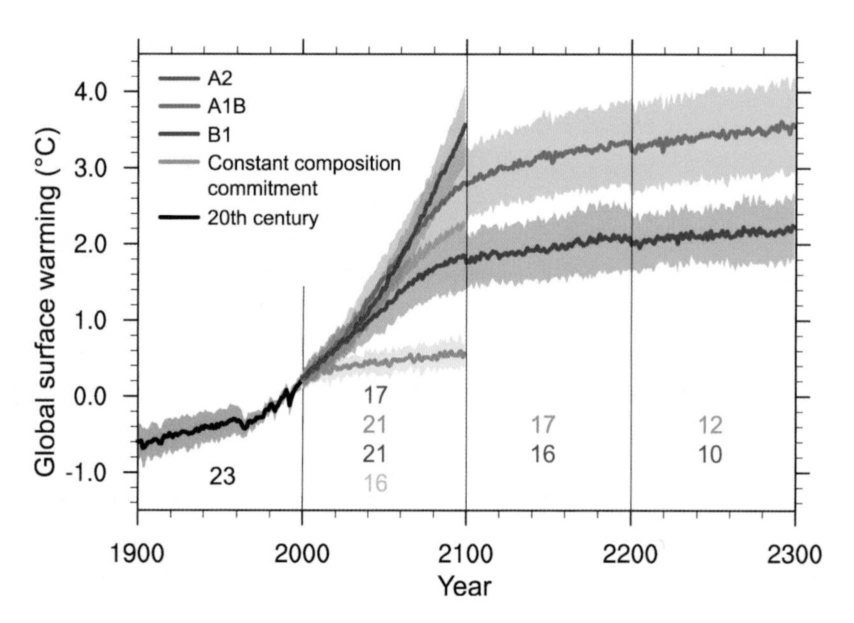

FIGURE 6.19 Model simulations of changes in global average temperature from 1900 to 2300. The black line and gray shading shows the average and spread1 of 23 model simulations of 20th-century climate using estimates of actual climate forcing. The colored lines and shading show average and spreads for projected global average temperatures for the 21st century under four different scenarios of future forcing: a "high-emissions" scenario (red), a "medium-high" scenario (green), a "low-emissions" scenario (blue), and a "commitment" scenario (orange), which assumes that GHG concentrations remain constant at year 2000 values. The green and blue curves also show commitment experiments for the 22nd and 23rd centuries (i.e., with the forcing at year 2100 held constant thereafter). Changes are relative to the 1960-1979 average. See text for additional discussion. SOURCE: Meehl et al. (2007a).

[1] The spreads in this figure indicate the 90 percent statistical confidence range of the model experiments (i.e., the annual average temperature traces from 90 percent of the included model experiments fall within the shaded bands). This spread is indicative of the uncertainty that the underlying models and forcing scenarios are able to resolve, but not the unresolved uncertainties discussed in the next section.

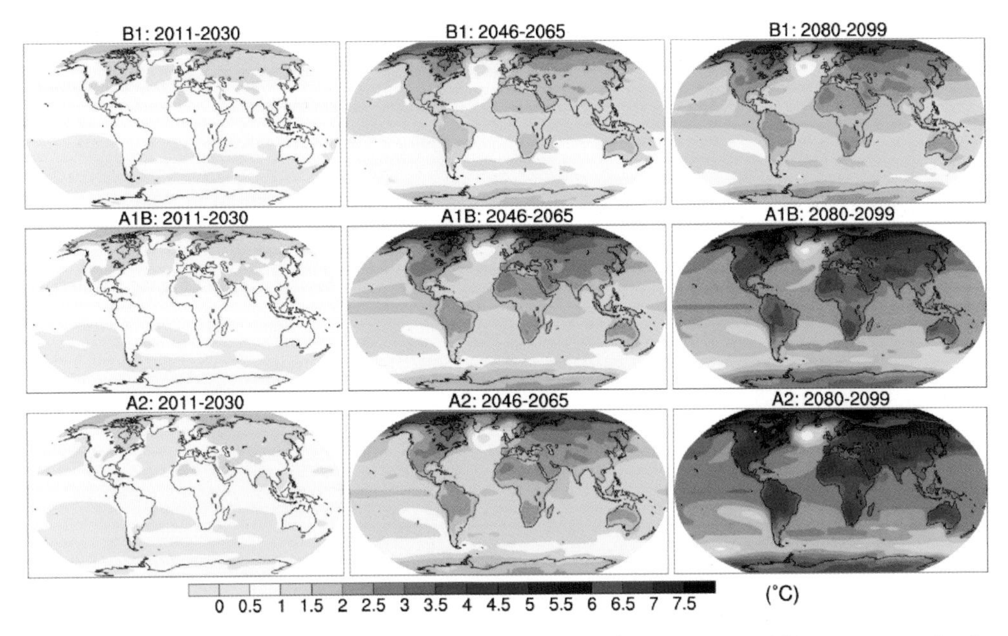

FIGURE 6.20 Worldwide projected changes in temperatures, relative to 1961-1990 averages, under three different emissions scenarios (rows) for three different time periods (columns). Projected warming is much stronger over land areas and high latitudes. SOURCE: Meehl et al. (2007a).

In addition to average temperature, a host of other climate variables are projected to experience significant changes over the 21st century, just as they have during the past century. For example, the frequency and intensity of heat waves is projected to continue to increase, both in the United States (Figure 6.21) and around the world. This projection is considered robust because a shift in the average value of a temperature distribution (or in another climate variable) typically entails an increase in the frequency of extreme and unprecedented events (see, e.g., Solomon et al., 2007). Similarly, there is considerable confidence that the frequency of cold extremes will decrease and that the number of frost days will decline in the middle and high latitudes, following current trends (Meehl et al., 2007a; USGCRP, 2009a). Projections of future climate also indicate that snow cover and sea ice extent will continue to decrease (Meehl et al., 2007a; USGCRP, 2009a; Zhang, 2010), while sea level will continue to rise (see Chapter 7).

Projections of precipitation change are generally more uncertain than projections of temperature and temperature-related changes. However, most models project increased precipitation in northern regions of the United States, while it is considered very likely that the southwestern United States will experience a net decrease in precipitation (USGCRP, 2009a). Another robust projection, which results from the fact

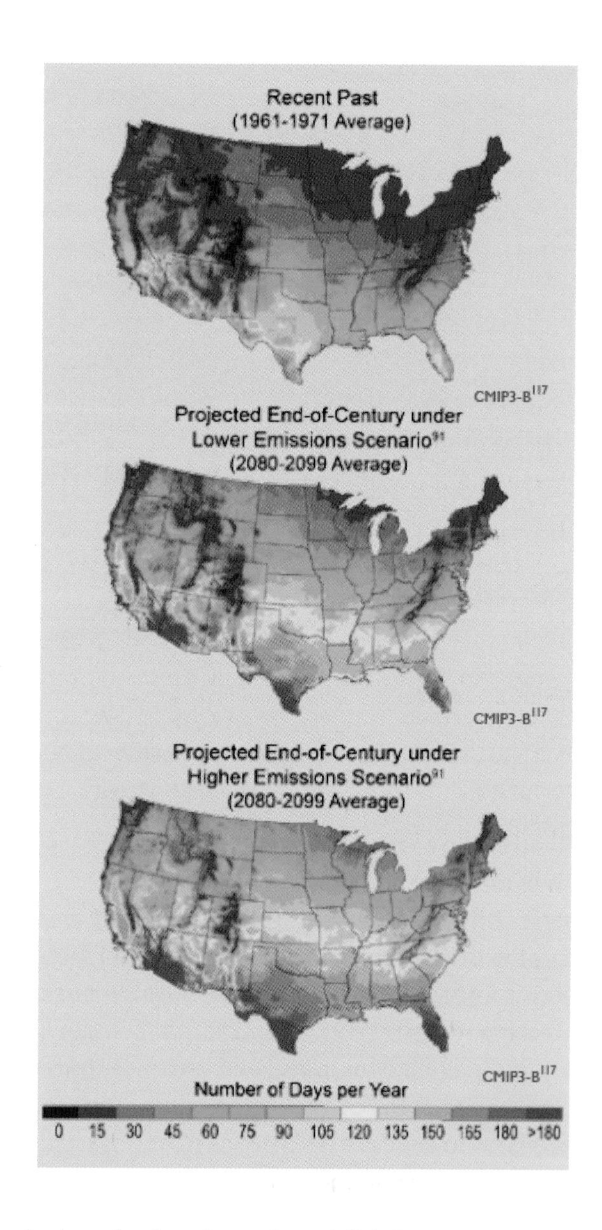

FIGURE 6.21 Projected changes in number of very hot days in the United States for lower- and higher-emissions scenario. The number of very hot days will increase substantially across virtually the entire country, in some places doubling or even trebling the number of days above 90°F. SOURCE: USGCRP (2009a).

that warmer air can hold more moisture, is that the fraction of rainfall falling in the form of heavy precipitation events will increase in many regions (Meehl et al., 2007a). These and other projected changes in precipitation, and the impact of these changes on freshwater resources, are explored in Chapter 8. Later chapters also explore how changes in temperature, precipitation, and other aspects of the physical climate system are likely to affect ecosystems (see Chapter 9), agriculture (Chapter 10), human health (Chapter 11), the urban environment (Chapter 12), transportation (Chapter 13) and energy systems (Chapter 14), and national security (Chapter 16).

Key Uncertainties in Projections of Future Climate

A great deal is known about past, present, and projected future climate change, especially at large (continental to global) scales. For example, there is high confidence that global temperatures will continue to rise, that the rate and magnitude of future temperature change depends strongly on current and future rates of GHG emissions, and that climate change—in interaction with other global and regional environmental changes—poses significant risks for a number of human and natural systems. Global climate models and, increasingly, regional techniques are also starting to provide useful information about future climate and climate-related changes on local to regional scales. Some of these projections—such as increases in extreme heat events and Arctic sea ice—are quite robust, while others are somewhat more speculative.

There are, however, several aspects of future climate change that remain more uncertain, and these represent some of the most important and active areas of current scientific research (see Research Needs at the end of this chapter). The uncertainties in climate projections can be categorized into two main sources: (1) uncertainties in future climate forcing and (2) uncertainties in how the climate system will respond to forcing, which includes both the known limitations of global climate models (such as an inability to resolve individual clouds) and the fact that the climate system is complex and might exhibit novel or unanticipated behavior in response to ongoing climate change.

The first of these categories, uncertainties in future climate forcing, was discussed in the Future Climate Scenarios section earlier in the chapter. The spread among the three colored curves in Figure 6.20 provides a rough indication of the importance of this uncertainty in terms of the magnitude of future climate change. As discussed above, and described in further detail in the companion report *Limiting the Magnitude of Future Climate Change* (NRC, 2010c), future climate forcing depends strongly on the choices that current and future human societies make, especially regarding energy production and use. However, actions that might be taken to limit the magnitude of future climate change, or adapt to its impacts, have not yet been fully and systematically integrated into climate forcing scenarios and evaluated across a range of different climate models to determine how they might ultimately affect both climate and other aspects of the Earth system.

As an illustration of some of the uncertainties present in climate model projections, Figure 6.22 shows projections of temperature change over North America from 21 different models, each using the same scenario of future climate forcing. Several robust features emerge from these projections—for example, all of the models project a

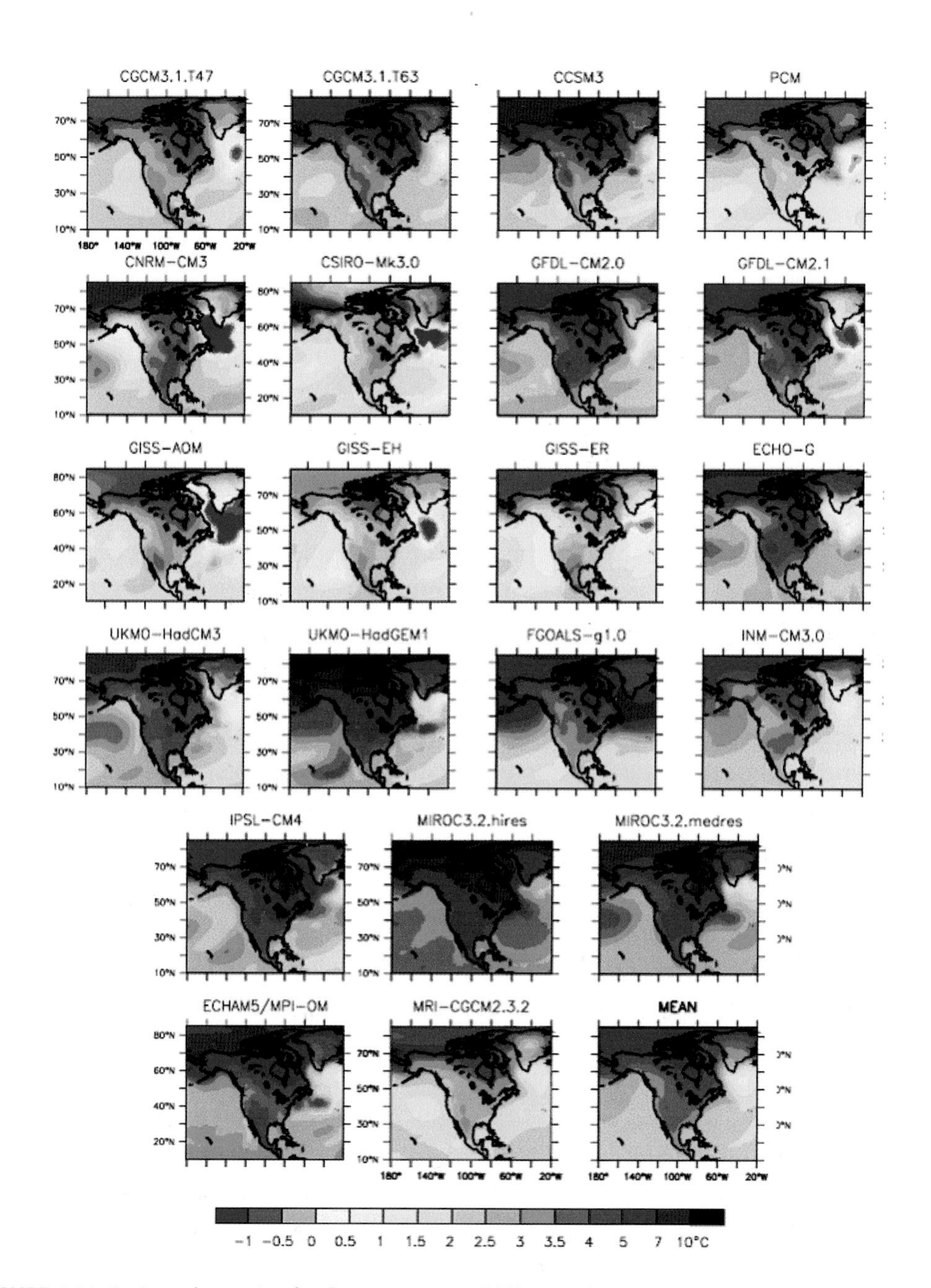

FIGURE 6.22 Projected warming for the 21st century (difference between 2080-2099 temperature and 1980-1999 temperature) for the North American region using 21 different climate models, all using the same scenario of future GHG emissions. The mean (average) of the 21 model experiments is also shown in the bottom right panel. Several robust features are evident, including enhanced warming over land areas and higher latitudes. Differences among the 21 projections are indicative of some of the uncertainties associated with model projections. SOURCE: Christensen et al. (2007).

substantial overall temperature increase, with stronger warming over land areas and at higher latitudes. Most of the models show somewhat less warming over the southeastern United States and a slight cooling, or at least less warming, over the western North Atlantic Ocean south of Greenland. In other regions, however, the exact pattern and magnitude of projected warming varies considerably among models. Typically, the average of many climate model simulations represents a more robust projection than any individual projection (Randall et al., 2007), so the average of these model calculations (shown in the bottom right panel) can be thought of as the most reliable prediction of future temperature change over North America. Differences among models indicate some (but not all) of the uncertainty in this "multimodel mean" projection. Analyses of the differences among models—such as CMIP3 and previous model intercomparison projects—are also a key tool for model development.

The other main type of "known" uncertainty in model-based projections of future climate change is associated with processes that are either not resolved or not very well simulated in the current generation of global climate models. These processes, which are discussed in further detail in the Research Needs section at the end of this chapter, include clouds and aerosols, the carbon cycle, ocean mixing processes, ice sheet dynamics, ecosystem processes, land use–related changes, and extreme weather events such as hurricanes, tornadoes, and droughts. Another key research area is the relationship between regional modes of variability and global climate change, including the possibility that regional variability modes may shift in response to either regional or global human activities.

Abrupt Changes and Other Climate Surprises

Confounding all projections of future climate is the possibility that abrupt changes or other climate "surprises" may occur. Abrupt changes in the climate system can occur when (1) there is a rapid change in forcing, such as a rapid increase in atmospheric GHG concentrations or reduction in aerosol forcing, or (2) thresholds for stability (or "tipping points") are crossed, such that small changes in the climate state are reinforced, leading to rapid shifts until the climate enters another stable state and stability is restored. Paleoclimate records indicate that the climate can go through abrupt changes in as little as a single decade (NRC, 2002a). For example, Greenland ice cores indicate that about 13,000 years ago, during the recovery from the last Ice Age, local temperatures fell more than 10°F (6°C) within a few decades and remained low for more than a millennium before jumping up more than 16°F (10°C) in about a decade (CCSP, 2007b). Since the Earth's temperature is now demonstrably higher than it has been for at least 400 years and possibly more than 1,000 years (NRC, 2006b), and GHG concentrations are now higher than they have been in many hundreds of thousands of years, it is possible that we may be nearing other stability thresholds. However, we have only a limited understanding of what those thresholds might be or when the climate system might be approaching them.

One example of a potential abrupt change mechanism is the possibility that GHGs stored in permafrost (frozen soils) across the Arctic could be released in large quantities as high-latitude warming continues. Permafrost contains huge amounts of carbon that have been locked away from the active carbon cycle for millennia, and it has been demonstrated that thawing permafrost releases some of this carbon to the atmosphere in the form of CH_4 and CO_2 (Shakova et al., 2010). If the release of these GHGs accelerates as the Arctic continues to warm, this could potentially accelerate the warming, leading to a positive feedback on the warming associated with GHGs released through human activities (Lawrence and Slater, 2005; Schuur et al., 2009; Zimov et al., 2006). In a related example, high-latitude warming can also alter the types of ecosystems covering the land (for instance, a shift from tundra to forest), which in turn changes the reflective characteristics of the land surface and thus potentially exerts a further positive feedback on warming (Field et al., 2007a).

Other potential abrupt changes include rapid disintegration of the major ice sheets (see Chapter 7), irreversible drying and desertification in the subtropics as a result of shifts in circulation patterns (see Chapter 8), changes in the meridional overturning circulation in the ocean (Broecker, 1997, 2002; Stocker, 2000; Stocker and Schmittner, 1997), or the rapid release of CH_4 from destabilized methane hydrates in the oceans (Archer and Buffet, 2005; Overpeck and Cole, 2006), all of which could dramatically alter the rate of both regional and global climate change. Other surprises that may be associated with future climate change include so-called "low-probability, high-impact" events, such as an unprecedented heat wave or drought, or when multiple climate changes interact with each other or with other environmental stresses to yield an unexpectedly severe impact on a human or environmental system. Some of these potential—or in some cases already observed—surprises are discussed in later chapters.

RESEARCH NEEDS

Advances in our understanding of the climate system have been and will continue to be a critical underpinning for evaluating the risks and opportunities posed by climate change as well as evaluating and improving the effectiveness of different actions taken to respond. Hence, even as actions are taken to limit the magnitude of future climate change and adapt to its impacts, it is important that continued progress be made in observing all aspects of the climate system, in understanding climate system processes, and in projecting the future evolution of the climate system, and as well as its interactions with other environmental and human systems (which are explored in the chapters that follow). The following are some of the most critical basic research needs in these areas.

Expand and maintain comprehensive and sustained climate observations to provide real-time information about climate change. Regular and sustained observations of climate variables are needed to monitor the progress of climate change, inform climate-related decision making, and to monitor the effectiveness of actions taken to respond to climate change. Observations are also critical for developing and testing climate models, projections of future climate forcing, and other tools for understanding and projecting climate change, as well as for supporting decision-support activities. As discussed in Chapter 3, a comprehensive climate observing system is needed to provide regular monitoring of biological, chemical, geological, and physical properties in the atmosphere, oceans, land, and cryosphere, as well as related biological, ecological, and socioeconomic processes. Expanded historical and paleoclimatic records would also be valuable for understanding natural climate variations on all time scales and how these modes of variability interact with global climate change. Finally, a comprehensive data assimilation system is also needed to bring these disparate observations into a common framework, so that the state of the whole Earth system can be assessed and impending feedbacks that could alter the rate of climate change can be identified. Research is especially needed on how to better integrate physical indicators with emerging indicators of ecosystem health and human well-being, as discussed in other chapters.

Continue to improve understanding of climate variability and its relationship to climate change. Great strides have been made in improving our understanding of the natural variability in the climate system over the past several decades. These improvements have translated directly into advances in detecting and attributing human-induced climate change, simulating past and future climate in models, and understanding the links between the climate system and other environmental and human systems. For example, the ability to realistically simulate natural climate variations, such as the El Niño-Southern Oscillation, is a critical test for climate models. Improved understanding of regional variability modes is also critical for improving regional climate projections, as discussed below. Understanding the impacts associated with natural climate variations also provides insight into the possible impacts of human-induced climate change. Continued research on the mechanisms and manifestations of natural climate variability in the atmosphere and oceans on a wide range of space and time scales, including events in the distant past, can be expected to yield additional progress.

Develop more informative and comprehensive scenarios of drivers of future climate forcing and socioeconomic vulnerability and adaptive capacity. Uncertainty in projections of the future is inevitable. However, the development of scenarios allows better understanding of the dynamics of the interconnected human-environment system and in particular how the dynamics will change depending on the choices we make. Scenarios are also critical for helping decision makers establish targets for future GHG emissions and concentration levels as well as helping make plans to adapt to the future projected impacts of climate change, topics addressed in many of the chapters that follow. Developing and improving assessments of the potential influence of various policy choices on emission profiles and adaptive capacity is particularly important in the context of supporting climate-related decision making—especially "overshoot" scenarios, which have the potential to cause irreversible changes to the climate system. Influences of shorter-lived forcing agents (including short-lived GHGs and aerosols) are also of high importance in the near term and could benefit from more near-term emphasis.

Developing enhanced scenarios and linking them to a variety of Earth system and socioeconomic models is an inherently interdisciplinary and integrative activity requiring contributions from many different scientific fields as well as processes that link scientific analysis with decision making and, ideally, public deliberation about desirable futures. The new "Representative Concentration Pathways" described earlier represent a few common, transparent, thoroughly documented representative scenarios of key variables over time. A number of research needs and developments are required to develop new socioeconomic scenarios that explore both mitigation and adaptation issues. It is particularly important to explore methods for coupling scenarios across geographic scales (from global to regional to local), to further develop methods for downscaling climate scenarios and providing regional climate information, and to develop data and information systems for pairing socioeconomic and climate scenarios for use in impacts research and to support the needs of particular decision makers.

Improve understanding of climate system forcing, feedbacks, and sensitivity. The past several decades have seen tremendous progress in quantifying human influences on climate and assessing the response of the climate system to these influences. This progress has been critical both in establishing the current level of confidence in human-induced climate change and in developing reliable projections of future changes. Key uncertainties remain, however, and continued research on the basic mechanisms and processes of climate change can be expected to yield additional progress. Some critical areas for further study include the following:

- Continued research to improve estimates of climate sensitivity, including theoretical, modeling, and observationally based approaches;
- Improved understanding of cloud processes, aerosols and other short-lived forcing agents, and their interactions, especially in the context of radiative forcing, climate feedbacks, and precipitation processes;
- Continued theoretical and experimental research on carbon cycle processes in the context of climate change, especially as they relate to strategies for limiting climate change (CCSP, 2007a; NRC, 2010j);
- Improve understanding of the relationship between climate change and other biogeochemical changes, especially acidification of the ocean (see Chapter 9);
- Improve understanding of the hydrologic cycle, especially changes in precipitation (see also Chapter 8);
- Improved understanding of the mechanisms, causes, and dynamics of changes in the cryosphere, especially changes in major ice sheets (see Chapter 7) and sea ice.

Overall, the need for improved understanding of climate forcing, feedbacks, and sensitivity was summarized well in the NRC report *Understanding Climate Change Feedbacks* (NRC, 2003b); these suggestions remain highly relevant today:

> The physical and chemical processing of aerosols and trace gases in the atmosphere, the dependence of these processes on climate, and the influence of climate-chemical interactions on the optical properties of aerosols must be elucidated. A more complete understanding of the emissions, atmospheric burden, final sinks, and interactions of carbonaceous and other aerosols with clouds and the hydrologic cycle needs to be developed. Intensive regional measurement campaigns (ground-based, airborne, satellite) should be conducted that are designed from the start with guidance from global aerosol models so that the improved knowledge of the processes can be directly applied in the predictive models that are used to assess future climate change scenarios.
>
> The key processes that control the abundance of tropospheric ozone and its interactions with climate change also need to be better understood, including but not limited to stratospheric influx; natural and anthropogenic emissions of precursor species such as NO_x, CO, and volatile organic carbon; the net export of ozone produced in biomass burning and urban plumes; the loss of ozone at the surface, and the dependence of all these processes on climate change. The

chemical feedbacks that can lead to changes in the atmospheric lifetime of methane also need to be identified and quantified.

Improve model projections of future climate change. Numerous decisions about climate change, including setting emissions targets and developing and implementing adaptation plans, require information that is underpinned by models of the physical climate system. There are a number of scientific and technological advances needed to improve model projections of future changes in the Earth system, especially changes over the next several decades and at the local and regional levels where many climate-related decisions occur. While this research should not be expected to eliminate uncertainties, especially given the inherent uncertainty in projections of future climate forcing, efforts to expand and improve model simulations of future climate changes can be expected to yield more, more robust, and more relevant information for decision making, including the effectiveness of various actions that can be taken to respond to climate change. It should also be noted that improvements in modeling go hand-in-hand with improvements in understanding and observation.

The core of the nation's climate modeling enterprise is the development and testing of global Earth system models, many of which already or are now beginning to incorporate some of the key forcing and feedback processes noted above, including an explicit carbon cycle, certain biogeochemical and ecological processes, and improved parameterizations for clouds, aerosols, and ocean mixing. While these important activities should continue, the nation should also initiate a strategy for developing the next generation of ultra-high-resolution global models; models that can explicitly resolve clouds and other small-scale processes, include explicit representations of ice sheets and terrestrial and marine ecosystems, and allow for integrated exploration of forcing and feedback processes from local to global scales (Shapiro et al., in press). It may be valuable to consider the merits of coordinating the development of climate models with the development of weather models through "seamless prediction" paradigms that could potentially improve the simulation of extreme events as well as lower development costs (Tebaldi and Knutti, 2007). Expanded computing resources and human capital are needed to support all of these activities.

Climate modelers in the United States and around the world have also begun to devise strategies for improving the utility of climate models. Decadal-scale climate prediction, in which climate models are initialized with present-day observations and run forward in time at fairly high resolution for three to four decades, is another emerging strategy to provide decision makers with information to support near-term decision making (Meehl et al., 2009b). Extending or coupling current models to models of human and environmental systems, including both ecosystems models and models of human activities, would foster the development of more robust and integrated assessments of key impacts of climate change (see Chapter 4). Finally, the usefulness of climate model experiments to decision making would be improved if they could be used to comprehensively assess a wider variety of climate response strategies, including specific GHG emissions-reduction strategies, adaptation strategies, and solar radiation management strategies (see Chapter 15).

Improve regional climate modeling, observations, and assessments. Given the importance of local and regional information to decision makers, and the fact that it might take decades to develop global models with sufficient resolution to resolve local-scale processes, it is essential to continue improving regional climate information, including observations and assessments of regional climate and climate-related changes as well as models that can project interannual, decadal, and multidecadal climate change, including extreme events, at regional to local scales across a range of future global climate change scenarios. Improvements in regional climate observations, modeling, and assessment activities often go hand in hand—for example, local and regional-scale observations are needed to verify regional models or downscaled estimates of precipitation. Models also require a variety of information, for example the regional climate forcing associated with aerosols and land use change, that is also useful to decision makers for planning climate response strategies and for other reasons (such as monitoring air quality). It will also be important to improve our understanding and ability to model regional climate dynamics, including atmospheric circulation in complex terrain as well as modes of natural climate variability on all time scales, especially how their intensity and geographic patterns may change under different scenarios of global climate change. Several strategies for improving regional climate models are described in this chapter, including statistical and dynamical approaches. As with the development of global climate models, further progress in regional modeling will require expanded computing resources, improvements in data assimilation and parameterization, and both national and international coordination.

Advance understanding of thresholds, abrupt changes, and other climate "surprises." Some of the largest potential risks associated with future climate change come not from the relatively smooth changes in average climate conditions that are reasonably well understood and resolved in current climate models, but from extreme events, abrupt changes, and surprises that might occur when thresholds in the climate system (or related human or environmental systems) are crossed. While the paleoclimate record indicates that abrupt climate changes have occurred in the past, and we have many examples of extreme events and nonlinear interactions among different components of the human-environment system that have resulted in significant impacts, our ability to predict these kinds of events or even estimate their likelihood is limited. Improving our ability to identify potential thresholds and evaluate the potential risks from unlikely but high-impact events will be important for evaluating proposed climate targets and developing adaptation strategies that are robust in the face of uncertainty. Sustained observations will be critical for identifying the signs of possible thresholds and for supporting the development of improved representations of extreme events and nonlinear processes in climate models. Expanded historical and paleoclimatic records would also be valuable for understanding the impacts associated with abrupt changes in the past. Finally, since some abrupt changes or other climate surprises may result from complex interactions among different components of the coupled human-environment system, improved understanding is needed on multiple stresses and their potential intersection with future climate shifts.

CHAPTER SEVEN

Sea Level Rise and the Coastal Environment

The coastlines of the United States and the world are major centers of economic, social, and cultural development, and coastal areas are home to critical ecological and environmental resources. Climate change poses a number of risks to coastal environments. Foremost among these is sea level rise, which threatens people, ecosystems, and infrastructure directly and also magnifies the impacts of coastal storms. Coastal environments face a variety of other stresses, such as pollution, development pressures, and resource harvesting, that can interact with climate-related changes and potentially increase the vulnerability of coastal areas.

Coastal managers, businesses, governments, and inhabitants are all concerned about current and future risks to coastal areas. In order to develop adequate responses to the risks posed by climate change and sea level rise, they require answers to questions such as the following:

- How much will sea level rise in the future and on what time scales?
- How will sea level rise and changing storm patterns translate into local problems such as erosion, flooding, damage to infrastructure, and loss of ecosystems?
- What coastal protection measures are physically and economically feasible and socially and environmentally acceptable in different locations, and how much time do we have to start implementing these measures?
- At what point is it more cost effective to retreat from the shoreline than to defend coastal land uses in place?
- How uncertain is the information about sea level rise and other coastal (physical, ecological, and socioeconomic) processes, and what are the implications of these uncertainties for decision making?

This chapter summarizes the information that is currently available regarding the history, causes, projections, and consequences of sea level rise and other climate-related changes that affect or may affect coastal environments. It includes updated projections of sea level rise and the impacts of sea level rise and other climate changes on coastal systems, and also discusses scientific knowledge to support responding to sea level rise. The final section of the chapter describes some of the additional scientific information that will be needed to better understand and develop more effective responses to sea level rise and other coastal management challenges.

OBSERVED SEA LEVEL CHANGES

Sea level has varied dramatically over Earth's history. For the past two to three million years, the ice age cycles—which are driven by periodic variations in Earth's orbit (see Chapter 6)—have led to regular fluctuations in sea level of several hundred feet. During an ice age, significant amounts of water are stored in continent-sized glaciers called ice sheets that are up to several miles thick. Much of this ice melts during warm interglacial periods, and the resulting water raises global sea level substantially when it enters the oceans. During the most recent ice age, which peaked about 26,000 years ago, global average sea level was approximately 400 feet (120 meters) lower than it is today.

By carefully analyzing the depths and dates of coral reefs, geologists have reconstructed the temporal history of sea level rise during the recovery from the last ice age (Fairbanks, 1989; Peltier and Fairbanks, 2006). This rise was not steady, but rather punctuated by periods of rapid rise of as much as 2 inches (5 centimeters) per year; it is inferred that these periods of rapid rise were driven by pulses of water from melting ice (Figure 7.1). By approximately 6,000 years ago—or around the time that agriculture expanded and larger-scale civilizations were first established—global average sea level had risen to close to its present-day value, and it subsequently remained relatively steady. Other direct and indirect observations have allowed oceanographers to estimate past sea levels going back a few thousand years. These historical records suggest that there was little net change in sea level from the first century A.D. to 1800 (Church et al., 2008; Sivan et al., 2004).

Intrumental Records of Sea Level Rise

Instrumental records for sea level date back about 140 years, when systematic measurements by tide gauges became available (NRC, 1990b). During the past few decades, tide gauge records have been augmented by satellite measurements that give precise sea level maps across the entire globe. Together, these modern records indicate that sea level has been rising since the mid-19th century, and that the rate of increase has been accelerating in recent years (Figure 7.2). For example, in the late 19th century, when tide gauge readings begin to provide accurate global sea level estimates, the rate of sea level rise was about 0.02 inches (0.6 millimeters) per year

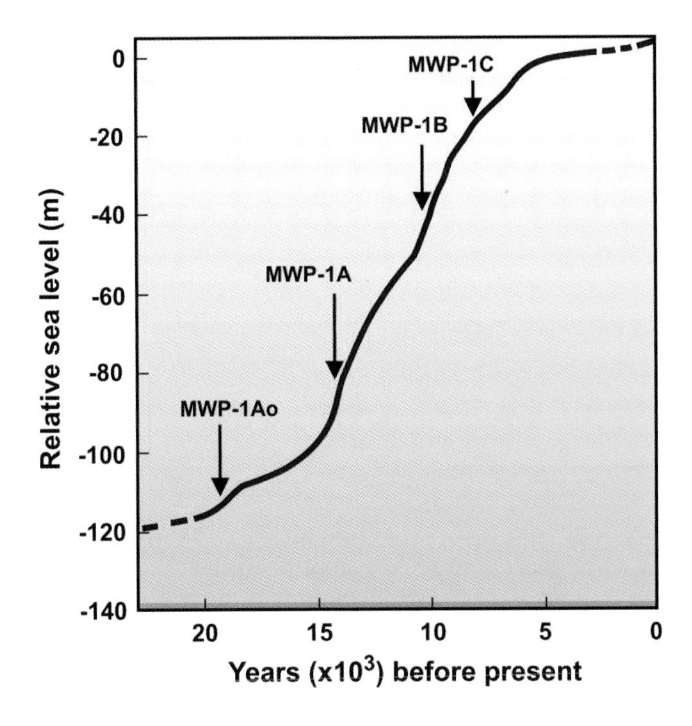

FIGURE 7.1 Illustration of relative sea level rise since the last ice age: 26,000 years ago, sea level is estimated to have been about 400 feet (120 meters) lower than it is today. This curve was assembled using analyses of coral reefs all over the world. The abbreviation MWP refers to various meltwater pulses, which caused sea level to rise relatively rapidly. MWP-1AO, ~19,000 years ago; MWP-1A, ~14,600 to 13,500 years ago; MWP-1B, ~11,500 to 11,000 years ago; MWP-1C, ~8,200 to 7,600 years ago. SOURCE: Gornitz (2009).

(Church and White, 2006); in the last half of the 20th century, this increased to approximately 0.07 inches (1.8 millimeters) per year (Miller and Douglas, 2004); and over the past 15 years, the rate of sea level rise has been in excess of 0.12 inches (3 millimeters) per year (Katsman et al., 2008; Vermeer and Rahmstorf, 2009). Ice core records show that atmospheric CO_2 concentrations have been rising since about 1830 (see Chapater 6), so sea level and CO_2 increases are generally coincident. Clear indications of interannual and decadal variability can also be seen in Figure 7.2. Distinguishing the effects of natural climate variability from human-caused warming is one of the challenges of understanding the details of past sea level rise and anticipating its future course.

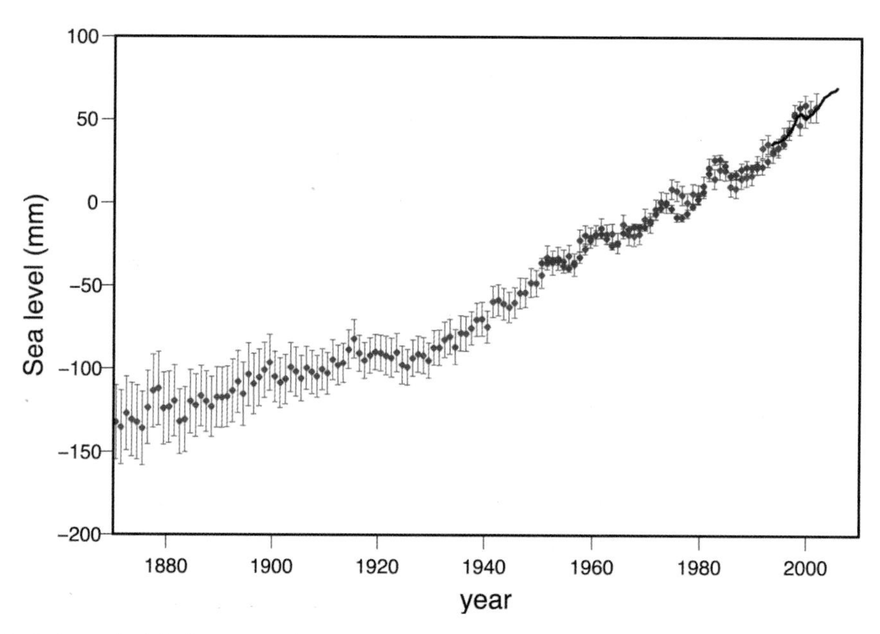

FIGURE 7.2 Annual, global mean sea level as determined by records of tide gauges (red curve with error bars, from Church and White [2006]; blue curve, from Holgate and Woodworth [2004]) and satellite altimetry (black curve, from Leuliette et al. [2004]). For the last half of the 20th century, the rate of sea level rise can be estimated as being about 0.07 in/yr (1.8 mm/yr), with the most recent decade exhibiting a rate of sea level rise over 0.12 in/yr (3 mm/yr). The red and blue curves show deviations in sea level relative to the 1961 to 1990 period; the black curve shows deviations from the average of the red curve relative to the 1993 to 2001 period. SOURCE: Bindoff et al. (2007).

CAUSES OF SEA LEVEL RISE

Past, present, and future changes in global sea level are mainly caused by two fundamental processes: (1) the thermal expansion of existing water in the world's ocean basins as it absorbs heat and (2) the addition of water from land-based sources—mainly ice sheets and glaciers, but also other smaller sources. Geological processes (subsidence and uplift), ocean circulation changes, and other processes are important for determining local and regional rates of sea level rise, but the total volume of the world's oceans—and hence global average sea level—is essentially controlled by thermal expansion and addition of water from land-based sources.

Ocean Thermal Expansion

The ocean is by far the most important heat reservoir in the climate system, with a heat storage capacity more than 1,000 times larger than that of the atmosphere. In fact, measurements of changes in ocean heat content show that 80 to 90 percent of the heating associated with human greenhouse gas (GHG) emissions over the past 50 years has gone into raising the temperature of the oceans (Levitus et al., 2001; Trenberth and Fasullo, 2010) (see Figure 7.3). One consequence of the large thermal capacity of the oceans is that it takes many years for the climate system to warm in response to GHG emissions; for example, as discussed in Chapter 6, global surface temperatures would continue to warm for many decades even if GHG concentrations and other climate forcings were stabilized at present values). Moreover, as heat is absorbed by the oceans, the volume of the water expands, causing sea levels to rise. Approximately 50 percent of the observed sea level rise since the late 19th century has been attributed to thermal expansion of the warming oceans (Gornitz et al., 1982).

Ocean expansion is neither spatially uniform nor steady in time (Levitus et al., 2009; Lozier et al., 2008). Over the last half century, ocean thermal expansion has varied from approximately one quarter of the total sea level contribution (1961-1993) to a

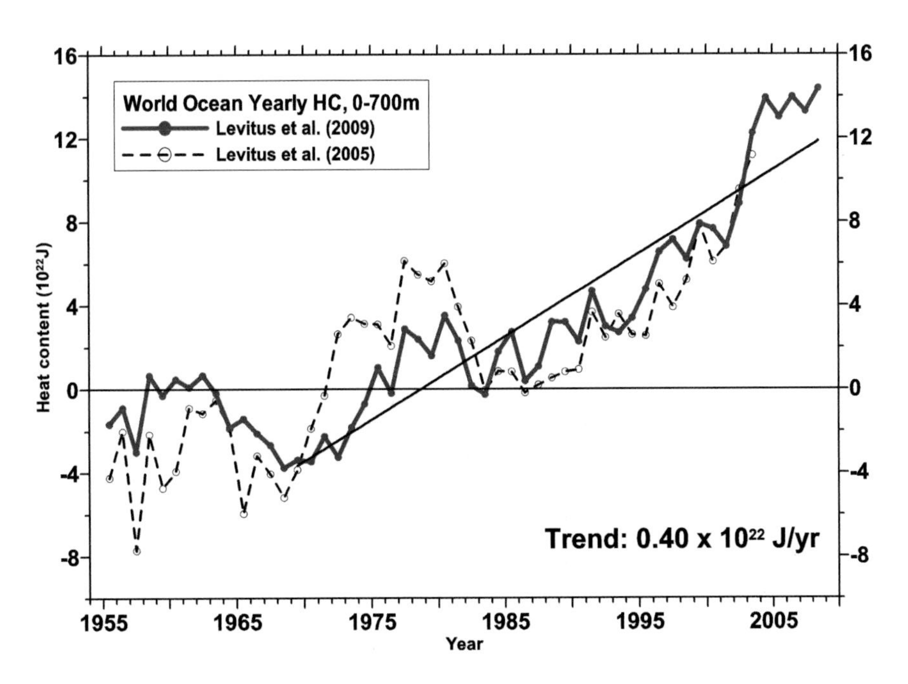

FIGURE 7.3 Increase in globally averaged ocean heat content (HC) for the topmost 700 m of the ocean. The dashed black line represents estimates from Levitus et al. (2005); the red line shows estimates from Levitus et al. (2009). For both lines, the values are calculated with respect to the 1957 to 1990 periods. The solid black line shows the positive trend in ocean heat content from 1969 to 2008. Units are 10^{22} Joules. SOURCE: Levitus et al. (2009).

little over one half (1993-2003; Bindoff et al., 2007). The absorption of heat energy by the oceans varies from place to place on interannual and decadal time scales, and the warmer waters of the tropics and near the ocean surface expand more in response to a given temperature increase than the cold waters at high latitude and at depth (Fofonoff, 1985). Monitoring spatial and temporal heat content changes of the ocean is thus important for predicting both the global average and spatial patterns of future sea level rise, as is developing a better understanding of mixing processes that distribute heat in the oceans.

Ice Sheets

Land ice contained in the world's glaciers and ice sheets contributes directly to sea level rise through melt or the flow of ice into the sea (Figure 7.4). In contrast, when sea ice, which is already floating on the ocean surface, melts, it contributes only a negligible amount to sea level rise (Jenkins and Holland, 2007; Noerdlinger and Brower,

FIGURE 7.4 Outlet glaciers in Northwest Greenland. SOURCE: Photo by K. Steffen.

2007). If all of the water currently stored as ice on land surfaces around the world were to melt, sea levels would rise up to 230 feet (70 meters; Bamber et al., 2001; Lythe and Vaughan, 2001). It is important to note, however, that the estimated time scale for complete melting of the major ice sheets is on the order of hundreds to thousands of years (Gregory et al., 2004; Lambeck and Chappell, 2001; Meehl et al., 2007a).

The major ice sheets of Greenland and Antarctica contain the equivalent of 23 and 197 feet (7 and 60 meters) of sea level rise, respectively. Recent observations of these ice sheets have revealed that not only are they shrinking (e.g., Lemke et al., 2007), but their rate of loss may have increased over the last decade (Lemke, et al., 2007; Rignot et al., 2004, 2008; Thomas et al., 2006; Velicogna, 2009). In Greenland, ice sheet melt has increased 30 percent over the past 30 years (Mote, 2007). In both Greenland and Antarctica, many outlet glaciers are accelerating their seaward flow, hastening the delivery of ice to the surrounding seas (Howat et al., 2007; Rignot and Kanagaratnam, 2006; Rignot et al., 2008). In many cases, when an outlet glacier reaches the sea, a large floating portion extends into the surrounding water, forming long, thin ice tongues or larger, thicker ice shelves that buttress the outlet glacier and restrain some of its discharge. Many of these ice shelves and ice tongues have retreated, thinned, and weakened—and in some cases, collapsed suddenly as seen in Figure 7.5—which has allowed the glaciers that discharge into the surrounding bodies of water to flow much more rapidly (Rignot and Kanagaratnam, 2006; Rignot et al., 2004, 2008; Scambos et al., 2004).

The implications of the loss of floating ice are particularly significant in West Antarctica, where ice shelves are enormous and much of the ice rests on a soft, deformable bed of rock that lies below sea level. The disappearance of Antarctic ice shelves and the retreat of the ice sheet at the continent's margins would allow the surrounding sea water to flow into the ice-bedrock interface, eroding the ice further from underneath and enhancing its discharge. The time scales of these processes are not well known, but, with the equivalent of 11 feet (3.3 meters) of sea level stored in the West Antarctic Ice Sheet (Bamber et al., 2009), this potential instability is of great importance to future sea level rise.

Two mechanisms contribute to the accelerating ice sheet loss to the ocean: (1) increased surface melt (in Greenland) and the associated lubrication of the ice-bedrock interface by surface meltwater during summer, and (2) increased calving processes and thinning at the glacial termini induced by a warming ocean, which in turn leads to faster ice flow and thinning upstream. For the Greenland ice sheet, increased surface melting is associated with earlier onset and longer length of the melt season (Mote, 2007). In addition to the increase in melt runoff, meltwater from the ice sheet surface

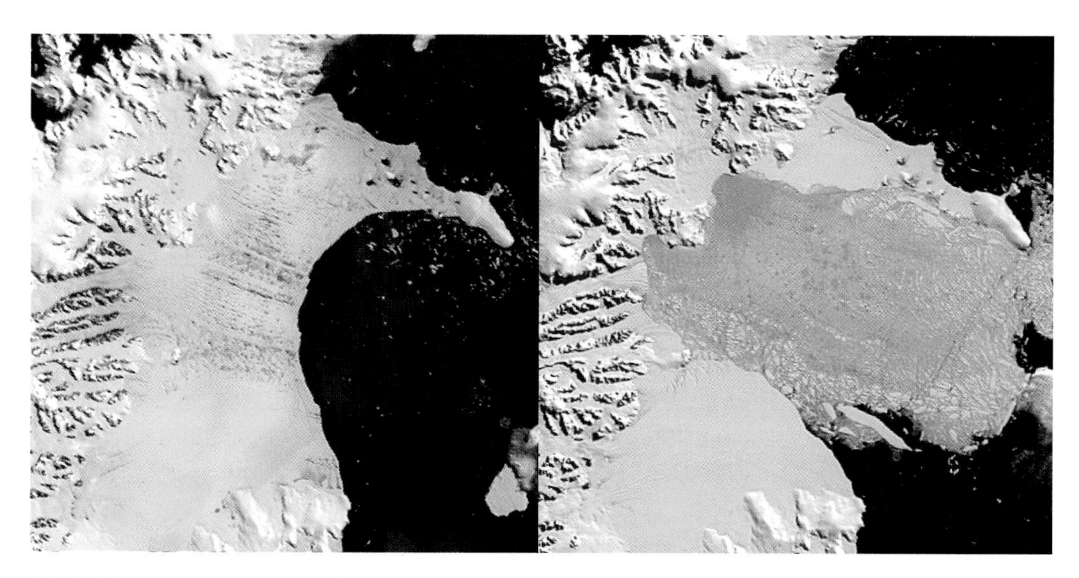

FIGURE 7.5 Larsen-B Ice Shelf (left) January 31, 2002, and (right) March 17, 2002. The 2,018-mile (3,250-km) section of ice shelf, estimated to be over 10,000 years old and 650 feet (200 meters) thick, disintegrated in 6 weeks. White areas correspond to the ice shelf and glaciers on the Antarctic Peninsula, and dark blue/black indicates ocean. The light blue streaks (left panel) correspond to melt ponds on the ice; the larger areas of light blue (right panel) indicate where the ice shelf has collapsed and formed icebergs. Some of the glaciers that fed the ice shelf accelerated by eightfold within months of the collapse. SOURCE: MODIS imagery courtesy of NASA and the National Snow and Ice Data Center.

can penetrate through crevasses or tunnels in the ice (moulins) to the bed, where it can lubricate the ice-bedrock interface, causing a summertime acceleration of glacier flow (Joughin et al., 2008; Zwally et al., 2002). This summer acceleration hastens the flow of ice toward the edges of the ice sheet, where it can melt or calve more rapidly. Recent paleoclimate reconstructions and modeling studies indicate that human GHG emissions have elevated Arctic air temperatures in recent decades by 2.5°F (1.4°C) above those expected from natural climate cycles (Kaufman et al., 2009), meaning that continued surface melting and melting of outlet glacier floating ice tongues can be expected.

Recent analysis of ICESat altimetry data (Pritchard et al., 2009) reveal that ice sheet thinning is mainly confined to the margins for both the Greenland and Antarctic ice sheets. This observation can be ascribed to ocean-driven melting, a mechanism supported by the recent discovery of a warming ocean around Greenland that appears to be contributing to year-round calving into the ocean (Hanna et al., 2009; Holland et al., 2008; Rignot et al., 2010; Straneo et al., 2010). An analysis of time-dependent changes in ice flow rates (Joughin et al., 2008) also suggests that ice-ocean interactions tend to dominate coastal ice losses. Numerical modeling (Nick et al., 2009) further supports this conclusion and suggests that tidewater outlet glaciers adjust rapidly to changing boundary conditions at the calving terminus. Expanded monitoring of both air and sea temperatures at high latitudes and an improved understanding of ice sheet dynamics will be needed to improve scientific knowledge of these processes.

Mountain Glaciers, Ice Caps, and Other Contributors to Sea Level Rise

The world's glaciers and ice caps contain the water equivalent of up to 2.4 feet (0.72 meters) of sea level (Dyurgerov and Meier, 2005). They have consistently been contributing about one quarter of the total sea level rise over the past 50 years, staying roughly proportional to the overall rate of sea level rise (Bindoff et al., 2007). Mountain glaciers are expected to continue to be a significant contributor to sea level rise during this century, and their retreat poses significant risks to populations that depend on glacial runoff as a water source (see Chapter 8). However, unlike the Greenland and Antarctic ice sheets, mountain glaciers are relatively small and do not carry the potential for large and sudden contributions to sea level rise.

There are additional contributions to sea level rise from other human activities such as wetland loss, deforestation, and the extraction of groundwater for irrigation and industrial use. While estimates of the size of these sources are somewhat uncertain, they are believed to be small relative to land ice melting and may be partially offset by the increased storage of water behind dams and in other surface reservoirs over the past century and a half (e.g., Chao et al., 2008). Moreover, the observed recent sea level rise rate of over 0.12 inches (3.3 ± 0.4 millimeters) per year (Cazenave et al., 2010) is consistent with what would be expected from the combination of thermal expansion of the oceans and melting of ice on land (Bindoff et al., 2007). Hence, the overall contribution of other land-based sources to global sea level rise is thought to be small. Nonetheless, small glaciers and ice caps remain important contributors to sea level rise, and their respective contributions need to be better understood.

PROJECTIONS OF FUTURE SEA LEVEL RISE

The Intergovernmental Panel on Climate Change (IPCC) estimated that sea level would rise by an additional 0.6 to 1.9 feet (0.18 to 0.59 meters) by 2100 (Meehl et al., 2007a). However, this projection was based only on current rates of change and was accompanied by a major caveat regarding the potential for substantial increases in the rate of sea level rise. The 2007 IPCC projections are conservative and may underestimate future sea level rise because they do not include one of the two major processes contributing to sea level rise discussed in this chapter: significant changes in ice sheet dynamics (Rahmstorf, 2010). While the growth of ice sheets—mainly through snow accumulation—is an inherently slow process, the processes that govern ice sheet losses, in particular discharge rates, can be strongly nonlinear, with the potential for sudden changes (Overpeck et al., 2006), as illustrated in Figure 7.5. Thus, there is a real potential for ice sheets to shrink rapidly, causing a rapid rise in sea levels. Unfortunately, we do not yet have a good understanding of the processes that control the flow rates; consequently, the potential for rapid ice sheet losses is not well understood at this time. This uncertainty prevented the IPCC from providing a quantitative estimate of how much ice sheet losses might contribute to sea level rise in the coming century.

Research on current and potential future rates of sea level rise has advanced considerably since the IPCC Fourth Assessment Report, which was based on data published in 2005 or earlier. Some research conducted during the past several years suggests that sea level rise during the 21st century could be several times the IPCC estimates, as shown in Figure 7.6. Empirical techniques (e.g., Grinsted et al., 2009; Rahmstorf, 2007; Vermeer and Rahmstorf, 2009) that relate sea level to historical average temperatures

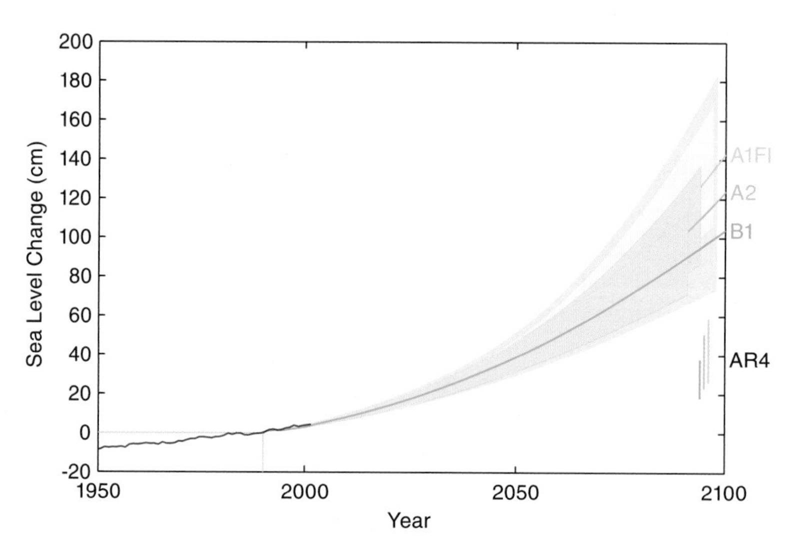

FIGURE 7.6 Projection of sea level rise from 1990 to 2100, based on IPCC temperature projections for three different GHG emissions scenarios (pastel areas, labeled on right). The gray area represents additional uncertainty in the projections due to uncertainty in the fit between temperature rise and sea level rise. All of these projections are considerably larger than the sea level rise estimates for 2100 provided in IPCC AR4 (pastel vertical bars), which did not account for potential changes in ice sheet dynamics and are considered conservative. Also shown are the observations of annual global sea level rise over the past half century (red line), relative to 1990. SOURCE: Vermeer and Rahmstorf (2009).

suggest that a sea level rise of up to nearly 5 feet (1.4 meters) is possible by 2100. By incorporating this empirical effect into models, Horton et al. (2008) estimates a sea level rise of 2 to 2.6 feet (0.62 to 0.88 meters) by 2100. In other work, Rohling et al. (2008) find that a rise rate of up to 5 feet (1.6 meters) per century is possible, based on paleoclimatic evidence from past interglacial periods (including the most recent interglacial period, 110,000 years ago, when global temperatures were 3.6°F [2°C] higher than today and sea levels were 13 to 20 feet [4 to 6 meters] higher). Kopp et al. (2009) estimate that sea level peaked at 22 to 31 feet (6.6 to 9.4 meters) higher than today during the last interglacial period and had a 1,000-year average rise rate between 1.8 and 3 feet (0.56 to 0.92 meters) per century. Pfeffer et al. (2008) used geophysical constraints of ice loss to suggest that a 2.5-foot (0.8-meter) sea level rise is more likely, with a 6.5-foot (2-meter) rise the maximum to be expected by 2100. Others (Siddall et al., 2009) suggest that a 2.5-foot (0.8-meter) rise is the most we could experience by 2100, based on a model that is fit to data only since the last glacial maximum.

The differences among these estimates highlight the uncertainties involved in sea level rise projections; however, there is widespread consensus that substantial long-term sea level rise will continue for centuries to come (Overpeck and Weiss, 2009). A considerable amount of sea level rise is to be expected simply from past CO_2 emissions as the ocean heat content catches up with radiative forcing (see Chapter 6); furthermore, the risk of ice sheet collapse, and the attendant large rates of sea level rise, will increase if GHG concentrations in the atmosphere continue to increase. The task of determining how much sea level rise to expect, when to expect it, and its regional character is a critical scientific challenge given the large numbers of people, assets, and economic activity at risk, and the substantially different planning and management challenges managers would face if they had to prepare for and adapt to a sea level rise of 2, 4, or 8 feet over the course of one century. While the risks cannot be quantified at present, the consequences of extreme and rapid sea level rise could be economically and socially devastating for highly built-up and densely populated coastal areas around the world, especially low-lying deltas and estuaries (Anthoff et al., 2010; Lonsdale et al., 2008; Nicholls et al., 2007; Olsthoorn et al., 2008; Poumadère et al., 2008; see further discussion below).

Regional Variability in Sea Level Rise

As noted above, sea level rise will not be uniform across the globe. Regional variations in the rate of sea level rise occur for a number of reasons. Some coasts are still adjusting to the disappearance of glaciers—the weight of glacial ice pushed them down, and they are still rising in response to the loss of ice. In other regions, coasts may be subsiding because of distant glacial rebound or subsurface fluid withdrawal due to water, oil, or gas extraction. Regional variation in sea level rise rates can also stem from changes in the Earth's rate of spin as water is redistributed from the poles as high-latitude ice melts. Several studies indicate that sea level rise will be particularly problematic for both coasts of the United States as a result of the altered global mass distribution; they may experience 20 percent greater sea level rise than the global average (Bamber et al., 2009; Mitrovica et al., 2001). Differing spatial patterns in sea level trends have already been observed with satellite altimetry (Wunsch et al., 2007).

Changes in the intensity of ocean currents could also produce regional variations in sea level rise. For example, Yin et al. (2009) suggest that a warming-induced slowdown of the Atlantic meridional overturning circulation would contribute to a 6- to 8-inch (15- to 20-centimeter) additional rise in sea level for New York and Boston. However, such changes in the ocean circulation are highly uncertain, since they depend on poorly known parameterizations of vertical mixing in ocean models. Other studies suggest that an intensification, rather than a slowdown, of the overturning circulation with global warming is possible (Huang, 1999; Nilsson et al., 2003), in which case sea levels would fall on the U.S. east coast. A critical factor needed to resolve these disparate projections is a better understanding of vertical mixing processes in the ocean, which are sensitive to changing stratification and govern the absorption of heat by the ocean at all latitudes.

Role of Ice Sheets in Producing Potential Climate Surprises

The same factors that can contribute to accelerated sea level rise over relatively short periods of time could also potentially lead to other abrupt climate changes or "climate surprises" (see Chapter 6). For example, if the Greenland ice sheet were to shrink substantially in a short period of time, freshwater delivery to key deep-water formation regions of the North Atlantic could alter the ocean structure and influence its circulation. Normally, the surface waters of the North Atlantic release large amounts of heat to the atmosphere, thereby becoming sufficiently dense to sink and return southward, making room to be replaced with more warm water from the south. This meridional overturning circulation is important for the oceanic redistribution of heat from the tropics to the Northern Hemisphere; it is confined to the North Atlantic because of its higher salinity and thus greater density than the North Pacific (Haupt and Seidov, 2007).

Compelling evidence has been assembled indicating that rapid freshwater discharges to the North Atlantic due to the breaking of ice dams and drainage of meltwater lakes during the termination of the ice ages caused abrupt circulation changes in the oceans, with significant impacts on regional climate (Boyle and Keigwin, 1987; Lehman and Keigwin, 1992; McManus et al., 2004). The paleoclimate record indicates that the strong meltwater pulses diluted the surface waters of the North Atlantic and rendered them too buoyant to sink, thus shutting down the meridional overturning circulation for centuries at a time (Alley et al., 2003; Broecker, 1987; NRC, 2002a). These shutdowns of the overturning circulation were associated with a dramatic cooling of European climate and also influenced global weather patterns (Vellinga and Wood, 2002). Whether human-caused warming will cause similar abrupt climate changes in the future is an important topic for research (Rahmstorf, 1995) . A freshening of the surface waters of the North Atlantic over the past 50 years has been well documented (Boyer et al., 2005; Curry et al., 2003; Dickson et al., 2002; Levitus, 1989) but it is unclear if climate change will ultimately lead to a gradual slowing or even an acceleration of the meridional overturning circulation (as discussed above). Many models suggest that some slowing of the meridional overturning circulation will result from the ice melting and increased Arctic river discharges that are already taking place, but these models have poor representation of oceanic mixing processes and coastal freshwater discharges. Thus, while the risk of these and other possible abrupt changes in climate should be taken seriously, much work remains to develop confident projections of future ocean circulation changes resulting from the ongoing freshening of the North Atlantic.

IMPACTS OF SEA LEVEL RISE AND OTHER CLIMATE CHANGES ON COASTAL ENVIRONMENTS

Coastal areas are among the most densely populated regions of the United States, and around the world. In 2003, 53 percent of the U.S. population lived in (1) counties with at least 15 percent of its total land area located within the nation's coastal watershed or (2) a county with a portion of its land that accounts for at least 15 percent of a coastal cataloging unit[1] bordering the ocean and the Great Lakes, and 23 of the 25 most densely populated counties in 2003 were coastal counties (Crosset et al., 2005). Considering only coastal counties that border the ocean or contain flood zones with at least a 1 percent chance every year of experiencing flooding from coastal storms and being impacted by wave action, the coastal population, excluding the Great Lakes counties, was 85,640,000, or 30 percent of the total U.S.

[1] The National Oceanic and Atmospheric Administration defines a coastal cataloging unit as "a drainage basin that falls entirely within or straddles an Estuarine Drainage Area or Coastal Drainage Area" (Crowell et al., 2007).

population in the 2000 Census (Crowell et al., 2007).[2] Such population concentration and growth are accompanied by a high degree of development and use of coastal resources for economic purposes, including industrial activities, transportation, trade, resource extraction, fisheries, tourism, and recreation. They also imply significant investments in infrastructure to support these human activities (see Chapters 12 and 13).

While humans have always made use of coastal resources and areas, permanent settlements with high levels of investment and infrastructure are a relatively recent phenomenon, as prehistoric peoples and even early settlers of the United States did not have the technology to protect themselves against storms. The modern concentration of people, human activities, development, and infrastructure is taking place in one of the most dynamic environments on Earth, where land, ocean, and climate are constantly changing. This interaction between a highly variable natural environment and the growing pressures from human use and development produces multiple stresses that make coastal areas particularly vulnerable to additional impacts from climate change.

The IPCC (Nicholls et al., 2007), the recent *Global Climate Change Impacts on the United States* report by the U.S. Global Change Research Program (USGCRP, 2009a), and other studies have documented that a growing number of well-studied coastal areas are already experiencing the effects of rising sea levels and related changes in climate. Physical damage and economic losses from coastal storms and related flooding, erosion, and cliff failures in highly developed regions are increasing; coastal wetlands, hemmed in by human development and deprived of river-borne sediment supplies, are being lost at an increasing rate; the frequency of coral bleaching and mortality events is increasing (see Chapter 9); water quality is declining from the combined impacts of effluent, higher water temperatures, and changes in runoff; saltwater is increasingly intruding into coastal groundwater resources (see Chapter 8); and coastal ecosystems are almost exclusively negatively impacted by the combination of all these climatic changes and human pressures, undermining fisheries, tourism, and long-term sustainability of coastal areas (Nicholls et al., 2007).

Increases in average sea level magnify the impacts of extreme events on coastal landscapes. Relatively small changes in average sea level can have dramatic impacts on storm surge elevation and on the inland extent and frequency of flooding events, depending on coastal topography and the existence of protective structures such as

[2] The selection of the most appropriate demographic data set for evaluating vulnerability to sea level rise (or any other impact of climate change) depends on the focus, scale, and purpose of the study (see, e.g., Crowell et al., 2010).

seawalls, levees, and dikes (e.g., Kirshen et al., 2008). For example, analyses for San Francisco Bay indicated that increases in average sea level as small as 1 foot (0.3 meter) would lead to floods as high as today's 100-year floods (that is, a flood that could be expected to occur once every one hundred years under current climate conditions) every 10 years (Field et al., 1999). Interestingly, in a number of locations along the U.S. coastline, average higher high water (the higher of the two high waters of any tidal day) is rising faster than average sea level, for reasons not yet fully understood; this increases the risk of extensive coastal flooding even more than the rise in average sea levels would suggest (Flick et al., 1999, 2003). In general, the direct losses of coastal habitat and built environments from gradual sea level rise can be greatly amplified by the far larger impacts of flooding, erosion, and wind damage caused by extreme events (Adams and Inman, 2009; Flick, 1998; Nicholls and Tol, 2006; Nicholls et al., 1999; Pendleton et al., 2009; Sallenger et al., 2002; Zhang et al., 2004).

The economic impacts of climate change and sea level rise on coastal areas are probably the second most frequently studied economic impacts in the United States after those on agriculture. Since the first study of this sort in 1980 (Schneider and Chen, 1980), economic impact assessment methodologies have become increasingly sophisticated, though they remain partial and subject to the commonly cited challenges of cost-benefit analyses (see Chapter 17). Analysts have examined the damage potential of gradual sea level rise on taxable real estate in coastal areas subject to inundation; expected impacts of extreme events (floods) on land loss, housing structures, property values, and building contents, as well as integrated impact analyses of combined sea level rise and extreme events; the wider impacts of sea level rise on economies dependent on coastal areas; and the cost of various response options (e.g., seawalls and other hard structures to prevent inundation or erosion loss, beach nourishment requirements as higher sea levels increase the rate of coastal erosion and sediment movement, and relocation or retreat from the shoreline; e.g., Bosello et al., 2007; Nicholls et al., 2007; Yohe et al., 1999). Simple conclusions about the nationwide magnitude of economic impacts cannot be drawn from these studies as metrics, modeling approaches, sea level rise projections, inclusions of coastal storms, and assumptions about human responses (e.g., the type and level of protection) vary considerably. The U.S. National Assessment's coastal sector assessment (Boesch et al., 2000) estimated the cumulative cost of an 18-inch (46-centimeter) sea level rise by 2100 at between $20 and $200 billion, and a 3-foot (roughly 1-meter) sea level rise produced roughly double this figure. The wide range of estimates illustrates the considerable uncertainties involved in the underlying assumptions and calculations. Consistent approaches across U.S. coastal regions would provide much improved understanding of the economic threats.

Steady progress is being made toward more interdisciplinary, integrated analyses of the impacts of sea level rise and other climate and climate-related changes on coastal areas (see Box 4.2). However, most analyses to date still have not assessed economic impacts on culturally or historically important sites, or on coastal infrastructure such as wastewater treatment plants, water supply (drinking water treatment and desalination facilities), utilities (natural gas, electricity, and telephone lines), roads, airports, harbors, and other transportation infrastructure although there are some notable exceptions for certain U.S. locations (e.g., Heberger et al., 2009; Larsen et al., 2008; NRC, 2008g). Impacts on nonmarket values such as the loss of natural habitat have been equally challenging to assess and therefore are still often omitted from economic impact assessments. Coastal ecosystems such as dunes, wetlands, seagrass beds, and mangroves provide numerous ecosystem goods and services, ranging from nursery habitat for certain fish and shellfish to habitat for bird, mammal, and reptilian species, including some endangered ones; protective or buffering services for coastal development against the impacts of storms; water filtering and flood retention; and the aesthetic, cultural, and economic value of beaches and coastal environments for recreation, tourism, and simple enjoyment (for detailed reviews of this literature see Darwin and Tol, 2001; Nicholls et al., 2007; West and Dowlatabadi, 1999).

As climate continues to change and sea level continues to rise through the twenty-first century, these physical, ecological, and socioeconomic impacts on coastal areas are expected to increase and intensify. Moreover, they can be expected to be exacerbated by continued growth in human pressures on coastal areas. Even if sea level rise were to remain in the conservative range projected by the IPCC (0.6 to 1.9 feet [0.18 to 0.59 meters])—not considering potentially much larger increases due to rapid decay of the Greenland or West Antarctic ice sheets—tens of millions of people worldwide would still become vulnerable to flooding due to sea level rise over the next 50 years (Nicholls, 2004; Nicholls and Tol, 2006). This is especially true in densely populated, low-lying areas with limited ability to erect or establish protective measures. In the United States, the high end of the conservative IPCC estimate would result in the loss of a large portion of the nation's remaining coastal wetlands. The impact on the east and Gulf coasts of the United States of 3.3 feet (1 meter) of sea level rise, which is well within the range of more recent projections for the 21st century (e.g., Pfeffer et al., 2008; Vermeer and Rahmstorf, 2009), is shown in pink in Figure 7.7. Also shown, in red, is the effect of 19.8 feet (6 meters) of sea level rise, which could occur over the next several centuries if warming were to continue unabated.

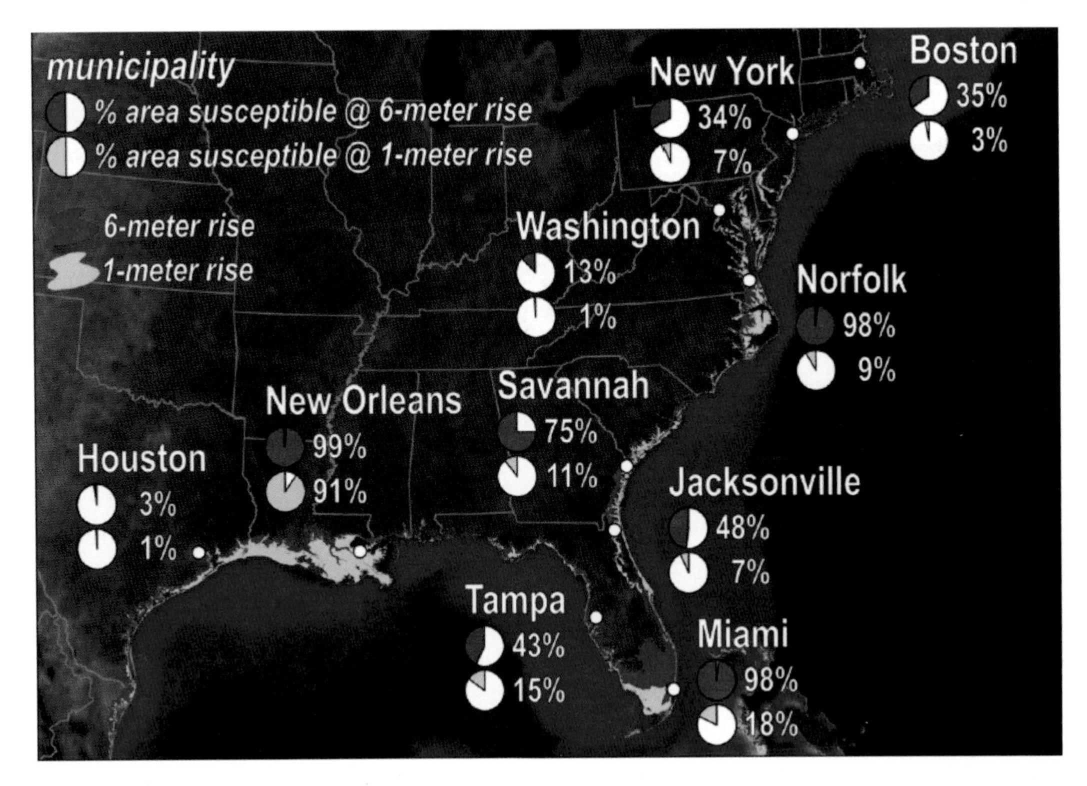

FIGURE 7.7 Areas of the east and Gulf coasts of the United States susceptible to coastal inundation following a 3.3-foot (1-meter; pink shading) or 19.8-foot (6-meter; red shading) sea level rise. Pie charts show the percentage of some cities that are potentially susceptible to 3.3-foot (1-meter; pink) or 19.8-foot (6-meter; red) sea level rise. SOURCE: Overpeck and Weiss (2009).

RESPONDING TO SEA LEVEL RISE

General scientific understanding of people's vulnerability and ability to adapt to sea level rise and other climate changes has increased substantially in recent years, though place-based, sector-specific knowledge remains extremely limited. Developing countries are expected to generally face greater challenges in dealing with the impacts of rising sea levels because of large exposed populations and lower adaptive capacity—which is largely a function of economic, technological, and knowledge resources, social capital, and well-functioning institutions (Adger et al., 2007; Nicholls et al., 2007). However, even in developed countries like the United States, significant gaps remain in our understanding of the impacts of sea level rise, especially for specific locations (Moser, 2009a), as well as considerable challenges in translating our greater adaptive capacity into real adaptation action on the ground (Adger et al., 2007, 2009b;

NRC, 2010a; O'Brien et al., 2006; Repetto, 2008). Certain technological options such as seawalls and levees are not feasible in all locations, and in many they could negatively impact the coastal ecology, beach recreation and tourism, and other social values (e.g., aesthetics). At the same time, a wide range of barriers and constraints make "soft" solutions equally challenging—these include changes in land use, planning, and, ultimately, retreat from the shoreline, with the associated costs and social and ecological consequences. Such constraints and limits on adaptation are increasingly recognized and researched (Adger et al., 2009b; Moser and Tribbia, 2006; Moser et al., 2008; NRC, 2010a). While there is extensive research about, and experience dealing with, coastal hazards, significant further research is required to determine the most appropriate, cost-effective, least ecologically damaging, and most socially acceptable adaptation options in the face of significantly faster rates of sea level rise than has been historically experienced. Past coastal hazards management approaches may not be ecologically sustainable or economically affordable in light of some of the high-end sea level rise projections.

While many research questions about managing coastal ecosystems and hazards remain (see, e.g., JSOST, 2007), the fundamental best practices are well known and include building new structures, elevating existing structures above flood elevation, and maintaining dunes as storm buffers. However, these measures are not frequently employed because underlying incentives and self-reinforcing factors favor continued development in at-risk areas, structural protection, and repeated emergency intervention (Burby, 1998; Kunreuther, 2008; Mileti, 1999; Platt, 1996, 1999). An additional challenge is how to evaluate and weigh near-term costs and benefits against long-term costs and benefits, given that neither is known with much precision and such evaluations are inherently place-specific. A critical research topic is how to foster adaptive coastal management actions with a long-term, systemic perspective while avoiding the worst economic, social, and ecological consequences for coastal areas (see also Chapter 4). Finally, little is known about local vulnerability to sea level rise in the context of multiple stresses, such as increased storm surge or rainfall rates, or about the feasibility and acceptability of various adaptation options. These issues are discussed in a recent synthesis of the impacts of climate change and vulnerability of coastal areas of the U.S. mid-Atlantic region (CCSP, 2009a; Najjar et al., 2009; Wu et al., 2009).

RESEARCH NEEDS

Significant sea level rise is expected for the foreseeable future, but the physical science of sea level rise and related climatic changes remains incomplete, making specific projections highly uncertain at this time. Moreover, place-specific social and ecological understanding of vulnerability and adaptation potential in coastal regions is severely lacking. Decision makers and other stakeholders should assume that changing scientific understanding as well as environmental and societal conditions will require considerable policy flexibility and make for potentially difficult tradeoffs. Thus, an adaptive risk management approach is indicated (see Box 3.1 and NRC, 2010b). Key research advances that would assist in improving both understanding and decision making include the following:

Reduce the scientific uncertainties associated with land-ice changes. Comprehensive, simultaneous, and sustained measurements of ice mass and volume changes and ice velocities are needed, along with measurements of ice thickness and bed conditions, both to quantify the current contributions of ice sheets to sea level rise and to constrain and inform ice sheet model development for future assessments. These measurements, which include satellite, aircraft, and in situ observations, need to overlap for several decades in order to enable the unambiguous isolation of ice melt, ice dynamics, snow accumulation, and thermal expansion. Equally important are investments in improving ice sheet process models that capture ice dynamics as well as interactions with the ocean and the ice bed. Efforts are already under way to improve modeling capabilities in these critical areas, but fully coupled ice-ocean-land models will ultimately be needed to reliably assess ice sheet stability, and considerable work remains to develop and validate such models—especially given the relatively small number of qualified researchers currently working in this area. Sustained observations and analysis programs are also needed for improving understanding and projections of glaciers and ice caps. Finally, additional paleoclimate data from ice cores, corals, and ocean sediments would be valuable for testing models and improving our understanding of the impacts of sea level rise.

Improve understanding of ocean dynamics and regional rates of sea level rise. Direct, long-term monitoring of sea level and related oceanographic properties via tide gauges, ocean altimetry measurements from satellites, and an expanded network of in situ measurements of temperature and salinity through the full depth of the ocean water column are needed to quantify the rate and spatial variability of sea level change and to understand the ocean dynamics that control global and local rates of sea level rise. A better understanding of the dependence of ocean heat uptake on vertical mixing and the abrupt change in polar reflectivity that will follow the loss of summer sea ice in the Arctic are some of the most critical improvements needed in ocean and Earth system models. In addition, oceanographic, geodetic, and coastal models are needed to predict the rate and spatial dynamics of ocean thermal expansion, sea level rise, and coastal inundation. The need for regionally specific information creates additional challenges. For example, coastal inundation models require better bathymetric data in the coastal seas, improved elevation data on land, the inclusion of wave and spillover effects, better data on precipitation rates and stream flows, ways of dealing with storm-driven sediment transport, and the ability to include the effects of built structures on coastal wind stress patterns.

Develop tools and approaches for understanding and predicting the vulnerability to, and impacts of, sea level rise on coastal ecosystems and coastal infrastructure, as well as for translating this understanding into decision-relevant information. The impacts of sea level rise on wetlands, coral reefs, marine fisheries, and estuarine bays and rivers need to be evaluated in concert with the impacts associated with increasing levels of CO_2 in the atmosphere and oceans, increasing nutrient inputs from land, and changes in use or management (see also Chapter 9). Likewise, the impacts of sea level rise on infrastructure, including ports, roads, cities, dikes, levees, and freshwater aquifers and storage facilities, should take into account potential shifts in storm patterns, rainfall rates, and other climate changes (see also Chapters 12 and 13). Improved valuation of nonmarket values, and development of decision-support tools to assess the trade-offs between physical, ecological, and social impacts and response options (see below) are needed to inform coastal management decisions that require long lead times.

Expand the ability to identify and assess vulnerable coastal regions and populations and to develop and assess adaptation strategies to reduce their vulnerability. With sea level rise acting in combination with other physical, social, and economic stressors, the ability to assess the social-ecological vulnerability of coastal regions, improve society's adaptive response options (through technological, economic, and land use changes), and identify constraints to adaptation (including legal, social, political, infrastructure-related, and economic issues) are all critical research needs (see also Chapter 4). This area of research has received very little attention to date, leaving many U.S. coastal communities without adequate place-specific information to inform their adaptation decisions.

Develop decision-support capabilities for all levels of governance. Methods for identifying preferences and weighing alternative adaptive responses will be needed as environmental and social conditions change over time. Frameworks and approaches need to be developed for the evaluation of market and nonmarket values of affected assets and habitats; of the economic costs and other consequences of different response options to sea level rise on both highly developed and less developed shorelines; and of the social and environmental feasibility of different adaptation options (including technological, economic, physical, ecological, social, or legal options) for different coastlines. This will require improved information of the kinds listed above, as well as financial and technical resources that enable decision makers to engage in adaptation planning and actions.

Build capacity. There is a significant shortage of expertise to conduct place-based vulnerability and adaptation needs assessments in coastal regions of the United States (as well as in other sectors and regions, as discussed in Chapter 4), making it extremely challenging to meet the rapidly increasing demand for such information by decision makers. Thus, a strong emphasis on training and capacity building is needed to generate human resources that can produce and also use the information essential for effective adaptation planning along U.S. coasts.

CHAPTER EIGHT

Freshwater Resources

Humans and ecosystems rely on water for life. The availability of water depends on both the climate-driven global water cycle and on society's ability to manage, store, and conserve water resources. Climate change is affecting both the quantity and quality of Earth's water supplies. Already, precipitation amounts and patterns are changing, and these trends are expected to continue or intensify in the future. This creates significant challenges for water resource management, especially where current water rights and consumption patterns were established under climate conditions different from the conditions projected for the future. Moreover, climate change is not the only problem putting demands on water supplies. Growing populations and consumptive use may cause shortages in some regions. Responding to these challenges will require better data and modeling as well as a better understanding of both the impacts of climate change and the role of water governance on water resources.

Questions water managers and other decision makers are asking, or will be asking, about climate change include the following:

- Given the relatively large uncertainties in model projections of future precipitation, what actions can we take now that we will not regret in 20 or 30 years?
- How robust are different long-term water management strategies under various scenarios of future climate change?
- Are there management and decision-support tools that can help us balance the water needs for urban, agricultural, energy, and in-stream environmental requirements, improve time-dependent decisions, and illuminate the relevant trade-offs?
- How can water institutions and legal mechanisms be modified to improve flexibility and fit changing baseline conditions? What can we learn from other regions and countries about flexible and fair water use?
- How can we develop tools to assess preparedness and develop capacity to respond to extreme water-related events such as flooding or drought?

Scientific research has steadily increased our understanding of how climate change is affecting freshwater resources. Changes in freshwater systems are expected to create significant challenges for flood management, drought preparedness, water supplies, and many other water resource issues. The research summarized in this chapter provides an overview of freshwater resources and what is known about how climate change will affect freshwater availability. We also indicate research needs and outline some of the fundamental challenges of making projections of climate impacts on water resources and governance strategies.

SENSITIVITY OF FRESHWATER RESOURCES TO CLIMATE CHANGE

Historically, the United States has relied heavily on surface water, and to a lesser extent groundwater, to meet its freshwater needs. It would be easy to assume that precipitation is the most critical factor in determining surface water availability, and thus future water supplies will be controlled almost entirely by changes in average annual precipitation. In reality, however, the relationship between climate change and water supplies is more complex. For example, climate change directly affects temperatures, and hence evaporation from soil and water surfaces, plant transpiration, and mountain snowmelt. The average intensity, seasonality, mode (i.e., rain or snow), and geographic distribution of precipitation are also important for water management decisions. All of these characteristics are closely connected to storm patterns, which are modulated by regional and global patterns of variability on a range of time scales, and both storm patterns and patterns of variability may shift as climate change progresses (e.g., Kundzewicz et al., 2007; Lemke et al., 2007; Trenberth et al., 2007). Moreover, water cycling through soils, land cover, and geologic formations, as well as rainfall intensity and amount, all affect the volume of surface runoff as well as infiltration rates and groundwater recharge, making the response of water resource systems to climate change complex. Changes in land cover and land use will complicate projections of water resource availability as well as the detection and attribution of climate-driven trends; for example, land degradation with accompanying vegetation changes can be a dominant driver of changes in stream flow (Wilcox et al., 2008). In many coastal regions, sea level rise (see Chapter 7) will affect surface and groundwater resources.

The complex processes involved in the water cycle, combined with uncertainties in model projections of future precipitation changes, prevent any easy conclusions about how climate change will affect regional water supplies. Even if model projections do not show any significant changes in total precipitation, for example, shifts in seasonal precipitation patterns or average storm intensity may be critical for water-dependent sectors like agriculture. As discussed in Chapter 6 and the next section below, a higher fraction of rainfall is expected to fall in the form of heavy precipitation events as temperatures increase, and in many locations such a shift has already been observed (see also CCSP, 2008f; Bates and Kundzewicz, 2008). Higher temperatures are also projected to increase soil and surface water evaporation, producing overall drier conditions even if total precipitation remains constant. Higher temperatures and runoff from intense rainfall can both negatively affect the physical and chemical characteristics of freshwater and thus water quality.

Despite considerable improvements in modeling, significant uncertainties remain in projections of precipitation—including its distribution, intensity, frequency, and other characteristics—as well as in related variables such as land use and land cover change. These uncertainties are compounded by uncertainties in our technical capacity to store, manage, and conserve water resources, as well as in socioeconomic, cultural, and behavioral issues that shape the use of water. Multisectoral planning and sophisticated decision-support tools can help water resource managers avoid the most undesirable consequences of climate change in their areas of responsibility (Bates and Kundzewicz, 2008; Gleick, 2000; Vorosmarty et al., 2000). Adaptive water management approaches at operational time scales will be particularly important (e.g., Georgakakos et al., 2005), and long-term strategic decisions need to be robust—that is, able to meet water management goals under a range of plausible future climate conditions (e.g., Dessai and Hulme, 2007; Lempert, 2002; Lempert and Collins, 2007; Lempert et al., 2003).

HISTORICAL AND FUTURE CHANGES IN FRESHWATER

Precipitation: Frequency, Intensity, Storminess

Observed changes in precipitation are broadly consistent with theoretical expectations and reasonably simulated by global climate models (Bates and Kundzewicz, 2008; Trenberth et al., 2007; Zhang et al., 2007a). While total precipitation in the United States has increased by about 5 percent over the past 50 years, there are significant regional differences, with generally wetter conditions in the Northeast and generally drier conditions in the Southeast and particularly the Southwest (Figure 8.1) (see also Field et al., 2007b). A wide range of climate models using different emissions scenarios predict that these regional trends will continue, with generally robust model results for the north and with high uncertainty for the south (Christensen et al., 2007; USGCRP, 2009a). Other factors in addition to temperature influence precipitation. Specifically, uncertainty remains in our understanding of the effects of aerosols on cloud formation and precipitation. For example, climate models underestimate the magnitude of the observed global land precipitation response to 20th-century volcanic forcing (Hegerl and Solomon, 2009) as well as human-induced aerosol changes (Gillett et al., 2004; Lambert et al., 2005).

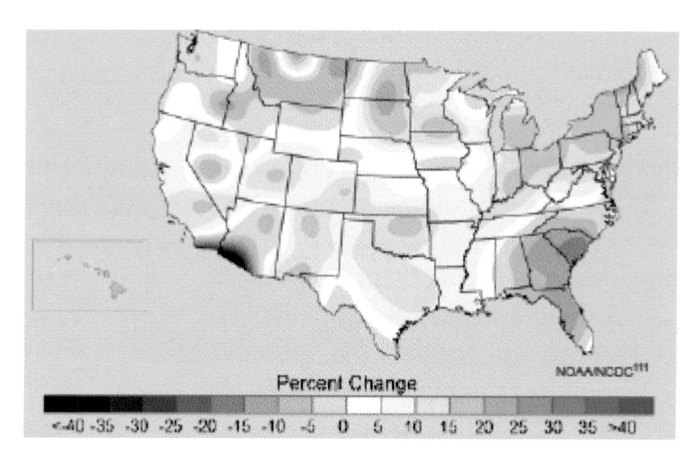

FIGURE 8.1 Observed annual average precipitation changes in the United States between 1958 and 2008. Blue indicates areas that have gotten wetter, while brown indicates areas that have gotten drier. SOURCE: USGCRP (2009a); data from NOAA/NCDC (2008).

Historical data also show an increase in precipitation intensity. In the United States, the fraction of total precipitation falling in the heaviest 1 percent of rain events increased by about 20 percent over the past century (Gutowski et al., 2008). Most climate models project that this trend will continue (Bates and Kundzewicz, 2008) and also project a strong seasonality, with notable summer drying across much of the Midwest, the Pacific Northwest, and California (Hesselbjerg and Hewitson, 2007).

Changes in major storm events are of interest both because a significant fraction of total U.S. precipitation is associated with storm events and because storms often bring wind, storm surges, tornadoes, and other threats. Tropical storms, which become hurricanes if they grow to a certain intensity, are of particular interest because of their socioeconomic impacts (e.g., Hurricane Katrina; see Box 4.3). Changes in the intensity of hurricanes have been documented and attributed to changes in sea surface temperatures (Emanuel, 2005; Trenberth and Fasullo, 2008), but the link between these changes and climate change remains uncertain (Knutson et al., 2010). Recent model projections indicate growing certainty that climate change could lead to increases in the strength of hurricanes, but how their overall frequency of occurrence might change is still an active area of research (Bender et al., 2010; Knutson et al., 2010). Extratropical storms, including snowstorms, have moved northward in both the North Pacific and the North Atlantic (CCSP, 2008f), but the body of work analyzing current and projected future changes in the frequency and intensity of these storms is somewhat inconclusive (Albrecht et al., 2009; Hayden, 1999). Historical data for thunderstorms and tornadoes are insufficient to determine if changes have occurred (CCSP, 2008f).

Snowpack, Glaciers, and Snowmelt

Worldwide, snow cover is decreasing, although substantial regional variability exists (Lemke et al., 2007; Slaymaker and Kelly, 2007). Since the 1920s, Northern Hemisphere snow cover has steadily declined (Figure 8.2), despite increased precipitation. Between 1966 and 2005, the total area of Northern Hemisphere snow cover shrank by approximately 1.4 percent per decade. In the Southern Hemisphere, there has been no significant trend in South American snow cover, and data are sparse and inconclusive in Australia and New Zealand.

In the United States, snowpack changes in the West currently represent the best-documented hydrological manifestation of climate change (e.g., Barnett et al., 2008; Pierce et al., 2008). About half of the observed decline in western snowpack, and resulting changes in the amount and seasonality of river discharge, can be linked to a warming climate. The largest losses in snowpack are occurring in the lower elevations of the mountains of the Northwest and California, because higher temperatures are causing more precipitation to fall as rain instead of snow. Moreover, snowpack is melting as much as 20 days earlier than the historical average in many areas of the West (Kapnick and Hall, 2009; Kim and Waliser, 2009; Stewart et al., 2005). Snow is expected

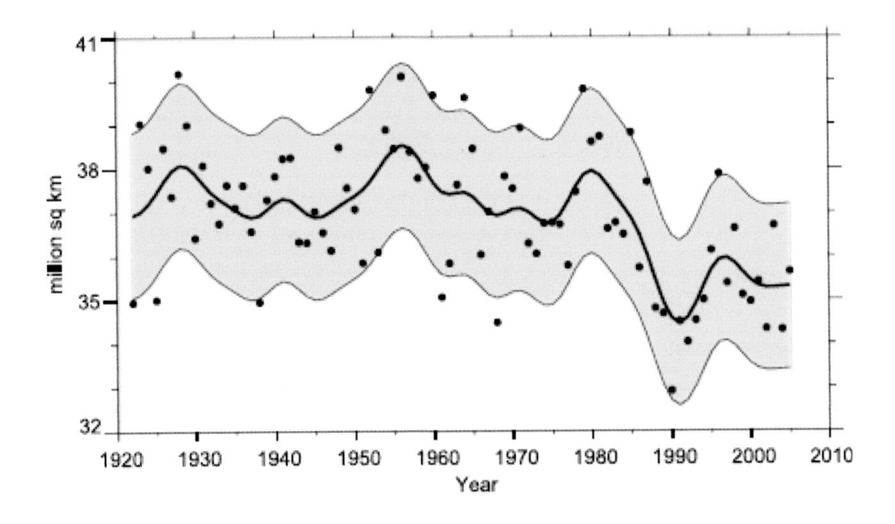

FIGURE 8.2 Area of Northern Hemisphere covered by snow in the spring. There is an overall trend toward a decrease in the area covered by snow for the entire period (1922-2005). The black dots correspond to individual years, the smooth black line shows decadal variations, and the yellow area indicates the 5 to 95 percent confidence range associated with decadal variations. SOURCE: Lemke et al. (2007).

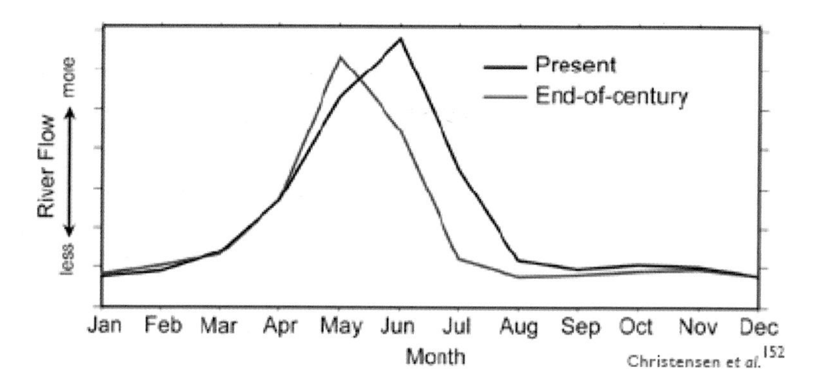

FIGURE 8.3 An example of how the timing and amount of runoff is projected to change following warming in the 21st century. The black line shows the amount of stream flow occurring in the Green River, which is part of the Colorado River Basin. The stream flow of the Green River is dominated by the timing and amount of snowmelt, and peak flows historically have occurred around June. Warming in the twenty-first century would tend to decrease snowfall during the winter and accelerate the timing and pace of snowmelt, leading to earlier peak flows and overall less stream flow (red line). SOURCE: USGCRP (2009a); data from Christensen et al. (2004).

to melt even earlier under projections of future climate change, resulting in reduced later-summer stream flows (Figure 8.3). This change would have major implications for ecosystems, hydropower, urban and agricultural water withdrawals, and requirements for other water uses. In regions where the summer growing season is the dry season, as in much of the western United States, this concentration of runoff in the spring and reduction in summer will stress water supply systems and could lead to summer water shortages (Barnett et al., 2005b; Cayan et al., 2009).

Finally, as discussed in Chapter 7, nearly all of the world's glacier systems are shrinking, and in many cases their rate of ice loss has been accelerating. Disappearing glaciers are ultimately expected to lead to reductions in river flows during dry seasons and lost water resources for the hundreds of millions of people who rely upon glacier-fed rivers worldwide (Barnett et al., 2005b). Changes in glacier-stream flow interactions are also expected to lead to changes in ecosystems and in water quality (Milner et al., 2009).

Elements of the Terrestrial Water Cycle: Surface and Groundwater Resources

Analyses of stream flow records for the United States over the past several decades show primarily increases, which is consistent with trends in precipitation (Lins and Slack, 2005). However, these observed changes in stream flow are due in large part to the aggregate effects of many human influences, of which climate change is only one (Gerten et al., 2008). Of the world's 200 largest rivers, 22.5 percent showed downward trends over the period 1948 to 2004, and 9.5 percent showed upward trends, both mostly as a result of climate variations (Dai et al., 2009). While projections of runoff changes generally mimic precipitation trends, such projections are uncertain in part because runoff is influenced by rates of evapotranspiration—the sum of evaporation of water from the surface and transpiration of water though the leaves of plants. The effects of temperature change and changes in CO_2 on plant processes can in turn affect evapotranspiration, and thus the magnitude of runoff (Gedney et al., 2006; Piao et al., 2007; Wolock and Hornberger, 1991).

Extreme conditions, namely floods and droughts, are generally of greatest concern to water managers. In addition to climate change, these events and can be magnified by human-influenced factors such as urbanization, streambed alterations, and deforestation. It is not clear whether the frequency of extreme runoff events has increased during the last several decades. Milly et al. (2002) reported a measurable increase in large floods, but Kundzewicz et al. (2005) found 27 increases, 31 decreases, and 137 with no significant trend in 195 catchments worldwide. These differences reflect both the regional nature of precipitation shifts as well as the multiple changes occurring in any individual region. For example, catchment-specific land use changes and streambed modifications may have occurred over the period of record and may mask or enhance the climate change signal. Such challenges suggest that adaptive water management decisions will require regional climate information and may differ in their specific application from one river basin to another. Given the observed increases in heavy precipitation events and the expectation that this intensification will continue, assessments indicate that generally, the risk from floods will increase in the future. However, local water, land use, and flood risk-management decisions can modify the actual flood vulnerability of communities and built infrastructure (Kundzewicz et al., 2007). Flood-control measures themselves can be a primary reason for changes in intensity of flooding (Pinter et al., 2008).

Long-term records do not exist for evapotranspiration. Trends in pan evaporation, a standard measurement of water loss to the atmosphere from an exposed pan of water at some meteorological stations, are actually negative for the past several decades in the United States (Golubev et al., 2001), which is the opposite of what would be expected under a warming climate. Several explanations are possible. Brutsaert and Parlange (1998) argue that pan evaporation reflects potential rather than actual evapotranspiration and that actual evapotranspiration and pan evaporation should have opposite signs due to feedbacks caused by the heat transferred during the transformation of water from liquid to vapor. An alternate explanation is that net surface radiation actually decreased in the United States during the past several decades due to increased cloudiness, and hence actual evapotranspiration decreased (Huntington, 2004). Discerning trends and making projections for evapotranspiration is complicated further by the indirect effect of increased CO_2 concentrations, which can alter plants' water-use efficiency (Betts et al., 2007). Thus, although evapotranspiration is a critically important process in the water cycle, our ability to understand trends and to predict the impacts of climate change on it is limited (Fu et al., 2009; Kingston et al., 2009).

Groundwater

Some regions of the United States rely partially—and others, such as Florida, mainly—on groundwater for drinking, residential use, and agriculture. According to the U.S. Geological Survey (USGS, 2004), total groundwater withdrawals in the country in 2000 amounted to 84,500 million gallons per day—about one quarter of total freshwater withdrawals. In the central United States, usage of the Ogalalla aquifer, mainly for agriculture, is withdrawing groundwater much faster than it can be recharged (Alley et al., 1999) and other aquifer systems are also being depleted (USGS, 2003). Significant changes in future rainfall rates will create additional vulnerabilities associated with groundwater usage.

The impacts of climate change on groundwater are far from clear; in fact, little research effort has been devoted to this topic. Changes in precipitation and evaporation patterns, plant growth processes, and incursions of seawater into coastal aquifers will all affect the rate of groundwater recharge, the absolute volume of groundwater available, groundwater quality, and the physical connection between surface and groundwater bodies (USGCRP, 2009a). Already, as climate change-driven impacts and other pressures on water resources unfold, water managers in drier regions of the United States find themselves confronted with the need to expand groundwater withdrawal and develop groundwater recharge schemes and infrastructure. The inconsistent regulation of groundwater and surface water from state to state and the lack of readily available legal mechanisms to link ground- and surface-water management—even where they are physically linked—makes comprehensive, integrated water management difficult.

Drought

Drought is a complex environmental impact and is affected strongly by the balance between precipitation and evapotranspiration and the concomitant effect on soil moisture. Global climate models predict increasing summer temperatures and decreasing summer precipitation in many continental areas, implying reductions in soil moisture. Long-term records of soil moisture are sparse, and the records that do exist do not show depletion of soil moisture, possibly due to reductions in solar radiation reaching the Earth's surface due to increased cloudiness (Robock et al., 2005). A surrogate indicator, derived from land-surface models, is the Palmer Drought Severity Index, which measures the duration and intensity of long-term drought-inducing patterns through thousands of data points such as rainfall, snowpack, stream flow, and other water supply indicators.[1] The historical record of the Palmer Index from 1870 to 2002 shows that very dry areas have more than doubled globally since the 1970s, and the expansion after the 1980s is associated with surface warming (Dai et al., 2004). However, there are considerable year-to-year variations in soil moisture associated with the El Niño-Southern Oscillation and other modes of climate variability, and model projections of soil moisture for the 21st century do not provide a consistent indication of future changes (Trenberth et al., 2007). This uncertainty in future soil moisture projections leads to uncertainties about ecosystem dynamics and projections of agricultural productivity and, thus, presents a challenge for farmers, natural resource managers, and others trying to plan adaptation measures.

Attributing increases in severe droughts to human causes using observed data is difficult (e.g., Seager et al., 2009) and cannot currently be done unambiguously (Seager et al., 2007; Sun et al., 2007). For the United States, trend analyses indicate that droughts decreased in intensity, duration, and frequency over the period from 1915 to 2003, except in the Southwest (Andreadis and Lettenmaier, 2006; Sheffield and Wood, 2008). However, other analyses (Groisman and Knight, 2008) suggest increases in extended dry periods over the past 40 years. Model projections indicate that the area affected by drought will probably increase in the decades ahead (Bates and Kundzewicz, 2008) and that the number of dry days annually will also increase (Kundzewicz et al., 2007). In snowmelt-dominated systems, the risk of drought is expected to increase (Barnett et al., 2005b).

[1] For an overview of the Palmer Drought Severity Index and limitations on its use, see Alley (1984).

Water Quality

Changes in the water temperature of lakes and rivers have consequences for freshwater quality (Bates and Kundzewicz, 2008). Increased temperatures generally have a negative impact on water quality, typically by stimulating growth of nuisance algae. Changes in heavy precipitation, runoff, and stream flow can impact a diverse set of water quality variables (Kundzewicz et al., 2007). Water quality will also be negatively affected by saline intrusion into coastal aquifers as sea levels rise (Kundzewicz et al., 2009; Alley et al., 1999; see also Chapter 7). In general, however, the water quality implications of climate change are less well understood than its impacts on water supply.

MANAGING FRESHWATER IN A CHANGING CLIMATE

In the face of the many, and sometimes uncertain, impacts on freshwater resources outlined above, water managers face a variety of challenges. For example, new infrastructure construction (e.g., large dams) is expected to be limited, so water managers will have to develop and implement approaches to increase the efficiency of water use (Gleick, 2003a,b). On the other hand, existing water infrastructure (e.g., reservoirs, conveyer pipes, sewage lines, and treatment plants) will need to be maintained and upgraded, which offers opportunities for taking account of current and projected impacts of climate change (e.g., ASCE, 2009; EPA, 2008; King County, 2008). Projections of freshwater supply as well as climate change impacts on water infrastructure itself are uncertain, so water managers will need more information about risks and about managing water in the face of uncertainty (Beller-Simms et al., 2008; CDWR, 2008; Delta Vision Blue Ribbon Task Force, 2007; EPA, 2008; Wilby et al., 2009). In addition to tools and models that expand their range of response options, managers and policy makers will need governance flexibility in order to increase adaptive capacity and resilience in water systems (Adger et al., 2007; Huitema et al., 2009; Zimmerman et al., 2008).

Two options for dealing with these challenges are governance frameworks such as Integrated Water Resources Management and adaptive management. Integrated Water Resources Management often involves reforming broader institutional structures of water governance including decentralization, integration, participatory/collaborative management, and social learning. Adaptive management involves organizational and management processes that maintain flexibility (see Box 3.1). While these frameworks could increase the adaptive capacity of freshwater management, there is still a need for more information about water institutions and governance structures and how they affect human and institutional behavior (Engle and Lemos, 2010; Huitema et al., 2009; Norman and Bakker, 2009; Urwin and Jordan, 2008).

RESEARCH NEEDS

Significant gaps remain in the knowledge base that informs both projections of climate change impacts on water resources and governance strategies that can curb demand and build adaptive capacity of water systems. Critical research needs include the following.

Improved projections of changes in the water cycle at regional and seasonal time scales. Because water most directly affects society at the watershed or regional level, improved regional-scale projections of changes in precipitation, soil moisture, runoff, and groundwater availability on seasonal to multidecadal time scales are needed to inform water management and planning decisions, especially decisions related to long-term infrastructure investments. Likewise, projections of changes in the frequency and intensity of severe storms, floods, and droughts are critical both for water management planning and for adapting the natural and human systems that depend on water resources. This will require new multiscale modeling approaches, such as nesting cloud-resolving climate models into regional weather models and then coupling these models to land surface models that are capable of simulating the hydrologic cycle, vegetation, multiple soil layers, ground water, and stream flow. These models will also need to reliably project changes in storm paths and modes of regional climate variability.

Long-term observations for measuring and predicting hydrologic changes and planning management responses. Improved physical observations are needed to monitor the impacts of climate change on water systems and to support model development and adaptation planning. Improved observations would also improve short-term hydrological forecasts. New technologies are needed to allow continuous high-precision measurements of inventories and fluxes of water, including precipitation, groundwater, soil moisture, snow, evapotranspiration, and stream flow. Time-series data related to human demographics, economic trends, vulnerabilities to changes in water quantity and quality, and human exposures and sensitivities to water contamination are also important, and should be made available in an integrated framework with physical observations to support integrated analysis and decision making.

Improved tools and approaches for decision making under uncertainty and complexity. Water resource managers are faced with making many important and complex decisions under uncertainty. To support more robust and effective decisions and strategies, further advances are needed in ensemble and integrated approaches to modeling, scenario building and comparison, and identification of no-regrets options. To improve the use and usability of climate knowledge in decision making, research is also needed on effective decision-support tools, such as forecasts, climate services, and methods for making complex trade-offs under uncertainty (see Chapter 4 for additional details).

Impacts of climate change on diverse water uses. Climate change will affect many water-related activities and sectors, including navigation, recreation, tourism, human health, drinking water, agriculture, hydroelectric power generation, and the ecological integrity of aquatic and terrestrial ecosystems. Continued and expanded research in all of these areas, and on the economics of water supply, demand, and costs of adaptation, is needed across and between different water-dependent sectors. The potential for local, state, and international disputes over water resources is also an area where further study is warranted (see Chapter 16). Another need is for better understanding of how institutions and behavior shape vulnerability and offer opportunities to adapt to changing water regimes.

Develop vulnerability assessments and integrative management approaches to respond effectively to changes in water resources. Changes in water resources are anticipated to affect coupled human-environment systems in a variety of ways and in interaction with many other environmental stresses. Assessing which water supplies and human-environment systems are most vulnerable to climate change will require analysis of place-based environmental conditions as well as social conditions and management needs. Frameworks need to be developed and tested for such assessments, and new integrative water resource management and adaptation approaches are needed for managing water in the context of climate change. Finally, the effects of actions taken to limit the magnitude of climate change (or adapt to other impacts)on water resources need to be more systematically assessed and accounted for in climate-related decision making.

Increase understanding of water institutions and governance, and design effective systems for the future. Water institutions of the future will have to deal with the complexity of multiple and interacting stresses as well as equity and economic issues related to water use. Reconciling water entitlements across different water systems, making water systems more flexible in the face of change, and shaping an institutional environment that encourages water conservation and reuse are only some of the challenges facing water resource institutions as climate change progresses. To improve our ability to design and deploy water institutions, more research is needed on governance mechanisms such as water markets, public-private partnerships, and community-based management. Evaluation of legacy effects of past infrastructure and management decisions will assist in understanding path-dependent effects, but only to the extent that such lessons are relevant to constantly evolving conditions.

Improve water engineering and technologies. Many water management systems are currently constrained by existing water infrastructures, many of which are old and need replacement. Thus, attention needs to be given to the development and implementation of more efficient water delivery systems. New technologies for water storage, supply, treatment, and recycling will also be needed, as will more efficient residential, commercial, and agricultural end-use technologies.

Evaluate effects of water resource use on climate. Changes in land and water use affect local and regional climate through effects on land-atmosphere interaction, particularly changes in evapotranspiration. The role of ecosystems in recycling precipitation, influencing stream flow, and mitigating droughts is particularly important. Improving our understanding of the effects of water and land use on regional climate will be an important component of developing local and regional integrated climate change responses.

CHAPTER NINE

Ecosystems, Ecosystem Services, and Biodiversity

Terrestrial and marine ecosystems supply the foundation for human well-being and livelihood through the food, water, timber, and other goods and services they provide. Advances over past decades have also revealed the importance of less visible but equally important services that ecosystems provide for society, such as water filtration, carbon storage, maintenance of biodiversity, protection from storm disturbance, and stabilization of local climates. Climate change has already led to a number of changes in both terrestrial and marine ecosystems, and future climate change will strongly influence biodiversity, ecosystem processes, and ecosystem services, adding to other stresses on ecosystems from human activities.

Some questions decision makers are asking, or will be asking, about ecosystems management in the context of climate change include the following:

- How is climate change—including changes in temperature, precipitation, and the chemistry of the atmosphere and oceans—altering the distribution of species?
- Will these changes have major economic and social consequences, such as the loss of pollination services or valuable fisheries?
- How does climate change relate to other ecosystem stresses, such as pollution and habitat loss?
- Can ecosystems be managed to improve their ability to adapt to anticipated changes?
- Is it possible to manage forests and other ecosystems in ways that can help limit the magnitude of future climate change?

Decades of focused research on terrestrial and marine ecosystems and their biodiversity have improved our understanding of their importance for society and their interactions with other components of the Earth system. The findings have been the subject of many authoritative syntheses and assessments, including those by the Pew Oceans Commission (2003), the Pew Center on Global Climate Change (Parmesan and Galbraith, 2004), the U.S. Commission on Ocean Policy (2004), the Millennium Ecosystem Assessment (MEA, 2005), the Intergovernmental Panel on Climate Change (IPCC; Fischlin et al., 2007), the Heinz Center (2008a), the National Research Council (NRC,

BOX 9.1
Glacier National Park

Glacier National Park is rapidly losing its namesake as summer temperatures rise and its glaciers disappear (see figure on facing page). The park, which straddles both the Continental Divide and the U.S.-Canada border in Montana, has lost about two-thirds of its glaciers since 1850 (Hall and Fagre, 2003). Plant and animal species are struggling to keep pace as suitable habitats retreat uphill as the climate warms. For example, pine trees are invading open grassland as the tree line migrates to higher elevation, in turn reducing fodder available for grazing mountain goats, bighorn sheep, and other ungulates.

Glacier National Park exemplifies some of the key questions that land and natural resource managers face with climate change (Pederson et al., 2006):

- As glaciers recede, will loss of scenic value reduce the millions of dollars that tourists spend there each year?
- How will populations of the grizzly bear and other species fare under climate change?
- Will dwindling glacial melt reduce populations of trout, a staple of the grizzly's diet and the fishing-tourism industry?
- Will droughts cause grizzlies and other large mammals to alter their seasonal movements in search of food, potentially exacerbating conflicts with human populations in and around the park?
- Will landslides increase, threatening animal and plant habitats and human-built infrastructure in the park?
- How will changes in water flowing out of the park into three major river systems—the Missouri/Mississippi, the Columbia, and the Saskatchewan/Nelson—alter availability downstream for irrigation and hydropower?
- Will fire become more frequent, and should more resources be allocated to fire-fighting or preemptive forest management?

All of these questions highlight the need for improved understanding of how plant and animal species will respond to climate change and other stresses. Projections of climate change on finer spatial scales would provide input for land managers to begin to assess the implications in their local context. Studies and models of the complex interactions among climate, biodiversity, ecosystem processes, and

2008b), and the U.S. Global Change Research Program (CCSP, 2009b), among others. This chapter outlines some of the key impacts of climate change on terrestrial and marine ecosystems (see Box 9.1), including the effects of ocean acidification, and also briefly summarizes current scientific knowledge about the potential role of ecosystems in limiting the magnitude of climate change and possible strategies for helping ecosystems adapt to climate change and other environmental stresses. The last section of the chapter outlines key research needs in all of these areas.

human decisions would provide a scientific basis for management decisions such as land use zoning, fire and forest management, animal population control, infrastructure maintenance, and habitat restoration appropriate for maintaining national treasures such as Glacier National Park.

Repeat photography showing the retreat of the Grinnell glacier in Glacier National Park. The top photos were taken around 1940, and the bottom photos show the glacier six decades later. SOURCE: USGS (2008).

TERRESTRIAL ECOSYSTEMS

Impacts of Climate Change on Land-Based Ecosystems and Biodiversity

A series of place-based observations, meta-analyses, and models indicate that climate shifts have already begun to change the geographical range of plants and animal species on land (IPRC, 2007c). In the extreme, some plants and animals have experienced maximum range shifts over the past 30 years that approach the magnitude of those witnessed in the transition from last glacial maximum to the present (NRC, 2008b; Parmesan and Yohe, 2003). In the Northern Hemisphere, range shifts are almost wholly northward and up in elevation as species search for cooler temperatures (NRC, 2008b). Special stress is being placed on cold-adapted species located on mountain tops and at high latitudes where boreal forests are invading tundra lands and where Arctic and Antarctic sea ice is rapidly diminishing (e.g., polar bears and various species of seals and penguins [NRC, 2008b]). Warming of streams, rivers, and lakes also potentially affects cold-water fish, such as economically important salmon and trout, through impacts on reproduction, food resources, and disease. The IPCC estimates with medium confidence that approximately 20 to 30 percent of plant and animal species assessed so far are likely to be at increasingly high risk of extinction as global average temperatures exceed a warming of 3.6°F to 5.4°F (2°C to 3°C) above preindustrial levels (Fischlin et al., 2007).

The phenology of species (seasonal periodicity and timing of life-cycle events) is also changing with warming. Biological indicators of spring (e.g., timing of flowering, budding, and breeding) arrive in the Northern Hemisphere as much as 3 days earlier each decade, and the growing season is longer (Walther et al., 2002). Such changes can disrupt the synchronicity between species and their food and water sources, pollinators, and other vital interactions. It also affects the timing and severity of insect and disease outbreaks, wildfire, and other disturbances, challenging the capacity of ecosystems and those charged with managing them to deal with new disturbance patterns. For example, large and long-duration forest fires have increased fourfold over the past 30 years in the American West; the length of the fire season has expanded by 2.5 months; and the size of wildfires has increased several-fold (NIFC, 2008; Westerling and Bryant, 2008; Westerling et al., 2006). Recent research indicates that earlier snowmelt, temperature changes, and drought associated with climate change are important contributors to this increase in forest fire (Westerling et al., 2006). Climate change in the western United States is also increasing populations of forest pests such as the spruce beetle, pine beetle, spruce budworm, and wooly adelgid (Logan et al., 2003) and expanding their range into forested areas previously protected from insect attack. Climate change thus increases the complexity and costs of forest and fire management practices (Chapin et al., 2003; Spittlehouse and Stewart, 2003), which in turn are strongly affected by policy. These policies and practices can be better informed by linking downscaled climate models with hydrologic and fire-vegetation models to determine, under different projections of climate change, which ecosystems will be most vulnerable to wildfires (Westerling, 2009).

Climate change, including the higher levels of CO_2 in the atmosphere that help to drive it, also affects the functioning of terrestrial ecosystems and their living communities (Loreau et al., 2001; Tilman et al., 1997); this, in turn, changes how ecosystems influence the atmosphere and climate system (Steffen et al., 2004). Experimental and modeling studies (e.g., Field et al., 2007b; Reich et al., 2006) reveal that, in general, exposure to elevated CO_2 and temperatures leads to increases in photosynthesis and growth rates in many plants, up to a point; thereafter, the trend may reverse owing to processes not yet fully understood (Woodward, 2002). Decomposition and associated release of CO_2 back to the atmosphere also increase as temperatures warm. However, ecosystem processes such as plant growth and decomposition are also determined by interactions with other factors such as nitrogen and carbon supplies, soil moisture, length of growing season, land use, and disturbance (Eviner and Chapin, 2003). Despite this complexity, projections suggest that forest productivity, especially in young forests on fertile soils where water is adequate, will increase with elevated CO_2 and climate warming. Where water is scarce and drought is expected to increase, however, forest productivity is projected to decrease (Janetos et al., 2008).

Climate warming alone is projected to drive significant changes in the range and species composition of forests and other ecosystems. Generally, tree species are expected to shift their ranges northward or upslope, with some current forest types such as oak-hickory expanding, others such as maple-beech contracting, and still others such as spruce-fir disappearing from the United States altogether (Figure 9.1). Importantly, however, whole forest communities or ecosystems will not shift their ranges intact. Plant and animal species will respond independently, according to their physiology and sensitivity to climate, resulting in the breakup of existing communities and ecosystem types and the emergence of new ones. The consequences of such reshuffling are not clear, either for the plants and animals that now exist together, or for the services those systems provide to humanity.

In addition to climate change, ecosystems and biodiversity are already being impacted by human activities. For example, human infrastructure such as farms, settlements, and road networks have directly or indirectly affected more than 50 percent of the ice-free, terrestrial surface of the Earth (Ellis and Ramankutty, 2008; Foley et al., 2005; Vitousek et al., 1997). As much as 41 percent of the vast expanse of the oceans has been affected by human activities, for example through eutrophication or fish stock depletion (Halpern et al., 2008). Considering indirect impacts, such as ocean acidification, ground-level air pollution, and climate change, virtually all ecosystems on Earth are being affected in some way by climate change, and other human pressures on ecosystems are also growing significantly (Auffhammer et al., 2006; Chameides et al., 1994; Orr et al., 2005).

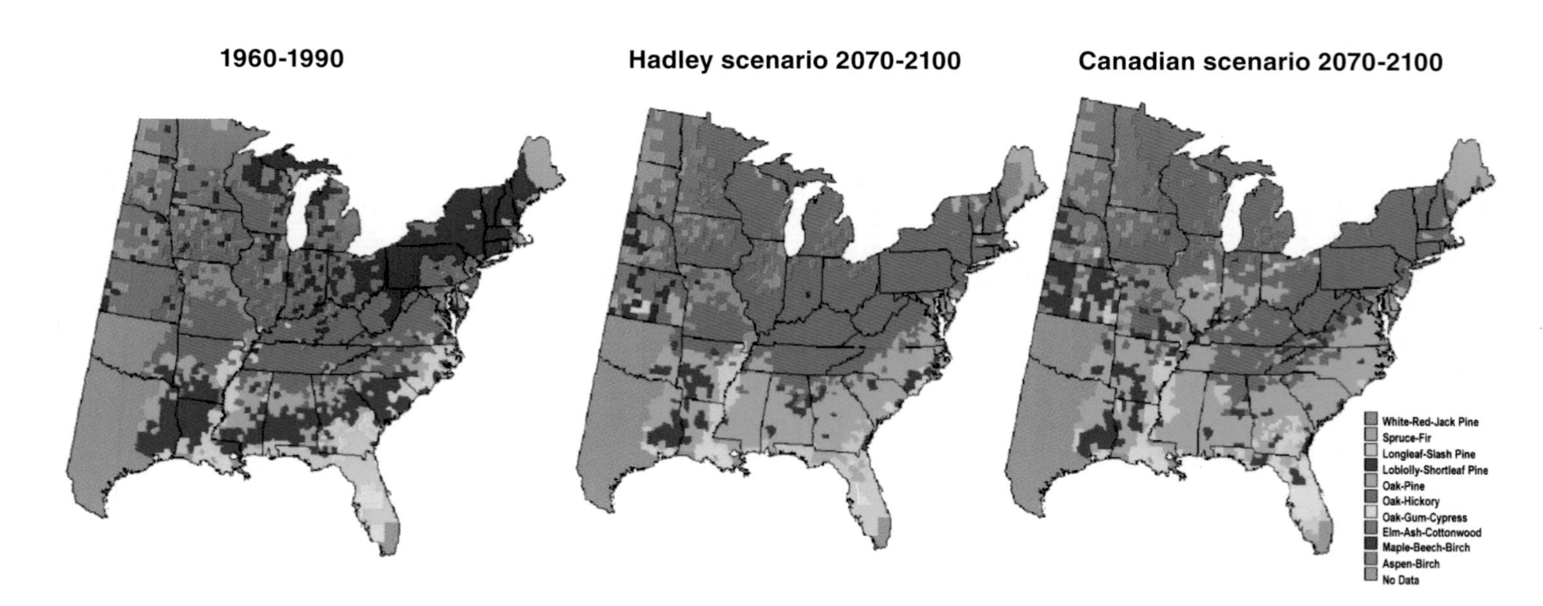

FIGURE 9.1 Potential changes in the geographic ranges of the dominant forest types in the eastern United States under projections of future climate change, based on the Hadley and Canadian climate models and a forest-type distribution model. Many forest types shift their ranges northward. Some types of forests, such as the loblolly-shortleaf pine in the Southeast (dark blue) or the maple-beech-birch forest type (red), shrink in area significantly or migrate to areas to the north and west. Oak-hickory (dark green) and oak-pine (light green) forest types expand their ranges. SOURCE: USGCRP (2001).

Managing the impacts of climate change on ecosystems and individual species already poses difficult challenges to land, resource, and conservation managers, and these challenges will undoubtedly increase. Past ecosystem conservation relied heavily on the assumption of a stable climate and focused on protecting individual species in place as well as preserving the habitat of entire species assemblages within protected areas. As climate change forces species to migrate to more suitable climates, ecosystems will be disassembled and reassembled in new locations, often outside the bounds of protection, and with new casts of characters. Some species will be lost, while other species will appear in new locations where they may become invasive and add to the pressures on existing species (NRC, 2008b).

Significant research is needed to better understand how climate change affects both individual species and entire ecosystems, and whether transitional or newly assembling ecosystems can continue to provide the ecosystem goods and services on which society depends (e.g., CCSP, 2008d; Fischlin et al., 2007). Moreover, social science research is needed to help land, resource, and conservation practitioners guide adaptive risk management in the face of altered species composition and a continually changing climatic and environmental baseline. In addition, very little is known yet about the social acceptability of new and evolving approaches to species conservation and land protection (including the Endangered Species Act under significant climate change, when many more species are at risk of extinction) or the social acceptability of a triage approach to species protection that may evolve as ecosystem functions are affected by climatic and species changes. Past experience with conservation management, however, indicates that societal values relative to species protection are significant to policy and practice. Integrated assessment and decision-support tools are also needed to help managers and the public understand and make wise judgments about the complex trade-offs that will be involved.

Role of Land-Based Ecosystems in Driving Climate Change

Modeling studies suggest that ecosystem responses to elevated CO_2 result in a net carbon sink (that is, some of the elevated carbon generated by human activities is taken up and stored in plant tissues and soils, and the amount stored exceeds the carbon released through plant respiration and decomposition) and that this sink will persist through the twenty-first century (Schimel et al., 2000, 2001). When the models include temperature change as well as elevated CO_2, however, they project that these carbon sinks could decrease, thereby increasing concentrations of CO_2 in the atmosphere and reinforcing climate warming (e.g., Field et al., 2007a). Indeed, recent analyses suggest that the reduction in efficiency of land ecosystem sinks may already be in decline (Canadell et al., 2007). Several major carbon sinks in terrestrial ecosystems face a high degree of risk from projected climate and land use changes (Fischlin et al., 2007). One of these is permafrost—frozen soil that covers vast areas of the northern latitudes and has locked away vast quantities of carbon. Permafrost temperatures are already rising due to high-latitude warming, creating a potential feedback that could drive further warming. Permafrost could also switch from a carbon sink to a source with thawing, releasing more carbon than in takes up (Dutta et al., 2006; Field et al., 2007a; McGuire et al., 2006; Norby et al., 2005; Zimov et al., 2006) and thus accelerating the pace of climate change. The potential for such a switch is one of several tipping points of concern in ecosystem-climate interactions (Barbier et al., 2008; Lenton et al., 2008; see also Chapter 6). Many other factors will ultimately determine whether terrestrial ecosystems provide a net feedback that enhances or slows the pace of climate change. Species redistributions, changes in major growth forms (e.g., from grass to woody plants, or from coniferous to deciduous trees), drought, length of growing seasons, air pollution, fire, insects and pathogens, deforestation and reforestation (Canadell and Raupach, 2008), and land use (Tilman et al., 2001) will influence uptake or release of CO_2 and other greenhouse gases (GHGs) such as N_2O and CH_4 (Canadell and Raupach, 2008; Swann et al., 2010; Tilman et al., 2001).

Globally, as much as 35 percent of human-induced CO_2 emitted over time to the atmosphere has had its origins in changes in land systems (both use and vegetative cover), principally deforestation (Foley et al., 2005). Biomass burning is also a major source of atmospheric aerosols (Andreae and Merlet, 2001). As discussed in Chapter 6, aerosols have direct effects on climate through scattering and absorbing solar radiation, and indirect changes in the properties and propensity for formation of clouds and hence precipitation, all of which can affect ecosystems (Lohmann and Feichter, 2005; Menon et al., 2002). Biomass burning is one of the largest sources of black carbon (soot) aerosols, a particularly potent warming agent that has been implicated in changing precipitation patterns and rapid ice melting in the Arctic (Flanner et al., 2007; McConnell et al., 2007; Wang, 2007). Finally, the emission of various trace gases by plants and from biomass burning leads to the formation of ground-level ozone, a gas that is both a climate-influencing agent and a pollutant that directly affects human and ecosystem health (Auffhammer et al., 2006; Chameides et al., 1994; Orr et al., 2005).

Land use change also influences climate by changing the reflective characteristics of the land surface and the exchange of water between the surface and the atmosphere. Deforestation, arid land degradation, and the transformation of ecosystems into built-up areas, for example, tend to increase reflectivity of the land surface and decrease evapotranspiration, leading to both local climate changes and, in combination with other land use changes, influencing large-scale climate forcing, feedbacks, and atmospheric circulation patterns (Chapin et al., 2002; Pielke et al., 1998; Zhao et al., 2001). Deforestation tends to lead to warmer and drier climate conditions in the humid tropics, apparently due to reductions in evapotranspiration, (Bounoua et al., 2002; DeFries and Bounoua, 2004). Reductions in vegetation at high latitudes, on the other hand, tend to exert a cooling effect because more snow cover is exposed, increasing the reflection of solar radiation back to space (see Chapter 6 and Bonan, 1999). Afforestation (planting trees where they do not naturally occur), replanting forests in previously deforested areas, or shifts in evergreen species into previously shrub or forb areas could lead to increased absorption of solar radiation and thus increases in temperature (Bala et al., 2007). All these factors are important to the critical question of whether changes in terrestrial ecosystems accelerate or decelerate climate change, yet their combined role has not been evaluated. Importantly, these and other facets of ecosystem change not only influence the global climate system but also generate large local to regional climate implications as well (Cook et al., 2009; Durieux et al., 2003; Li et al., 2006; Malhi et al., 2008; Pielke et al., 1998).

Science to Support Managing Terrestrial Ecosystems to Limit the Magnitude of Climate Change

Managing land ecosystems provides opportunities to both limit the magnitude of climate change and ameliorate its negative consequences for society. Tropical deforestation and degradation, for example, contributed approximately 17 percent of anthropogenic carbon emissions in 2004 (Barker et al., 2007a). The opportunity to reduce emissions from deforestation and degradation (REDD) has been recognized within the United Nations Framework Convention on Climate Change as a relatively low-cost option to limit climate change (Gullison et al., 2007; Stern, 2007). Research is needed to support and improve such policies. While it is now a feasible goal to monitor changes in forest area by satellite throughout the tropics (DeFries et al., 2007; GOFC-GOLD, 2009), substantial uncertainties remain about the amount and distribution of biomass (carbon contained organic plant material such as leaves, branches, and roots). Accurate biomass estimates are critical for improving estimates of GHG emissions generated by deforestation (Houghton, 2005). Both ground-based measurements and new satellite technologies (e.g., Asner, 2009) for estimating above- and below-ground carbon are needed to improve these estimates.

Understanding the socioeconomic and ecological drivers of deforestation and degradation is also critically important for developing effective policies to reduce deforestation. Global-scale drivers, from international trade in agricultural products to subsistence needs by small-scale farmers, are complex and vary in different locations

(Nepstad et al., 2006; Rudel, 2005). Research focused on ecosystems needs to include intertwined climatic, ecological, and socioeconomic factors. For example, more clearing and more fires occur during relatively dry years in the tropical forests of southeast Asia, creating a positive feedback between emissions and climate change (van der Werf et al., 2008). The synergies and trade-offs between REDD and biodiversity conservation, watershed protection, and livelihood needs for local people require more rigorous analysis. (Research needs are discussed at the end of this chapter.)

Ecosystems also provide the opportunity to limit climate change through the enhancement of carbon storage or surface reflectivity. In forest ecosystems, protection from fire, insect damage, and forest thinning through logging and other human use can enhance carbon storage as can secondary regrowth of forests in abandoned croplands, tree plantations, and agroforestry (Rhemtulla et al., 2009; Gough et al., 2008). The extent to which these strategies might be able to offset GHG emissions on a global scale is poorly known. As noted above, land use and land cover changes also alter the reflectivity of the land surface, and this fact could potentially be exploited to limit the magnitude of climate change. Research is needed to evaluate these many interacting factors and quantify the potential and of these strategies relative to costs of adapting to climate change (Bala et al., 2007; Bonan, 2008; Jackson et al., 2008; Ollinger et al., 2008).

Ecosystems management is also a potential strategy to ameliorate some of the societal impacts of climate change. Restoration of wetlands in the Gulf of Mexico, for example, can reduce damage from hurricanes by damping wave action and diminishing wind penetration (Day et al., 2007). Mangroves protected people from a 1999 Asian cyclone (Das and Vincent, 2009) and will potentially provide some protection against storm surges that will move further inland with sea level rise. Quantitative and rigorous analysis of these ecosystem management opportunities, including their effectiveness and costs, is needed to assess their potential in different locations.

MARINE ECOSYSTEMS

Marine ecosystems are fundamental to the large role the oceans play in regulating the climate system. For example, the oceans contain many times more carbon than the atmosphere and terrestrial ecosystems combined, and are thus critical in regulating the amount of CO_2 in the atmosphere. Climate change will have broad effects on marine ecosystems, their capacity to take up CO_2 from the atmosphere, and the diverse ecosystem services they provide to society. These ecosystem effects will be driven by projected changes in ocean temperature, circulation (Bryden et al., 2005), storms, and chemistry (Doney et al., 2009). Unlike on land, the majority of food that humans derive from the sea is still harvested from wild populations (FAO, 2008). Therefore, the oceans' capacity to provide seafood, a major protein source for more than a billion people, will be directly affected by climate impacts on marine ecosystems (see Chapter 10).

Climate Change Impacts on Ocean Ecosystems

Over recent decades, marine scientists have detected widespread poleward shifts in species distributions that are consistent with patterns of a warming ocean (Alheit and Hagen, 1997; Holbrook et al., 1997; Mueter and Litzow, 2008; Sagarin et al., 1999; Southward et al., 1995). Marine species can be highly mobile, both as adults and as microscopic young drifting in the plankton (Kinlan and Gaines, 2003). This mobility can lead to larger and faster geographic shifts than in terrestrial ecosystems. For example, two-thirds of the 36 most common bottom-dwelling fish in the North Sea have shifted the geographical center of their range north toward the pole over just 25 years (Perry et al., 2005) (Figure 9.2). Such shifts, if they continue, could move the fish beyond the range of national fisheries. More broadly, because species move at different rates depending on their unique life histories, such shifts could lead to rapid rearrangements of the species composition of some ocean ecosystems (Cheung et al., 2009). The unpredictability of responses by different species is a key barrier to anticipating and adapting to the resulting ecosystem rearrangements.

Given the prominent role of oceans in storing carbon, climate impacts on ocean productivity could also alter their role in the carbon cycle. Overall, oceans contribute roughly half of the globe's net primary productivity (NPP; Field et al., 1998), defined as the net carbon gain by ecosystems over a specific time period, typically annually. Some ocean habitats (polar seas, coastal upwelling systems) may see increased productivity under projected climate change (Arrigo et al., 2008; Bakun, 1990; Behrenfeld et al., 2006; Pabi et al., 2008; Polovina et al., 1995; Snyder et al., 2003). Most of the ocean, however, is permanently stratified with shallow, warm, nutrient-depleted water isolated from cold, nutrient-rich water below. In these seas, warmer surface temperatures generally decrease phytoplankton productivity (Figure 9.3). Given the prominence of these stratified seas, a substantially warmer ocean would "inevitably alter the magnitude and distribution of global ocean net air-sea carbon exchange, fishery yields, and dominant … biological regimes" (Behrenfeld et al., 2006).

Just as on land, high-latitude marine ecosystems may experience more stress than lower-latitude marine ecosystems, since rates of warming are higher (Gille, 2002; Hansen et al., 2006) and the opportunity for poleward range shifts is limited. Sea ice

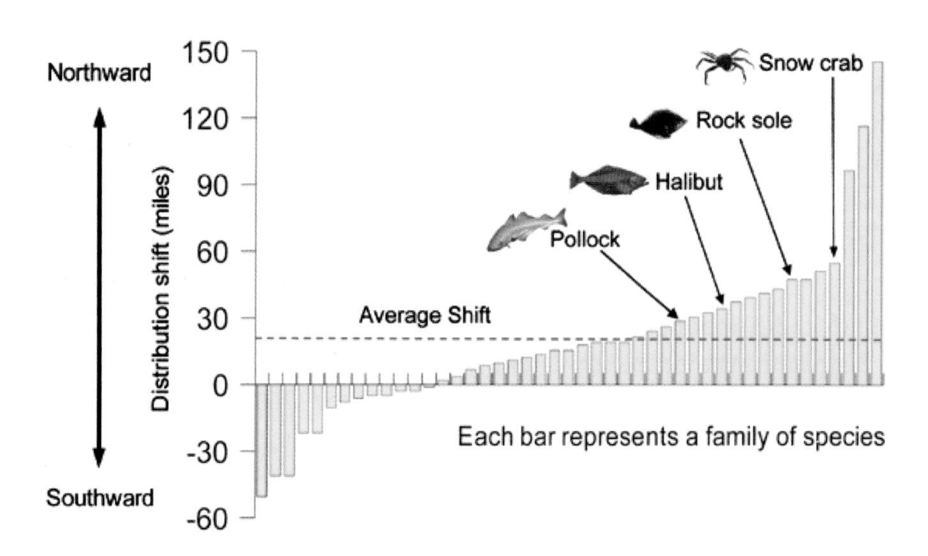

FIGURE 9.2 Observed northward shift of marine species in the Bering Sea between the years 1982 and 2006. Length of the yellow bars indicates the distance that the center of a species range has shifted. The average shift among the species examined was approximately 19 miles north of its 1982 location (red line). The northward shift is primarily linked to warming of the Bering Sea during this period. SOURCES: Mueter and Litzow (2008) and USGCRP (2009a).

creates critical habitat for a diverse array of marine species, including many mammals and birds (Hunt and Stabeno, 2002). Major declines in sea ice thickness and extent have been observed in the Arctic (see Chapter 6) and are projected for the next few decades (Overland and Stabeno, 2004; USGCRP, 2009a). Ice dynamics, which are highly sensitive to climate, drive dynamics of ocean primary productivity, which in turn has impacts throughout the marine food web in ways that are not clearly understood (Moore and Huntington, 2008; Smetacek and Nicol, 2005). Declines in sea ice can lead to large blooms in phytoplankton (e.g., Arrigo et al., 2008; Pabi et al., 2008) and declines in production from benthic (seafloor) habitats. These changes alter both the food webs of animals that ultimately depend on these different sources of productivity, including humans (Grebmeier et al., 2006; Mueter and Litzow, 2008; USGCRP, 2009a), and the role of high-latitude ocean ecosystems in the carbon cycle. Although the details are highly uncertain, many high-latitude ocean ecosystems appear to be at the threshold of major ecosystem changes (USGCRP, 2009a), especially since climate-induced changes may soon be joined by new human uses and stresses (e.g., oil and mineral exploration, expanded maritime use, and new fisheries in the Arctic) made possible by reductions in sea ice.

Some of the most productive ocean ecosystems are coastal regions where winds push

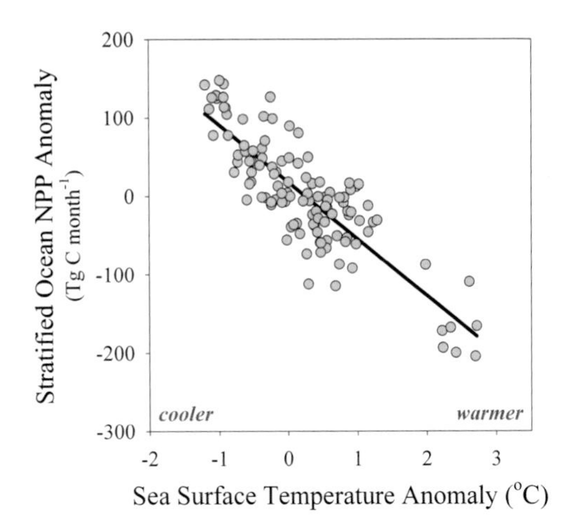

FIGURE 9.3 Relationship between changes in sea surface temperature and net primary productivity (NPP) from 1999 to 2004 based on satellite observations. Warmer ocean temperatures typically lead to reduction in the productivity of phytoplankton, which means that they remove less carbon from the atmosphere. SOURCE: Updated from Behrenfeld et al. (2006).

surface waters offshore and draw deep, cold, nutrient-rich waters to the surface (e.g., the west coast of North America). The nutrients fuel plankton blooms that support diverse and abundant food webs and fisheries. These upwelling regions may become even more productive under climate change if forecasts of increasing upwelling and favorable winds hold true (Bakun, 1990). Substantial increases in upwelling, however, can also have catastrophic consequences if the system crosses key thresholds (Chan et al., 2008; Helly and Levin, 2004). Deep ocean waters are typically extremely low in oxygen (hypoxic). Strong upwelling of deep cold waters can pull such hypoxic water onto shallow ocean shelves with devastating impacts on many marine species (Grantham et al., 2004). Hypoxia of coastal waters is more commonly associated with nutrient-laden runoff from land (NRC, 2000; Rabalais and Turner, 2001), but climate-driven changes in winds, ocean temperature, and circulation can cause similar devastation even in areas without runoff from land (Bakun and Weeks, 2004; Chan et al., 2008). The system can rapidly switch from high productivity to "dead zones," where most species cannot live. For example, this transition has recently occurred in summers off the coasts of Oregon and Washington (Chan et al., 2008). Over more than 50 years of observations in the 20th century, hypoxia was rare or absent from these near-shore waters. In the past decade, however, hypoxia has become common and caused major die-offs of coastal species. By 2006, these once highly productive waters were oxygen-depleted along much of the coastline as upwelling winds increased (see Figure 9.4).

In the tropics, warm temperatures pose a "bleaching" threat to corals. Coral reef ecosystems have been compromised by a diverse set of activities including overfishing, damaging fishing practices, eutrophication, and sedimentation, among others (USGCRP, 2009a). On top of these human-caused stresses, recent decades have brought an

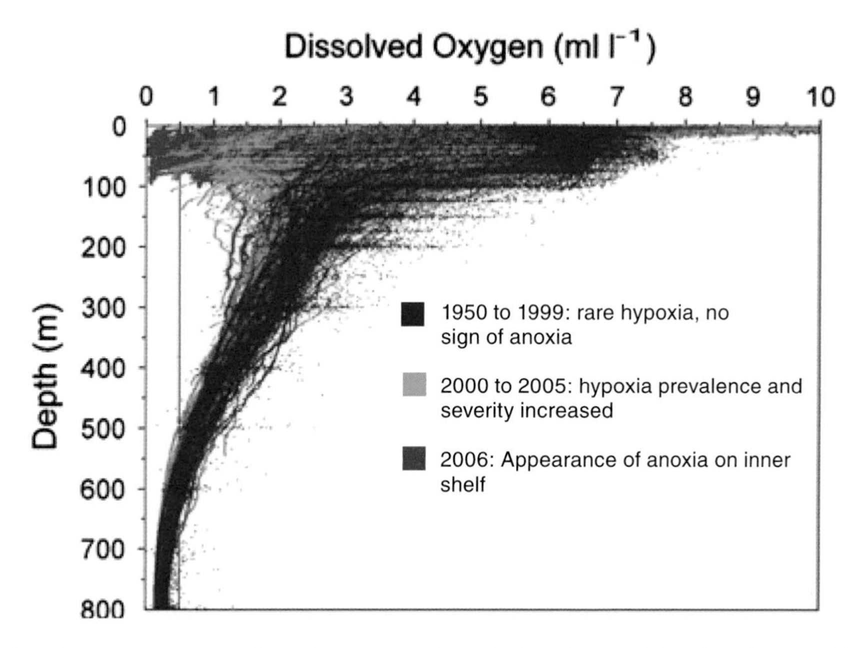

FIGURE 9.4 Hypoxia and anoxia in shallow waters. Values below 0.5 ml/l (left of black vertical line) represent severe hypoxia. Over the latter of half of the 20th century, hypoxia was only found in deep waters. In recent years (red and green), hypoxia has extended into waters close to the surface. SOURCE: Modified from Chan et al. (2008).

increase in widespread bleaching events, where corals eject their symbiotic algae in the face of extreme temperatures (Figure 9.5). In some cases, the bleached corals recover with new symbionts (Lewis and Coffroth, 2004). In other cases, the coral is killed. Periods of mass bleaching have occurred globally since the late 1970s (Glynn, 1991; Hoegh-Guldberg, 1999), with the most severe event in 1998, an El Niño year in which an estimated 16 percent of the world's reef corals died (Wilkinson, 2000). The extent of bleaching varies greatly among species and locations. Some of the variability is tied to the level of other human stresses, which argues for managing reefs for greater resilience to climate change by reducing other stressors (Hughes et al., 2003). The next subsection discusses ocean acidification, which serves as an additional and potentially devastating stressor to corals. Recent models of coral-symbiont dynamics suggest that adaptation could greatly reduce coral bleaching catastrophes if the pace of climate warming is not too rapid (Baskett et al., 2005).

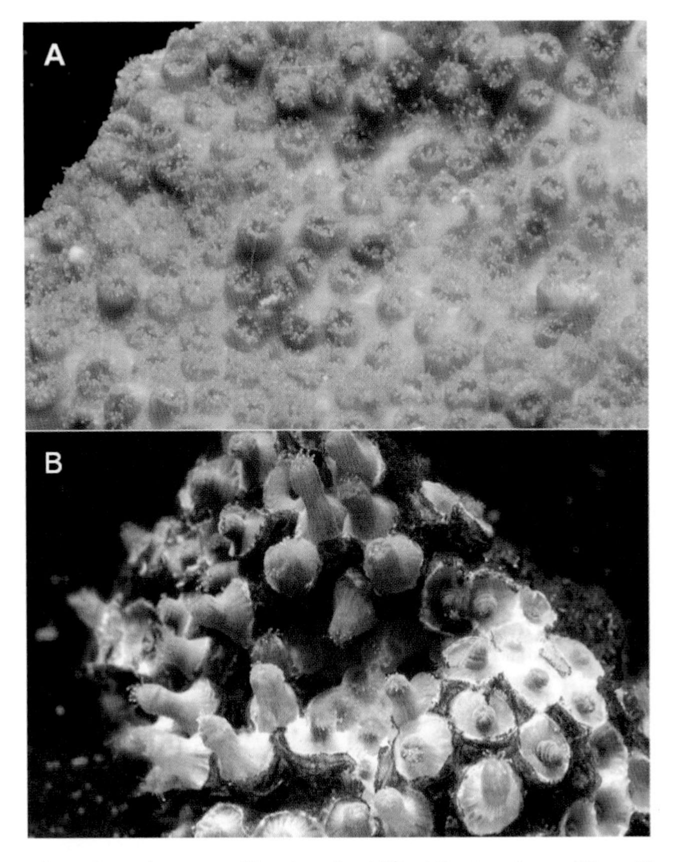

FIGURE 9.5 Photos of corals under normal (top) and acidified (bottom) conditions. The bottom coral lack a protective skeleton (appearing as light yellow in the top panel) and are sometimes called "naked coral." SOURCE: Doney et al. (2009).

Ocean Acidification

In addition to its climate impacts, CO_2 released by human activities can influence ecosystem dynamics in aquatic systems by altering water chemistry—in particular, the reaction of CO_2 with water to form carbonic acid (H_2CO_3), which lowers (acidifies) ocean pH. Roughly one-third of all CO_2 released by human activities since preindustrial times has been absorbed by the sea (Doney et al., 2009; Sabine and Feely, 2005; Sabine et al., 2004; Takahashi et al., 2006); consequently, ocean pH has decreased by approximately 0.1 units since preindustrial times. While this might not seem like a large change, it actually represents a 25 percent increase in acidity, because pH is measured on a logarithmic scale. By the end of this century, the oceans are projected to acidify by an

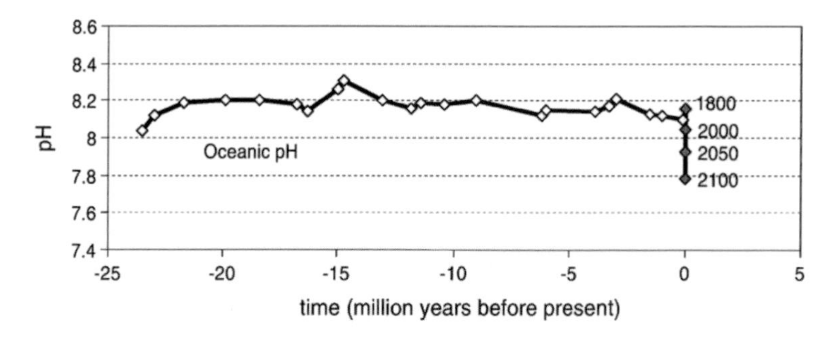

FIGURE 9.6 Estimates of ocean pH over the past 23 million years (white diamonds) and for contemporary times (gray diamonds). Projections are made for the future using IPCC projections of atmospheric concentrations of CO_2. The projected changes in pH are extremely large and rapid, considering the relative stability of oceanic pH in the past. SOURCE: Blackford and Gilbert (2007).

additional 0.3 to 0.4 units (Orr et al., 2005) under the highest IPCC emissions scenario (Figure 9.6).

Because pH interacts with temperature to determine saturation levels for various related chemical species, cold-water ocean areas are projected to become undersaturated with calcium carbonate ($CaCO_3$)—a key building block for the shells of many marine organisms—as early as 2050 (Orr et al., 2005). A broad array of marine species produce $CaCO_3$ skeletons during at least part of their life cycle, so ocean acidification threatens nearly all ocean ecosystems by altering calcification rates while simultaneously increasing the rate of $CaCO_3$ dissolution (Yates and Halley, 2006). Physiological studies suggest wide variations in the ability of organisms to cope with such changes (Doney et al., 2009). Acidification is especially challenging for coral reefs, which are defined by the $CaCO_3$ skeletons of corals. Acidification, in tandem with elevated temperatures and other human stresses, decreased calcification rates on the Great Barrier Reef by 21 percent between 1988 and 2003 (Cooper et al., 2008). Numerous controlled experiments under elevated pH now complement these field observations (e.g., Doney et al., 2009). Projections of future ocean chemistry and climate change indicate that, by the time atmospheric CO_2 content doubles over its preindustrial value, there will be virtually no place left in the ocean that can sustain coral reef growth (Cao and Caldeira, 2008; Silverman et al., 2009). Ocean acidification could also have dramatic consequences for polar food webs since several prominent species at the base of the food web may be unable to form shells—including species that salmon and other iconic species depend on for survival. Overall, ocean acidification has the potential to alter marine ecosystems catastrophically, but the details and consequences of these impacts are only beginning to be understood (see also NRC, 2010f).

The Role of Ocean Ecosystems in Managing Carbon and Climate Change

The ocean contains far more carbon than the atmosphere or land ecosystems. Storage of carbon in the ocean occurs by several mechanisms whose rates can be altered by human activities. In the ocean, CO_2 dissolves directly in sea water; CO_2 is sequestered when marine plants photosynthesize, and organic carbon ultimately sinks to great depths; and CO_2 is also sequestered by conversion to $CaCO_3$ by plankton, invertebrates, and fish (Wilson et al., 2009), $CaCO_3$ that either forms sediments or sinks to deep water after the organism dies. None of these forms of storage is permanent, but sequestration rates can be modified greatly by a variety of factors (e.g., water temperature, pH, and the abundance of fish and plankton), ultimately affecting how much CO_2 remains locked away or returns to the sea.

Because the oceans provide such an enormous reservoir for carbon storage, it may be possible to manipulate (i.e., geoengineer—see Chapter 15) ocean ecosystems to cause a transfer of CO_2 from the atmosphere to the oceans. Several different approaches have been proposed to achieve this end, most of them involving the introduction of some kind of fertilizer to the upper ocean. The basic hypothesis is that fertilization may stimulate the incorporation of dissolved CO_2 into organic matter through phytoplankton blooms, which could then sink to the deeper ocean. Some of the carbon that sinks out of the upper ocean should be replaced by CO_2 from the atmosphere, thus reducing atmospheric CO_2 concentrations.

Most of the attention given to the ocean fertilization hypothesis has focused on iron (Martin and Fitzwater, 1988; see also *Limiting the Magnitude of Future Climate Change* [NRC, 2010c]). In some parts of the ocean, especially the Southern Ocean and parts of the equatorial Pacific Ocean, marine biological productivity is limited by the availability of iron. The ratios of carbon to iron in marine phytoplankton typically exceed 10,000 to 1, so there is the potential that small amounts of iron could lead to substantial carbon uptake in the form of phytoplankton blooms. While there is still considerable uncertainty, the prevailing view is that this approach could store some carbon, but maximum achievable sustainable rates might be only a small fraction of the total carbon emitted due to fossil fuel emissions (Buesseler et al., 2008). There have been various proposals to fertilize the ocean with other nutrients, such as phosphate or nitrogen, or to fertilize the oceans by bringing up nutrients from the deep ocean, but these approaches have received even less study and attention on either their potential efficacy in reducing atmospheric CO_2 or their broader environmental impacts.

In general, significant uncertainties remain about the effectiveness of ocean fertilization at removing CO_2 from the atmosphere, as well as the length of time this CO_2

would stay isolated from the atmosphere. Furthermore, there is considerable uncertainty about the impact of these manipulations on marine ecosystems and the services they provide to society, particularly since CO_2 causes ocean acidification, which is expected to harm marine ecosystems. Much effort has been focused on trying to protect marine ecosystems by keeping CO_2 out of the ocean, whereas ocean fertilization proposals seek to do the opposite. Because large parts of the oceans are a global commons, regulation of such activities represents a significant issue that has yet to be addressed. Furthermore, verification of amounts of carbon stored by ocean fertilization activities would be challenging, at best.

In summary, it is feasible that human manipulation of marine ecosystems could store at least some extra CO_2 in the oceans. While maximum storage rates are projected to be at most a few percent of total human-generated GHG emissions, significant questions remain regarding exactly how much carbon could be stored, and for how long, using these approaches. Furthermore, considerations such as ocean acidification and the difficulty of predicting responses of marine ecosystems make it doubtful whether such manipulations could contribute to overall environmental risk reduction.

RESEARCH NEEDS

Improve understanding of the effects of climate change and impacts of enhanced CO_2 on ecosystems, ecosystem services, and biodiversity. Given the complexity of the impacts of different scenarios of climate change and elevated CO_2 levels on ecosystem function, services, and biodiversity, further research is needed to evaluate the consequences of multiple interacting changes. For example, movement of species, changes in phenology and synchronicity, changes in productivity and carbon cycling processes, and changes in disturbance regimes in response to temperature, moisture, and CO_2 have not been well assessed, especially at regional scales. Enhanced capacity for linking models of physical change in the climate system to species response models would help meet these challenges. Research is also needed to identify those ecosystems, ecosystem services, species complexes, and people reliant on them that are most resilient or most vulnerable (see Box 9.2).

Evaluate the climate feedbacks from changes in ecosystems and biodiversity. Changes in ecosystem biogeochemical processes (including GHG emissions) and biodiversity (including changes in reflectance characteristics) have the potential to exacerbate or offset certain aspects of climate change (i.e., act as feedbacks). Models and experiments that integrate knowledge about ecosystem processes, plant physiol-

BOX 9.2
National Marine Sanctuary

Ocean ecosystems face growing threats globally from overfishing, habitat damage, pollution, and especially acidification (Halpern et al., 2008). As a result, the persistence of several marine species is at risk, and ecosystem services provided by intact coastal ecosystems could be compromised. Compared to the land, a minute fraction of the sea is set aside for protection. In response to growing threats, a number of nations, including the United States, are establishing networks of new marine protected areas (MPAs) with special protections (Airame et al., 2003; Fernandes et al., 2005). In the United States, the largest network of MPAs is being established along the coast of California, where dozens of new protected areas are currently being designed and implemented.

Although MPAs can be dramatically successful at restoring depleted ocean ecosystems (Lester et al., 2009), many questions remain:

- Will the effectiveness of MPAs be compromised by climate change, ocean acidification and/or the migration of marine species outside the boundaries of protected areas?
- MPA network design is based on where species occur today, not where they will be driven by future climate shifts—will the expected conservation gains from MPA networks go unrealized as the seascape shifts?
- Alternatively, could large networks of MPAs along entire coastlines provide protected havens to aid species driven poleward by shifting climate?
- Does uncertainty about future climate change increase the need for MPAs as a hedging strategy?

ogy, vegetation dynamics, and disturbances such as fire need to be further developed and included in advanced Earth systems models.

Assess the potential of land and ocean ecosystems to limit or buffer impacts of climate change. How can specific land uses (including managed and unmanaged forests and grasslands, agricultural systems, fisheries, urban systems, and aquatic systems) be managed for provisioning services as well as for their effects on GHG emissions, carbon storage, reflectivity, and evapotranspiration? What ecosystem management strategies can provide co-benefits that meet multiple goals, including carbon storage, biodiversity conservation, and watershed protection? To address these questions, new tools and approaches need to be developed for evaluating different land and ocean uses for their potential in helping to limit the magnitude of climate change. Such research needs to address the trade-offs between alternative land management options, including economic costs and impacts on ecosystem services that are difficult to quantify in economic terms. The efficiency and efficacy of overlapping systems of governance and management structures to address trade-offs and determine management strategies is also a critical area of research.

Assess vulnerabilities of ecosystems and the benefits society derives from them to climate change. Ecosystems on land and in the ocean, and the services they provide, are key components of the maintenance of environmental functions and human well-being. Climate change affects this maintenance, with potentially significant societal consequences. Identifying critical linkages and feedbacks among changing ecosystems, their services, and human outcomes (e.g., crop yields, water supply) is essential. To do this requires analytical frameworks and methods for assessing vulnerability of coupled human-environment systems, and the ability of the social and environmental components of such systems to adapt to change. Complicating these assessments is the need to address climate change in the context of other changes, such as land use, acid rain, and nitrogen deposition.

Improve observations and modeling. There is a great need for global-scale, long-term, and continuous observations of land and ocean ecosystems and ongoing changes within them. Such observations will enable measures of ecological processes at relatively fine spatial and temporal scales, which are needed both to provide critical inputs to Earth system models and to track gradual and abrupt change in Earth system processes. The development of indicators of ecosystem health and ecosystem vulnerability is also needed as part of an early warning system (see NRC, 2009i). As mentioned earlier, new Earth system models that address multiple drivers and feedbacks from climate-ecosystem interactions are needed, and they will be most effective if linked to climate models that function at regional scales.

CHAPTER TEN

Agriculture, Fisheries, and Food Production

Meeting the food needs of a still-growing and more affluent global population—as well as the nearly one billion people who already go without adequate food—presents a key challenge for economic and human security (see Chapter 16). Many analysts estimate that food production will need to nearly double over the coming several decades (Borlaug, 2007; FAO, 2009). Recent trends of using food crops for fuel (e.g., corn ethanol) or displacing food crops with fuel crops, along with potential opportunities for reforesting land for carbon credits, may amplify the food security challenge by increasing competition for arable land (Fargione et al., 2008). Climate change increases the complexity of meeting these food needs because of its multiple impacts on agricultural crops, livestock, and fisheries. The potential ability of agricultural and fishery systems to limit climate change adds yet another dimension to be considered.

Questions that farmers, fishers, and other decision makers are asking or will be asking about agriculture, fisheries, and food production in the context of climate change include the following:

- How will climate change affect yields?
- How will climate change affect weeds and pests, and will I need more pesticides or different technology to maintain or increase yields?
- Will enough water be available for my crops? Will the risk of flooding or drought increase?
- Should I change to more heat-resistant or slower-growing crop varieties?
- What new market opportunities should I take advantage of? How will competitors in other regions be affected?
- What adjustments do I need to make to guarantee the sustainability of the fisheries under my management?
- How will climate change affect my catch? Will I need new equipment and technology? Will regulations change?
- How will climate change affect the availability of food in domestic and international markets? Will food become more expensive? Will food security increase or decrease?
- How can changes in agricultural production and practices contribute to reduc-

tions in greenhouse gas emissions or dampen regional-scale impacts related to climate change?

The scientific knowledge summarized in this chapter illustrates how agriculture will be influenced by climate change, and it explores the less well understood impacts of climate change on fisheries. The chapter also indicates how agricultural management may provide opportunities to reduce net human greenhouse gas (GHG) emissions, and it offers insight into the science needed for adaptation in agriculture systems as well as food security issues. Finally, the chapter provides examples of a broad range of research that is needed to understand the impacts of climate change on food production systems and to develop strategies that assist in both limiting the magnitude of climate change through management practices and reducing vulnerability and increasing adaptive capacity in regions and populations in the United States and other parts of the world.

CROP PRODUCTION

Crop production will be influenced in multiple ways by climate change itself, as well as by our efforts to limit the magnitude of climate change and adapt to it. Over the past two decades, numerous experimental studies have been carried out on crop responses to increases in average temperature and atmospheric CO_2 concentrations (often referred to as carbon fertilization), and mathematical models depicting those relationships (singly or in combination) have been developed for individual crops. Fewer experiments and models have evaluated plant responses to climate-related increases in air pollutants such as ozone, or to changes in water or nutrient availability in combination with CO_2 and temperature changes. A recently published report of the U.S. Climate Change Science Program (CCSP, 2008e) summarized the results from experimental and modeling analyses for the United States. Results of experimental studies, for example, indicate that many crop plants, including wheat and soybeans, respond to elevated CO_2 with increased growth and seed yield, although not uniformly so. Likewise, elevated CO_2 also reduces the conductance of CO_2 and water vapor through pores in the leaves of some plants, with resulting improvements in water use efficiency and, potentially, improved growth under drought conditions (Leakey et al., 2009). On the other hand, studies carried out in the field under "free air CO_2 enrichment" environments indicate that growth response is often smaller than expected based on more controlled studies (e.g., Leakey et al., 2009; Long et al., 2006). The response of crop plants to carbon fertilization in field environments hence remains an important area of research (see Research Needs section at the end of the chapter).

Some heat-loving crop plants such as melons, sweet potatoes, and okra also respond positively to increasing temperatures and longer growing seasons; but many other crops, including grains and soybeans, are negatively affected, both in vegetative growth and seed production, by even small increases in temperature (Figure 10.1). Many important grain crops tend to have lower yields when summer temperatures increase, primarily because heat accelerates the plant's developmental cycle and reduces the duration of the grain-filling period (CCSP, 2008b; Rosenzweig and Hillel, 1998). In some crop plants, pollination, kernel set, and seed size, among other variables, are harmed by extreme heat (CCSP, 2008b; Wolfe et al., 2008). Studies also indicate that some crops such as fruit and nut trees are sensitive to changes in seasonality, reduced cold periods, and heat waves (Baldocchi and Wong, 2008; CCSP, 2008e; Luedeling et al., 2009).

Most assessments conclude that climate change will increase productivity of some crops in some regions, especially northern regions, while reducing production in others (CCSP, 2008b; Reilly et al., 2003), an expected result given the range of projected climate changes and diversity of food crops around the world. The Intergovernmental Panel on Climate Change (IPCC) suggests, with medium confidence, that moderate warming (1.8°F to 5.4°F [1°C to 3°C]) and associated increases in CO_2 and changes in precipitation would benefit crop and pasture lands in middle to high latitudes but decrease yields in seasonally dry and low-latitude areas (Easterling et al., 2007). This response to intermediate temperature increases would generate a situation of midlatitude "winners" in developed countries and low-latitude "losers" in developing coun-

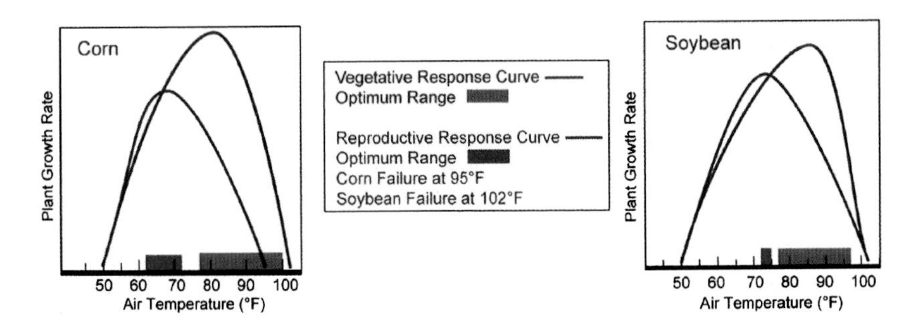

FIGURE 10.1 Growth rates (green) and reproductive response (purple) versus temperature for corn (left) and soybean (right). The curves show that there is a temperature range (colored bars) within which the plants can optimally grow and reproduce, and that growth and reproduction are less efficient at temperatures above this range. The curves also show that, above a certain temperature, the plants cannot reproduce. SOURCE: USGCRP (2009a).

tries, thus magnifying rather than reducing existing inequities in food availability and security. The IPCC also concludes with medium to low confidence that, on the whole, global food production is likely to decrease with increases in average temperatures above 5.4°F (3°C).

Regional assessments of agricultural impacts in the United States (e.g., CCSP, 2008b, and references therein) suggest that over the next 30 years, the benefits of elevated CO_2 will mostly offset the negative effects of increasing temperature (see below for limits in modeling conducted to date). In northern regions of the country, many crops may respond positively to increases in temperature and atmospheric CO_2 concentrations. In the Midwest corn belt and more southern areas of the Great Plains, positive crop responses to elevated CO_2 may be offset by negative responses to increasing temperatures; rice, sorghum, and bean crops in the South would see negative growth impacts (CCSP, 2008b). In California, where half the nation's fruit and vegetable crops are grown, climate change is projected to decrease yields of almonds, walnuts, avocados, and table grapes by up to 40 percent by 2050 (Lobell et al., 2007). As temperatures continue to rise, crops will increasingly experience temperatures above the optimum for growth and reproduction. Adaptation through altered crop types, planting dates, and other management options is expected to help the agricultural sector, especially in the developed world (Burke et al., 2009; Darwin et al., 1995). However, regional assessments for other areas of the world consistently conclude that climate change presents a serious risk to critical staple crops in sub-Saharan Africa, where adaptive capacity is expected to be less than in the industrialized world (Jones and Thornton, 2003; Parry et al., 2004). Parts of the world where agriculture depends on water resources from glacial melt, including the Andean highlands, the Ganges Plain, and portions of East Africa, are also at risk due to the worldwide reduction in snowpack and the retreat of glaciers (Bradley et al., 2006; Kehrwald et al., 2008; also see Chapter 8).

While models of crop responses to climate change have generally incorporated shifts in average temperature, length of growing season, and CO_2 fertilization, either singly or in combination, most have excluded expected changes in other factors that also have dramatic impacts on crop yields. These critical factors include changes in extreme events (such as heat waves, intense rainfall, or drought), pests and disease, and water supplies and energy use (for irrigation). Extreme events such as heavy downpours are already increasing in frequency and are projected to continue to increase (CCSP, 2008b; Rosenzweig et al., 2001). Intense rainfalls can delay planting, increase root diseases, damage fruit, and cause flooding and erosion, all of which reduce crop productivity. Drought frequency and intensity are likely (Christensen et al., 2007) to increase in several regions that already experience water stress, especially in developing countries where investments have focused on disaster recovery more than adaptive capacity (e.g., Mirza, 2003).

Changes in water quantity and quality due to climate change are also expected to affect food availability, stability, access, and utilization. This will increase the vulnerability of many farmers and decrease food security, especially in the arid and semiarid tropics and in the large Asian and African deltas (Bates and Kundzewicz, 2008). As noted in Chapter 8, freshwater demand globally will grow in coming decades, primarily due to population growth, increasing affluence, and the need for increased production of food and energy. Climate change is exacerbating these issues, and model simulations under various scenarios indicate that many regions face water resource challenges, especially in regions that depend on rainfall or irrigation from snowmelt (Hayhoe et al., 2007; Kapnick and Hall, 2009; Maurer and Duffy, 2005). As a result, many regions face critical decisions about modifying infrastructure and pricing policies as climate change progresses.

Many weeds, plant diseases, and insect pests benefit from warming (and from elevated CO_2, in the case of most weed plants), sometimes more than crops; as temperatures continue to rise, many weeds, diseases, and pests will also expand their ranges (CCSP, 2008b; Garrett et al., 2006; Gregory et al., 2009; Lake and Wade, 2009; McDonald et al., 2009). In addition, under higher CO_2 concentrations, some herbicides appear to be less effective (CCSP, 2008b; Ziska, 2000; Ziska et al., 1999). In the United States, aggressive weeds such as kudzu, which has already invaded 2.5 million acres of the southeast, is expected to expand its range into agricultural areas to the north (Frumhoff, 2007). Worldwide, animal diseases and pests are already exhibiting range extensions from low to middle latitudes due to warming (CCSP, 2008b; Diffenbaugh et al., 2008). While these and other changes are expected to have negative impacts on crops, their impact on food production at regional or national scales has not been thoroughly evaluated.

Similar to crop production, commercial forestry will be affected by many aspects of climate change, including CO_2 fertilization, changes in length of growing season, changing precipitation patterns, and pests and diseases. Models project that global timber production could increase through a poleward shift in the locations where important forest species are grown, largely as a result of longer growing seasons. Enhanced growth due to carbon fertilization is also possible (Norby et al., 2005). However, experimental results and models typically do not account for limiting factors such as pests, weeds, nutrient availability, and drought; these limiting factors could potentially offset or even dominate the effects of longer growing seasons and carbon fertilization (Angert et al., 2005; Kirllenko and Sedjo, 2007; Norby et al., 2005).

LIVESTOCK PRODUCTION

Livestock respond to climate change directly through heat and humidity stresses, and they are also affected indirectly by changes in forage quantity and quality, water availability, and disease. Because heat stress reduces milk production, weight gain, and reproduction in livestock, production of pork, beef, and milk is projected to decline with warming temperatures, especially those above 5.4°F (3°C; Backlund et al., 2008) (Figure 10.2). In addition, livestock losses due to heat waves are expected to increase, with the extreme heat exacerbated by rising minimum nighttime temperatures as well as increasing difficulties in providing adequate water (CCSP, 2008b).

Increasing temperatures may enhance production of forage in pastures and rangelands, except in already hot and dry locations. Longer growing seasons may also extend overall forage production, as long as precipitation and soil moisture are sufficient; however, uncertainty in climate model precipitation projections makes this difficult to determine. Although CO_2 enrichment stimulates production on many rangelands and pastures, it also reduces forage quality, shifts the dominant grass species toward those with lower food quality, and increases the prevalence of nonforage weeds (CCSP, 2008b; Eakin and Conley, 2002). In northern Sonora, Mexico, for example, buffelgrass, which was imported from Africa and improved in the United States, is increasingly planted as livestock pasture in arid conditions. However, the grass has become an

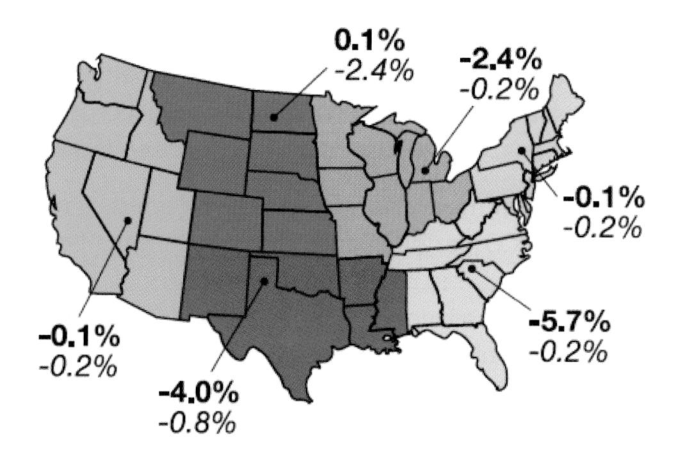

FIGURE 10.2 Percent change in milk yield from 20th-century (1850 to 1985) climate conditions to projected 2040 climate conditions made using two different models of future climate (bold versus italicized numbers) in different regions of the United States. The bold values are associated with the model that exhibits more rapid warming. SOURCE: CCSP (2008e).

aggressive invader, spreading across the Sonoran Desert landscape and into Arizona and overrunning important national parks and reserves (Arriaga et al., 2004). Overall, changes in forage are expected to lead to an overall decline in livestock productivity.

FISHERIES AND AQUACULTURE PRODUCTION

Over one billion people around the world rely on seafood as their primary source of protein, and roughly three billion people obtain at least 15 percent of their total protein intake from seafood (FAO, 2009). Global demand for seafood is growing at a rapid rate, fueled by increases in human population, affluence, and dietary shifts (York and Gossard, 2004). While demand for seafood is increasing, the catch of wild seafood has been declining slightly for 20 years (Watson and Pauly, 2001). Meeting the growth in demand has only been possible by rapid growth in marine aquaculture. The United States consumes nearly five billion pounds of seafood a year, ranking it third globally behind China and Japan. This large consumption, however, comes primarily from fish caught outside the nation's boundary waters. Nearly 85 percent of U.S. consumption is imported, and that fraction is increasing (Becker, 2010). Therefore, consumption of food from the sea links the United States to nearly all the world's ocean ecosystems.

Marine Fisheries

The impacts of climate change on marine-based food systems are far less well known than impacts on agriculture, but there is rapidly growing evidence that they could be severe (see Chapter 9). This is especially problematic given that a sizeable fraction of the world's fisheries are already overexploited (Worm et al., 2009) and many are also subject to pollution from land or under stress from the decline of critical habitats like coral reefs and wetlands (Halpern et al., 2008; Sherman et al., 2009).

Year-to-year climate variability has long been known to cause large fluctuations in fish stocks, both directly and indirectly (McGowan et al., 1998; Stenseth et al., 2002), and this has always been a challenge for effective fisheries management (Walters and Parma, 1996). Similar sensitivity to longer time-scale variations in climate has been documented in a wide range of fish species from around the globe (Chavez et al., 2003; Steele, 1998), and this portends major changes in fish populations under future climate change scenarios. Successful management of fisheries will require an improved ability to forecast population fluctuations driven by climate change; this in turn demands significant new investments in research, including research on various management options (e.g., Mora et al., 2009). Fundamental shifts in management practices may be needed. For example, restoration planning for depleted Chinook salmon populations in the Pacific Northwest needs to account for the spatial shift in salmon habitat (Battin et al., 2007). An added complexity is that, because most of the fish catch comes from open oceans under international jurisdiction, any management regime will need to be negotiated and accepted by multiple nations to be effective.

Fished species tend to be relatively mobile, either as adults or young (larvae drifting in the plankton). As a result, their distributions can shift rapidly compared to those of land animals. In recent decades, geographical shifts toward the poles of tens to hundreds of kilometers have been documented for a wide range of marine species in different areas (Grebmeier et al., 2006; Lima et al., 2006; Mueter and Litzow, 2008; Sagarin et al., 1999; Zacherl et al., 2003). Model projections for anticipated changes by 2050 suggest a potentially dramatic rearrangement of marine life (Cheung et al., 2009). Although such projections are based upon relatively simple models and should be treated as hypotheses, they suggest that displacements of species ranges may be sufficiently large that the fish species harvested from any given port today may change dramatically in coming decades. Fishers in many Alaskan ports are already facing much longer commutes as distributions of target species have shifted (CCSP, 2009b).

Such projected shifts in fisheries distributions are likely to be most pronounced for U.S. fisheries in the North Pacific and North Atlantic, where temperature increases are likely to be greatest and will be coupled to major habitat changes driven by reduced sea ice (CCSP, 2009b). Abrupt warming in the late 1970s, which was associated with a regime shift in the Pacific Decadal Oscillation, greatly altered the marine ecosystem composition in the Gulf of Alaska (Anderson and Piatt, 1999). Rapid reductions in ice-dominated regions of the Bering Sea will very likely expand the habitat for subarctic piscivores such as arrowtooth flounder, cod, and pollock. Because there are presently only fisheries for cod and pollock, arrowtooth flounder may experience significant population increases with broad potential consequences to the ecosystem (CCSP, 2009b).

The effects of ocean acidification from increased absorption of CO_2 by the sea (see Chapters 6 and 9) may be even more important for some fisheries than other aspects of climate change, although the overall impact of ocean acidification remains uncertain (Fabry et al., 2008; Guinotte and Fabry, 2008). Many fished species (e.g., invertebrates such as oysters, clams, scallops, and sea urchins) produce shells as adults or larvae, and the production of shells could be compromised by increased acidification (Fabry et al., 2008; Gazeau et al., 2007; Hofmann et al., 2008). Many other fished species rely on shelled plankton, such as pteropods and foraminifera, as their primary food source. Projected declines in these plankton species could have catastrophic impacts on fished species higher in the food chain. Finally, acidification can disrupt a variety of physiological processes beyond the production of shells. Hence, the potential impacts of acidification—especially in combination with other climate changes on marine fisheries—is potentially enormous, but the details remain highly uncertain (NRC, 2010f).

Aquaculture and Freshwater Fisheries

Today, approximately a third of seafood is grown in aquaculture, and that number rises to half if seafood raised for animal feed is included. As the fastest growing source of animal protein on the planet, aquaculture is widely touted as critical for meeting growing demands for food. Although aquaculture avoids some of the climate impacts associated with wild fish harvesting, others (e.g., ocean acidification) are equally challenging. Indeed, the current predominance of aquaculture facilities in estuaries and bays may exacerbate some of the impacts of ocean acidification (Miller et al., 2009). In addition, since different forms of aquaculture may require a variety of other natural resources such as water, feed, and energy to produce seafood, there may be much broader indirect impacts of climate change on this rapidly growing industry.

Freshwater fisheries face most of the same challenges from climate change as those in saltwater, as well as some that are unique. Forecasting the consequences of warming on fish population dynamics is complicated, because details of future climate at relatively small geographic scales (e.g., seasonal and daily variation, regional variation across watersheds) are critical to anticipating fish population responses (Littell et al., 2009). Yet, as noted in Chapter 6, regional and local aspects of climate change are the hardest to project. Expected effects include elevated temperatures, reduced dissolved oxygen (Kalff, 2002), increased stratification of lakes (Gaedke et al., 1998; Kalff, 2002), and elevated pollutant toxicity (Ficke et al., 2007). Although the consequences of some of these changes are predictable when taken one at a time, the complex nature of interactions between their effects makes forecasting change for even a single species in a single region daunting (Littell et al., 2009). In addition to altering these physical and chemical characteristics of freshwater, climate change will also alter the quantity, timing, and variability of water flows (Mauget, 2003; Ye et al., 2003; Chapter 8). Climate-driven alterations of the flow regime will add to the decades or even centuries of alterations of stream and river flows through other human activities (e.g., urbanization, water withdrawals, dams; Poff et al., 2007). Finally, changes in lake levels that will result from changed patterns of precipitation, runoff, groundwater flows, and evaporation could adversely affect spawning grounds for some species, depending on bathymetry. While the full ramifications of these changes for freshwater fish require further analysis, there is evidence that coldwater fish such as salmon and trout will be especially sensitive to them. For example, some projections suggest that half of the wild trout population of the Appalachians will be lost; in other areas of the nation, trout losses could range as high as 90 percent (Williams et al., 2007).

Globally, precipitation is expected to increase overall, and more of it is expected to occur in extreme events and as rain rather than snow, but anticipated regional changes in precipitation vary greatly and are highly uncertain (see Chapter 8). As a result, major alterations of stream and lake ecosystems are forecast in coming decades, but the details remain highly uncertain (Ficke et al., 2007). Although freshwater fish and invertebrates are typically as mobile as their marine counterparts, their ability to shift their range in response to climate change may be greatly compromised by the challenges of moving between watersheds. In contrast to the rapid changes in species ranges in the sea (Perry et al., 2005), freshwater fish and invertebrates may be much more constrained in their poleward range shifts in response to climate change, especially in east-west stream systems (Allan et al., 2005; McDowall, 1992).

In the United States, per capita consumption of fish and shellfish from the sea and estuaries is more than 15 times higher than consumption of freshwater fish (EPA, 2002); nevertheless, freshwater fish are important as recreation and as food for some U.S. populations. Globally, however, freshwater and diadromous fish (fish that migrate between fresh- and saltwater) account for about a quarter of total fish and shellfish consumption (Laurenti, 2007) and in many locations serve as the predominant source of protein (Bayley, 1981; van Zalinge et al., 2000). Given the large uncertainty in how climate change impacts on freshwater ecosystems will affect the fisheries they support, this important source of food and recreation is at considerable risk.

SCIENCE TO SUPPORT LIMITING CLIMATE CHANGE BY MODIFYING AGRICULTURAL AND FISHERY SYSTEMS

Food production systems are not only affected by climate change, but also contribute to it. Agricultural activities release significant amounts of CO_2, methane (CH_4), and nitrous oxide (N_2O) to the atmosphere (Cole et al., 1997; Paustian et al., 2004; Smith et al., 2007). CO_2 is released largely from decomposition of soil organic matter by microorganisms or burning of live and dead plant materials (Janzen, 2004; Smith, 2004); decomposition is enhanced by vegetation removal and tillage of soils. CH_4 is produced when decomposition occurs in oxygen-deprived conditions, such as wetlands and flooded rice systems, and from digestion by many kinds of livestock (Matson et al., 1998; Mosier et al., 1998). N_2O is generated by microbial processes in soils and manures, and the flux of N_2O into the atmosphere is typically enhanced by fertilizer use, especially when applied in excess of plant needs (Robertson and Vitousek, 2009; Smith and Conen, 2004). The 2007 IPCC assessment concluded, with medium certainty, that agriculture accounts for about 10 to 12 percent of total global human-caused emissions of GHGs, including 60 percent of N_2O and about 50 percent of CH_4 (Smith et al., 2007). The Environmental Protection Agency (EPA) estimates that about 32 percent of CH_4 emissions and 67 percent of N_2O emissions in the United States are associated with agricultural activities (EPA, 2009b).

Typically, the projected future of global agriculture is based on intensification—increasing the output per unit area or time—which is typically achieved by increasing or improving inputs such as fertilizer, water, pesticides, and crop varieties, and thereby potentially reducing agricultural demands on other lands (e.g., Borlaug, 2007). Given this projected intensification, global N_2O emissions are predicted to increase by about 50 percent by 2020 (relative to 1990) due to increasing use of fertilizers in agricultural systems (EPA, 2006; Mosier and Kroeze, 2000). If CH_4 emissions grow in direct proportion to increases in livestock numbers, then global livestock-related CH_4 production is expected to increase by 60 percent up to 2030 (Bruinsma, 2003); in the United States, the EPA (2006) forecasts that livestock-related CH_4 emissions will increase by 21 percent between 2005 and 2020. Projected changes in CH_4 emissions from rice production vary but are generally smaller than those associated with livestock (Bruinsma, 2003; EPA, 2006).

The active management of agricultural systems offers possibilities for limiting these fluxes and offsetting other GHG emissions. Many of these opportunities use current technologies and can be implemented immediately, permitting a reduction in emissions per unit of food (or protein) produced, and perhaps also a reduction in emissions per capita of food consumption. For example, changes in feeds and feeding practices can reduce CH_4 emissions from livestock, and using biogas digesters for manure management can substantially reduce CH_4 and N_2O emissions while producing energy. Changes in management of fertilizers, and the development of new fertilizer application technologies that more closely match crop demand—sometimes called precision or smart farming—can also reduce N_2O fluxes. It may also be possible to develop and adopt new rice cultivars that emit less CH_4 or otherwise manage the soil-root microbial ecosystem that drives emissions (Wang et al., 1997). Alternatively, organic agriculture or its fusion into other crop practices may reduce emissions and other environmental problems. To date, however, there has been little research on the willingness of farmers and the agricultural sector in general to adopt practices that would reduce emissions, or on the kinds of education, incentives, and institutions that would promote their use.

Beyond limiting the trace gases emitted in agricultural practice, there are opportunities for offsetting GHG emissions more broadly by managing agricultural landscapes to absorb and store carbon in soils and vegetation (Scherr and Sthapit, 2009). For example, minimizing soil tillage yields multiple benefits by increasing soil carbon storage, improving and maintaining soil structure and moisture, and reducing the need for inorganic fertilizers, as well as reducing labor, mechanization, and energy costs. Such practices may also have beneficial effects on biodiversity and other ecosystem services provided by surrounding lands and can be made economically attractive to farmers (Robertson and Swinton, 2005; Swinton et al., 2006). Incorporating biochar (charcoal from fast-growing trees or other biomass that is burned in a low-oxygen environment) has also been proposed as a potentially effective way of taking carbon out of the atmosphere; the resulting biochar can be added to soils for storage and improvement of soil quality (Lehmann and Joseph, 2009), although there has been some debate about the longevity of the carbon storage (Lehmann and Sohi, 2008; Wardle et al., 2008). Shifting agricultural production systems to perennial instead of annual crops, or intercropping annuals with perennial plants such as trees, shrubs, and palms, could also store carbon while producing food and fiber. Biofuel systems that depend on perennial species rather than food crops could be an integral part of such a system. Research is needed to develop these options and to test their efficacy. Most important, a landscape approach would be required in order to plan for carbon storage in conjunction with food and fiber production, conservation, and other land uses and the ecosystem services they provide.

Land clearing and deforestation have been major contributors to GHG emissions over the past several centuries, although as fossil fuel use has grown, land use contributions have become proportionally less important. Still, tropical deforestation alone accounted for about 20 percent of the carbon released to the atmosphere from human activities from 2000 to 2005 (Gullison et al., 2007) and 17 percent of all long-lived GHGs in 2004 (Barker et al., 2007). Reducing deforestation and restoring vegetation in degraded areas could thus both limit climate change and provide linked ecosystem and social benefits (see Chapter 9). It is not yet clear, however, how such programs would interact with other forces operating on agriculture to affect overall land uses and emissions. Finally, as with all proposed emissions-limiting land-management approaches, it is critical that attention be paid to consequences for all GHGs, not just a single target gas (Robertson et al., 2000), and to all aspects of the climate system, including reflectivity of the land surface (Gibbard et al., 2005; Jackson et al., 2008), as well as co-benefits in conservation, agricultural production, water resources, energy, and other sectors.

SCIENCE TO SUPPORT ADAPTATION IN AGRICULTURAL SYSTEMS

The ability of farmers and the entire food production, processing, and distribution system to adapt to climate change will contribute to, and to some extent govern, the ultimate impacts of climate change on food production. Adaptation strategies may include changes in location as well as in-place changes such as shifts in planting dates and varieties; expansion of irrigated or managed areas; diversification of crops and other income sources; application of agricultural chemicals; changes in livestock care, infrastructure, and water and feed management; selling assets or borrowing credit (Moser et al., 2008; NRC, 2010a; Wolfe et al., 2008). At the broadest level, adaptation also includes investment in agricultural research and in institutions to reduce vulnerability. This is because the ability of farmers and others to adapt depends in important ways on available technology, financial resources and financial risk-management instruments, market opportunities, availability of alternative agricultural practices, and importantly, access to, trust in, and use of information such as seasonal forecasts (Cash, 2001; Cash et al., 2006a). It also depends on specific institutional arrangements, including property rights, social norms, trust, monitoring and sanctions, and agricultural extension institutions that can facilitate diversification (Agrawal and Perrin, 2008). Not all farmers have access to such strategies or support institutions, and smallholders—especially those with substantial debt, and the landless in poor countries—are most likely to suffer negative effects on their livelihoods and food security. Smallholder and subsistence farmers will suffer complex, localized impacts of climate change (Easterling et al., 2007).

Integrated assessment models, which combine climate models with crop models and models of the responses of farmers and markets, have been used to simulate the impacts of climate changes on productivity and also on factors such as farm income and crop management. Some modeling studies have included adaptations in these integrated assessments (McCarl, 2008; Reilly et al., 2003), for example by adjusting planting dates or varieties and by reallocating crops according to changes in profitability. For the United States, these studies usually project very small effects of climate change on the agricultural economy, and, in some regions, positive increases in productivity and profitability (assuming adaptation through cropping systems changes). As noted earlier with regard to climate-crop models, assessments have not yet included potential impacts of pests and pathogens or extreme events, nor have they included site- and crop-specific responses to climate change or variations. Moreover, even integrated assessment models that include adaptation do not include estimates of rates of technological change, costs of adaptation, or planned interventions (Antle, 2009). Thus, our understanding of the effects climate change will have on U.S. agriculture and on international food supplies, distribution, trade, and food security remains quite limited and warrants further research.

As they have in the past, both autonomous adaptations by farmers and planned interventions by governments and other institutions to facilitate, enable, and inform farmers' responses will be important in reducing potential damages from climate change and other related changes. Investments in crop development, especially in developing countries, have stagnated since the 1980s (Pardey and Beintema, 2002), although recent investments by foundations may fill some of the void. Private-sector expenditures play an important role, especially in developed countries, and some companies are engaging in efforts to develop varieties well suited for a changing climate (Burke et al., 2009; Wolfe et al., 2008).

Government investments in new or rehabilitated irrigation systems (of all sizes) and efficient water use and allocation technologies, transportation infrastructure, financial infrastructure such as availability of credit and insurance mechanisms (Barnett et al., 2008; Gine et al., 2008; World Bank, 2007), and access to fair markets are also important elements of adaptation (Burke et al., 2009). Likewise, investments in participatory research and information provision to farmers have been a keystone of past agricultural development strategies (e.g., through extension services in both developed and developing countries) and no doubt will remain so in the future. Finally, the provision of social safety nets (e.g., formal and informal sharing of risks and costs, labor exchange, crop insurance programs, food aid during emergencies, public works programs, or cash payments), which have long been a mainstay of agriculture in the developed world, will remain important (Agrawal, 2008; Agrawal and Perrin, 2008). These considerations need to be integrated into development planning.

It is important that agriculture be viewed as an integrated system. As noted above, the United States and the rest of the world will be simultaneously developing strategies to adapt agriculture to climate change, to utilize the potential of agricultural practices and other land uses to reduce the magnitude of climate change, and to increase agricultural production to meet rising global demands. With careful analysis and institutional design, these efforts may be able to complement one another while also enhancing our ability to improve global food security. However, without such integrated analysis, various practices and policies could easily work at cross purposes, moving the global food production system further from, rather than closer to, sustainability. For example, increased biofuel production would decrease reliance on fossil fuels but could increase demand for land and food resources (Fargione et al., 2008).

FOOD SECURITY

Food security is defined as a "situation that exists when all people, at all times, have physical, social, and economic access to sufficient, safe, and nutritious food that meets their dietary needs and food preferences for an active and healthy life" (Schmidhuber and Tubiello, 2007). The four dimensions of food security are availability (the overall ability of agricultural systems to meet food demand), stability (the ability to acquire food during income or food price shocks), access (the ability of individuals to have adequate resources to acquire food), and utilization (the ability of the entire food chain to deliver safe food). Climate change affects all four dimensions directly or indirectly; all can be affected at the same time by nonclimatic factors such as social norms, gender roles, formal and informal institutional arrangements, economic markets, and global to local agricultural policies. For example, utilization can be affected through the impact of warming on spoilage and foodborne disease, while access can be affected by changing prices in the fuels used to transport food. Most studies have focused on the first dimension—the direct impact of climate change on the total availability of different agricultural products. Models that account for the other three dimensions need to be developed to identify where people are most vulnerable to food insecurity (Lobell et al., 2008; see also Chapter 4).

Because the food system is globally interconnected, it is not possible to view U.S. food security, or that of any other country, in isolation. Where food is imported—as is the case for a high percentage of seafood consumed in the United States—prices and availability can be directly affected by climate change impacts in other countries. Climate change impacts anywhere in the world potentially affect the demand for agricultural exports and the ability of the United States and other countries to meet that demand. Food security in the developing world also affects political stability, and thereby U.S. national security (see Chapter 16). Food riots that occurred in many countries as prices soared in 2008 are a case in point (Davis and Belkin, 2008). Over the past 30 years, there has been dramatic improvement in access to food as real food prices have dropped and incomes have increased in many parts of the developing world (Schmidhuber and Tubiello, 2007). Studies that project the number of people at risk of hunger from climate change indicate that the outcome strongly depends on socio-economic development, since affluence tends to reduce vulnerability by enlarging coping capacity (Schmidhuber and Tubiello, 2007). Clearly, international development strategies and climate change are inextricably intertwined and require coordinated examination.

RESEARCH NEEDS

Given the challenges noted in the previous section, it is clear that expanded research efforts will be needed to help farmers, development planners, and others engaged in the agricultural sector to understand and respond to projected impacts of climate change on agriculture. There may also be opportunities to limit the magnitude of future climate change though changes in agricultural practices; it will be important to link such strategies with adaptation strategies so they complement rather than undermine each other. Identifying which regions, human communities, fisheries, and crops and livestock in the United States and other parts of the world are most vulnerable to climate change, developing adaptation approaches to reduce this vulnerability, and developing and assessing options for reducing agricultural GHG emissions are critical tasks for the nation's climate change research program. Focus is also needed on the developing world, where the negative effects of climate change on agricultural and fisheries production tend to coincide with people with low adaptation capacity. Some specific research areas are listed below.

Improve models of crop response to climate and other environmental changes. Crop plants and timber species respond to multiple and interacting effects—including temperature, moisture, extreme weather events, CO_2, ozone, and other factors such as pests, diseases, and weeds—all of which are affected by climate change. Experimental studies that evaluate the sensitivity of crops to such factors, singly and in interaction, are needed, especially in ecosystem-scale experiments and in environments where temperature is already close to optimal for crops. Many assessments model crop response to climate-related variables while assuming no change in availability of water resources, especially irrigation. Projections about agricultural success in the future need to explicitly include such interactions. Of particular concern are assumptions about water availability that include consideration of needs by other sectors. The reliability of water resources for agriculture when there is competition from other uses needs to be evaluated in the context of coupled human-environment systems, ideally at regional scales. Improved understanding of the response of farmers and markets to production and prices and also to policies and institutions that affect land and resource uses is needed; incorporation of that information in models will aid in designing effective agricultural strategies for limiting and adapting to climate change.

Improve models of response of fisheries to climate change. Sustainable yields from fisheries require matching catch limits with the growth of the fishery. Climate variation already makes forecasting the growth of fish populations difficult, and future climate change will increase this critical uncertainty. Studies of connections between climate and marine population dynamics are needed to enhance model frameworks for fisheries management. In addition, there is considerable uncertainty about differences in sensitivity among and within species to ocean acidification (NRC, 2010f). This inevitable consequence of increasing atmospheric CO_2 is poorly understood, yet global in scope. Most fisheries are subject to other stressors in addition to warming, acidification, and harvesting, and the interactions of these other stresses need to be analyzed and incorporated into models. Finally, these efforts need to be linked to the analysis of effective institutions and policies for managing fisheries.

Expand observing and monitoring systems. Satellite, aircraft, and ground-based measures of changes in crops yields, stress symptoms, weed invasions, soil moisture, ocean productivity, and other variables related to fisheries and crop production are possible but not yet carried out systematically or continuously. Monitoring of the environmental and social dynamics of food production systems on land and in the oceans is also needed to enable assessments of vulnerable systems or threats to food security. Monitoring systems will require metrics of vulnerability and sustainability to provide early warnings and develop adaptation strategies.

Assess food security and vulnerability in the context of climate change. Effective adaptation will require integration of knowledge and models about environmental as well as socioeconomic systems in order to project regional food supplies and demands, understand appropriate responses, to develop institutional approaches for adapting under climate variability and climate change, and to assess implications for food security (NRC, 2009k). Scenarios that evaluate implications of climate change and adaptation strategies for food security in different regions are needed, as are models that assess shifting demands for meat and seafood that will influence price and supply. Approaches, tools, and metrics are needed to assess the differential vulnerability of various human-environment systems so that investments can be designed to reduce potential harm (e.g., through interventions such as the development of new crop varieties and technologies, new infrastructure, social safety nets, or other adaptation measures). A concerted research effort is needed both for conducting assessments and to support the development and implementation of options for adaptation. Surprisingly, relatively little effort has been directed toward identification of geographic areas where damages to agriculture or fisheries could be caused by extreme events (hurricanes, drought, hypoxia); where there is or will be systematic loss of agricultural area due to sea level rise, erosion, and saltwater intrusion; or where there will be changes in average conditions (e.g., extent of sea ice cover, and warming of areas that are now too cold for agriculture) that could lead to broad-scale changes—positive or negative—in the type and manner of agricultural and fisheries production.

Evaluate trade-offs and synergies in managing agricultural lands. Improved integrated assessment approaches and other tools are needed to evaluate agricultural lands and their responses to climate change in the context of other land uses and ecosystem services. Planning approaches need to be developed for avoiding adaptation responses that place other systems (or other generations) at risk—for example, by converting important conservation lands to agriculture, allocating water resources away from environmental or urban needs, or overuse of pesticides and fertilizers. Integrated assessments would help to evaluate both trade-offs (e.g., conservation versus agriculture) and co-benefits (e.g., increasing soil carbon storage while also enhancing soil productivity and reducing erosion) of different actions that might be taken in the agricultural sector to limit the magnitude of climate change or adapt to its impacts.

Evaluate trade-offs and synergies in managing the sea. The oceans provide a wide range of services to humans, but conflicts over use of the oceans are often magnified because of the absence of marine spatial planning and relatively weak international marine regulatory systems. Efforts to limit the magnitude of climate change are causing society to consider the sea for new sources of energy (e.g., waves, tides, thermal gradients), while the opening of ice-free areas in the Arctic is encouraging exploration of offshore reserves of minerals and fossil fuels. Without analyses of the looming trade-offs between these emerging uses and existing services, such as fisheries and recreation, conflicts will inevitably grow. New approaches for analyses of such trade-offs are needed as an integral component of marine spatial planning.

Develop and improve technologies, management strategies, and institutions to reduce GHG emissions from agriculture and fisheries and to enhance adaptation to climate change. Research on options for reducing emissions from the agricultural sector is needed, including new technologies, evaluation of effectiveness, costs and benefits, perceptions of farmers and others, and policies to promote implementation. Technologies such as crop breeding and new cropping systems could dramatically increase the sector's adaptive capacity. Research on the role of entitlements and institutional barriers in influencing mitigation or adaptation responses; the effectiveness of governance structures; interactions of national and local policies; and national security implications of climate-agriculture interactions are also needed.

CHAPTER ELEVEN

Public Health

Extreme heat can be fatal, and hurricanes and tornadoes cause injuries and damage infrastructure. Air pollution can be linked to respiratory illness, and drought can lead to malnutrition. These are just a few examples of how weather and climate can influence human health. Climate change has the potential to affect any health outcome that is sensitive to environmental conditions. However, the causal chain linking climate change to shifting patterns of health threats and outcomes is complicated by factors such as wealth, distribution of income, status of public health infrastructure, provision of preventive and acute medical care, and access to and appropriate use of health care information. As with many other consequences of climate change, concurrent changes in nonclimatic factors, such as combustion-related air pollution, will influence the severity of future health impacts.

Questions decision makers are asking, or will be asking, about climate change and public health include the following:

- What climate change effects are potentially the most dangerous to human health, and who is most at risk?
- What kinds of preventative measures and response systems can be put in place to manage these risks?
- Are there lessons that can be learned from other threats to human health?
- What kinds of monitoring systems are available to track the health impacts of climate change?
- How do we help ensure that actions taken to limit or adapt to climate change do not result in unintentional adverse health impacts?
- What actions to limit and adapt to climate change will yield public health co-benefits?

This chapter summarizes the current understanding of the health effects of climate change from stressors such as temperature, severe weather, infectious disease, and air quality. It also reviews how health may be affected—in negative or positive ways—by many of the strategies societies use to limit, prepare for, and adapt to climate change (Figure 11.1). More extensive discussions of the relationship between climate change

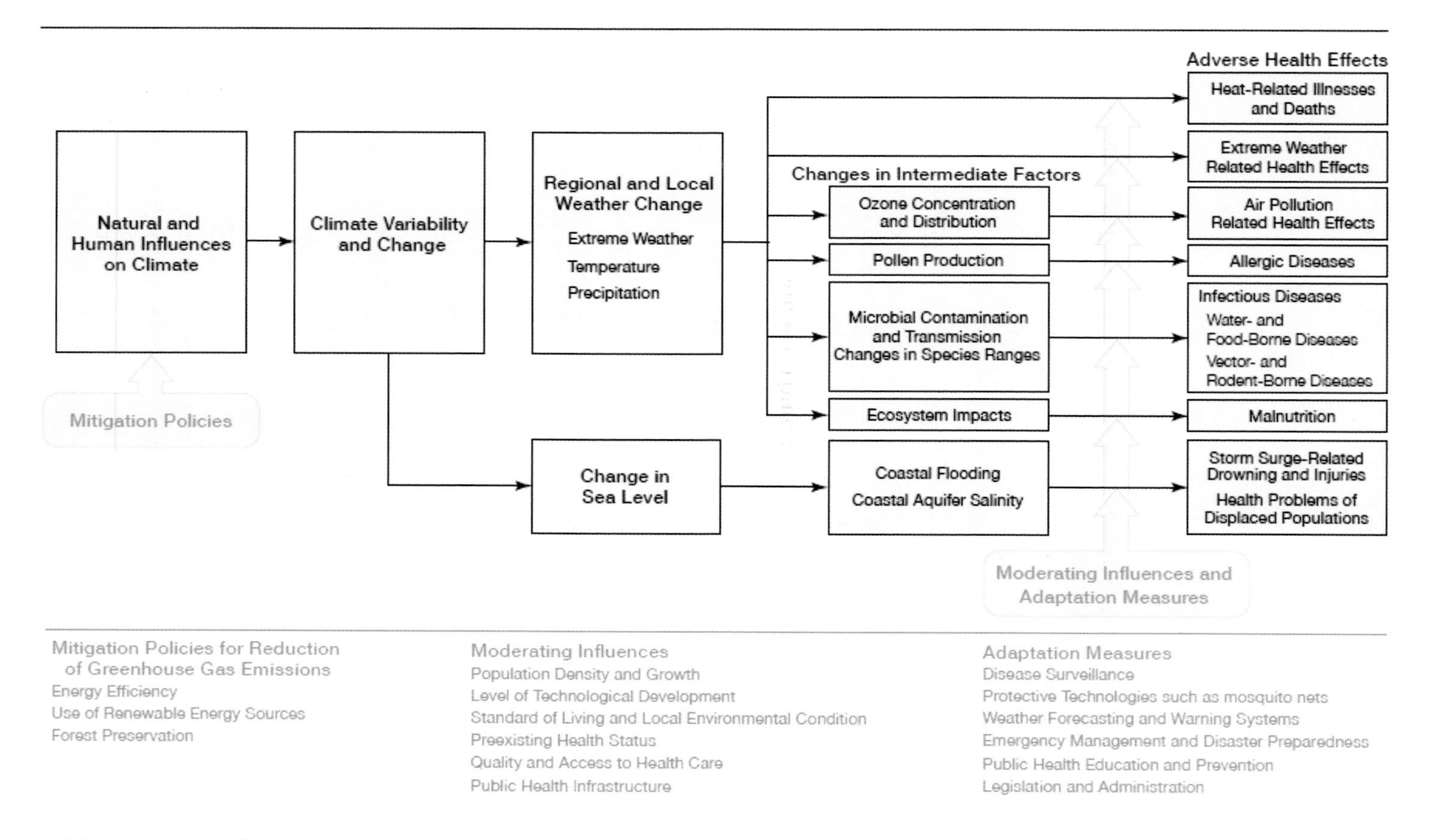

FIGURE 11.1 Simplified illustration of the mechanisms through which climate change can affect health outcomes. The blue text provides lists of policies or factors that can influence these health outcomes. Mitigation policies can potentially reduce the magnitude of climate change, while moderating influences (such as access to quality health care) or adaptation measures (such as improved public health education) could lessen the impact of adverse health effects. Mitigation and adaptation strategies will vary in the time necessary to implement them and realize their effects. SOURCE: Modified from Haines and Patz (2004).

and public health can be found in the recent synthesis of the U.S. Climate Change Science Program (CCSP, 2008a) and in other recent reports and syntheses (e.g., Confalonieri, 2007; Confalonieri et al., 2007). Additionally, this chapter identifies research needed to clarify exposure-response relationships, better quantify the impacts of climate change on human health, and identify efficient adaptation options.

EXTREME TEMPERATURES AND THERMAL STRESS

Heat waves are the leading causes of weather-related morbidity and mortality in the United States (CDC, 2006; Changnon et al., 1996). Between 1979 and 1999, some 8,015 deaths in the United States were heat related, and 3,829 of these were linked to weather conditions (Donoghue et al., 2003). As with other extreme events, the risk of heat waves is not evenly distributed across the country; for example, populations in the Midwest are at increased risk for illness and death during heat waves (CCSP, 2008a; Jones et al., 1982; Palecki et al., 2001; Semenza et al., 1996). Heat stress and heat waves are significant factors for increased morbidity and mortality in other parts of the world as well. A typical U-shaped curve (Figure 11.2) illustrates temperatures beyond which human mortality rates are observed to rise, depending on latitude.

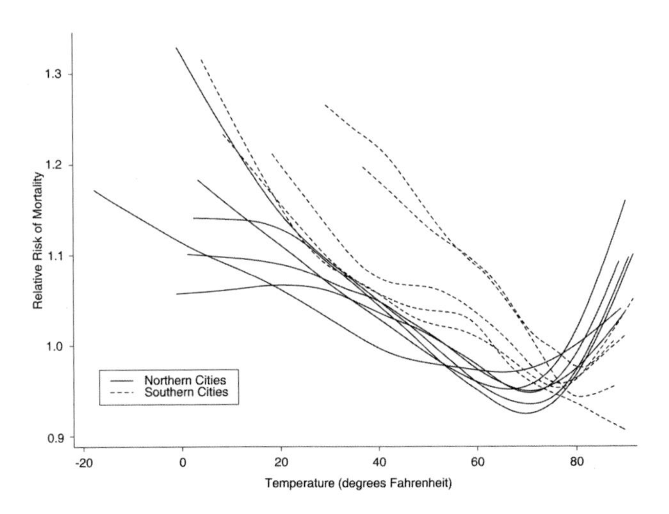

FIGURE 11.2 Temperature-mortality relative risk functions for 11 eastern U.S. cities for the period 1973 to 1994. Each city has its own line. Many northern cities (solid lines) exhibit a U-shaped curve, indicating that higher rates of mortality are exhibited at relatively cold and relatively warm temperatures. In the southern cities (dashed lines), mortality risks bear a stronger relationship with relatively cold temperatures. SOURCE: Curriero et al. (2002).

Hot days and hot nights have become more frequent in recent decades (Trenberth et al., 2007), and the frequency, intensity, and duration of heat waves are projected to increase in the decades ahead, especially under higher warming scenarios (CCSP, 2008a). By applying the magnitude of the 2003 European heat wave (see Box 12.1 in Chapter 12) to five major U.S. cities, Kalkstein et al. (2008) concluded that a heat wave of the same magnitude could increase excess heat-related deaths by more than five times the average. Projected excess deaths in New York City associated with such a heat wave, for example, would exceed the current average number of heat-related deaths nationwide each summer, with a death rate approaching that for all accidents.

There is also the potential, however, for warming temperatures to reduce exposure and health impacts associated with cold winter temperatures, although this potential is projected to vary by location (CCSP, 2008a). For example, research has shown that regions with milder winters actually have higher mortality rates during cold weather than regions with colder winters (Curriero et al., 2002; Davis et al., 2004). Seasonal variations in death rates in the United States are well documented, with more deaths occurring during winter than during summer months (Curriero et al., 2002; Mackenbach et al., 1992). However, mortality rates are influenced by a range of factors other than temperature, including housing characteristics and personal behaviors, which have not been extensively studied in the context of future climate projections. Thus, determining whether warming temperatures could alter winter temperature mortality relationships is complex and requires understanding all of the factors involved.

There have been several attempts to project future heat-related health impacts of climate change, and this is an active, albeit not large, area of current research. Figure 11.3 shows a schematic illustration of the expected impacts of warming temperatures and increased number of hot days on human health. Figure 11.4 shows a projection of total increases in heat-related deaths for a major U.S. city (Chicago) experiencing a

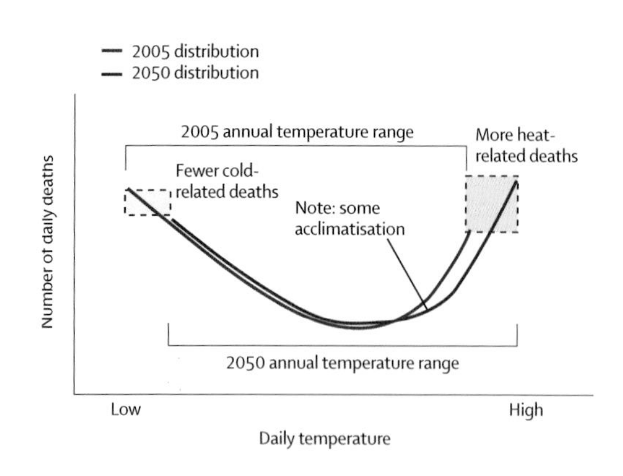

FIGURE 11.3 Schematic representation of the relationship of temperature-related deaths and daily temperature assuming no adaptation measures. The 2050 range of daily temperature (red curve) is shifted to the right of the 2005 range of daily temperature (blue curve), indicating that there could be an increase in heat-related deaths and a decrease in cold-related deaths. SOURCE: McMichael et al. (2006).

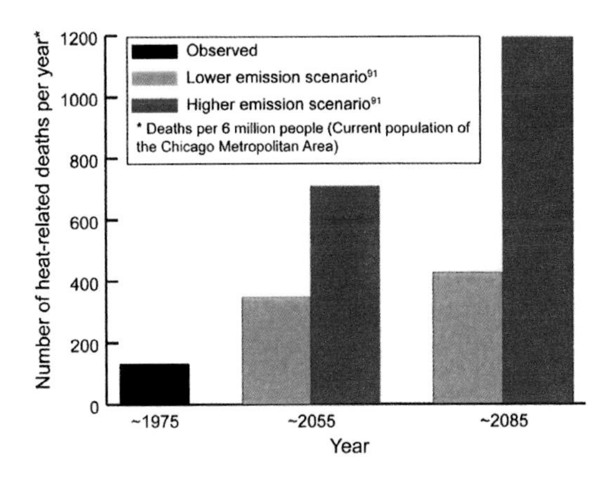

FIGURE 11.4 Potential increases in heat-related deaths in Chicago as a result of temperature increases over the 21st century. The graphs correspond to three-decade averages, centered on 1975, 2055, and 2085. Orange corresponds to climate projections with lower emissions and relatively less warming, and red corresponds to higher emissions and relatively more warming. SOURCES: USGCRP (2009a) and Hayhoe et al. (2010).

typical heat wave at three different future dates. One challenge with projecting future health impacts of heat waves is uncertainty about the extent to which people will acclimatize to higher temperatures. Uncertainty regarding adaptation strategies is also a challenge for projecting health impacts. For example, implementation of early warning systems and alteration of infrastructure to reduce urban heat islands could help minimize the impacts of heat waves by increasing adaptive capacities in communities.

SEVERE WEATHER

Deaths and physical injuries from severe weather events such as hurricanes, tornados, floods, and wildfire occur annually across the United States. Direct morbidity and mortality increase with the intensity and duration of such an event and can decrease with advance warning and preparation (CCSP, 2008a). While uncertainties remain, the general trend is that climate change will lead to an increase in the intensity of several types of severe weather events, such as flooding (see Chapter 8). Severe weather events may also lead to increases in diarrheal disease and increased incidence of respiratory symptoms, particularly in developing countries (CCSP, 2008a; Haines and Patz, 2004). Extreme events can also affect health indirectly. The mental health impacts (e.g., posttraumatic stress disorder, anxiety, and depression) of extreme events could be especially important, but they are difficult to assess (CCSP, 2008a; Haines and Patz, 2004).

INFECTIOUS DISEASES

The ranges and impacts of a number of important pathogens may change as a result of changing temperatures, precipitation, and extreme events (Confalonieri et al., 2007; Gage et al., 2008; Paaijmans et al., 2009; Pascual et al., 2006; Patz et al., 2008), resulting in greater human exposures in many parts of the world. Increasing temperatures may expand or shift the ranges of disease vectors, including mosquitoes, ticks, and rodents. Mosquito-borne diseases that may be affected by climate change include malaria, dengue fever, the West Nile virus, and the Saint Louis encephalitis virus. The West Nile and encephalitis viruses have both been associated with drought conditions brought on by extended periods of high temperatures (CCSP, 2008a; Haines and Patz, 2004). The range of the dog tick *Rhipicephalus sanguineus* that carries Rocky Mountain spotted fever may also expand due to increasing temperatures (Parola et al., 2008). Other tickborne diseases that may be impacted by increasing temperatures include Lyme disease and encephalitis. Rodent-borne diseases such as the hantavirus and leptospirosis may also be impacted by climate change. Aside from climate change impacts on vector-borne diseases, several pathogens that cause food- and waterborne diseases are sensitive to changes in temperature, with faster replication rates at higher temperatures. In addition, waterborne disease outbreaks (e.g., cholera outbreaks in developing countries) are also associated with heavy rainfalls and flooding (Confalonieri et al., 2007).

While vector-borne diseases will all be affected by climate-related changes in temperature, humidity, rainfall, and sea level rise, the geographical range of disease vectors depends on a variety of other factors, including population movement, land use change, public health infrastructure, and emergence of drug resistance (CCSP, 2008a; Haines and Patz, 2004). In addition, while there is wide range of vulnerability to disease within and between populations, this is also dependent on multiple, interacting factors (e.g., preexisting conditions such as malnutrition). Greater understanding of the factors contributing to the spread of infectious diseases, and the role that a changing climate will play in that spread, is needed.

AIR QUALITY

Many constituents of the atmosphere that impact public health also play a significant role in influencing climate. Of concern are aerosols, including black carbon, organic carbon, and sulfates. As discussed in Chapter 6, aerosols can have a net cooling effect on climate if they increase the Earth's reflectivity, such as inorganic carbon released during biomass burning, or a net warming effect if they absorb outgoing infrared radiation, such as the black carbon released during incomplete combustion of diesel fuel and biomass burning. Aerosols are of concern for human health due to their impacts on lung function and on respiratory and cardiac disease (Smith et al., 2009).

Tropospheric ozone is not only a greenhouse gas (GHG); it is also classified as a criteria air pollutant. Ozone is a secondary pollutant formed from the action of sunlight on ozone precursors such as carbon monoxide, nitrogen oxides, and volatile organic compounds (see Chapter 6). Human-caused emissions of ozone precursors have led to large increases in tropospheric ozone over the past century (Marenco et al., 1994; Wang and Jacob, 1998). When increased ozone events occur simultaneously with heat waves, the mortality rate can rise by as much as 175 percent (Filleul et al., 2006). Acute exposure to elevated concentrations of ozone is associated with increased hospital admissions for pneumonia, chronic obstructive pulmonary disease, asthma, allergic rhinitis, and other respiratory diseases, and also with premature mortality (e.g., Bell et al., 2005, 2006; Gryparis et al., 2004; Ito et al., 2005; Levy et al., 2005; Mudway and Kelly, 2000). A National Research Council committee concluded that "the association between short-term changes in ozone concentrations and mortality is generally linear throughout most of the concentration ranges. If there is a threshold, it is probably at a concentration below the current ambient air standard" (NRC, 2008e).

Although projected increases in temperatures across the United States in the decades ahead may raise the occurrence of high ozone concentrations (see Figure 11.5), ozone concentrations also depend on a wide range of other factors, including the rate and amount of ozone precursor emissions, human actions taken to limit ozone precursors, and meteorological factors. For example, extremely hot days tend to be associated with stagnant air circulation patterns that can concentrate ground-level ozone, exacerbating respiratory diseases and short-term reductions in lung function (USGCRP, 2001). Under one scenario of climate change for 50 U.S. cities, the increase in temperature projected to occur by the 2050s due to climate change, and a subsequent rise in tropospheric ozone, could exacerbate ozone-related health effects such as cardiovascular, respiratory, and total mortality, as well as hospital admissions for asthma, chronic obstructive pulmonary disease, and respiratory diseases of the elderly (Bell et al., 2007).

Climate change could also affect local and regional air quality through temperature-induced changes in chemical reaction rates, changes in boundary-layer heights that affect vertical mixing of pollutants, and changes in airflow patterns that govern pollutant transport. Responses to climate change can also affect air quality, most notably through changes in emissions associated with efforts to limit the magnitude of climate change. Sources of uncertainty include the degree of future climate change, future emissions of air pollutants and their precursors, and how population vulnerabil-

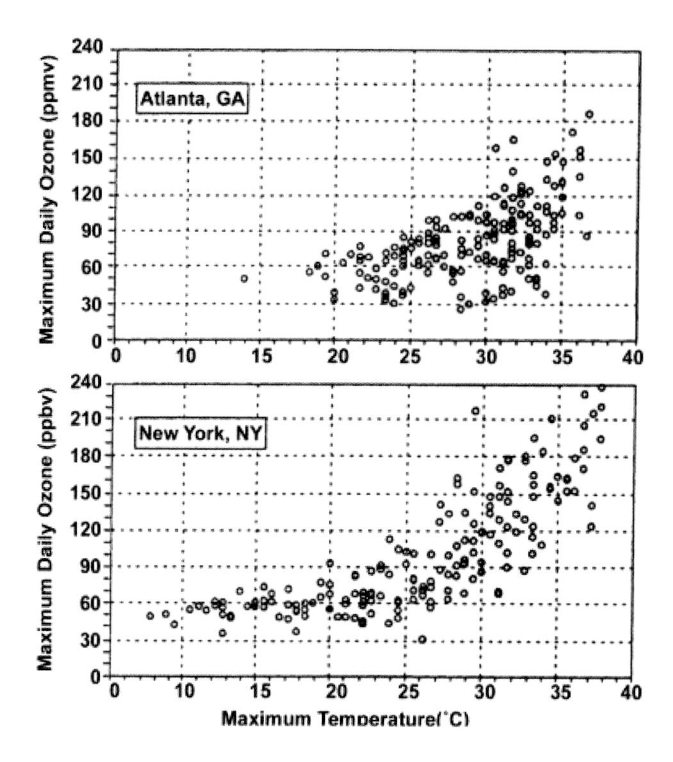

FIGURE 11.5 Ground-level ozone concentrations and temperature in Atlanta and New York City, measured between May and October, 1988 to 1990. The plots show that ozone concentrations are generally higher at warmer temperatures. SOURCE: USGCRP (2009a); based on EPA data.

ity may change in the future. When precursor emissions are held constant, projections suggest climate change will increase concentrations of tropospheric ozone across many regions, increasing morbidity and mortality (Ebi and McGregor, 2008). Increases in urban ozone pollution alone could be as much as 10 parts per billion (ppb) over the next few decades, which would make it difficult for many cities to meet air quality standards (Ebi and McGregor, 2008; Jacob and Winner, 2009). The evidence is less robust for other air pollutants, although several studies have found increased mortality associated with simultaneous rise in temperature and surface aerosols, including both particulate matter and sulfur dioxide (Hu et al., 2008; Katsouyanni et al., 1997; Smith et al., 2009). However, research is needed to understand how concentrations of these pollutants could change with climate change.

There are several examples of how the health impacts of climate change intersect with ecosystem and agricultural impacts in the context of air quality. For example, higher ozone concentrations would be detrimental not only to human health but also to crop production. Losses in crop yields due to increasing ozone and other climate-related factors over the next two to three decades in some rapidly developing regions are expected to have a major impact on the food supply (see Chapter 10), possibly leading to malnutrition and other negative public health impacts (CCSP, 2008a; Epstein, 2005; Haines and Patz, 2004; The Royal Society, 2009). Another example is that the frequency and intensity of wildfires is enhanced in a warming climate (see Chapter 9), and this would be expected to lead to increases in the atmospheric concentration of fine particulate matter, which would have adverse health consequences (Epstein, 2005; Haines and Patz, 2004).

The potential synergies and trade-offs between climate change policies and public health policies are complex. For example, reducing some aerosols such as organic carbon or sulfates would reduce air pollution-related health impacts but increase the rate of climate change (Forster et al., 2007; see also Chapter 6). Conversely, some of the technologies and policy mechanisms that might be used to control climate change may also be complementary to measures adopted to control air pollution; for example, reducing commuter traffic by encouraging mass transit and carpooling would reduce both transportation-related GHG emissions and ozone precursors. Walking or biking for transportation would have the added benefit of increasing physical activity, potentially lowering the incidence of obesity and its related negative health outcomes. Policies designed to reduce or offset climate change may thus have a variety of intended and unintended consequences on public health, and vice versa.

OTHER HEALTH EFFECTS OF CLIMATE CHANGE

Allergies and asthma are influenced by the growth and toxicity of numerous plant species like ragweed, poison ivy, and stinging nettle; based on limited evidence, these plants increase growth and toxicity at higher temperatures and/or concentrations of CO_2 (Hunt et al., 1991; Mohan et al., 2006; USGCRP, 2009a; Ziska, 2003). Drought, changes in water resources (Chapter 8), and climate impacts on agricultural production (Chapter 10) all may have consequences for human health and nutrition (CCSP, 2008a; Epstein, 2005; Haines and Patz, 2004). There could also be an increase in psychiatric disorders, such as anxiety and depression, occurring after severe weather events that cause a disruption of the home environment and economic losses (CCSP, 2008a; Haines and Patz, 2004). Shifts in migration patterns and refugee pressures may result from changes in sea level, food production, severe weather, and drought, resulting in additional human health challenges in some areas.

PROTECTNG VULNERABLE POPULATIONS

Not everyone is equally at risk from the health impacts of climate change (CCSP, 2008a; Confalonieri, 2007). For example, in the United States, cities with cooler climates generally experience more heat-related mortality than cities with warmer climates. This difference is attributed to the ability of populations to acclimatize to different levels of temperature through physiological, behavioral, and technological mechanisms (Haines and Patz, 2004). The heat island effect can result in residents of high-density urban areas being more vulnerable to heat-related health effects (CCSP, 2008a). Residents of low-lying coastal areas could be particularly vulnerable to the health impacts associated with sea level rise, coastal erosion, and more intense storms (CCSP, 2008a). Other reasons for geographic differences in vulnerability include differences in physical, ecological, and activity-related exposure to the risks within and across countries; differences in sensitivity due to the overlap with other changes and stresses in particular regions or populations; and widely varying adaptive capacities. In addition to geographic variations, certain subpopulations could also be more susceptible to the health impacts of climate change. These groups include infants and children, pregnant women, older adults, impoverished populations, people with chronic conditions, people with mobility and cognitive restraints, certain occupational groups, and recent migrants and immigrants. The specific vulnerabilities of these population groups are outlined in Table 11.1.

Responses to recent extreme weather and climate events such as Hurricane Katrina show that, even in the United States, current levels of adaptation are insufficient (USGCRP, 2009a). Substantial inequities exist in access to public heath infrastructure, both in the United States and elsewhere (Pellow and Brulle, 2007), so health risks will be disproportionately high for the poor, elderly, and otherwise disadvantaged. Additionally, analysis has shown that, without further investment, the public health infrastructure most important for addressing the challenges of climate change could be insufficient (Ebi et al., 2009). Concerted efforts will be needed to reduce the vulnerability of populations in both this country and the world, particularly the poorest and most marginalized.

The companion report *Adapting to the Impacts of Climate Change* provides a summary of potential adaptation strategies for human health (NRC, 2010a). For example, public health systems need to be strengthened to enable rapid monitoring, identification of, and response to new climate change-related health risks as they arise. Other societal stresses such as poverty or economic disadvantages, chronic work-related risks or exposure to otherwise unhealthy environmental conditions, lack of access to preventive and ongoing health care, insufficient emergency preparedness, and related

TABLE 11.1 Population Groups with Specific Vulnerabilities to Climate-Sensitive Health Outcomes

Groups with Increased Vulnerability	Climate-Related Exposures
Infants and children	Heat stress, ozone air pollution, water- and foodborne illnesses, psychological consequences of extreme events
Pregnant women	Heat stress, extreme weather events, water- and foodborne illnesses
Older adults	Heat stress, air pollution, extreme weather events, water- and foodborne illnesses
Impoverished populations	Heat stress, extreme weather events, air pollution, vector-borne illnesses
People with chronic conditions and mobility and cognitive restraints	Heat stress, extreme weather events, air pollution
Outdoor workers	Heat stress, ozone air pollution, vector-borne illnesses
Recent migrants and immigrants	Heat stress, vector-borne illnesses, extreme weather events

SOURCE: Modified from NRC (2010a) with information from CCSP (2008a).

institutional gaps or lack of effective collaboration (as was apparent in the response to Hurricane Katrina) will make effective preparation for and adaptation to climate change impacts on health more difficult. However, adaptive capacity and preparedness can be enhanced by addressing those underlying chronic problems where they persist. In addition, explicit consideration of climate change is needed within federal, state, and local programs (including nongovernmental services) and research activities to ensure that they have maximum effectiveness.

RESEARCH NEEDS

Systematically investigate current and projected health risks associated with climate change. Research is needed to develop a more complete understanding of the health effects of weather and climate events (temperature, heat waves, and severe weather) within the context of other drivers of climate-sensitive health outcomes (age, wealth, fitness, and location). This area of research has seen significant progress during the past decade but needs to be expanded more systematically in the United States and around the world. Key to this analysis is the development of reliable methods to link and quantify the relationships between climate change, and changes in food systems, water supplies, air pollution, and health outcomes.

Advance research on how air quality, heat waves, and the transmission of vector-borne diseases will change. Although several efforts have been made to project future morbidity and mortality effects of climate change-related ozone concentrations, there are currently few efforts to model the impact of climate change on other air pollutants (CCSP, 2008a). Refining projections of the frequency and occurrence of hot days and the range of disease-spreading species is necessary for effective adaptation planning and decision making. New science is needed to provide information for dealing with the impacts of climate change on public health, both nationally and internationally, keeping in mind the transboundary transport of air pollutants and disease vectors.

Characterize the differential vulnerabilities and adaptive capacity levels of particular populations to climate-related impacts, and the multiple stressors they already face or are likely to encounter in the future. The likelihood that various people and regions will suffer adverse health impacts related to climate change depends on (1) their exposure to climatic and other changes; (2) their sensitivity to these stressors, some of which are population- or person-specific (e.g., age, race) and some of which are modified (often magnified) by concurrent, nonclimatic multiple stresses; and (3) their capacity to cope, respond, and adapt to extreme events and health-related climatic changes. The latter in particular is affected by the status of and access to local public health infrastructure, including early warning systems, and human, social, and financial capital. Specific features of the local geographic and environmental situation also can affect the capacity to adapt.

Identify effective, efficient, and fair adaptation measures. Incorporating proactive adaptation into public health and health care planning would increase resilience to the health impacts of climate change. Improvements in health care interventions, access to health care, medical technologies, disease-vector surveillance systems, comprehensive heat-health warning systems, and raising awareness among health care providers are examples of such proactive measures. There are many other opportunities, however, for developing additional adaptation options. For example, seasonal and finer-scale forecasts can be used to develop early warning systems that could increase resilience to climate variability and extreme events.

Evaluate and develop effective information, education, and outreach strategies. Linking knowledge to action through partnerships with private, public, and nongovernmental organizations, and faith communities, and carefully building effective information, education, and outreach strategies that bring credible health information to potentially affected populations will be a critical element of increasing adaptive capacity and responses in the health sector. In addition, local and state governments can be instrumental in building awareness of climate-related health impacts and adaptation options among health care providers, caregivers, and potentially affected populations. The effectiveness of various outreach efforts in affecting human behavior requires careful research and testing in place-based contexts that take advantage of local knowledge and perspectives and the particulars of social networks.

Cities and the Built Environment

The world is rapidly urbanizing. Cities now house slightly more than half the world's population, and 70 percent of the global population will live in urban areas by 2050 (UN, 2007). An unprecedented reorganization is occurring in where people live and how they are restructuring their physical environment. Such growth has led to the emergence of urban conglomerations, or "megalopolises," in which one built environment stretches to another (urban to suburban to "exurban" infrastructure and design), covering entire ecosystems, landscapes, and watersheds (Figure 12.1). The majority of growth in global population over the next several decades is projected to take place in the cities of the developing world (Cohen, 2006), with much of it focusing on emerging urban conglomerations. Given these factors, cities and the built environment are becoming a major focus area for understanding and responding to climate change.

Questions decision makers are asking, or will be asking, about cities, the built environment, and climate change include the following:

- What is the potential for cities to contribute to limiting the magnitude of climate change in ways that also improve air quality and reduce overall environmental impact?
- Which cities and urban conglomerations are most vulnerable to the negative impacts of climate change, including sea level rise, water supply changes, heat waves, and extreme precipitation events?
- What are the most feasible and efficient adaptation actions that cities can take to reduce the stresses associated with climate change?
- How can cities enhance ecosystem services and human well-being in the face of climate change and other environmental stresses?

This chapter summarizes research on how the concentration of people, industry, and infrastructure in cities and built environments plays a major role in driving climate change. It also outlines current scientific knowledge regarding the impacts of climate change on cities, adaptation options, and the potential of cities to limit the magnitude of future climate change. Finally, it details some of the research needed to address the impacts, adaptation, and special vulnerabilities of urban environments with respect to climate change.

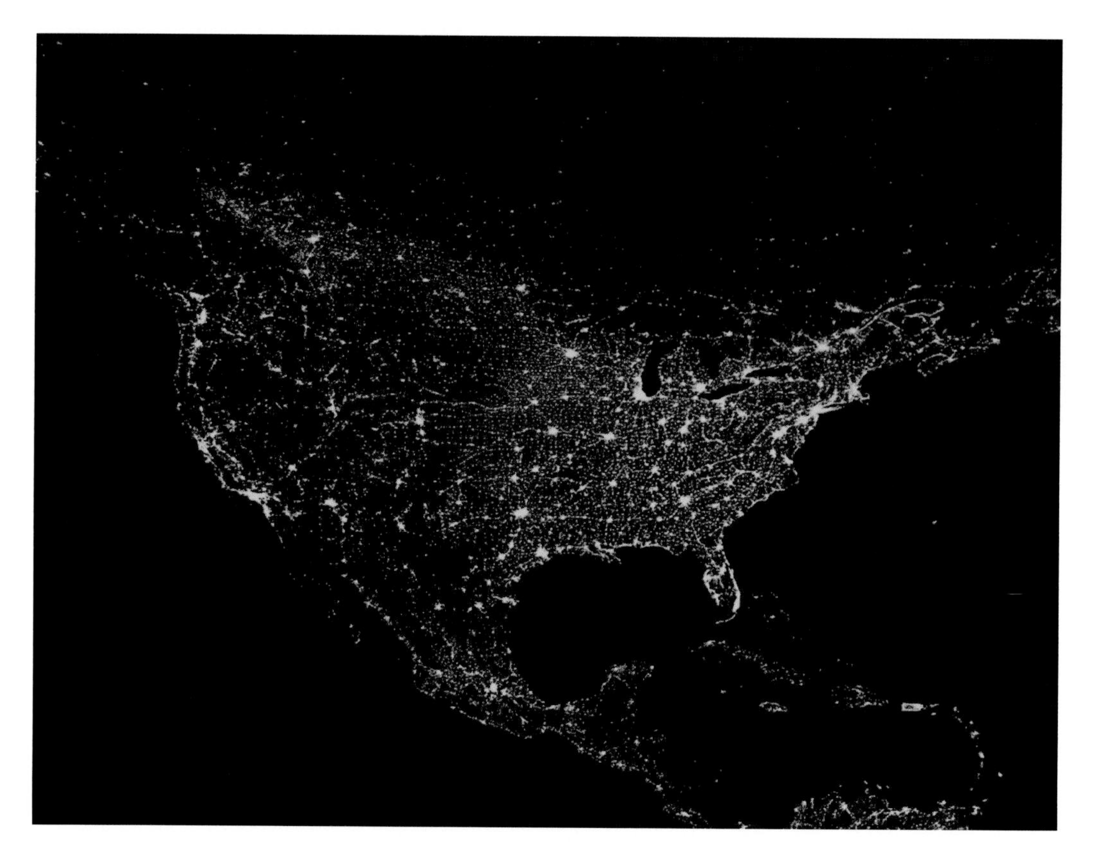

FIGURE 12.1 Lights of North America at night. Note the continuous lighting of extended concentrations of large cities (urban conglomerations), such as Washington to Boston, San Diego to Santa Barbara, and southwestern Lake Michigan. SOURCE: NASA (2001).

ROLE OF CITIES IN DRIVING CLIMATE CHANGE

Urbanized areas play an increasingly important role in driving climate change. For example, energy production and use generate about 87 percent of U.S. greenhouse gas (GHG) emissions; of this amount, the majority is associated with electricity, heat, industrial production, transportation, and waste located in cities and other built-up areas (Folke et al., 1997). The concentration of emissions from urban areas also commonly generates major problems for urban air quality (e.g., Mage et al., 1996). The economies of scale associated with concentrating people in cities generally result in lower per capita emissions relative to nonurban settlements (Dodman, 2009; Satterthwaite, 2008). However, especially in developing economies, the shift to an urban economy and lifestyle increases expectations of consumption and triggers rapid urban expansion (Angel et al., 2005; Guneralp and Seto, 2008), thus enlarging the urban ecological footprint (Rees and Wackernagel, 2008). This footprint involves land use changes in, and resource extraction from, not only the immediate city hinterland but also in distant areas as a result of globalization (DeFries et al., 2010). Thus, energy consumption, indirect land use change (e.g., deforestation), and ecosystem impacts (e.g., ground-level air pollution) beyond the city's boundaries play important roles in climate change (e.g., Auffhammer et al., 2006).

Urbanized or built-up areas directly change reflectivity (Sailor and Fan, 2002), especially through the concentration of roads and other dark surfaces, and so can affect global radiative forcing even though they cover only 1 to 2 percent of the land surface of the Earth (Akbari et al., 2009). The urban heat island effect is relatively well understood (see Figure 12.2) and also has consequences for regional and global climate (e.g., Jin et al., 2005; Lin et al., 2008); for example it may have amplified the effects of the 2003 heat wave in western Europe (Stott et al., 2004). Sustained research demonstrates that urbanization also affects precipitation, including its variability and intensity over and on the leeward side of cities (e.g., Changnon, 1969; Jauregui and Romales, 1996; Shem and Shepherd, 2009). In addition, large built-up areas affect the global carbon balance via their configuration, which affects vegetation and soils (Pickett et al., 2008), and their almost inevitable spread over prime croplands (Angel et al., 2005; Seto and Shepherd, 2009).

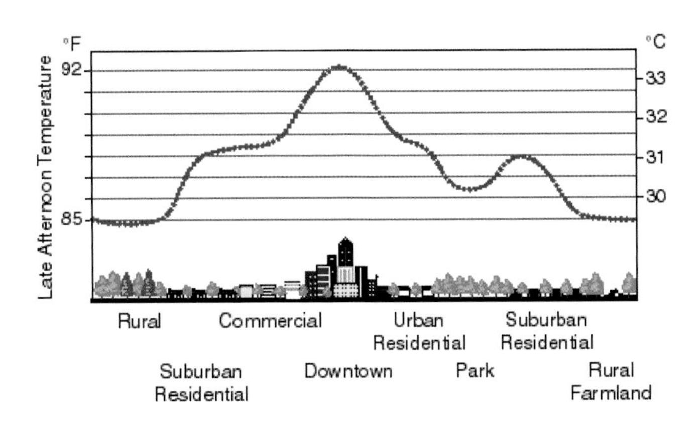

FIGURE 12.2 Schematic representation of an urban heat island, showing how urbanized areas can be several degrees warmer than the surrounding rural areas. The effect can be especially strong on warm summer days. SOURCE: Heat Island Group, Lawrence Berkeley National Laboratory (*http://heatisland.lbl.gov/HighTemps/*).

IMPACTS OF CLIMATE CHANGE ON CITIES

Given their concentration of people, industry, and infrastructure, cities and built environments are generally expected to face significant impacts from climate change. Some of the most important impacts will be associated with changes in the frequency and intensity of extreme weather. Hurricane Katrina in 2005 illustrated the potential for extreme events to cause catastrophic damage to human well-being as well as urban infrastructure; likewise, temperature extremes in cities increasingly cause severe human and environmental impacts, even in the developed world (see Box 12.1). The impacts of warming are amplified in large urban conglomerations because of the heat island effect and the interaction of other environmental stressors (Grimmond, 2007; Hayhoe et al., 2004; Rosenzweig et al., 2005; Solecki et al., 2005). For example, the urban heat island of Phoenix raises the minimum nighttime temperature in parts of the city by as much as 12.6°F (7°C), generating serious water, energy, and health consequences (Brazel et al., 2000). The growth of the southwestern U.S. "sunbelt" as well as that of megacities throughout other arid regions of the world increases the populations at risk from extreme heat as well as their demand for energy and water (Rosenzweig et al., 2005).

In addition, CO_2, nitrogen oxides, volatile organic compounds, particulate matter, and other pollutants and pollutant precursors react in the urban airshed to produce high levels of surface ozone and other potential health hazards (see Chapter 11). In a warmer future world, stagnant air, coupled with higher temperatures and absolute humidity, will lead to worse air quality even if air pollution emissions remain the same (e.g., Cifuentes et al., 2001a,b In many cases, air pollution plumes extend well beyond the urban area per se, affecting people and agriculture over large areas, such as the Ganges Valley (e.g., Auffhammer et al., 2006). In the developing world, such decreases in outdoor air quality come on top of poor indoor air quality—for example, from wood fuel heating (Zhang and Smith, 2003).

As discussed in Chapter 11, certain groups (such as the elderly) are especially vulnerable to intensive heat waves in cities worldwide, especially in temperate climates. Groups with preexisting medical problems, without air-conditioned living quarters, who are socially isolated, or who live on top floors are particularly vulnerable (Naughton et al., 2002; Patz et al., 2005; Semenza et al., 1996). The elderly, as well as portions of the population with asthma and related problems, are also susceptible to poor air quality (e.g., Hiltermann et al., 1998). The U.S. population over age 65 is expected to reach 50 million (20 percent of the total U.S. population) by 2030, with the overwhelming majority living in cities. Cities throughout the nation and the world are differentially prepared (CCSP, 2008a), as illustrated by the relative success of Marseille in the

BOX 12.1
Urban-Climate Interactions and Extreme Events

In the summer of 2003, a persistent anticyclone anchored above western Europe triggered temperatures in excess of 95°F-99°F (35°C-37°C) for as long as 9 days (see figure below). Temperatures were especially high in cities, where urban heat islands amplified the maximum temperatures (Beniston, 2004) and ground-level ozone concentrations climbed to 130 to 200 $\mu g/m^3$ (equivalent to the Environmental Protection Agency's code orange alert; Pirard et al., 2005). It is estimated that this heat wave and the associated poor air quality caused more than 50,000 excess deaths, mostly among elderly urbanites (Brüker, 2005). In France alone, where the housing infrastructure from Paris to Marseille commonly does not include air conditioning or insulation between roofs and rooms, more than 14,800 excess deaths occurred during that period, and the number of deaths is positively correlated with the number of consecutive hot days (Pirard et al., 2005). The rash of deaths, including over 2,200 excess deaths on a single day in August, overwhelmed emergency rooms and morgues.

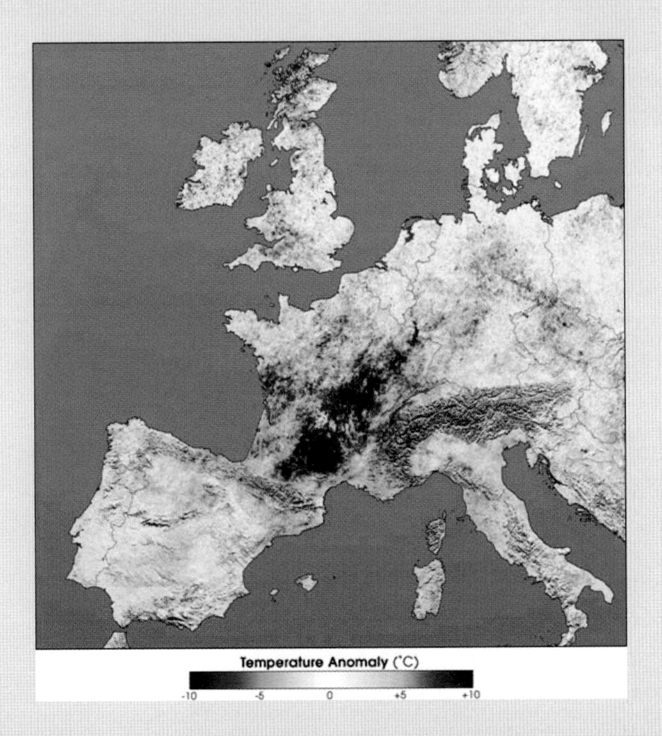

The 2003 summer heat wave in Europe. Colors indicate differences in daytime surface temperature between July 2003 and July 2001. Dark red areas across much of France indicate that temperatures in 2003 were as much as 10°C (18°F) higher than in 2001. SOURCE: Earth Observatory, NASA (*http://earthobservatory.nasa.gov/NaturalHazards/view.php?id=11972*).

2003 heat wave over France (Box 12.1; Pirard et al., 2005) versus the 700 excess deaths in Chicago's 1995 heat wave (Semenza et al., 1996). As noted in Chapter 11 and consistent with the findings of the panel report *Adapting to the Impacts of Climate Change* (NRC, 2010a), research on health infrastructure and preparedness, especially in urban complexes, is needed to inform practice.

Other climate change impacts will also affect cities. Many of the 635 million people occupying coastal lands worldwide live less than 33 feet (10 meters) above sea level and are thus threatened by sea level rise (McGranahan et al., 2007; Wu et al., 2002, 2009; see Chapter 7). Existing tensions over water withdrawal between rapidly growing urban areas and agricultural sectors will be exacerbated by decreasing snowpack in the American West and other regions as a result of climate change and variability (NRC, 2007b). Water vulnerabilities in general are expected to pose major problems for cities in the developing world (Vorosmarty et al., 2000). Expected increases in the frequency of extreme events (Milly et al., 2002), such as intense and prolonged rain storms (see Chapter 8) that stress drainage and flood protection systems, also threaten aging urban infrastructure. Climate change impacts on the megalopolises will also stress regional ecosystem function, water withdrawal, and movement of biota, among other environmental issues (Folke et al., 1997; Grimm et al., 2008; IHDP, 2005).

Cities are centers of economic, cultural, educational, research, social, and political activity, and as such they experience a myriad of nonclimatic changes and stresses that affect their institutional, technological, and economic capacities, the social capital available within and among different population groups, and the relationships between urban centers and their surroundings. Climate change impacts cannot be fully appreciated and addressed without understanding the complex nature of multiple stressors and interacting climatic and nonclimatic factors that affect the vulnerability and adaptive capacity of cities (e.g., Campbell-Lendrum and Corvalán, 2007; Pelling, 2003).

SCIENCE TO SUPPORT LIMITING FUTURE CLIMATE CHANGE

Just as cities loom large in driving and being affected by climate change, they also have important roles to play in limiting the magnitude and ameliorating the impacts of climate change (Grove, 2009). The largest opportunities for reducing GHG emissions from urban centers lie in the transportation, construction, commercial, and industrial sectors, which typically lead in energy consumption and GHG emissions. Reducing industrial and transportation emissions provides a potential for multiple co-benefits to cities in limiting future climate change, reducing the urban heat island effect, and also improving air quality (e.g., NRC, 2009e).

The design and geometry of cities and metropolitan areas afford various means for reducing emissions as well as surface reflectivity. The urban form of most cities has grown in an ad hoc way, through piecemeal planning, development, and control under multiple, independent decision-making units (Batty, 2008). Many, if not all, of these decision-making entities respond foremost to considerations other than climate change, and they rarely consider environmental spillovers beyond their area of control or concern. Yet, the development of cities has profound impacts on infrastructure, travel behavior, and energy consumption (e.g., Ewing and Rong, 2008; Filion, 2008; NRC, 2010g), all of which offer opportunities for interventions that could offset the role of cities in driving climate change. These interventions are only beginning to be explored and appreciated.

One potential response option is altering the reflectivity of surface structures by whitening roofs and road surfaces or employing green rooftop and landscaping options (Akbari et al., 2001; Betsill, 2001). Roofs and paved surfaces typically comprise about 25 and 35 percent, respectively, of dense urban areas (Akbari et al., 2009), so increasing the reflectivity of these surfaces offers the potential to offset some of the urban heat island effect and influence global climate (see Chapter 15). Green rooftops and landscaping options not only reduce urban and regional heat islands but can also improve local and regional air quality (Taha et al., 1997) and provide recreational opportunities and other nonclimate benefits. Alternative city designs or configurations can also lower the heat island effect (Eliasson, 2000; Unger, 2004), although with varying impacts on water and energy consumption that introduce a new suite of trade-offs to consider. "Smart" or "green" redesigns of cities that foster less use of automobiles, among other factors, could reduce GHG emissions from urban areas (Ewing et al., 2007).

SCIENCE TO SUPPORT ADAPTING TO CLIMATE CHANGE

Options to adapt to the impacts of climate change in cities and built-up areas encompass a wide array of potential actions. To date, most of the options considered have fallen into the category of structural or engineering strategies such as protecting existing development and infrastructure from sea level rise (e.g., NYCDEP, 2008); improving water supply, drainage, and water treatment infrastructure; and reducing urban heat island effects. In some cases, local and regional entities sharing a common problem thought to be amplified by climate change, such as water in the American

West, have begun planning to address adaptation beyond infrastructure per se, including more efficient water markets. Although noninfrastructural strategies, such as improving emergency preparedness and response (above), have also been considered, in general there is insufficient concern with, or scientific understanding of, the underlying social-ecological vulnerabilities that cities and the people within them face (see Chapter 4). Many more ways to reduce vulnerability and enhance adaptive capacity may become available when the vulnerabilities of cities are better understood, particularly the vulnerability of subpopulations (e.g., the urban poor, minority groups, children, the elderly, or manual laborers; Campbell-Lendrum and Corvalán, 2007) and the differences between large and smaller urban areas in different regions (e.g., Bartlett, 2008; Hardoy and Pandiella, 2009; Hess et al., 2008; Porfiriev, 2009; Thomalla et al., 2006). Urban areas adjacent to ecological reserves or bordering on forested areas or wildlands may also have to take preventive and preparatory measures to reduce wildfire risks and find ways to protect urban ecology (Collins, 2005).

In general, urban areas face all the climate-related problems faced in other sectors described in this report, but focused on a particular spatial scale. While lessons and techniques on adaptation to climate change from one urban area may be transferrable to others, many will be location specific, and clusters of municipalities in close proximity will have to devise integrated responses across extended metropolitan areas. These considerations raise both institutional and economic opportunities and challenges for adaptation (see the companion report *Adapting to the Impacts of Climate Change* [NRC, 2010a]). They also open up the opportunity to develop sustainable solutions to climate change that integrate actions to limit the magnitude of climate change with those taken to adapt to its impacts—a challenge that some cities around the world are already exploring (e.g., Heinz Center, 2008b). Important scientific questions remain, however, about how to analyze these dual strategies in an integrated fashion (e.g., Hamin and Gurran, 2009; Wilbanks, 2005).

RESEARCH NEEDS

Because the majority of the U.S. and world population already lives in urban areas, and existing or new urban centers will continue to grow in size and economic importance, research on reducing the climate change and accompanying environmental impacts of urban areas is critical. This includes assessing the differential vulnerability of urban areas and populations to climate change impacts as well as the full range of options for limiting and adapting to climate change. Opportunities for integrated, multidisciplinary, and use-inspired research abound, but better connections are needed particularly to the applied science, engineering, and planning professions.

Characterizing and quantifying the contributions of urban areas to both local and global changes in climate. The role of large built environments and how they vary in terms of GHG emissions (including per capita emissions), aerosols, ground-level air pollution, and surface reflectivity need to be examined in a systematic and comparative way. Such research should include the extended effect of urban areas on surrounding areas (such as deposition of urban emissions on ocean and rural land surfaces) as well as interactions between urban and regional heat islands and urban vegetation-evapotranspiration feedbacks on climate. Examination of both local and supralocal institutions, markets, and policies will be required to understand the various ways urban centers drive climate change and identify leverage points for intervention.

Understanding the impacts of climate change on cities. Improving assessments of the impacts of extreme events (e.g., heat waves, drought, floods, and storms) and sea level rise on cities will require improved regional climate models, improved monitoring systems, and better understanding of how extreme events will change as climate change progresses. Evaluations of climate change impacts on urban heat islands and local-regional precipitation should extend to the analysis of their combined impacts on urban and periurban ecosystem and landscape function, ecosystem services, and demands on water and energy consumption.

Assessing the vulnerability of cities to climate change. Improved understanding is needed of who and what are threatened by climate change in the urban context, in both developed and developing countries. This includes human cohorts, neighborhoods, infrastructure, and coupled human-environment systems, as well as implications for food and water security. Most of the world's largest cities are in developing nations and have difficulty achieving global standards for clean air and other healthy environmental qualities. At the same time, very few U.S. cities have received concerted attention from climate researchers. As a result, the relative vulnerability of different urban forms (e.g., design, geometry, and infrastructure) and urban configurations relative to other settlement forms is largely unknown and deserves further study. In addition, given the large population adjacent to coastlines, attention to the vulnerability of coastal cities to sea level rise deserves special attention.

Developing and testing methods and approaches for limiting and adapting to climate change in the urban context. Limiting the magnitude of climate change and adapting to its impacts in the urban context raises a wide range of issues, including the relationships among urban land use, heat islands, water and energy use, and air quality. Additional research is needed, for example, on the efficacy and sociological considerations involved in adoption and implementation of white and green roofs, landscape architecture, smart growth, and changing rural-urban socioeconomic and political linkages. Additional questions include the following: What legacy or lock-in effects, including infrastructure and governance, serve as impediments to responses to climate change? What co-benefits can be gained in the reconfiguration of cities? Which adaptation strategies synergistically benefit the goal of limiting climate change, which potentially counteract it, and how can the trade-offs be adjudicated effectively?

Linking air quality and climate change. Research is needed to provide information for decision making about air quality in the face of climate change. This includes measurements, understanding, modeling, and analyses of socioeconomic benefits and trade-offs associated with different GHG emissions-reduction strategies, including those that simultaneously benefit both climate and air quality (see also Chapter 11) and those that could exacerbate one issue while monitoring the other.

Developing effective decision-support tools. What do we know about effective decision making under uncertainty, especially when multiple governance units may be involved? Much research is needed in comparing the results of city action plans for climate change and identifying similarities and differences between and among small and large cities. Questions that need answers include which qualities of these different plans break or create path dependencies (lock-in, e.g., through infrastructure design, tax policies, or other institutions), and which lead to more flexible, adaptive responses to the risks of climate change.

CHAPTER THIRTEEN

Transportation

The transportation sector encompasses all movement of people and goods. Almost 28 percent of U.S. greenhouse gas (GHG) emissions can be attributed to this sector, and the overwhelming share of these emissions are from CO_2 emitted as the result of burning transportation fuels derived from petroleum (EPA, 2009c). Between 1970 and 2007, U.S. transportation energy use and hence GHG emissions nearly doubled.[1] Consequently, transportation is a major driver of climate change, and a sector with a potentially large role in limiting the magnitude of climate change.

Reducing transportation-related GHG emissions, and understanding the impacts of climate change on transportation systems, are concerns of many decision makers. Questions they are asking, or will be asking, about transportation and climate change include the following:

- How much do various modes of transportation contribute to climate change?
- What technologies and strategies can be used to reduce GHG emissions by the largest transportation contributors?
- How will transportation systems in my area be affected by climate change?
- What steps can be taken to make transportation systems less vulnerable to the impacts of climate change, and how can I apply them in current systems and incorporate them in the design and development of new infrastructure and policy?

This chapter summarizes how reducing the total amount of transportation activity, shifting some of the activity to less energy- and emissions-intensive modes, increasing energy efficiency, and reducing the GHG intensity of transportation fuels could help in lowering GHG emissions from this sector. Additionally, the chapter outlines how climate change will affect the transportation sector and describes the scientific and engineering knowledge regarding adaptation options. The last section of the chapter indicates research that is needed to better understand the impacts of climate change on transportation and ways to reduce GHG emissions in the transportation sector.

[1] The almost exclusive reliance on a single fuel source, petroleum, in the transportation sector means that relative energy expenditures can be interpreted as relative GHG emissions.

ROLE OF TRANSPORTATION IN DRIVING CLIMATE CHANGE

A large proportion of GHG emissions can be attributed to transportation, specifically from the burning of gasoline, diesel, and other fuels derived from petroleum. In fact, the transportation sector is responsible for 70 percent of U.S. petroleum use, which exceeds the percentage of oil that is imported (Davis et al., 2008). Reducing transportation's dependence on petroleum, much of it imported from politically unstable regions of the world, is one of the most direct connections between the issues of climate change, energy security, and national security (see Chapter 16). Transportation's use of petroleum fuels also leads to emissions of particulate matter, sulfur dioxide (which forms sulfate aerosols and ultimately leads to acid rain), and substances that are precursors to photochemical smog (nitrogen oxides [NO_x] and carbon monoxide [CO]) and to various forms of pollution in freshwater and marine systems. Hence, efforts to reduce GHG emissions in the transportation sector will also confer other benefits to the environment and public health (see Chapter 11).

Transportation activity is typically divided into two categories: the movement of people and the movement of goods. The movement of people, usually expressed in passenger-miles, accounts for 70 percent of the transportation sector's energy use and GHG emissions (Davis et al., 2008). The principal vehicles involved in the movement of people are light-duty personal vehicles—automobiles and light trucks—and commercial aircraft, which together account of almost 99 percent of passenger-miles (Davis et al., 2008). The movement of goods, usually expressed in ton-miles, is dominated by trucks, railroads, and ships. These freight modes account for the remaining 30 percent of transportation-related emissions (Davis et al., 2008). Table 13.1 shows the relative importance of different modes of personal and goods transport to total transport energy use and, by implication, its approximate contribution to GHG emissions.

In the United States between 1970 and 2007, energy intensity—the amount of energy required to produce a unit of transport activity—declined for nearly all transportation modes (for example, energy intensity declined by 0.3 percent per year on average for medium and heavy freight trucks, 0.8 percent per year for passenger cars, 1.5 percent per year for light trucks, 1.8 percent per year for freight rail, and 3.3 percent per year for domestic passenger air travel). However, these increases in efficiency were more than offset by an increase in total transportation activity (for example, the number of passenger-miles flown grew by 4.9 percent per year), leading to the overall growth in energy use and GHG emissions.

TABLE 13.1 Energy Use and Activity Characteristics of Various Transportation Modes in 2006

	Energy Use			Passenger Transport Activity			Goods Transport Activity		
Mode	Trillion BTU	%	Cum %	Passenger -miles (millions)	%	Cum %	Ton-miles (millions)	%	Cum %
Light-duty personal vehicles	16,824	65%	65%	4,546,618	87.5%	87.5%			
Medium/heavy trucks	5,188	20%	85%				1,294,492	34.8%	34.8%
Domestic air transport	1,834	7%	92%	590,633	11.4%	98.9%	15,860	0.4%	35.2%
Pipeline	842	3%	95%						
Freight rail	585	2%	98%				1,852,833	49.7%	84.9%
Domestic waterborne freight	304	1%	99%				561,629	15.1%	100%
Transit (all modes)	164	1%	99%	52,154	1.0%	99.9%			
School bus	73	0%	100%						
Intercity bus	30	0%	100%						
Motorcycles	28	0%	100%						
Intercity passenger rail	14	0%	100%	5,381	0.1%	100%			
Total of above	25,886			5,194,786			3,724,814		

SOURCE: Based on data from Davis et al. (2008) and DOT (2008).

REDUCING TRANSPORTATION-RELATED GREENHOUSE GAS EMISSIONS

There are four possible strategies that could be employed to reduce GHG emissions from the transportation sector:

- Reduce the total volume of transportation activity;
- Shift transportation activity to modes that emit fewer GHGs per passenger-mile or ton-mile;
- Reduce the amount of energy required to produce a unit of transport activity (that is, increase the energy efficiency of each mode); or
- Reduce the GHG emissions associated with the use of each unit of energy.

Each of these strategies is briefly discussed below. Additional details can be found in the companion report *Limiting the Magnitude of Future Climate Change* (NRC, 2010c), and the Transportation Research Board report *Potential Energy Savings and Greenhouse Gas Reductions from Transportation* (NRC, 2010f). The *Limiting* report concludes that "near-term opportunities exist to reduce GHGs from the transportation sector through increasing vehicle efficiency, supporting shifts to energy efficient modes of passenger and freight transport, and advancing low-GHG fuels." Achieving large (that is, on the order of 50 to 80 percent) long-term reductions in GHG emissions in the transportation sector, however, would require major technological and behavioral changes (e.g., Fawcett et al., 2009); this in turn implies a need for additional research to support the development and deployment of new and improved transportation modalities.

Reducing the Volume of Transport Activity

The most basic—but perhaps most difficult—way to reduce transportation-related GHG emissions is to reduce the total amount of transportation activity. While there has been some attention devoted to reducing total freight transport volumes—by, for example, promoting consumption of locally produced food and goods—most of the attention in this area has focused on reducing personal transportation activity, especially activity by light-duty vehicles. Since 1980, the number of light-duty vehicle passenger-miles has grown at an average rate of 2.3 percent per year (FHA, 2008). This growth has been spurred by, among other factors, the suburbanization of America. As recently as the 1960s, the majority of daily commutes were from downtown to downtown or from close-in suburbs to downtown. Now, the majority of commutes are from suburb to suburb, with the attendant traffic and pollution issues (NRC, 2006a; see also Chapter 12). Suburbanization has also stimulated the increased use of light-duty vehicles for trips other than commuting—for example, according to the National Household Travel Survey, in 2001 commuting accounted for 27 percent of all vehicle trips per household while "household-serving" travel (e.g., shopping errands, chauffeuring family members) accounted for most of the remainder (BTS, 2001).

Both logic and empirical evidence suggest that developing at higher population and employment densities results in trip origins and destinations that are closer to one another, on average, leading to shorter trips on average and less vehicle travel. Shorter trips can also reduce vehicle travel by making walking and bicycling more viable as alternatives to driving, while higher densities make it easier to support public transit. A recent National Research Council report, *Driving and the Built Environment* (NRC, 2009e), examined the relationships between land use patterns and vehicle miles traveled and concluded "[l]ooking forward to 2030 and, with less certainty, to 2050, it appears that housing preferences and travel patterns may change in ways that support higher-density development and reduced [vehicle miles traveled], although it is unclear by how much." While the study concluded that significant increases in more compact, mixed-use development result in only modest short-term reductions in energy consumption and CO_2 emissions, these reductions will grow over time. The implications of this and other findings for limiting GHG emissions from the transportation sector can be found in the companion report *Limiting the Magnitude of Future Climate Change* (NRC, 2010c).

Another trend that has led to increased travel activity has been the reduction over time in the average number of people traveling in each automobile and light truck. In 1977, the average vehicle carried 1.9 people; by 2001, this had declined by 14 percent, to 1.6 people. For travel to and from work, the average declined from 1.3 to 1.1 (Hu and Reuscher, 2005). Increasing the average vehicle occupancy could lead to reductions in total vehicle miles traveled and thus GHG emissions, even considering small offsets due to the need to pick up and drop off the additional passengers. Many municipalities have instituted policies to encourage carpooling; however, few of these policies were developed based on research on patterns and determinants of human behavior or effective mechanisms for informing such behavior, and there is a need for more evaluation of effectiveness.

Because commuting only accounts for about a quarter of passenger trips, carpooling strategies have limited potential for reducing transportation-related GHG emissions. However, it may be possible to increase the prevalence of ridesharing through more effective conveyance of information and the provision of incentives, both in monetary and convenience terms. New technologies could help in this regard; for instance, it is already possible to use personal telecommunications devices and computers to connect drivers with prospective riders to create casual forms of carpooling. Such opportunities will increase. Indeed, it is conceivable that in some locations public transit services will evolve away from the large fixed-route systems into smaller van-type vehicles that employ dynamic routing technologies to offer transportation services similar to that of private cars but with higher average occupancy (WBCSD, 2004). While such concepts are in limited use in Europe, they have not been explored in the United States.

Shifting Transportation Modes

Because there are significant differences in the energy expended per passenger-mile or ton-mile among the major modes of transportation, a second candidate strategy for reducing transportation-related GHG emissions is to shift people or freight to more energy efficient modes. The two most widely discussed options are (1) inducing people to substitute some of their driving with public transportation service, bicycling, and walking; and (2) shifting more freight from truck to rail.

The viability of public transportation (as well as walking and biking) as an alternative to driving hinges in part on there being favorable urban land use patterns, as discussed in the preceding subsection and in the recent report *Driving and the Built Environment* (NRC, 2009e). For public transportation to be an energy efficient alternative to the private vehicle, however, requires that the services be heavily used. At present, except in a few very dense urban areas such as New York City, public transportation load factors are not high enough to make these services more energy- and GHG-efficient than driving. Because demand is especially low outside of rush hours, transit systems often operate with very low levels of occupancy for much of the day (NRC, 2009c). As a consequence, buses—the most prevalent form of transit—used 24 percent more energy per passenger-mile than private cars in 2006 (Davis et al., 2008). Subways and commuter rail systems, in contrast, used about 20 percent less energy per passenger-mile than private cars, but these systems accounted for a minority of total public transportation ridership.

There is also significant geographic variability in the availability of public transportation: 97 percent of all subway and transit rail trips occurred in metropolitan areas with a population of over 5 million, and the New York metropolitan area alone was responsible for 38 percent of all national transit use for travel to and from work (NRC, 2006a). Bicycling and walking do not emit any GHGs and are associated with health co-benefits, but they currently constitute a very small share of all miles traveled by people when compared with motorized modes. Strategies designed to facilitate and promote these modalities could yield multiple benefits.

There has also been interest in using passenger rail for medium-distance (500 miles or less) intercity travel in the United States, which is currently dominated by automobiles and, to a lesser extent, air travel. In Europe and Japan, high-speed rail is succeeding in winning substantial market share away from automobiles and air transport for city-to-city travel at distances of up to 500 miles (FRA, 2009). There are many challenges, however, to duplicating such a system in the United States. While high gasoline and deisel fuel taxes and road tolls tend to discourage intercity travel by private car in Europe and Japan, the ease and low out-of-pocket cost for automobile travel in the United States favors their use. Automobiles also offer flexibility for local travel once at the final destination, which is particularly important for families and leisure travelers who make trips between suburbs rather than center cities. A large share of business travel also takes place in suburban areas, which are poor locations for high-speed rail terminals. Another challenge is that there are relatively few large U.S. metropolitan areas located within 500 miles of one another, especially when compared with Europe and Japan. Because of the long distances between cities, aviation is the only practical alternative for timely intercity travel in the United States. Moreover, U.S. airlines, operating in vast networks that funnel passengers through hubs, have the passenger volumes required to offer large numbers of flights between city pairs. This ability to offer a dense schedule of flights—which is highly valued by time-sensitive business travelers—cannot be matched by high-speed rail. The recent uptick in intercity bus travel in the United States, which has been attributed both to the recent economic downturn and to higher fuel prices, is another longer-distance travel option that could potentially be promoted to reduce overall energy use and GHG emissions, particularly among leisure travelers.

The practicality and benefits of shifting additional freight traffic from truck to rail has been studied and debated for years. In 1939, 64 percent of freight ton-miles moved by rail, while trucks carried only 9 percent, with most of the remainder moved on waterways (Department of Commerce, 1975). In 2006, rail's share had declined to 40 percent, dominated by heavy, bulk commodities such as coal, while trucking had increased its share to 28 percent (Margreta et al., 2009). Although moving freight by rail is generally more energy efficient than moving freight by truck, it is not clear that a significantly larger share of freight could be practically moved by rail. For example, because many rail sidings have been abandoned, most freight traffic, and especially manufactured goods, are moved by truck for at least a portion of the journey. On the other hand, the containerization of freight—especially for imported goods—increases the potential for movement by rail, and the recent sharp increases in the price of fuel seem to have shifted some containers from truck to rail and some truck trailers to rail (in "piggyback" service) for the line-haul segment of the trip. Observers who study freight movements contend that rail container and trailer movements such as these are generally not economically viable until line-haul distances reach 700 miles and, with the exception of the longest moves (over 1,500 miles), between the most heavily traveled markets having lane traffic densities in excess of 400,000 tons annually (Wittwer, 2006).

Reducing Energy Intensity

Increasing the efficiency of transportation—especially light-duty vehicles—has been a major strategy for reducing U.S. petroleum consumption. The companion report *Limiting the Magnitude of Future Climate Change* (NRC, 2010c) includes a summary of changes in fuel economy standards over the past 30 years, the effectiveness of these standards, and their implications for climate policy. For example, the fuel economy potential of new passenger cars and light trucks (measured in terms of ton-miles per gallon) has improved at a rate of about between 1 and 2 percent per year since 1975 (EPA, 2009c), mainly through a series of technological advances in engines and aerodynamics. However, this potential has not been reflected in actual new vehicle fuel economy; since the mid-1980s, the fuel economy of new automobiles and light trucks as tested by the Environmental Protection Agency (EPA) has essentially been stable. Instead, vehicles have become heavier (by about 900 pounds on average [Davis et al., 2008]) and have improved their acceleration performance (average 0 to 60 mph times have declined from just over 14 seconds to about 9.5 seconds [Davis et al., 2008]). The EPA estimates that if the potential improvements in fuel economy had been realized, model year 2008 cars would have averaged 33 to 34 mpg instead of the 30 mpg they did average, and new light trucks would have averaged 27 to 28 mpg instead of 22 mpg.

Congress has called for a fleetwide combined fuel economy for cars and light trucks that reaches 35 mpg by model year 2020, representing a 30 percent increase over current levels (Energy Independence and Security Act of 2007, P.L. 110-140). In addition, new EPA GHG-performance standards for cars and light trucks will acclerate these fuel economy improvements by 3 or 4 years (EPA, 2009c). Tapping the reservoir of unrealized fuel economy potential with continued modest improvements in the efficiency of conventional gasoline and diesel engines would be the easiest way for motor vehicle manufacturers to meet these new efficiency standards. Doing so, however, would require consumers to sacrifice certain desired performance attributes such as acceleration capabilities. In order to meet the new standards under these constraints, manufacturers will need to increase the use of hybrid-electric propulsion systems, make cars and trucks lighter (typically through the use of materials such as fiberglass and carbon fibers), and develop next-generation propulsion systems—batteries and fuel cells being the two main candidates (see next subsection). It will be important with respect to some of these vehicle technologies to consider the life-cycle energy costs associated with producing more efficient vehicles; for example, some of the materials used for lightweight and hybrid vehicles are associated with significant energy production costs, which may offset some fuel savings.

To advance the technologies required to enable the production of more fuel-efficient light vehicles, the federal government has over the years funded cooperative research and development programs such as the Program for a New Generation of Vehicles. In addition to such federal actions, some states, led by California, have set their own fuel economy standards and taken other actions, such as requirements to sell a certain number or fraction of low-emissions vehicles.

In addition to improving the efficiency of the vehicle fleet, there are behavioral changes that may be able to increase the energy efficiency of the operations of existing vehicles in the light-duty fleet, such as maintaining properly inflated tires, reducing time spent idling, and removing excess weight from trunks. Each of these alone is a minor factor for the individual driver, but small changes multiplied across the U.S. passenger vehicle fleet could have an impact (Dietz et al., 2009b). More information is needed on the prevalence and effectiveness of these behaviors as well as on how they might be further encouraged.

It merits noting that Congress has called for fuel efficiency standards for medium- and heavy-duty trucks (P.L. 110-140). EPA may also develop GHG performance standards for trucks and other transportation vehicles (EPA, 2010b). Developing efficiency standards for trucks presents a particular challenge, because these vehicles are used in so many different ways that a single metric for efficiency is impractical (e.g., using miles per gallon as a metric would encourage smaller trucks with less payload and would reduce ton-miles per gallon). A recent NRC report examines the issues surrounding the development of such standards (NRC, 2010i). As this report and others have pointed out, trucking and the other long-distance freight and passenger modes of transportation already have powerful economic incentives to care about energy efficiency, since they are highly competitive and cost-conscious industries in which fuel is a main operating cost.

Reducing the GHG Intensity of Transportation Fuels

A final strategy for reducing transportation GHG emissions is reducing the GHG emissions associated with the use of each unit of transport energy, typically through the development and deployment of vehicles that run on electricity or liquid or gaseous transportation fuels not based on petroleum, such as biofuels or hydrogen. In addition to propulsion and energy storage technologies themselves, this requires the development of ways to manufacture and distribute the new fuel or energy sources. While some of these vehicle and fuel combinations would significantly reduce or completely eliminate tailpipe GHG emissions, the GHG emissions generated as a result of fuel production and distribution could be significant and offset all or some of these benefits. Indeed, in some circumstances, the resulting "well-to-wheels" GHG emissions—emissions resulting from the extraction, production, and distribution of fuel plus the emissions resulting from its use by the vehicle—can exceed the well-to-wheels emissions generated by current transport vehicles using petroleum-based fuels. For example, some biofuels, especially corn-based ethanol but also certain forms of biodiesel, may not yield a net reduction in GHG emissions (Campbell et al., 2009; Searchinger et al., 2008).

In its analysis of the well-to-wheels impacts of alternative liquid transportation fuels, the *America's Energy Future* panel on this topic found that CO_2 emissions from corn grain ethanol are only slightly lower than those from gasoline (NRC, 2009b). In contrast, CO_2 emissions from cellulosic ethanol (biochemical conversion) are much lower (NRC, 2009b). However, cellulosic processes are not yet economical and production of corn-based ethanol may be encouraged for other reasons, such as bolstering domestic agricultural markets and building the market for biofuels (see NRC, 2009b). Similar concerns have been raised about battery- and hydrogen-powered vehicles, especially if the feedstock used to make the hydrogen or electricity that charges the batteries comes from GHG-intensive energy sources. In addition, the production of alternative fuel sources may carry unintended negative consequences for other resources, environmental concerns, trade issues, and human security issues, and the trade-offs and life-cycle costs and benefits of these alternatives have to be evaluated (see Chapter 14).

IMPACTS OF CLIMATE CHANGE ON TRANSPORTATION

In 2008 the Transportation Research Board released a report titled *Potential Impacts of Climate Change on US Transportation* (NRC, 2008g). The report assesses some of the possible impacts of climate change on various transportation systems, with an emphasis on four categories of climate change impacts: increases in very hot days and heat waves, increases in arctic temperatures, rising sea levels, and increases in hurricane intensity. These impacts are summarized in Table 13.2. While not an exhaustive or quantitative list, this analysis provides an overview of the types of impacts that could be experienced in the transportation sector.

TABLE 13.2 Potential Climate Change Impacts on Transportation

Potential Change in Climate	Impact on Operations	Impact on Infrastructure
Increases in very hot days and heat waves	Impact on liftoff load limits at high-altitude or hot-weather airports, resulting in flight cancellations or limits on payload or both Limits on periods of construction activity due to health and safety concerns	Thermal expansion on bridge joints and paved surfaces Concerns regarding pavement integrity, traffic-related rutting, and migration of liquid asphalt Rail-track deformities
Increases in Arctic temperatures	Longer ocean transport season and more ice-free ports in northern regions Possible availability of a northern sea route or a northwest passage	Thawing of permafrost, causing subsistence of roads, railbeds, bridge supports, pipelines, and runway foundations Shorter season for ice roads
Rising sea levels, combined with storm surges	More frequent interruptions to coastal and low-lying roadway travel rail service due to storm surges More severe storm surges, requiring evacuation or changes in development patterns Potential closure or restrictions at airports that lie in coastal zones, affecting service to the highest-density U.S. population centers	Inundation of roads, rail lines, and airport runways in coastal areas More frequent or severe flooding of underground tunnels and low-lying infrastructure Erosion of road base supports Reduced clearance under bridges Change in harbor and port facilities to accommodate higher tides and storm surges
Increases in intense precipitation events	Increase in weather-related delays and traffic disruptions Increased flooding of evacuation routes Increase in airline delays due to convective weather	Increase in flooding of roadways, rail lines, runways, and subterranean tunnels Increase in road washout, damages to railbed support structures, and landslides and mudslides that damage roads and tracks Increases in scouring of pipeline roadbeds and damage to pipelines
More intense or more frequent hurricanes	More frequent interruptions in air service More frequent and potentially more extensive emergency evacuations More debris on roads and rail lines, interrupting travel and shipping	Greater probability of infrastructure failures Increased threat to stability of bridge decks Impacts on harbor infrastructure from wave damage and storm surges

SOURCE: NRC (2008g).

SCIENCE TO SUPPORT ADAPTING TO CLIMATE CHANGE IN THE TRANSPORTATION SECTOR

The report *Potential Impacts of Climate Change on US Transportation* (NRC, 2008g) identifies a number of potential engineering options for strengthening and protecting transportation facilities such as bridges, ports, roads, and railroads from coastal storms and flooding as a short-term adaptation measure. The report also identifies a number of research needs and potential actions that will be necessary to support climate-related decision making in the transportation sector, including improved communication processes among transportation professionals, climate scientists, and other relevant scientific disciplines; a clearinghouse for transportation-relevant information on climate change; developing climate data and decision-support tools that incorporate the needs of transportation decision makers; developing and implementing monitoring technologies for major transportation facilities; developing mechanisms for sharing best practices; reevaluation of existing and development of new design standards; and creating a federal-level interagency working group focused on adaptation. Many of these initiatives would require federal action, while others would require action by professional organizations and university researchers.

Potential options and considerations for adaptation to climate change in the transportation sector are discussed in the companion report *Adapting to the Impacts of Climate Change* (NRC, 2010a). The report also notes that planning for adaptation in the transportation sector will require new modeling tools, the establishment of standards consistent with future climate risks (as opposed to those based on historical conditions), and improved communication between the climate science and transportation decision making communities.

RESEARCH NEEDS

Improve understanding of how transportation contributes to climate change. As society moves from vehicles propelled by internal combustion engines using petroleum-based fuels to vehicles using more varied types of propulsion systems and fuels, it will be increasingly important to understand the full life cycle of GHG emissions generated by various vehicle and fuel combinations, including the emissions and energy implications associated with vehicle production. The move from tank-to-wheels to well-to-wheels emissions analyses represents an important step in this understanding. For example, our understanding of the true life-cycle emissions from various biofuels is still incomplete, as is understanding of trade-offs and consequences for other resources and environmental issues. Also, the construction and maintenance of transportation infrastructure is an additional source of GHG emissions, but little is known of the relative emissions associated with different transportation modes or infrastructure types even as large investments are being planned for constructing new systems such as high-speed rail.

Improve understanding of what controls the volume of transportation activity. While there is potential for tempering growth in vehicle miles traveled by increasing land development densities, a recent NRC report (NRC, 2009e) found a lack of sound research on the potential for increasing metropolitan densities to affect travel, energy use, and emissions. Further research is needed on the relationships among household location, workplace location, trip-making activity, and light-duty vehicle travel, and on the effectiveness of various policy mechanisms to influence these relationships. Technological improvements such as online shopping, telecommuting, and virtual conferencing also have the potential to significantly reduce total transportation activity, but further research is needed on how to facilitate and promote expanded use of these technologies (and this research will require data on current levels of usage of these technologies—an example of a climate-relevant observation that falls outside the rubric of traditional climate observations).

Conduct research on the most promising strategies for encouraging the use of less fuel-intensive modes of transportation. Any increase in fuel prices, whether a result of climate or energy policy or other factors, can be expected to promote a shift toward more fuel-efficient modes of transportation, both at the personal level and through major private-sector transportation providers. However, as noted earlier in this chapter, there are a variety of strategies that might be employed to encourage less energy-intensive modes. As with overall reductions in travel volume, additional research is needed on the factors that influence travel mode choice—understanding how, for example, intermodal service can be made more attractive to shippers or public transit more attractive to passengers. Research is also needed on potential large-scale changes in the built environment and infrastructure that would encourage less energy-intensive modes, and the policy mechanisms that might be used to facilitate these changes.

Continue efforts to improve energy efficiency. In addition to the continued improvement of more efficient vehicle designs and propulsion systems, there could potentially be major energy efficiency gains in other transportation modes. For example, there is room for improvement in medium- and heavy-duty truck aerodynamics and means of reducing idling (NRC, 2010i). Ultralight materials such as carbon fiber are already beginning to see widespread application in new commercial aircraft (e.g., the Boeing 787), and additional research by both public and private sectors may help accelerate this and other efficiency improvements, such as "blended wings" and open fan propulsion systems.

In addition to technology development and deployment, there is a wide range of research needed on human behavior as it relates to transportation use and on the best policies for influencing both technology development and human behavior. For example, there are behavioral changes that increase the efficiency of existing vehicles, such as maintaining properly inflated tires, but we lack basic data on the prevalence of these behaviors as well as on how they might be effectively encouraged. Further research is also needed on factors that encourage the purchase of more efficient vehicles—fuel prices are certainly one factor, but, as with the adoption of any new technology, prices are only part of the explanation and a more nuanced understanding might lead to the design of effective policies. There may actually be substantial proprietary information on what influences consumer choice and technology adoption, but there is little open literature on this subject or on how policies, programs, and institutions might influence vehicle or mode choice. Finally, the history of U.S. fuel economy over the last 35 years, where efficiency improvements were offset by consumer demands for larger, more powerful vehicles (with little resulting fuel consumption penalty, because efficiency had increased), suggests a need for better understanding of how to design regulatory policies that have the intended results.

Accelerate the development and deployment of alternative propulsion systems, fuels, and supporting infrastructure. New, less carbon-intensive fuels and alternative propulsion systems will ultimately be needed to make major reductions in GHG emissions from the transportation sector. The two primary candidates for replacing internal combustion engines are batteries and hydrogen fuel cells, and major technological advances are still needed to make these methods competitive with current propulsion systems. Moreover, while these alternative propulsion systems would reduce petroleum consumption, they will only reduce GHG emissions significantly if the needed electricity or hydrogen is produced using low-emissions fuels and processes. As discussed in the companion report *Limiting the Magnitude of Climate Change* (NRC, 2010c) and elsewhere, widespread adoption of these technologies also implies a major restructuring of the nation's transportation infrastructure, and reasearch will play an important role in optimizing that design.

Advance understanding of how climate change will affect transportation systems and how to reduce the magnitude of these impacts. One of the most difficult tasks for transportation planners in addressing climate change is obtaining relevant information in the form they need for planning and design (NRC, 2008g). Improved regional-scale climate information is needed, but so is a better understanding of how projected climate changes, such as changes in temperature and precipitation, will affect different kinds of infrastructure in different regions, and improved methods of providing information to transportation decision makers. Practical research on adaptation measures, both for current transportation systems and for the design of new systems and infrastructure, is needed to better inform all kinds of transportation-related decisions as climatic conditions continue to exit the range of past experience.

CHAPTER FOURTEEN

Energy Supply and Use

Energy is essential for a wide range of human activities, both in the United States and around the world, yet its use is the dominant source of emissions of CO_2 and several other important climate forcing agents. In addition to total demand for energy, the type of fuel used and the end-use equipment affect CO_2 emissions. The diversity of ways in which energy is supplied and used provides ample opportunities to reduce energy-related emissions. However, achieving reductions can be very difficult, especially because it involves considerations of human behavior and preferences; economics; multiple time frames for decision making and results; and myriad stakeholders.

Questions decision makers are asking, or will be asking, about energy supply and consumption in the context of climate change include the following:

- What options are currently available for limiting emissions of greenhouse gases (GHGs) and other climate forcing agents in the energy sector, and what are the most promising emerging technologies?
- What are the major obstacles to widespread adoption of new energy technologies that reduce GHG emissions?
- What are the best ways to promote or encourage the use of energy-conserving and low-GHG energy options?
- What impacts will climate change have on energy production, distribution, and consumption systems, and how should possible impacts be accounted for when designing and developing new systems and infrastructure?
- What are the possible unintended consequences of new energy sources for human and environmental well-being?

This chapter focuses on what is already known about energy and climate change and about what more needs to be known. Strategies to limit emissions of CO_2 and other GHGs through changes in agriculture practices, transportation, urban planning, and other approaches are addressed in other chapters, and policy approaches that span these strategies are discussed in Chapter 17. Because *America's Energy Future* was the focus of a recent suite of National Research Council reports (NRC, 2009a,b,c,d), and energy-related GHG emissions reductions are a major point of emphasis in the companion volume *Limiting the Magnitude of Future Climate Change* (NRC, 2010c), this chapter provides only a brief summary of critical knowledge and research needs in the energy sector.

ENERGY CONSUMPTION

Globally, total energy consumption grew from 4,675 to 8,286 million tons of oil equivalent between 1973 and 2007 (IEA, 2009). The United States is still the world's largest consumer of energy, responsible for 20 percent of world primary energy consumption. The next largest user, China, currently accounts for about 15 percent. Energy consumption in the United States has increased by about 1 percent per year since 1970, although there is no longer a direct relationship between energy use and economic growth. Between 1973 and 2008, for example, U.S. energy intensity, measured as the amount of energy used per dollar of gross domestic product (GDP), fell by half, or 2.1 percent per year (EIA, 2009). Despite this trend, the United States still has higher energy use per unit of GDP and per capita than almost all other developed nations. For example, Denmark's per capita energy use is about half that of the United States (NRC, 2009c).

A nation's energy intensity reflects population and demographic and environmental factors as well as the efficiency with which goods and services are provided, and consumer preference for these goods and services. Comparison of the energy intensity of the United States with that of other countries indicates that about half of the difference is due to differences in energy efficiency (NRC, 2009c). The differences also reflect structural factors such as the mix of industries (e.g., heavy industry versus light manufacturing[1]) and patterns of living, working, and traveling, each of which may have developed over decades or even centuries.

Today, about 40 percent of U.S. energy use is in the myriad private, commercial, and institutional activities associated with residential and commercial buildings, while roughly 30 percent is used in industry and the same amount in the transport of goods and passengers (see Chapter 13). Most significantly for GHG emissions, 86 percent of the U.S. energy supply now comes from the combustion of fossil fuels—coal, oil, and

[1] In accounting for the energy or environmental implications of shifts in the mix of products produced and consumed in the economy, it is important to consider trade flows. For example, if a reduction in domestic production of steel is offset by an increase in steel imports, domestic GHG emissions may appear to decline but there may be no net global reduction in GHG emissions (and emissions may even increase, given the possibility of differences in production-related emissions and the energy expended in transporting the imported product). This concept is an important factor in negotiations over international climate policy (see Chapter 17).

natural gas (Figure 14.1). The transportation sector is 94 percent reliant on petroleum, 56 percent of which is imported (EIA, 2009).

There are important economic and national security issues related to the availability of fossil fuel resources, as well as significant environmental issues associated with their use—including, but not limited to, climate change. For example, the recent report

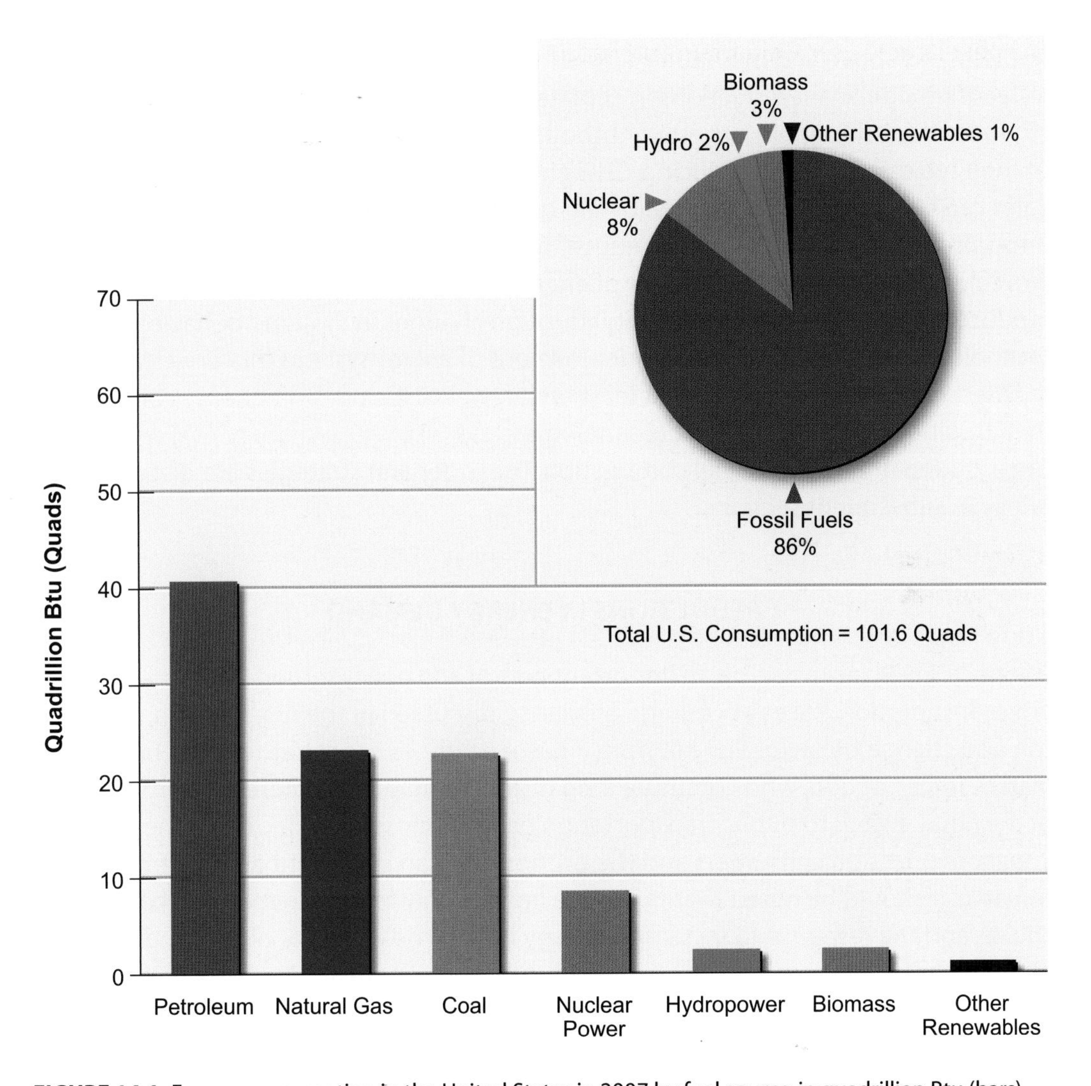

FIGURE 14.1 Energy consumption in the United States in 2007 by fuel source, in quadrillion Btu (bars) and as a percentage of total energy consumption (pie chart). Fossil fuels serve as the primary source of energy. SOURCE: NRC (2009d).

Hidden Costs of Energy: Unpriced Consequences of Energy Production and Use (NRC, 2009f) estimated that the damages associated with energy production and use in the United States totaled at least $120 billion in 2005, mostly through the health impacts of fossil fuel combustion (and not including damages associated with climate change or national security, which are very difficult to quantify in terms of specific monetary damages). While this is undoubtedly a small fraction of the benefits that energy brings, it reinforces the message that there are significant benefits associated with reducing the use of energy from fossil fuels.

As discussed above and in Chapter 6, limiting the magnitude of future climate change will require significant reductions in climate forcing, and GHGs emitted by the energy sector are the single largest contributor. Hence, many strategies to limit climate change typically focus on reducing GHG emissions from the energy sector. These strategies can be grouped into four major categories: (1) *reductions in demand*, typically through changes in behavior that reduce the demand for energy; (2) *efficiency improvements*, or reducing the amount of energy needed per unit of goods and services produced (also called energy intensity) through changes in systems, behaviors, or technologies; (3) development and deployment of *energy systems that emit few GHGs or other climate forcing agents*, or at least emit fewer GHGs per unit energy consumed than traditional fossil fuel-based technologies; and (4) *direct capture* of CO_2 or other GHGs during or after fossil fuel combustion. These general strategies are discussed briefly in subsequent sections.

REDUCTIONS IN ENERGY DEMAND

The price mechanism can be an important part of any policy intended to reduce energy consumption. Prices encourage efficiency, discussed in the next section, but they can also change behavior. For example, if gasoline prices rise, whether from taxes or market forces, people who commute long distances may buy a more efficient vehicle or they may switch to public transportation or move closer to work. Nevertheless, the impact of prices on consumers and the economy are an important area for further research. It should be noted that prices are not the only feature involved in consumer choice, and the response to increased energy prices (the elasticity of demand) is often modest. There are many possible explanations for this: modest changes in price are not noticed, consumers cannot easily change some aspects of their consumption (for example, it is not always feasible to sell a car with low gas mileage to buy one with higher mileage when gas prices rise, at least in the short run), and there are many other factors that influence decisions that affect energy consumption and in some circumstances may have more influence than prices (Carrico et al., 2010; Stern et al., in press; Wilson and Dowlatabadi, 2007).

ENERGY EFFICIENCY IMPROVEMENTS

Although energy intensity has declined in the United States over the past 30 years (EIA, 2009; NRC, 2009d), per capita consumption in the United States still exceeds that of almost all other developed countries. In addition, a considerable fraction of the intensity improvements in the United States may be due to the changing nature of demand (e.g., the shift away from manufacturing toward a service- and information-based economy) as well as increased imports of energy-intensive products and materials, which simply shift emissions to other locations. The recent report *Real Prospects for Energy Efficiency in the United States* (NRC, 2009c), part of the *America's Energy Future* suite of activities, carried out a comprehensive review of methods to improve energy efficiency in industry, buildings, and transportation sectors. The report concludes that energy efficient technologies in those sectors exist today that could be implemented without major changes in lifestyles and could reduce energy use in the United States by 30 percent by 2030. The companion report *Limiting the Magnitude of Future Climate Change* (NRC, 2010c) also discusses energy efficiency at length.

The building sector offers the greatest potential for energy savings through efficiency; options range from simple approaches like insulation and caulking, to the use of more efficient appliances and lighting, to changing patterns of building use. Investments in these areas could reduce energy use in residences by one-third, although systematic estimates that take account of both technological and behavioral changes have not been made. For example, participation in programs that subsidize weatherization with identical financial incentives can differ by an order of magnitude depending on how the programs are presented to the public (Stern et al., 1986). Efficiency improvements can be made through the development and use of more efficient devices, with more efficient systems for managing devices, and with changing patterns of use—all of which require both technological innovation and a better understanding of human behavior and institutions.

While implementation of current technologies holds immediate opportunities for reducing energy use and GHG emissions, new technological and scientific advances are likely to yield longer-term benefits. For example, the development of new materials for insulation, new kinds of lighting, fundamental changes in heating and cooling systems, computational technologies for energy systems management, and landscape architecture and materials for natural cooling could all contribute to major improvements in energy efficiency. As noted in Chapter 13, energy efficiency advances are also possible in the next decade in the transportation sector due to improved vehicle technologies and behavior changes. However, simply developing and making a new technology available is not sufficient to ensure its adoption; to be effective, research on all energy technologies, including efficiency technologies, needs to include analysis of the barriers to adoption of innovation and of public acceptance of new technology.

ENERGY SOURCES THAT REDUCE EMISSIONS OF GREENHOUSE GASES

Technologies that reduce the amount of GHGs emitted during the production of usable energy include renewable energy sources such as solar, wind, bioenergy, geothermal, hydropower, as well as nuclear power and carbon capture and storage (CCS) applied to fossil fuels or biomass. Even switching among fossil fuels can reduce carbon emissions per unit of energy produced. The *America's Energy Future* study (NRC, 2009d) evaluated the near- and intermediate-term potential of each of these technologies and concluded that fossil fuels are likely to retain their dominant position in energy production over the next several decades; however, the study also identified numerous areas where investments in technologies and policy changes could hasten the transition to a low-GHG energy economy. Some of these areas are briefly summarized below, with an emphasis on the research needed to accelerate technology development and deployment.

Fuel Switching

Natural gas is the cleanest of the fossil fuels, with the lowest GHG emissions per unit of energy, emitting about half of the CO_2 of coal when burned for electricity generation, as well as generally lower emissions of other pollutants. Shifting electric generation from coal to natural gas could significantly reduce emissions. Such a shift would be useful but would not by itself reduce emissions sufficiently for a low-emissions future to minimize climate change. Thus, natural gas is more likely to be a bridge than a final solution. Additionally, the feasibility of natural gas as a bridge fuel will depend on the stringency of any emissions-limiting policies that are adopted.

Until recently, resources of natural gas were thought too small to support a transition. Recent improvements in technology have made economic unconventional gas resources, such as shale, leading to higher resource estimates. If these estimates are confirmed, natural gas could be a long-term option. However, there is some concern that shale gas development may have negative impacts on the local freshwater resources and land resources (DOE, 2009a). Another possible future source is natural gas hydrates found on the ocean floor, which are estimated to contain from one to a hundred times the world resource of conventional natural gas. Methods for recovery of hydrates are under investigation, but it is unlikely that hydrates would contribute significantly to the production of natural gas in the near term without major breakthroughs in the recovery process (NRC, 2010h).

Solar Energy

The total solar energy incident on the surface of the earth averages about 86,000 terawatts (TW), which is more than 5,000 times the 15 TW of energy currently used by humans (of which roughly 12 TW now comes from fossil fuels) and more than 100 times larger than the energy potential of the next largest renewable source, wind energy (Hermann, 2006). Hence, the potential resource of solar energy is essentially limitless, which has led many to conclude that it is the best energy resource to rely on in the long run. Currently, this resource is exploited on a limited scale—total installed worldwide solar energy production totaled 15 gigawatts (GW) in 2008,[2] or just 0.1 percent of total energy production, with similar penetration in the United States (EIA, 2009). Solar energy can be used to generate electricity and heat water for domestic use. Passive solar heating can be used in direct heating and cooling of buildings.

There are two main classes of solar energy technology used to generate electricity: concentrating solar power (CSP) and photovoltaics (PVs). CSP technologies use optics (lenses or mirrors) to concentrate beam radiation, which is the portion of the solar radiation not scattered by the atmosphere. The radiation energy is converted to high-temperature heat that can be used to generate electricity or drive chemical reactions to produce fuels (syngas or hydrogen). CSP technologies require high-quality solar resources, and this restricts its application in the United States to the southwest part of the country. However, CSP technologies are commercially available and there are a number of upcoming projects in the United States, particularly in California. The CSP industry estimates 13.4 GW could be deployed for service by 2015 (WGA, 2006). In the short term, incremental design improvements will drive down costs and reduce uncertainty in performance predictions. With more systems installed, there will be increased economies of scale, both for plant sites and for manufacturing. However, new storage technologies, such as molten salt, will be needed in the longer term to make wide-

[2]Energy production is generally reported as the "nameplate capacity" or the maximum amount of energy that could be produced from a given source. For energy sources such as solar or wind, which are intermittent in nature, the actual output is often lower than the nameplate capacity.

spread CSP deployment feasible. The global research community is studying the use of concentrated solar energy to produce fuels through high-temperature chemical processing (Fletcher, 2001; Perkins and Weimer, 2004, 2009; Steinfeld, 2005). At the international scale, the SolarPACES organization is working to further the development and deployment of CSP systems.[3] This organization brings experts from member countries together to attempt to address technical issues associated with commercialization of these technologies.

While incremental improvements in CSP performance are anticipated, there is the potential for large improvements in PV electricity generation technologies. Over the past 30 years, the efficiency of PV technologies has steadily improved, though commercial modules achieve, on average, only about 10 to 15 percent efficiency (that is, only 10 to 15 percent of the solar energy incident on the cell is converted into electricity), which is 50 percent or less of the efficiency of the best research cells (NRC, 2009d). Most current PV generation is produced by technologies that rely on silicon wafers to convert photons to electrons (Green, 2003; Lewis, 2007). Recent shortages of polycrystalline silicon have increased prices for PV modules and spurred increases in the use of thin-film solar PV technologies that do not require as much or any silicon. Thin-film solar PV technologies have about a 40 percent market share in the United States (EIA, 2009). In the short term, research is continuing on PV technologies; most of the work on improving these cells has focused on identifying new materials, new device geometries (including thin films), and new manufacturing techniques (Ginley et al., 2008).

The overall costs of a PV system, not just the costs of PV cells, determine its competitiveness with other sources of electricity. For example, approximately 50 percent or more of the total installed cost of a rooftop PV system is not in the module cost but in the costs of installation, and of the inverter, cables, support structures, grid hookups, and other components. These costs must come down through innovative system-integration approaches, or this aspect of a PV system will set a floor on the price of a fully installed PV system. In the medium term, new technologies are being developed to make conventional solar cells by using nanocrystalline inks as well as semiconducting materials. Thin-film technologies have the potential for substantial cost reduction over current wafer-based crystalline silicon methods because of factors such as lower material use, fewer processing steps, and simpler manufacturing technology for large-area modules. Thin-film technologies have many advantages, such as high throughput and continuous production rate, lower-temperature and nonvacuum processes, and ease of film deposition. Even lower costs are possible with plastic organic solar

[3] See *http://www.solarpaces.org/inicio.php.*

cells, dye-sensitized solar cells, nanotechnology-based solar cells, and other new PV technologies.

If next-generation solar technologies continue to improve and external costs associated with emissions from fossil fuel-based electricity are incorporated into the cost of electricity, it is possible that solar technologies could produce electricity at costs per kilowatt-hour competitive with fossil fuels. This transition could be accelerated through carefully designed subsidies for solar energy, as several other countries have done, or by placing a price on carbon emissions (Crabtree and Lewis, 2007; Green, 2005). Modifications to the energy distribution network along with energy storage would also improve the ability to exploit solar energy resources (see the section Energy Carriers, Transmission, and Distribution in this chapter). However, it should be noted that a bifurcated market for PV systems exists, depending on whether the system is installed on a customer's premises (behind the meter) or as a utility-scale generation resource. Behind-the-meter systems compete by displacing customer-purchased electricity at retail rates, while utility-scale plants must compete against wholesale electricity prices. Thus, behind-the-meter systems can often absorb a higher overall system cost structure. In the United States, much of the development of solar has occurred in this behind-the-meter market (NRC, 2009d).

There are several potential adverse impacts associated with widespread deployment of solar technologies. Utility-scale solar electricity technologies would require considerable land area. When CSP is used with a conventional steam turbine, the water requirements are comparable to fossil fuel-fired plants, making water availability a concern and, in some cases, a limiting factor. For PV technology, there are also concerns associated with the availability of raw materials (particularly a few rare earth elements; NRC, 2008f) and with the potential that some manufacturing processes might produce toxic wastes. Finally, the energy payback time, which is a measure of how much time it takes for an energy technology to generate enough useful energy to offset energy consumed during its lifetime, is fairly long for silicon-based PV.

In addition to electricity generation, nonconcentrating solar thermal technologies can displace fossil fuels at the point of use, particularly in residential and commercial buildings. The most prevalent and well-developed applications are for heating swimming pools and potable water (in homes and laundries). Systems include one or more collectors (which capture the sun's energy and convert it into usable heat), a distribution structure, and a thermal storage unit. The use of nonconcentrating solar thermal systems to provide space heating and cooling in residential and commercial buildings could provide a greater reduction of fossil fuels than do water heaters, but at present it is largely an untapped opportunity. Recently there has been limited deployment of liquid-based solar collectors for radiant floor-heating systems and solar air heaters, but the challenge with these applications is the relatively large collector area required in the absence of storage. Solar cooling can be accomplished via absorption and desiccant cycles, but commercial systems are not widely available for residential use.

Wind Energy

Wind electricity generation is already a mature technology and approximately cost competitive in many areas of the country and the world, especially with electricity generated from natural gas. The installed capacity for electricity generated from wind at the end of 2009 was approximately 159 GW, or about 2 percent of worldwide energy usage (WWEA, 2010). Wind turbine size has been increasing as technology has developed, and offshore wind farms are being constructed and proposed worldwide. As with solar power, wind energy alone could theoretically meet the world's energy needs (Archer and Jacobson, 2005), but a number of barriers prevent it from doing so, including dependence on location, intermittency, and efficiency. Other estimates of the resource base are not as large, but also indicate the United States has significant wind energy resources. Elliott et al. (1991) estimate that the total electrical energy potential for the continental U.S. wind resource in class 3 and higher wind-speed areas is 11 million GWh per year. As noted in NRC (2009d), this resource estimate is uncertain, however, and the actual wind resource could be higher due to the low altitude this estimate was developed at, or lower due to the inaccuracy of point estimates for assessing large-scale wind-power extractions (Roy et al., 2004). Assuming an estimated upper limit of 20 percent extraction from this base, an upper value for the extractable wind electric potential would be about 2.2 million GWh/yr, equal to more than half of the total electricity generated in 2007. This estimate does not incorporate the substantial offshore wind resource base. Development of offshore wind power plants has already begun in Europe, but progress has been slower in the United States. Though offshore wind power poses additional technical challenges, these challenges are being addressed by other countries. However, political, organizational, social, and economic obstacles may continue to inhibit investment in offshore wind power development in the United States, given the higher risk compared to onshore wind energy development (Williams and Zhang, 2008).

The key technological issues for wind power focus on continuing to develop better turbine components and to improve the integration of wind power into the electricity system, including operations and maintenance, evaluation, and forecasting. Goals appear relatively straightforward: taller towers, larger rotors, power electronics, reducing the weight of equipment at the top and cables coming from top to bottom, and ongoing progress through the design and manufacturing learning curve (DOE, 2008a; Thresher et al., 2007). Basic research in materials and composites is expected to lead to improved and more efficient wind energy systems, for example by improving the efficiency of turbines for use in low-windspeed areas (DOE, 2009c). Research on materials reliability and stabilizing control systems could help reduce maintenance requirements and further enable wind machines to survive extreme weather events. Continued research on forecasting techniques, operational and system design, and optimal siting requirements would improve the integration of wind power into the electricity system. As with solar energy technologies, modifications to the electricity transmission and distribution system along with energy storage capacity would also improve the ability to exploit wind energy resources (see the section Energy Carriers, Transmission, and Distribution in this chapter).

Along with technology advances, research on policy and institutional factors affecting the widespread implementation of wind systems is needed, as well as continued assessment of the potential adverse impacts of wind energy systems—for example, past research has shown that adverse impacts on flying animals, especially birds and bats, can be reduced both with advanced turbine technologies and by considering migration corridors when siting wind farms (NRC, 2007e). Siting is also critical in order to reduce potential negative effects on the viewscape, effects on noise, and unintended consequences on local wind and perhaps weather patterns (Keith et al., 2004). Concerns with the adverse effects of wind farms have led to substantial public opposition on some areas (Firestone et al., 2009; Swofford and Slattery, 2010). Further research and analysis of these factors would help decision makers evaluate wind energy plans and weigh alternative land uses—for agriculture, transportation, urbanization, biodiversity conservation, recreation, and other uses—to maximize co-benefits and reduce unintended consequences.

Bioenergy

Bioenergy refers to liquid or solid fuels derived from biological sources and used for heat, electricity generation, or transportation. Electricity generation using biomass is much the same as that from fossil fuels; it generally involves a steam turbine cycle. The key difference is that typical output for a wood-based biomass power plant is about 50 MW, while conventional coal-fired plants generally produce anywhere from 100 to 1,500 MW (NRC, 2009a).

In the United States, interest in biomass for energy production is usually in the form of liquid transportation fuels. Such biofuels currently take several forms, including biodiesel, the sugarcane-based ethanol systems used widely in Brazil, and the corn-based ethanol system that has been encouraged through subsidies in the United States. While the sugarcane system has an energy output that is more than five times greater than the energy input, corn ethanol has an energy output that on average is slightly greater than its input, and thus does not significantly reduce GHG emissions (Arunachalam and Fleischer, 2008; Farrel et al., 2006). Ongoing research into cellulosic feedstocks, algae-based fuels, and other next-generation biofuel sources could lead to more favorable bioenergy effects and economics. Other areas of research include improving the productivity of current bioenergy crops through genetic engineering (Carroll and Sommerville, 2009), reducing the environmental impact of bioenergy crops by growing native species on marginal lands (McLaughlin et al., 2002; Schmer et al., 2008), and developing biofuels that can be used within the current, petroleum-based fuel infrastructure (NRC, 2009b).

Many different disciplines are contributing to the development of new bioenergy strategies, including biochemistry, bioenergetics, genomics, and biomimetics research. For example, research in plant biology, metabolism, and enzymatic properties will support the development of new forms of biofuel crops that could potentially have high yields, drought resistance, improved nutrient use efficiency, and tissue chemistry that enhances fuel production and carbon sequestration potential. Significant research is also being directed toward strategies for cellulose treatment, sugar transport, and the use of microbes to break down different types of complex biomass, as well as on advanced biorefineries that can produce biofuels, biopower, and commercial chemical products. Many developments in biofuels have been recently summarized (see DOE, 2009c; NRC, 2008a, 2009b).

Widescale development of bioenergy crops could have significant unintended negative consequences if not managed carefully. Conversion of solar energy to chemical energy by ecosystems is typically less than 0.5 percent efficient, yielding less than 1 W/m^2, so relatively large land areas would be required for biomass to be a major source of energy (Larson, 2007; Miyamoto, 1997; NRC, 1980a). If the land required to grow bioenergy crops comes from deforesting or converting natural lands, there could be a net increase in GHG emissions as well as losses of biodiversity and ecosystem services. If grown on marginal lands, increased emissions of N_2O, a potent GHG, may result as a side effect of nitrogen fertilizer use (Wise et al., 2009b). If bioenergy crops are grown on existing agricultural areas, food prices and food security could be compromised (Crutzen et al., 2008; Searchinger et al., 2008). Production of bioenergy crops also has the potential to negatively impact water quality and availability for other uses (NRC, 2008i), and methods are needed to more fully assess their potential impacts on ecosystem services (Daily and Matson, 2008). The recent report *Liquid Transportation*

Fuels from Coal and Biomass (NRC, 2009b) contains a more detailed discussion of the potential environmental and ecosystem impacts and provides recommendations for sustainable methods for increased bioenergy use. Focused interdisciplinary research efforts are needed to develop such methods and more fully assess the full spectrum of possible benefits and side effects associated with different bioenergy production strategies.

Geothermal Energy

There are three components to the geothermal resource base: (1) geothermal heating and cooling, or direct heating and cooling by surface or near-surface geothermal energy; (2) hydrothermal systems involving the production of electricity using hot water or steam accessible within approximately 3 km of Earth's surface; and (3) enhanced geothermal systems (EGS) using hydraulic stimulation to mine the heat stored in low-permeability rocks at depths down to 10 km and use it to generate electricity. Currently, geothermal heating provides approximately 28 GW of energy (mainly for heating and industrial applications). For example, municipalities and smaller communities provide district heating by circulating the hot water from aquifers through a distribution pipeline to the points of use. The barriers to increased penetration of direct geothermal heating and cooling systems are not technical, but with the high initial investment costs and the challenges associated with developing appropriate sites. The resource for direct heating is richest in the western states, and geothermal heat pumps have extended the use of geothermal energy into traditionally nongeothermal areas of the United States, mainly the Midwestern and eastern states. A geothermal heat pump draws heat from the ground, groundwater, or surface water and discharges heat back to those media instead of into the air. The electric heat pump is standard off-the-shelf equipment available for installation in residences and commercial establishments. There are no major technical barriers to greater deployment. The United States currently has 700,000 installed units and the rate of installation is estimated to be 10,000 to 50,000 units per year (NRC, 2009d). One barrier to growth is the lack of sufficient infrastructure (i.e., trained designers and installers) and another is the high initial investment cost compared to conventional space-conditioning equipment.

In terms of electricity generation, hydrothermal systems are mature systems relying on conventional power-generating technologies. Technology is not a major barrier to developing conventional hydrothermal resources, but improvements in drilling and power conversion technologies could result in cost reductions and greater reliability. There is some potential for expanding electricity production from hydrothermal resources and thus providing additional regional electricity generation. For example, a study of known hydrothermal resources in the western states found that 13 GW of electric power capacity exists in identified resources within this region (WGA, 2006). However, in general the potential for major expansion of electricity produced from hydrothermal resources in the United States is relatively small and concentrated in the western states.

Enhanced geothermal systems represent the much larger resource base—the theoretical potential EGS resource below the continental United States is over 130,000 times the total 2005 U.S. energy consumption (MIT, 2006). Though this resource is vast, it exists at great depths and low fluxes. Accessing the stored thermal energy would first require stimulating the hot rock by drilling a well to reach the hot rock, and then using high-pressure water to create a fractured rock region. Drilling injection and production wells into the fractured region would follow next, and the stored heat would then be extracted, using water circulating in the injection well. The heat extraction rate would depend on the site. EGS reservoirs can cool significantly during heat-mining operations, reducing extraction efficiency with time and requiring periodic redrilling, fracturing, and hydraulic stimulation. Even so, the MIT report assumes that the individual reservoirs would only last around 20 to 30 years. Other challenges include a general lack of experience in drilling to depths approaching 10 km, concerns with induced seismicity, the need to enhance heat transfer performance for lower-temperature fluids in power production, and improving reservoir-stimulation techniques so that sufficient connectivity within the fractured rock can be achieved. Further research and demonstration projects will thus be needed before EGS is deployed on large scales.

Hydropower

Technologies for converting energy from water to electricity include conventional hydroelectric technologies and emerging hydrokinetic technologies that can convert ocean tidal currents, wave energy, and thermal gradients into electricity. Conventional hydroelectricity or hydropower, the largest source of renewable electricity, comes from capturing the energy from freshwater rivers and converting it to electricity. Hydroelectric power supplies about 715,000 megawatts (MW), or 19 percent, of world electricity. In the United States, conventional hydropower provides approximately 7 percent of the nation's energy (USGS, 2009). Hydropower is regionally important, providing about 70 percent of the energy used in the Pacific Northwest (PNWA, 2009).

Since this resource has been extensively exploited, most prime sites are no longer available. Furthermore, there is increasing recognition of negative ecosystem consequences from hydropower development. Future hydropower technological developments will relate to increasing the efficiency of existing facilities and mitigating the dams' negative consequences, especially on anadromous fish. Existing hydropower capacity could be expanded by increasing capacity at existing sites; installing electricity-generating capabilities at flood-control, irrigation, or water supply reservoirs; and developing new hydropower sites (EPRI, 2007a). Turbines at existing sites also could be upgraded to increase generation. None of these strategies require new technologies.

Because use of the conventional hydroelectric resource is generally accepted to be near the resource base's maximum capacity in the United States, further growth will largely depend on nonconventional hydropower resources such as low-head power[4] and on microhydroelectric generation.[5] A 2004 Department of Energy (DOE) study of total U.S. water-flow-based energy resources, with emphasis on low-head/low-power resources, indicated that the total U.S. domestic hydropower resource capacity was 170 GW of electric power (DOE, 2004). However, these numbers represent only the identified resource base that was undeveloped and was not excluded from development. A subsequent study assessed this identified resource base for feasibility of development (DOE, 2006). After taking into consideration local land use policies, local environmental concerns, site accessibility, and development criteria, this value was reduced to 30 GW of potential hydroelectric capacity (DOE, 2006). A report from the Electric Power Research Institute (EPRI) determined that 10 GW of additional hydroelectric resource capacity could be developed by 2025 (EPRI, 2007). Of the 10 GW of potential capacity, 2.3 GW would result from capacity gains at existing hydroelectric facilities, 2.7 GW would come from small and low-power conventional hydropower facilities, and 5 GW would come from new hydropower generation at existing non-powered dams.

New technologies to generate electricity from ocean water power include those that can harness energy from currents, ocean waves, and salinity and thermal gradients. There are many pilot-scale projects demonstrating technologies tapping these sources, but only a few commercial-scale power operations worldwide at particularly favorable locations. In general, there is no single technological design for converting energy in waves, tides, and currents into electricity. For example, approaches for tapping wave energy include floating and submerged designs that tap the energy in the impacting wave directly or that use the hydraulic gradient between the top and bottom of a wave (MMS, 2006). One such device concentrates waves and allows them

---

[4] Vertical difference of 100 feet or less in the upstream surface water elevation (headwater) and the downstream surface water elevation (tailwater) at a dam.

[5] Hydroelectric power installations that produce up to 100 kW of power.

to overtop into a reservoir, generating electricity as the water in the reservoir drains out through a turbine. Other approaches include long multisegmented floating structures that use the differing heights to drive a hydraulic pump that runs a generator or subsurface buoys that generate electricity through their up-down motion. Over the next 10 years, many large-scale demonstration projects will be completed to help assess the capabilities of these technologies, though it will take at least 10 to 25 years to know whether these technologies are viable for the production of significant amounts of electricity (NRC, 2009d). Over the longer term, other significant potential technologies that use ocean thermal and salinity gradients to generate electricity may also be investigated. However, these technologies currently only exist as conceptual designs, laboratory experimentation, and field trials. In general, even though waves, currents, and gradients contain substantive amounts of energy resources, there are significant technological and cost issues to address before such sources can contribute significantly to electricity generation. Storms and other metrological events also pose significant issues for hydrokinetic technologies.

Nuclear Power

Nuclear power is an established technology that could meet a significant portion of the world's energy needs. France obtains roughly 78 percent of its electricity from nuclear sources and Japan obtains 27 percent (EIA, 2007). About 20 percent of U.S. electricity comes from nuclear reactors, by far the largest source of GHG-free energy (EIA, 2009).[6] The reliability of U.S. reactors has increased dramatically over the past several decades, but no nuclear power plants had been ordered for over 30 years, largely because of high costs, uncertain markets, and public opposition. Improved availability and upgrades have kept nuclear power's share of generation constant at 20 percent despite the growth of other generation technologies. A nuclear revival has been initiated recently, largely because of concerns over limiting the magnitude of climate change. The U.S. government is providing loan guarantees for the first set of plants now being planned to compensate for uncertainties in costs and regulation. If these plants are successful in coming online at reasonable cost, their numbers could grow rapidly.

While nuclear power does not emit GHGs, there are other serious concerns associated with its production, including radioactive wastes (especially long-term storage of certain isotopes), safety, and security concerns related to the proliferation of nuclear

[6] Total generation of electricity from nuclear power in the United States is greater than in France or Japan.

weapons (MIT, 2003). The absence of a policy solution for the disposal of long-lived nuclear wastes, while not technically an impediment to the expansion of nuclear power, is still a concern for decision makers. New reactor construction has been barred in 13 U.S. states as a result, although several of these states are reconsidering their bans. Safety concerns stem from the potential for radioactive releases from the reactor core or spent fuel pool following an accident or terrorist attack. Nuclear reactors include extensive safeguards against such releases, and the probability of one happening appears to be very low. Nevertheless, the possibility cannot be ruled out, and such concerns are important factors in public acceptance of nuclear power. Proliferation of nuclear weapons is a related concern, but after 40 years of debate, there is no consensus as to whether U.S. nuclear power in any way contributes to potential weapons proliferation. A critical question is whether there are multilateral approaches that can successfully decouple nuclear power from nuclear weapons (Socolow and Glaser, 2009). Finally, public opinion is less skeptical of nuclear power in the abstract than it once was, but a majority of Americans oppose the location of nuclear (and coal or natural gas) power plants near them (Ansolabere and Konisky, 2009; Rosa, 2007). Some evidence suggests that the lack of support for nuclear power is based in part on a lack of trust in the nuclear industry and federal regulators (Whitfield et al., 2009).

Current U.S. nuclear power plants were built with technology developed in the 1960s and 1970s. In the intervening decades, ways to make better use of existing plants have been developed, along with new technologies that improve safety and security, decrease costs, and reduce the amount of generated waste—especially high-level waste. These technological innovations include improvements or modification of existing plants, alternative new plant designs (e.g., thermal neutron reactor and fast neutron reactor designs), and the use of alternative (closed) nuclear fuel cycles. The new technologies under development may allay some of the concerns noted above, but it will be necessary to determine the functionality, safety, and economics of those technologies through demonstration and testing.

Finally, research on nuclear fusion has been funded at several hundred million dollars per year since the 1970s. Fusion promises essentially unlimited, non-GHG energy, but harnessing it has proved to be extremely difficult. Most research addresses magnetic confinement (e.g., Tokamak reactors), but laser fusion (inertial confinement) also has promise. While fusion research and development is still worthwhile, it is uncertain whether a workable, cost-effective, power-producing reactor can be developed.

CARBON DIOXIDE REMOVAL APPROACHES

Fossil fuel sources are likely to remain an important part of the U.S energy system for the near future (NRC, 2009d), in part because of their abundance and the legacy of infrastructure investments. Hence, it makes sense to consider options for capturing the GHGs emitted during or after fossil fuel combustion. Virtually all of these approaches have focused on removing CO_2, as it is by far the most abundant GHG contributing to human-caused global warming. While there have been pilot projects and small commercial-scale projects to demonstrate the feasibility of some of these approaches, for the most part they remain in the research stage, and many involve important legal, practical, and governance concerns, as well as further technical research. Additional details about these approaches can be found in the companion report *Limiting the Magnitude of Future Climate Change* (NRC, 2010c).

Carbon Capture and Storage

Approaches for capturing the CO_2 released from coal- and gas-fired power plants and compressing and storing it underground (either in geological formations or via mineralization) are an important subject of research. While many of the component processes needed for this form of CCS are already used—for example, CO_2 injection is often used to improve yield or extend the lifetime of oil fields—there is currently only one demonstration CCS facility integrated with electrical power production in the United States,[7] and there are only a handful worldwide. As a result, many questions remain about the technological feasibility, economic efficiency, and social and environmental impacts of this approach.

Much of the needed research to support further development and, if proven feasible, widespread deployment of CCS has been outlined by the Intergovernmental Panel on Climate Change (IPCC, 2005). Research on the storage component focuses on the assessment of potential geologic reservoirs where CO_2 could be stored safely for long amounts of time, on the efficacy of carbon adsorption in geologic formations, and on monitoring techniques that would allow tracking of CO_2 once underground. Research on carbon capture focuses on improved methods for separating CO_2 from power plant waste, including analysis and development of approaches to feasibly (both

[7] In late 2009, the Mountaineer Plant in West Virginia began capturing and storing CO_2 from a 20-MW portion of the 1,400-MW plant using the chilled ammonia process. A project to scale up to a commercial-scale capture and sequestration demonstration has just been awarded. The DOE expects sequestration of 1.5 million tons per year of CO_2 to begin in 2015 (DOE, 2010). The FutureGen project, if built, would gasify coal, burn the gases in a combined turbine/steam cycle plant, and then capture and sequester the CO_2.

technologically and economically) retrofit existing plants with new technology. In addition, research is needed on environmental and social impacts of CCS (for example, its potential impacts on freshwater resources) and on the issues of adoption of new technology and public resistance to technologies that are perceived to be hazardous, all of which are critical to sound decision making about CCS.

The *America's Energy Future* committee highlighted the need for technical, cost, risk, environmental impact, legal, and other data to assess the viability of CCS in conjunction with fossil fuel-based power generation. It judged that the period between now and 2020 could be sufficient for acquiring the needed information, primarily through the construction and operation of full-scale demonstration facilities (NRC, 2009d).

Direct Air Capture

While conventional CCS is an attractive option for centralized power stations, there may be opportunities for other CCS technologies that may be more economic or environmentally preferable in certain situations (e.g., Rau et al., 2007) or could be used to remove CO_2 released by many small sources (e.g., Lackner et al., 1999). There have been many initial forays into the possibility of capturing GHGs directly from the atmosphere via technological means, but research in this area is generally only in preliminary stages. The only strategy for direct air capture that has emerged thus far involves physical or chemical absorption from airflow passing over some recyclable sorbent such as sodium hydroxide. A few research groups are developing and evaluating prototypes of such systems (Rau, 2009; Stolaroff et al., 2006). Major challenges remain in making such systems viable in terms of cost, energy requirements, and scalability. Direct capture approaches must also deal with the same challenges of long-term storage of the captured CO_2 as conventional CCS.

Other proposed approaches to direct capture from air involve fertilizing the ocean or modifying agricultural or ecosystem management practices (see Chapters 9 and 10). Further details and discussion about direct air capture approaches can be found in the companion report *Limiting the Magnitude of Future Climate Change* (NRC, 2010c). As noted in Chapter 15, sometimes direct air capture approaches are grouped together with solar radiation management approaches under the rubric of geoengineering (e.g., The Royal Society, 2009).

ENERGY CARRIERS, TRANSMISSION, AND STORAGE

Fossil fuels have come to dominate our energy system because they are dense energy sources that can be transformed into easily transportable and storable fuels and have historically been readily available at relatively low market prices. Moving to an energy system that produces fewer GHG emissions will require examination of issues involving integrating intermittent renewable energy sources from remote sites, smarter transmission and distribution grids, storage, and flexible/manageable loads, among others. As the America's Energy Future committee noted, the U.S. electricity transmission and distribution system is in urgent need of modernization to meet growing demand and to accommodate ever-larger amounts of intermittent sources of energy, especially wind and solar power. Moreover, many of the best areas for wind and solar generation are far from centers of energy demand and, on the other end, there is likely to be an increased need for accommodating distributed generation and two-way metering (e.g., for homes with PV panels). Finally, many of the renewable technologies discussed above have higher direct land use requirements than fossil fuels. These land use impacts have led to (and will presumably continue to generate) instances of local opposition to the siting of renewable electricity-generating facilities and associated transmissions lines.

Improvements in energy transmission efficiency and "intelligence" are needed for these resources to most effectively meet energy needs. Linking together many stable, intermittent, and distributed resources as well as grid-based storage in an extensive "smart" grid is needed to smooth out the fluctuations experienced at individual installations and improve the overall efficiency of transmission (Arunachalam and Fleischer, 2008). Grid intelligence involves extensive use of advanced measurement, communications, and monitoring devices together with decision-support tools. Taken together, the elements of a smart grid would also increase grid resilience, reducing the risk of widespread collapse following a local disruption or damage from natural events (such as storms and flooding) as well as physical and cyber attacks. Improved two-way information flows form the foundation of new ways for consumers to understand and control their electricity consumption (Denholm et al., 2010).

Improving energy storage technology and finding new ways to store energy is critical for addressing the intermittency of many renewable energy sources. Storage in compressed air systems has been under development, as well as improved battery technologies, focusing on improvements in storage capacity, charge time, power output, and cost. For further discussion of the role of storage, see Denholm et al. (2010).

SCIENCE TO SUPPORT TECHNOLOGY DEPLOYMENT

Substantial reductions in CO_2 emissions from the energy sector will require integrated deployment of multiple technologies: energy efficiency, renewables, coal and natural gas with CCS, and nuclear. Widespread deployment is expected to take on the order of years to decades. Such system-level implementation and integration require not only technology research and development but also research on potential hidden costs of implementation, the barriers to deployment, and the infrastructure and institutions that are needed to support implementation. All technologies have multiple impacts that require analysis and trade-offs in making choices among them. For example, impacts associated with the manufacturing and ultimate disposal of technologies can be substantial, even in comparison to the impacts of the operation of the technology. Life-cycle analysis and other analytical approaches (discussed in Chapter 4) can help identify the full set of impacts associated with a technology and thus can be an important tool for technology-related decision making.

Research is also needed to understand and address barriers to implementation. A full discussion of the strategies for, and barriers of, deployment of the technologies outlined above is beyond the scope of this chapter; however, Tables 14.1 and 14.2 provide a summary of issues as outlined by the U.S. Climate Change Technology Program. Analyses and approaches that identify and address these issues will be critical to implementation strategies (see *America's Energy Future* [NRC, 2009a,b,c,d] and *Limiting the Magnitude of Future Climate Change* [NRC, 2010b]). Finally, for some deployment challenges, full-scale demonstrations are critical precursors to implementation. *America's Energy Future* (NRC, 2009d) identified two kinds of demonstrations that should be carried out in the next decade: assessing the viability of CCS for sequestering CO_2 from coal and natural gas-fired electricity generation, and demonstrating the commercial viability of evolutionary nuclear plants in the United States. Such demonstration projects can provide research testbeds for understanding and evaluating the full suite of issues related to implementation.

LIKELY IMPACTS OF CLIMATE CHANGE ON ENERGY SYSTEM OPERATIONS

In addition to producing climate-forcing agents, the U.S. energy sector itself is expected to be affected by climate change and will need to adapt to the accompanying changes. Research on the possible impacts on energy system operations is still in its infancy; therefore, the examples noted below are merely illustrative of the ways climate change could affect energy systems (see the companion report *Adapting to the Impacts of Climate Change* [NRC, 2010a]).

TABLE 14.1 Summary of Activities for Deploying New Energy Technologies and Strategies

CCTP Goal Area	Technology Strategies	Education, labeling and information, dissemination	Tax policy and other financial incentives
Energy End-Use and Infrastructure	Transportation	54	29
	Buildings	58	21
	Industry	45	14
	Electric Grid and Infrastructure	19	7
Energy Supply	Low-Emission, Fossil-Based Fuels and Power	23	15
	Hydrogen	11	6
	Renewable Energy & Fuels	48	30
	Nuclear Fission	7	4
Carbon Sequestration	Carbon Capture	5	5
	Geologic Storage	4	4
	Terrestrial Sequestration	18	12
Non-CO_2 Greenhouse Gases	Methane Emissions from Energy and Waste	14	3
	Methane and Nitrous Oxide Emissions from Agriculture	8	7
	Emissions of High Global-Warming Potential Gases	17	3
	Nitrous Oxide Emissions from Combustion and Industrial Sources	14	9
Totals		345	169

NOTE: Column totals represent the number of deployment activities impacting the 15 technology strategies. Totals are indicative measures of relative frequency of application. Double counting occurs because a single deployment activity may impact multiple technology strategies. The count does not include activities that are authorized but not implemented.
SOURCE: DOE, 2009c.

Coalitions & partnerships	International cooperation	Market conditioning, including government procurement	Technology demonstration	Codes and standards	Legislative act of regulation	Risk mitigation
24	15	16	12	10	7	1
22	15	20	5	14	5	3
28	13	4	6	2	1	2
11	12	4	6	1	3	1
8	14	5	6	2	1	1
2	5	3	4	3	0	1
19	19	18	11	7	7	2
3	7	2	2	0	0	2
4	6	2	4	0	0	1
4	7	2	3	1	1	1
7	8	5	2	0	0	1
7	9	1	1	0	2	1
1	6	1	0	0	0	2
15	6	1	0	2	0	1
10	7	2	3	6	5	1
165	149	86	65	48	32	21

TABLE 14.2 Summary of Major Barriers to Deployment of New Energy Technologies

CCTP Goal Area	External Benefits and Costs	High Costs	Technical Risks	Market Risks	Incomplete and Imperfect Information	Lack of Specialized Knowledge	Infrastructure Limitations	Industry Structure	Policy Uncertainty	Competing Fiscal Priorities
Energy End-Use and Infrastructure	✓	✓	✓	✓	✓	✓		✓	✓	✓
Energy Supply	✓	✓	✓	✓		✓	✓	✓		
Carbon Capture and Sequestration	✓	✓	✓	✓	✓		✓		✓	
Non-CO_2 Greenhouse Gases	✓	✓	✓	✓	✓	✓				

NOTE: Checks indicate that a barrier is judged to be a critical or important obstacle to the deployment of two or more technology strategies within a particular CCTP goal area.
SOURCE: DOE, 2009c.

- Increases in energy demands for cooling and decreases in energy demands for heating can be expected across most parts of the country. These changes could drive up peak electricity demands, and thus capacity needs, but could also reduce the use of heating oil and natural gas in winter.
- Even as electricity demand increases in many regions, climate change may affect energy production. For example,
  - Water availability for cooling is a critical resource at thermal electric power plants (e.g., gas, coal, oil, CSP, bioenergy, and nuclear plants). Water limitations in parts of the country, and increased demand for water for other uses, may result in less water for use in energy production.
  - Increased water temperatures may reduce the cooling capacity of available water resources.
  - Water flows at hydropower sites may increase in some areas and decrease in others.
- Changes in river flows and sea levels may affect ship and barge transportation of coal, oil, and natural gas (as well as hydrokinetic energy sources).
- Changes in circulation and weather patterns may change the efficiency of electricity generation by solar and wind farms. For example, increased cloudiness could reduce solar energy production, and wind energy production could be reduced if wind speeds increase above or fall below the acceptable operating range of the technology. Not all of the possible impacts on intermittent renewable energy sources are well understood.
- Large-scale deployment of bioenergy may cause large new stresses on water supplies for growing the biofuel crops and processing them into usable liquid, gaseous, or solid fuels.
- Changes in the severity and frequency of extreme weather events—including hurricanes, floods, droughts, and ice storms—may disrupt a wide range of energy system operations, including thermal power plants, transmission lines, oil and gas platforms, ports, refineries, wind farms, and solar installations. Changes in sea levels (together with subsidence) could also threaten coastal energy system operations.

As with the other impacts of climate change discussed in this report, most of these impacts on energy production and use will be highly variable and place-dependent.

SCIENCE TO SUPPORT ADAPTING TO CLIMATE CHANGE

Potential actions to help the energy sector adapt to the effects of climate change include increasing electric power generating capacity, accounting for changing patterns of demand (summer-winter, north-south); increasing the energy efficiency of heating and cooling technologies; hardening infrastructures to withstand increased floods, wind, lightning, and other storm-related stressors; developing electric power generation strategies that use less water; instituting contingency planning for reduced hydropower generation; and increasing resilience of fuel and electricity delivery systems and of energy storage capacity. For more details, see the companion report *Adapting to the Impacts of Climate Change* (NRC, 2010a).

RESEARCH NEEDS

The remainder of this chapter focuses on what we still need to know—what we need research to tell us—in order to optimize strategies to both reduce emissions and adapt to climate changes in energy supply and use.

Develop new energy technologies and implementation strategies. Numerous scientific and engineering disciplines will need to contribute to the development of energy technology options and their effective implementation. Some key areas include materials science, electrochemistry and catalysis, biological sciences, and social and behavioral sciences. For example, materials science research could lead to advanced materials that could increase efficiency and offer other improvements in energy use, while research into photochemistry could provide the basis for engineering systems that mimic photosynthesis at higher efficiencies and rates. Technology assessment and portfolio analysis methods based on sequential decision making and risk-management paradigms need to be improved to help better set research priorities. Environmental, behavioral, and institutional analyses are essential to address obstacles and avoid unintended negative consequences. Of particular importance will be assessments of economic and technical performance of new technologies as well as full life-cycle environmental impacts.

Develop improved understanding of behavioral impediments to adopting new technologies, at both individual and institutional levels. New methods and increased research efforts are needed to develop understanding of the determinants of consumer choice and institutional decision making. Factors such as market failures and hidden costs could have important consequences on energy use and adoption of new energy technologies. Understanding possible impediments, and developing behavioral and policy interventions that circumvent them at both the individual and institutional level, are critical to rapid adjustments in energy consumption

Research on development of analytical frameworks for evaluating trade-offs and avoiding unintended consequences. Analytical frameworks are needed for evaluating trade-offs and synergies among efforts to limit the magnitude of climate change and efforts to adapt to climate change. There are many possible co-benefits associated with some of the technologies and strategies discussed in this chapter and the companion reports (NRC, 2010a,c). For example, along with the benefits of reducing GHG emissions and climate change, use of almost every energy efficiency or lower-emissions energy alternative will yield co-benefits in terms of reduced air pollution and associated health impacts. Some approaches may also yield co-benefits through increasing national energy security or conserving water resources. On the other hand, negative effects or interactions are also possible. For example, energy efficiency programs could disadvantage the poor or marginalized communities if they are not carefully included, and biofuels programs or large-scale deployment of other renewable energy sources could lead to food insecurity, loss of biological diversity, competition for land and water resources, and other impacts. It is also possible that carbon pricing could disproportionately affect the poor. Further research is needed on the interactions between the broad range of such benefits and consequences.

Develop new integrated approaches that evaluate energy supply and use within a systems context and in relation to climate change and other societal concerns. To date, scientists from many disciplines have investigated and developed some understanding of new energy technologies and strategies, individual and institutional choices related to energy consumption and adoption of new technologies, and the benefits and unintended consequences of limiting and adaptation policies. As described in the previous three research needs, further research is still needed to advance our understanding of all these areas. It is critical that such research is not conducted in an isolated manner but rather using integrated approaches and analyses that investigate energy supply and use within the greater context of efforts to achieve sustainable development goals and other societal concerns.

CHAPTER FIFTEEN

Solar Radiation Management

For over 45 years, proposals for deliberate, large-scale manipulation of Earth's environment—or geoengineering (see Box 15.1 and Figure 15.1)—have been put forward as ways to potentially offset some of the consequences of climate change. For example, whitening clouds, injecting particles into the stratosphere, or putting sunshades in space could increase Earth's reflectivity, thereby reducing incoming solar radiation and offsetting some of the warming associated with increasing GHG concentrations. Although few if any voices are promoting geoengineering as a near-term option to limit the magnitude of climate change, the concept has recently been gaining more serious attention as a possible backstop measure to be used if traditional strategies to limit emissions fail to yield significant emissions reductions or if climate trends become disruptive enough to warrant extreme and risky measures.

Questions decision makers are asking, or will be asking, about solar radiation management and other geoengineering approaches include the following:

- Can the negative impacts associated with increasing atmospheric greenhouse gas (GHG) concentrations be reduced or offset by intentionally intervening in the climate system? If so, how?
- What undesirable, unintended consequences might result from such interventions? How could these consequences be anticipated or detected?
- Who should decide, whether, when, and how to intentionally intervene in the climate system?
- What institutional mechanisms would be needed to initiate, carry out, monitor, and respond to the impacts—foreseen and unforeseen—of such an effort?
- Which types of interventions might be most socially acceptable and what frameworks for evaluation, governance, and compensation should be used?

In this chapter, we briefly review what is known about proposed solar radiation management (SRM) approaches and related governance and ethical issues and conclude with a discussion of the research needed to better understand SRM. Carbon dioxide removal approaches are addressed in Chapters 9, 10, and 14 and in the companion report *Limiting the Magnitude of Climate Change* (NRC, 2010c). Note that SRM research is in its infancy and that most conclusions should be regarded as preliminary.

BOX 15.1
Geoengineering: Solar Radiation Management and GHG Removal

The term *geoengineering* refers to deliberate, large-scale manipulations of the Earth's environment designed to offset some of the harmful consequences of GHG-induced climate change (see AGU, 2009; AMS, 2009; NRC, 1992b; The Royal Society, 2009). Geoengineering encompasses two very different classes of approaches: carbon dioxide removal (CDR) and solar radiation management (SRM). Figure 15.1 depicts the most commonly discussed options in both these categories.

CDR approaches (also referred to as post-emission GHG management or carbon sequestration methods) involve removal and long-term sequestration of atmospheric CO_2 (or other GHGs) in forests, agricultural systems, or through direct air capture with geological storage. These techniques and their implications are discussed in the companion report *Limiting the Magnitude of Future Climate Change* (NRC, 2010c) and are also mentioned in several previous chapters. There is no consensus regarding the extent to which the term *geoengineering* should be applied to various widely accepted practices that remove CO_2 from the atmosphere (e.g., reforestation).

SRM approaches, the focus of this chapter, are those designed to increase the reflectivity of the Earth's atmosphere or surface in an attempt to offset some of the effects of GHG-induced climate change.

HISTORY OF SOLAR RADIATION MANAGEMENT PROPOSALS

In November of 1965, the Environmental Pollution Panel of the President's Science Advisory Council (PSAC) for the first time informed a president of the United States about the threats posed by increasing atmospheric CO_2 concentrations. Their report stated:

> The climatic changes that may be produced by the increased CO_2 content could be deleterious from the point of view of human beings. The possibilities of bringing about countervailing climatic changes therefore need to be thoroughly explored. A change in the radiation balance in the opposite direction to that which might result from the increase of atmospheric CO_2 could be produced by raising the albedo, or reflectivity, of the earth (PSAC, 1965).

The topic of SRM was also taken up in the National Research Council's 1992 report *Policy Implications of Greenhouse Warming* (NRC, 1992b). That report noted:

> [W]e are at present involved in a large project of inadvertent "geoengineering" by altering atmospheric chemistry [i.e., by increasing GHG concentrations], and it does not seem inappropriate to inquire if there are countermeasures that might be implemented to address adverse impacts.... Our current project of "geoengineering" involves great uncertainty and risk. Engineering coun-

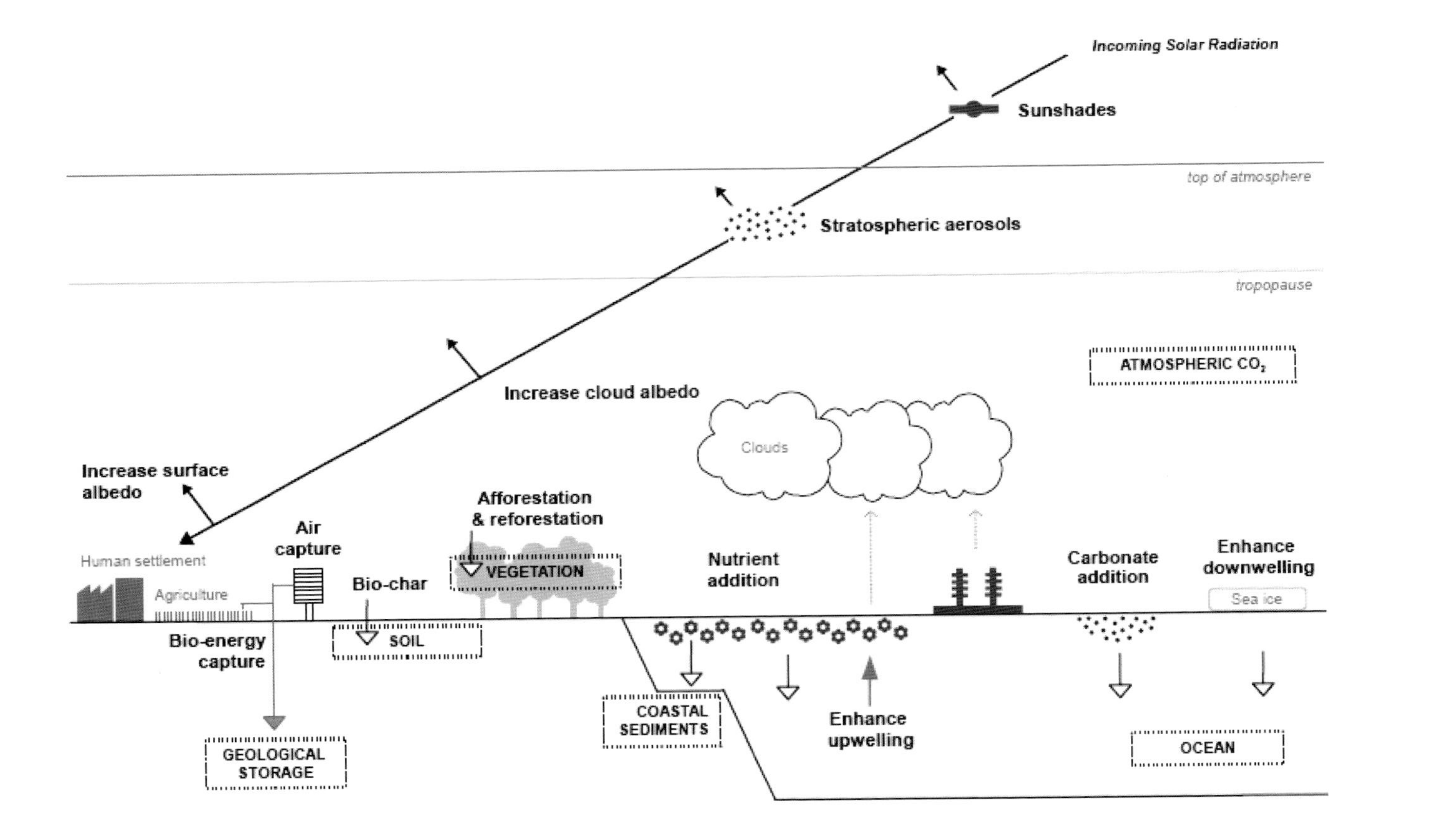

FIGURE 15.1 Various geoengineering options, including both solar radiation management and carbon dioxide removal. Dashed boxes represent carbon reservoirs (e.g., soil, ocean); black arrowheads represent shortwave radiation and are associated with solar radiation management; white and gray arrowheads pointing down correspond to a variety of natural and engineered processes, respectively, for removing CO_2 from the atmosphere; the thicker, gray arrowhead pointing up represents enhanced ocean upwelling, which could conceivably help to remove CO_2 from the atmosphere by enhancing biological activity at the ocean's surface; and the thinner gray arrowheads correspond to increased cloud condensation nuclei sources. SOURCE: Lenton and Vaughn (2009).

termeasures need to be evaluated but should not be implemented without broad understanding of the direct effects and potential side effects, the ethical issues, and the risks.

The PSAC (1965) and NRC (1992b) reports suggested that proposals to increase the reflectivity of the Earth (and to remove GHGs from the atmosphere) be thoroughly examined. This sentiment was echoed by many participants at the geoengineering workshop held in June 2009 as part of the suite of activities for the *America's Climate Choices* study (Appendix F), as long as such research does not undermine other critical climate research efforts (see the discussion of ethical issues below), including research on adapting to the impacts of climate change and on conventional strategies for limiting the magnitude of future climate change (i.e., reducing fossil fuel consumption, deforestation, and other activities that contribute to climate forcing). Critically, these evaluations should explore the intended effects of geoengineering approaches and their potential unintended side effects, as well as the ethical, institutional, social, and political aspects of intentional manipulation of the climate system.

PROPOSED SOLAR RADIATION MANAGEMENT APPROACHES

A number of different SRM methods have been proposed. This subsection briefly outlines some of the approaches that have been discussed in the literature (Keith, 2000; Rasch et al., 2008) and briefly summarizes their potential to reduce total radiative forcing. Other sources, including a recent report by the Royal Society (2009), provide a more comprehensive description. The relative advantages and disadvantages, potential for unintended consequences, and governance and ethical issues associated with these approaches are discussed in the next subsection. It should be noted that, unlike many other areas of research discussed in this report, these issues have undergone relatively little scientific scrutiny, with most of the relevant research done by a few small groups of scientists working with limited resources. Thus, many of the conclusions presented here must be regarded as preliminary and subject to revision.

Space-Based Options

A variety of options have been proposed for placing vast satellites in space, typically at the L1 point1 between Earth and the Sun (Early, 1989). However, to compensate for the increase in GHGs, nearly 4,000 square miles (10,000 square kilometers) of reflective

[1] "Lagrange Point 1" refers to a point roughly 1.5 million km above the surface of the Earth and between the Earth and the Sun. An object at the L1 point appears stationary from the perspective of Earth, as the net

surface would need to be constructed and put into orbit each year—or approximately an additional 10 square miles per day each and every day—for as long as CO_2 emissions continue increasing at rates comparable to today's (Govindasamy and Caldeira, 2000). Due to the magnitude of spaced-based deployment required for such an undertaking, and the enormous cost of putting objects into orbit, these options appear impractical for addressing threats posed by climate change this century.

Stratosphere-Based Options

One of the most widely discussed options for SRM involves the injection of sulfate aerosols into the stratosphere, although other types of particles could potentially serve the same function. As discussed in Chapter 6, particles can reflect solar radiation back to space, offsetting some of the warming associated with GHGs. The amount of sulfur that would need to be supplied to the stratosphere to offset the radiative forcing associated with GHG emissions could be delivered through a variety of means, including aircraft and artillery shells, with relatively small direct costs (Crutzen, 2006; NRC, 1992b; Robock et al., 2009; The Royal Society, 2009). Since sulfate particles are also injected into the stratosphere by volcanic eruptions, cooling following recent eruptions serves at least as a general "proof of concept" for this approach. For example, in the year following the eruption of Mount Pinatubo in June 1991, global temperatures cooled by approximately 0.9°F (0.5°C; Trenberth and Dai, 2007). Process understanding could be developed through small-scale tests, but an understanding of global climate effects would require either reliance on models or tests that would be of global scale and at least one-tenth the size of a full deployment. Full deployment would require a long-term, uninterrupted commitment to continued injection at the scale of tens of kilograms of material per second injected quasi-continuously. A sudden cessation after a sustained deployment could result in rapid temperature increases over a period of a few years, causing potentially severe impacts on ecological and social systems (Matthews and Caldeira, 2007).

Cloud-Based Options

A range of options have been proposed to "whiten" clouds, or make them more reflective, by increasing the number of water droplets in the clouds. The most widely discussed proposal involves whitening low clouds over remote parts of the ocean by

gravitational forces of the Earth and Sun are balanced by the centripetal force associated with that object's orbit of the Sun.

spraying a fine seawater spray in the air (Latham, 2002). This approach may be able to offset some or most of the radiative forcing associated with a doubling of atmospheric CO_2 (Bower et al., 2006; Latham, 2002). Process understanding relevant to this approach (e.g., cloud physics) can be tested at relatively small scales (Salter et al., 2008), although such tests would not permit direct inference of climate consequences of large-scale deployment. Another proposed cloud-based approach involves the seeding of high cirrus clouds with heterogeneous ice nuclei to reduce their coverage, potentially using commercial airplanes (Mitchell and Finnegan, 2009). While this method is not technically an example of SRM, it could potentially increase the amount of longwave (infrared) radiation emitted to space, which would cool the Earth.

Surface-Based Options

It has been proposed that global warming could be slowed by whitening roofs to reflect more sunlight back to space (Akbari et al., 2009). Under certain circumstances, whiter roofs could both reduce heating costs and help keep the Earth cool by reflecting sunlight back to space. Others have proposed growing more reflective crops (Ridgwell et al., 2009). Both approaches, if applied on a global scale, could potentially yield a modest cooling effect (The Royal Society, 2009), and white roofs also have the potential for co-benefits such as reducing urban heat islands (see Chapter 12). To date, studies indicate limited potential for such approaches, and the efficacy and environmental consequences of these approaches have yet to be carefully studied.

POSSIBLE UNINTENDED CONSEQUENCES

The overall climatic and environmental responses to SRM approaches are not well characterized. All proposed approaches have the potential for unintended negative consequences for both environmental and human systems. While the magnitude of the consequences is generally proportional to the scale on which the approach is deployed (painting an individual home white would yield fewer impacts—and be more easily reversible—than injecting millions of tons of sulfur into the stratosphere), several issues associated with large-scale deployment merit discussion.

First, none of the SRM approaches would stem ocean acidification (see Chapter 9) associated with enhanced atmospheric CO_2 levels. This is a key difference between SRM approaches and the CDR approaches discussed in Chapters 9 and 14 and in the companion report *Limiting the Magnitude of Future Climate Change* (NRC, 2010c).

Second, despite the potential for SRM approaches to offset warming in a globally averaged sense, local imbalances in radiative forcing could still lead to regional climate shifts, and the impact of SRM on precipitation and the hydrologic cycle is not very well understood. Short-term volcanic eruptions are not a good direct analog of long-term deployments, yet they provide valuable tests of our process understanding and ability to simulate the climate response to such forcings. Currently climate models underestimate the magnitude of the observed global land precipitation response to 20th-century volcanic forcing (Hegerl and Solomon, 2009) as well as human-induced aerosol changes (Gillett et al., 2004; Lambert et al., 2005), suggesting that these models may not reliably predict the simultaneous effect of SRM approaches on both precipitation and temperature (Caldeira and Wood, 2008). Some modeling studies (Robock et al., 2008) indicate that sulfate aerosol injection could decrease rainfall in the Asian and African monsoons, thereby affecting food supplies. Observational studies also reported that the Ganges and Amazon rivers both experienced very low flows immediately following the eruption of Mount Pinatubo (Trenberth and Dai, 2007). With regard to cloud-based options, it is also unclear if changes to cloud properties in one region could lead to "downwind" changes in the hydrologic cycle, including changes to precipitation.

For the injection of sulfate aerosols, an additional concern exists: the potential for increased concentrations of stratospheric aerosols to enhance the ability of residual chlorine, left from the legacy of chlorofluorocarbon use, to damage the ozone layer, especially in the early spring months at high latitudes. A sudden increase in stratospheric sulfate aerosol could strongly enhance chemical loss of stratospheric polar ozone for several decades, especially in the Arctic (Tilmes et al., 2008). There is also some evidence, however, that sulfate injection, by scattering some of the sunlight that does reach the Earth's surface, could actually boost ecosystem productivity and crop yields—this could disturb natural ecosystems but be an unintended co-benefit for agricultural systems (Gu et al., 2003; Roderick et al., 2001).

Finally, many SRM approaches require continuous intervention with the climate system in order to offset the forcing associated with GHGs. At some point in the future, if geoengineering were abandoned following its deployment, the adjustment of the climate system to the accumulated GHGs could involve warming on the order of several degrees Fahrenheit per decade (Matthews and Caldeira, 2007), a rate far greater than that estimated for the planet in the absence of geoengineering.

GOVERNANCE ISSUES

The deployment of SRM approaches has been discussed as a means of buying time for society to develop more effective ways to reduce GHG emissions, to avoid having to reduce emissions, and to produce a global cooling within years and decades in order to avert or reduce damage from a "climate emergency" (Lane et al., 2007) such as ice sheet collapse, rapid GHG degassing from melting permafrost, or other abrupt shifts in climate (see Chapter 6). Regardless of the ability of an SRM intervention to effectively buy time or avert crisis, several governance issues are associated with the decision to test or deploy SRM.

Due to the global nature of SRM, and especially considering some of the potential unintended consequences discussed in the preceding subsections, most analyses suggest that some sort of international framework—whether a series of bilateral or global, multilateral treaties—will be needed for governing SRM (e.g., Virgoe, 2009). Currently, no widely agreed-upon international governing body or legal or regulatory framework exists to govern the testing or deployment of SRM methods. Recent conferences on this topic have discussed how such a framework might be developed; the Council on Foreign Relations' Workshop on Unilateral Planetary-Scale Geoengineering (Ricke et al., 2008) suggested the application of standards such as "encapsulation" (the degree to which SRM releases material into the environment) and "reversibility" (the ability to terminate and reverse the effects of SRM) (The Royal Society, 2009), and another recent conference recommended voluntary governance mechanisms and basic principles to guide future geoengineering research (Asilomar Scientific Organizing Committee, 2010), but international endorsement and formal adoption by relevant research institutions and governments have not been undertaken. Because some research groups may be ready to test SRM approaches in the near term, there is also a near-term need to define what kinds of field experiments might be permitted in the near term while a broader regulatory framework is developed. Without a clear international agreement and relevant international and complementary national institutions, the probability of unilateral testing or deployment of SRM is elevated. Such unilateral action could potentially result in international tension, distrust, or even conflict (Virgoe, 2009), which could compromise the physical feasibility of SRM or increase the economic cost (Gardiner, 2010).

ETHICAL ISSUES

Intentional climate alteration, including SRM, raises important issues with respect to ethics and responsibility (Gardiner, 2010; Jamieson, 1996). First and foremost is the issue of equity. Issues of inequities include unequal representation in relevant decision-making bodies in relationship to benefits, and intergenerational equity, where future generations inherit the long-term commitment to certain types of interventions, or face the consequences involved in phasing out past SRM interventions. Second, consideration of SRM may pose a "moral hazard," where focus on SRM as a solution to climate change may detract from efforts to reduce GHG emissions or adapt to the consequences of climate change, or create an institutional inertia that essentially commits us to its deployment (Gardiner, 2010). Finally, there is the question—probably impossible to discern scientifically but nonetheless powerful in coloring public debate—about the "appropriate" place of the human species in the global ecology and whether human attempts to control the complex Earth system are a matter of hubris or a desirable evolution (e.g., Jamieson, 1996; Keutartz, 1999; Lovelock, 2008; Schneider, 1996, 2008).

Issues of ethics are likely to affect the social acceptability and political feasibility of planetary-scale, intentional manipulation of the climate system. Judging from past experience with siting and deployment of potentially fear-invoking technologies, these issues may dominate the political process (e.g., Douglas, 1985; Erikson, 1994; Fischhoff, 1981; Freudenburg and Pastor, 1992; Kates et al., 1984). Little if anything is known at present, however, about how U.S. citizens or other countries perceive SRM or other geoengineering options, and improved understanding of these perceptions may be critical inputs to governance discussions.

RESEARCH NEEDS

Improve understanding of the physical potential and technical feasibility of approaches. None of the SRM approaches have proceeded beyond the level of relatively simple analyses, small-scale laboratory experiments, and preliminary computer simulations. Hence, only a little is known about how effective proposed approaches would be at achieving their stated goals, or how possible it would be to actually deploy them. For example, in the case of stratospheric sulfur aerosol injection options, modeling and experiments to improve understanding of how particles aggregate in the stratosphere are needed. Because this and similar basic research questions relevant to climate engineering would also improve fundamental knowledge about the atmosphere, they could contribute more broadly to understanding the physical climate system. Engineering and cost analyses of different approaches are also likely to be useful as options are explored.

Focus research attention on the potential consequences of srm approaches

on other aspects of the earth system, including ecosystems on land and in the oceans. Because the coupled human-environment system is large and complex, it is impossible to fully anticipate all consequences of a geoengineering intervention in advance, or any other type of intervention for that matter. Nevertheless, it is possible to predict and anticipate some of these consequences through a combination of analysis; small-scale de minimis experiments; and climate, Earth system, and integrated assessment modeling. Again, in the case of stratospheric sulfur aerosol injection options, experiments that evaluate how increases in diffuse solar radiation would affect ecosystem productivity or how stratospheric particles might affect the ozone layer could be carried out. Similarly, modeling studies and analysis of observations around volcanic eruptions may provide insight into the changes to be expected in the hydrologic cycle from SRM.

Develop metrics and methods for informing discussions and decisions related to "climate emergencies." There are at least two components to this research need. For use of SRM as a potential "backstop option" in the case of an emerging "climate emergency," improved observations and understanding of climate system thresholds, reversibility, and abrupt changes (see Chapter 6)—for example, observations to let us know when an ice sheet or methane hydrate field may become unstable (e.g., Khvorostyanov et al., 2008; Shakhova et al., 2010)—could inform societal debate and decision making about needs for deployment of a climate intervention system. Second, there is no consensus on what constitutes a "climate emergency," nor is there a consensus regarding when an SRM deployment might be warranted. The notion of an "emergency" is not simply a scientific concept, but one that involves both scientific facts and human values—quite similar to discussions about "dangerous interference in the climate system" (e.g., Dessai et al., 2004; Gupta and van Asselt, 2006; Hansen, 2005; Lorenzoni et al., 2005; Oppenheimer, 2005; Smith et al., 2009). To some people, losing Arctic ecosystems constitutes a climate emergency, whereas to others the declaration of an "emergency" might require widespread loss of human life. Therefore, to inform a broader discussion of how society wants to address issues of risk, climate intervention cannot be studied in isolation but must be placed in a broader context considering, for example, drivers of climate change, climate consequences, sociopolitical systems, and human values.

Develop and evaluate systems of governance that provide models for decision making about whether, when, and how to intentionally intervene in the climate system. Because decisions about intentional alteration in the climate system will have widespread consequences, options for governance, including different types of institutions, assigned decision makers, procedures, norms, and rules and regulations, will be needed and can be provided through analysis. Much can be learned, for example, by studying past environmental and national security agreements, the siting and deployment of large-scale technology, and the conditions under which cooperation or conflict develops. Further research can help elucidate when and what type of governance might be useful not only for deployment but also for field experiments that can be reasonably expected to involve risks of negative consequences. Decisions about intentional interventions in the climate system require not only an understanding of the physical climate system response but also how these climate responses affect differentially vulnerable people and things people need or care about such as food and water security.

Improve detection and attribution of climate change so as to provide an adequate baseline of observations of the "nonengineered" system with which to compare observations of the "engineered" system. Just as it is a nontrivial exercise to quantitatively attribute observed climate change among different climate forcing agents, distinguishing the effects of intentional climate intervention from other causes of climate change to ascertain the effectiveness of SRM approaches is a nontrivial task. Detection and attribution of climate change, and evaluation of all actions taken to respond, including initial testing, will require enhanced observing systems and analyses covering a wide array of climate and other environmental variables, especially more complete observations of energy flows in Earth's climate system. In particular, preparations are needed to carefully observe the effects of the next major volcanic eruption.

Measure and evaluate public attitudes and test communication approaches to effectively inform and engage the public in decision making. Past experience with large and potentially dangerous technologies (or technologies perceived as dangerous) shows the importance of involving the public in advancing ideas and deliberations regarding testing or deployment of climate engineering approaches (see references above). However, little is known at this time about how different publics would perceive such large-scale interventions, what their attitudes are, how they should be engaged, and how best to communicate the complex issues concerning climate engineering. Also, attitudes and communicative approaches are likely to change over time and require periodic reassessment.

Develop an integrated research effort that considers the physical, ecological, technical, social, and ethical issues related to srm. Much of the research and observations needed to advance the scientific understanding of SRM approaches are also needed to advance general understanding of the climate system and related human and environmental systems. Examples of dual-purpose research include studies of the climate effects of aerosols, cloud physics, and how ecosystems, ocean circulation, permafrost, and ice sheets respond to changes in temperature and precipitation.

There is, however, additional research that would be needed to support full evaluation of SRM approaches (just as there is with other options for limiting the magnitude of future climate change), including a variety of social, ecological, and physical sciences (see Chapter 4). Such an effort would no doubt draw on many of the experts already engaged in climate change research, but would also need to engage new disciplines and expertise to aid in issues related to governance, public acceptance, and ethics.

CHAPTER SIXTEEN

National and Human Security

Over the past three decades, a number of concerns have emerged about potential interactions between global environmental change and security. Changes in temperature, sea level, precipitation patterns, and other aspects of the climate system can add substantial stresses to infrastructure and especially to the food, water, energy, and ecosystem resources that societies require. Several recent reports have argued that responding to climate change is a critical part of the U.S. national security agenda (Table 16.1). This assessment arises from concerns about how climate change directly affects military operations and regional strategic priorities, as well as the possible links between environmental scarcity and violent conflict, the role of environmental conservation and collaboration in promoting peace, and relationships between environmental quality, resource abundance, and human security.[1]

Questions decision makers are asking, or will be asking, about climate change and security include the following:

- How will changes in the physical environment, natural resources, and human well-being influence human security, interactions, and conflicts among nations, and the national security of the United States?
- Through what measures and interventions can we increase human security?
- What are the most critical implications of climate change for U.S. military operations and their supporting infrastructure?
- How will international GHG treaties be verified, what are the treaty provisions for onsite inspections in all signatory countries, and how will potential violations be detected and investigated in denied territories?
- What role should the U.S. intelligence community and the remote sensing infrastructure it supports contribute to these efforts, and what can be learned from previous treaty verifications efforts?

This chapter summarizes how climate change and our responses to it may affect U.S. military operations and international relations. The chapter also outlines the role of climate science in verifying international treaties and in analyzing human security. The last section lists research needs for studying the relationships between environmental change and security.

[1] Human security is defined as freedom from violent conflict and physical want (see Khagram and Ali [2006] for one recent review and synthesis).

TABLE 16.1 A Summary of Recent Studies Related to National Security and Climate Change Commissioned by Congress or Undertaken by Nonprofit and University Research Centers

Study	Year	Author	Synopsis
National Security and the Threat of Climate Change	2007	The CNA Corporation	A board of U.S. Military retired flag and general officers provide a perspective on the potential national security implications of climate change.
The Age of Consequences, Foreign Policy and National Security Implications of Global Climate Change	2007	Campbell et al., Center for New American Security	A projection and discussion of three (expected, severe, and catastrophic) potential climate scenarios as viewed through the eyes of national security and foreign policy.
The Arctic Climate Change and Security Policy Conference, Final Report and Findings	2008	Yalowitz et al., Dartmouth College	Results and findings from a December 2008 conference on the subject of Arctic climate change and security policy, as addressed through an international group of academics, scientists, government officials, and representatives of indigenous peoples.
Impact of Climate Change on Colombia's National and Regional Security	2009a	Catarious and Espach, The CNA Corporation	Projected impacts of climate change on Colombia's natural systems and resources and potential follow-on regional effects.
Climate-Related Impacts on National Security in Mexico and Central America, Interim Report	2009	The Royal United Services Institute	An examination of potential climate change impacts in Mexico and Central America, and their projected political, social, and security implications.
Socioeconomic and Security Implications of Climate Change in China, Conference Paper	2009	The CNA Corporation	An examination of the security implications of climate change in China from Chinese, American, and British Perspectives.
National Security Implications of Climate Change for U.S. Naval Forces: Letter Report	2010e	National Research Council	First component of a study to assess the implications of climate change for the U.S. Naval Services.

TABLE 16.1 Continued

Study	Year	Author	Synopsis
Lost in Translation: Closing the Gap between Climate Science and National Security Policy	2010	Rogers and Gulledge, Center for New American Security	Explores the gap between the science and policy communities and offers recommendations for collaboration to ensure the United States can effectively plan for the national security implications of climate change.
The following National Intelligence Council (NIC) Conference Research Reports are intelligence community documents summarizing the security and geopolitical implications of climate change from the perspective of a specific country.			
India: The Impact of Climate Change to 2030—Geopolitical Implications	2009c	NIC-CR 2009-07 May 2009	
China: The Impact of Climate Change to 2030—Geopolitical Implications	2009a	NIC-CR 2009-09 June 2009	
Russia: The Impact of Climate Change to 2030—Geopolitical Implications	2009f	NIC-CR 2009-16 September 2009	
The following NIC Commissioned Research Reports are intelligence community examinations of the security and geopolitical implications of climate change from the perspective of a specific country. Analysis includes impacts on stability of the governments and the economic vulnerability of each country.			
China: The Impact of Climate Change to 2030	2009b	NIC-2009-02D	
India: The Impact of Climate Change to 2030	2009d	NIC-2009-03D	
Russia: The Impact of Climate Change to 2030	2009g	NIC-2009-04D	
Southeast Asia and Pacific Islands: The Impact of Climate Change to 2030	2009h	NIC-2009-06D	
North Africa: The Impact of Climate Change to 2030	2009e	NIC-2009-07D	

THE RELATIONSHIP BETWEEN CLIMATE AND NATIONAL SECURITY

Identified concerns about climate-national security linkages, and associated research areas and needs, can be divided into two categories.

First, both climate change and efforts to respond to it may have significant effects on the operations, assets, and missions of the U.S. military. Many U.S. bases are located in areas that may be affected by sea level rise and tropical storms, and some future military operations may take place in areas subject to extreme high temperatures and droughts, compounding logistic problems. U.S. military operations are also substantial consumers of fossil fuels and thus will be affected by shifts in fuel prices and availability, as well as new technologies intended to displace fossil fuels.

Second, the impacts of climate change on specific assets and resources of international significance may affect multiple issues in bilateral and multilateral relations, shifting national strategic interests or perceptions thereof, or providing new bases for international conflict or cooperation. For example, declines in sea ice thickness and extent could result in increased access to and conflict over offshore resources in the Arctic Ocean associated with the opening of the Northwest and Northeast passages. Other examples include the effects of sea level rise and extreme events on coastal ports, navigable waterways, runways, roads, canals, or pipelines of international significance; changes in precipitation regimes that affect international river systems and ground vehicle mobility; and increases in humanitarian aid/disaster response stemming from changes in climate extremes (NRC, 2010e).

Military Operations

lCimate change and responses to it may affect the U.S. military in several ways. The Department of Defense (DOD) was directed in 2009 by the U.S. Congress to include the potential impacts of climate change in their 2010 Quadrennial Defense Review (QDR). The QDR is a legislatively mandated review of DOD strategy and priorities that sets the long-term course for DOD by assessing the threats and challenges the nation faces and rebalancing the Department's strategies, capabilities, and forces to address today's conflicts and tomorrow's threats. The QDR recognized climate change as one of many factors that has the potential to impact all facets of the DOD mission:

> The rising demand for resources, rapid urbanization of littoral regions, the effects of climate change, the emergence of new strains of disease, and profound cultural and demographic tensions in several regions are just some of the trends whose complex interplay may spark or exacerbate future conflicts (DOD, 2010).

The QDR focused on four specific issues where reform is imperative: security assistance, defense acquisition, the defense industrial base, and energy security and climate change. It stated the need for "incorporating geostrategic and operational energy considerations into force planning, requirements development, and acquisition processes."

Climate change may affect military assets and operations directly, for example through physical stresses on military systems and personnel, severe weather constraints on operations due to increased frequency and intensity of storms and floods, or increased uncertainty about the effects of Arctic ice and ice floes on navigation safety both on and below the ocean surface. U.S. military bases and associated infrastructure both inside the United States and overseas, particularly in coastal areas, will face risks from continuing sea level rise, extreme weather events, and interactions with other environmental stresses. For example, low-lying military bases in South Carolina, Guam, and Diego Garcia are particularly vulnerable to sea level rise. Changes in energy supply systems, including both fuel and electricity, as a result of either climate change or policies to limit greenhouse gas (GHG) emissions, would also have major impacts on military readiness and operations since the military is a major energy consumer and many military bases get their electricity from the national grid. In 2009, the Chief of Naval Operations directed the Navy's Task Force on Climate Change to assess the Navy's preparedness to respond to emerging requirements and to develop a science-based timeline for future Navy actions regarding climate change.

Sea level rise, reductions in sea ice, and changes in precipitation patterns may also affect key navigation routes of military as well as commercial importance, such as the Panama and Suez canals. Summer melting of Arctic sea ice will also make the Arctic Ocean more navigable, albeit with considerable seasonal ice floes, and the U.S. Coast Guard currently has just three commissioned icebreakers, only two of which are active (Borgerson, 2008; NRC, 2007g). Congress has asked the military to assess its preparedness for climate change, and the assessment process is now under way (e.g., Dabelko, 2009); the military has also requested input from a number of outside organizations, including the National Research Council (NRC), to gauge its preparedness for climate change and provide advice on prudent adaptation strategies. The Navy in particular, as directed by the Chief of Naval Operations, chartered the NRC's Naval Studies Board to conduct a study to explore the potential climate change impacts on naval forces (NRC, 2010e).

In general, there is substantial overlap between military and civilian needs with regard to climate change planning, such as the need for expanded and more interdisciplinary impact and vulnerability assessments and improved observation and modeling capabilities; hence, expanded collaboration between the defense, intelligence, and climate change research communities may yield benefits to all.

The U.S. military is also a major user of fossil fuels. Consequently, the military could play an important role in reducing the U.S. contribution to global GHG emissions, both through direct reductions and by providing a large market and consumer base for low-emission technology. Since supply chains that provide fuel to military equipment are a point of vulnerability during military operations, there are obvious co-benefits to strategies that increase the energy efficiency of the U.S. military and reduce its reliance on fossil fuels. Research to advance this goal will have many points of overlap with the broader research agenda to reduce emissions from transportation and energy use (see Chapters 13 and 14).

Climate change may also affect the U.S. military through new and changed missions. The military has substantial logistical, engineering, and medical capabilities that have been used to respond to emergencies both in the United States and abroad (for example, the 2005 Indian Ocean tsunami, the 2008 Burma/Myanmar typhoon, and the 2010 Haiti earthquake). Because climate change is expected to increase the severity and possibly the number of storms, floods, droughts, and other climate-related natural disasters in many parts of the world, military preparedness planning and the role of the military in responding to such disasters needs to be considered as part of adaptation planning (NRC, 2010e). Again, much of the research that will be needed to support analysis of military involvement in disaster support overlaps with that needed for impact and vulnerability studies in other sectors.

International Relations

While most discussions of climate change and security have examined the role of general environmental stress and resource scarcity on vulnerable populations and the risk of conflict, climate change also has the potential to disrupt international relations and raise security challenges through impacts on specific assets and resources. Such effects may arise as climate change increases or decreases the strategic value of resources of international significance, disrupting the basis for existing arrangements of ownership, control, or benefit sharing, or changing perceptions of national interests and threats to those interests. Perhaps the most obvious example is the loss of Arctic sea ice and the resultant increased value of Arctic navigation routes and offshore Arctic resources. Both the Northwest and Northeast passages will shorten major navigation routes during summer—for example, the Northwest passage through the Canadian Arctic archipelago would shorten the voyage from Rotterdam to Yokahama by 40 percent—and new orders will double the global fleet of ice-capable ships able to take advantage of these routes during other seasons (Borgerson, 2008). Other implications of an ice-free Arctic include increased tourism, expanded operating demands on the U.S. Coast Guard, and changes in the operating environment for surface and subsurface naval vessels. The legal status of the Northwest Passage in particular has long been contested, but the prospect of its becoming more widely usable raises the stakes substantially (NRC, 2010e).

Similarly, the prospect of substantial mineral reserves under the Arctic Ocean has prompted new offshore claims in the region by Canada, Denmark, Norway, and Russia. The U.S. Senate has not ratified the U.N. Convention on the Law of the Sea, so the United States cannot formally assert rights associated with the roughly 1,000 miles of Arctic coastline in U.S. territory. Climate change will also affect shorelines and in some cases "exclusive economic zones" and baselines used for projecting national boundaries seaward (Paskal, 2007). This may create or revive conflicts over resources in the offshore exclusive economic zone. Areas that may be affected include boundaries in the South China Sea and the boundary between the United States and Cuba. Changes in precipitation may also affect flow regimes in international river systems, risking new or intensified conflict in cases where claims over flows are already disputed or are subject to agreements not sufficiently robust to accommodate the flow changes that will occur. These and other challenges associated with climate change have been hypothesized but have not yet received thorough analysis (e.g., Liverman, 2009; Salehyan, 2008).

TREATY VERIFICATION

The prospect of binding international agreements with specific targets for GHG emissions from signatory countries will require methods and protocols for treaty verification and compliance. Whereas the measurement of GHGs has previously been in the research domain, the advent of climate treaties will require operational monitoring to meet the needs of verification and compliance. Remote sensing systems and other DOD and National Intelligence Council systems could play an important role in providing coverage of large regions of the globe and in monitoring local or point sources in remote or hostile locations.

This operational monitoring may require onsite visits to signatory countries by international observers with the ability to take direct in situ measurements to characterize, quantify, and validate sources and sinks of GHGs. Historical precedence for robust and potentially intrusive verification regimes can be found in the START I and START II treaties. START I Provisions included data exchanges, notifications, inspections, national technical means, and cooperative measures. The START II agreement built upon the START I verification regime and added several in situ inspection protocols to address issues that could only be verified through onsite inspections.

Reliable measurements of GHG concentrations and emissions are needed to effectively inform national and international policy aimed at regulating emissions, to verify compliance with emissions policies, and to ascertain their effectiveness. A system of measurement that is the basis for international agreement or financial transactions (e.g., carbon trading systems) needs to meet a higher level of scrutiny than a system used exclusively for research because of its legal, liability, and compliance implications. Consideration must therefore be given to data security, authentication, reliability, and transparency. In addition, as noted in Chapter 15, concerns about the possibility of unilateral implementation of solar radiation management schemes or other geoengineering approaches raise the need for improved monitoring of both GHG emissions reduction efforts and other climate intervention methods.

At present, there is no single U.S. agency that has the lead responsibility for operational GHG monitoring, and recent experiences with joint civilian and military satellite design and operation (e.g., the National Polar-orbiting Operational Environmental Satellite System; see NRC, 2008d) have highlighted some potential pitfalls of attempting to merge research and operational applications across multiple agencies. However, the defense, intelligence, and diplomatic communities have considerable experience with designing both technology and institutional arrangements to monitor treaty compliance. It would be valuable to complement this experience with the knowledge of the scientific community in designing and building monitoring and verification systems that have the appropriate resolution and accuracy to fulfill treaty verification requirements (NRC, 2009h). Improved interaction and engagement between the military, intelligence, diplomatic, and scientific communities would also be expected to advance the pace at which the science of monitoring evolves and enhance decision making around international treaties.

THE RELATIONSHIP BETWEEN CLIMATE AND HUMAN SECURITY

Climate changes are part of a set of interacting stresses that affect human welfare. Climate change may decrease human security and increase risks of domestic and international conflict in many parts of the world over the next several decades. The fact that climate change impacts may increase the probability of conflict has become a prominent argument for considering climate change in security analyses (e.g., Busby, 2007; Dalby, 2009; DOD, 2010).

As global changes and their potential consequences are becoming more evident, both in the United States and internationally, and as the global sustainability agenda has expanded (Brundtland, 1987; UN, 2009), the security agenda has been broadened (Sorensen, 1990). For example, a 1994 United (UNDP) report argued:

> The concept of security has for too long been interpreted narrowly: as security of territory from external aggression, or as protection of national interest in foreign policy or as global security from the threat of a nuclear holocaust. It has been related more to nation-states than to people (UNDP, 1994).

As threats associated with sustainable human development and global environmental changes became more prominent, UNDP's formulation of human security began to include "safety from such chronic threats as hunger, disease, and repression, [as well as] protection from sudden and hurtful disruptions in the patterns of daily life—whether in homes, in jobs or in communities" (UNDP, 1994). The concept of "human security" continues to gain prominence in both academic and policy arenas and is expected to be featured in the forthcoming Intergovernmental Panel on Climate Change (IPCC) Fifth Assessment Report (IPCC, 2009).

While these developments have been intended to encourage an integrative conception of security and threats that reflect the lived realities that individuals and communities face, there are still multiple ways of thinking about human security and no agreement on a policy agenda (Dalby, 2009). Most scholars understand human security as some combination of freedom from fear, want, harm, and violence. To some it is simply the converse of "vulnerability" (e.g., Barnett, 2001; Brauch, 2005; Dalby, 2009; Khagram and Ali, 2006; O'Brien et al., 2009). The International Human Dimensions Programme on Global Environmental Change (IHDP) core project on Global Environmental Change and Human Security has stressed the complementary nature of security threats and people's capacity to respond, focusing on "the ways that environmental changes contribute to (or exacerbate) pervasive threats and critical situations, while at the same time undermining the capacity to respond to these threats" (IHDP, 2009).

Human security scholars have examined the potential impacts of many types of environmental change, including food and water security, disaster vulnerability, land use and land degradation, urbanization and migration, the spread of infectious disease, and the associated challenges of building sustainable economic pathways out of poverty and deprivation. Insecurity can result, for example, when infrastructure developments (such as hydroelectric dams) put in place to meet other needs result in dislocations and displacements (Khagram, 2004), or as a negative consequence of rapid urbanization and the associated pollution, intergroup struggles, and crime (Evans, 2002). Through myriad interregional and international linkages, via political, economic, financial, sociocultural, military, public health, and environmental systems, human insecurities in one part of the world will affect the security of communities and economies in other parts (e.g., Adger et al., 2009b).

Research on human security and the environment has highlighted issues of equity, fairness, and human dignity, and especially the condition of women, because interacting socioeconomic and environmental stresses are experienced most severely by those who are most vulnerable (e.g., Adger et al., 2006; O'Brien et al., 2008). Related research has helped advance understanding of barriers and limits to adaptation (e.g., Adger et al., 2009b). Efforts are currently under way to synthesize a 10-year research effort on Global Environmental Change and Human Security (IHDP, 2009). Already this effort has identified conditions needed to maintain or restore human security, including effective governance systems, healthy and resilient ecosystems, comprehensive and sustained disaster risk-management efforts, empowerment of individuals and local institutions, and supportive values. Existing scientific insight is available, and further use-inspired social science research is needed, to inform the establishment of international mechanisms for effective, verifiable, accountable, and just efforts to limit and adapt to climate change. Such mechanisms have been found to be critical to the ability of communities anywhere to pursue sustainable livelihoods, meet fundamental human needs, secure human rights, and ultimately to ensure that climate change does not disrupt the natural environment so severely that it can no longer support the adequate and safe provision of ecosystem goods and services essential to human life and well-being (MEA, 2005).

RESEARCH NEEDS

Many plausible mechanisms of environment-security interaction have been proposed, but few have been carefully tested through research. Security scholars have cautioned against facile assumptions of cause and effect, and suggested that more focused and critical research into causal links between climate and environmental stresses and human conflict is needed (e.g., Barnett, 2003, 2009; Dabelko, 2009; Dalby, 2009; Liverman, 2009). Research in the following areas would help to solidify understanding of these linkages and project the future security implications of climate change with greater confidence.

Research on the relationship between climate change and national security. There has been little detailed scientific research on the direct and indirect impacts of climate change on national, international, or human security. Such research will require broadly interdisciplinary efforts, enhanced understanding (through observations and modeling) of the effects of climate change around the globe, empirical studies of impacts on both natural and human systems, and analyses of effective mechanisms for developing response strategies—in short, virtually all of the research that is needed to support other aspects of improved understanding discussed in previous chapters. To connect this improved understanding of general climate change impacts with security-specific concerns, research will also be needed on the relationships among environmental changes, social instability, and other threat multipliers. Such research is methodologically challenging because data are often limited in quality and quantity and the analysis needs to take account of thresholds, nonlinearities, and contextual and interaction effects. Nonetheless, given the prominence of climate change in recent discussions of national security and vice versa, it seems appropriate to dedicate additional resources to develop a coordinated program of research in the area.

Development of improved observations, models, and vulnerability assessments for regions of importance in terms of military infrastructure. There is an opportunity for considerable cooperation and synergy between the climate change research and national security communities. Improved regional climate projections and risk-management approaches are two important needs that these communities could work together to address. The needed research ranges from hydrological cycles at high latitudes and their implications for military operations to "war game" scenarios with climate-related crises.

Research on monitoring requirements for treaty verification. While considerable progress has been made in monitoring GHG emissions for climate research purposes, less is known about the operational observation standards that may be needed to meet treaty monitoring and verification requirements, and this is an active area of assessment, research, and planning (NRC, 2009h). Additional research and cooperation among communities is needed to determine the optimal mix of in situ and space-based civilian, military, and intelligence assets and the best data assimilation and analysis techniques to translate collected data into robust and reliable verification tools.

Identification of potential human insecurity in response to climate change impacts interacting with other social and environmental forces. Vulnerability analyses and better metrics are needed to identify people and places that might be expected to suffer the greatest harm from climate-related impacts—both individually and collectively. Of particular value would be metrics or observational approaches that can provide leading indicators of areas at risk, to help support preventive measures or anticipatory provision of humanitarian aid, or contribute to increased resilience. Moreover, new methods are needed to understand and predict interactions of climate change impacts, associated environmental changes, and social vulnerabilities, and how they are linked across regions.

CHAPTER SEVENTEEN

Designing, Implementing, and Evaluating Climate Policies

Governments broadly recognize the risks posed by climate change. From the local to the international level, many governments have considered and adopted policies designed to limit the magnitude of climate change and adapt to its expected impacts. Consequently, better understanding of climate policies is paramount to inform public- and private-sector decisions regarding climate change. Policy options are many and complex. For instance, the Intergovernmental Panel on Climate Change (IPCC) identifies six basic forms of policy instruments intended to directly reduce greenhouse gas (GHG) emissions and 53 different proposals for structuring international agreements to limit climate change (Gupta et al., 2007; see also Aldy and Stavins, 2007). Climate policies for adaptation are less well developed and mostly codified at the international level through the United Nations Framework Convention on Climate Change (UNFCCC) and the Kyoto Protocol (see also NRC, 2010a). Understanding how well implemented policies are working, or how proposed policies will work, requires scientific research on both current and possible future climate policies.

In general, how a policy actually works depends to some degree on all aspects of institutional design and on interaction with other policies and actors in the decision environment (Hill and Hupe, 2009; Mazmanian and Sabatier, 1981; Pressman and Wildavsky, 1984; Sabatier, 1986; Scheberle, 2004; Victor et al., 1998). Thus, questions that decision makers are asking, or will be asking, about climate policy include the following:

- What are the potential consequences of different GHG emissions-reduction targets—both in terms of climate change-related impacts and in terms of costs, feasibility, and other socioeconomic factors?
- What are the advantages and disadvantages of different policy instruments designed to pursue emissions targets, including their expected effectiveness, cost, robustness, adaptability, administrative burden, and distributional effects across different sectors, regions, and groups?
- What insights can scientific research and analysis provide about interactions (and especially the potential for conflict) among different climate-related policies? Are there policies that can contribute to both limiting climate change and adapting to its impacts? How can we avoid the potential for one type of policy to undermine another?

- How should different preferences across different sections of society be weighed? Who stands to gain and who stands to lose under different kinds of climate policies? How will climate policies interact with other policy objectives, such as moving toward sustainability?
- What does science tell us about building political support for policy implementation?

This chapter summarizes the scientific aspects of climate policy, including how science can contribute to policy design as well as its implementation and evaluation. Strengths and weaknesses of different policy approaches have been examined substantially in the scientific literature, and this chapter provides an overview of the general conclusions that have been reached by the IPCC and others as a prelude to identifying key areas for further research. For an actual assessment of current policies being considered in the United States to limit the magnitude of future climate change, see the companion report *Limiting the Magnitude of Future Climate Change* (NRC, 2010c); for a more detailed description of potential policy approaches related to adaptation to climate change, see the companion report *Adapting to the Impacts of Climate Change* (NRC, 2010a). The companion report *Informing Effective Decisions Related to Climate Change* (NRC, 2010b) also contains a detailed treatment and analysis of various policy mechanisms, as well as other approaches for improving climate-related decision making. The last section of the chapter summarizes research that is needed to support understanding of the interaction of climate change with natural and social systems, as well as policy design and implementation.

TYPES OF CLIMATE POLICIES AND AGREEMENTS

While there is a great deal of complexity and nuance involved in policy assessment, the IPCC (Gupta et al., 2007) concludes that there is "high agreement" and "much evidence" to support a number of conclusions about the major kinds of national policies that have been proposed and in some cases implemented to limit climate change. The IPCC also points out (see Table 17.1):

- Direct regulation, when enforced, can reduce emissions.
- Taxes are cost effective but do not guarantee a particular level of emissions reductions and are hard to adapt and adjust.
- The environmental effectiveness and cost effectiveness of tradable permits depend on the structure of the policy, including the number of permits issued, how they are distributed, and whether permits can be banked.

- Voluntary agreements between industry and government have played a role in the evolution of national policies and have accelerated adoption of best available technology but have not achieved significant emissions reductions.
- Subsidies, support for public research and development, or other incentives to develop and adopt new, low-emitting technologies, when used alone, have higher costs than other approaches; however, these strategies can complement policies targeting emissions directly via market mechanisms and enhance their overall environmental and cost effectiveness (this is particularly important when markets alone fail to achieve needed reductions in emissions; see also Jaffe et al., 2005).
- While information programs alone do not seem to lead to substantial emissions reductions, they can improve the effectiveness of other programs.
- A well-designed mix of policy types can be more effective than a single "pure form."

The companion report *Limiting the Magnitude of Future Climate Change* (NRC, 2010c) contains an extensive analysis of the advantages and disadvantages associated with different climate change policy options for reducing U.S. GHG emissions.

At the international level, climate policies have been codified in the UNFCCC and the Kyoto Protocol. Policies for limiting the magnitude of climate change are implemented through a variety of mechanisms such as the Clean Development Mechanism (CDM) and Joint Implementation. While experience with climate change treaties is limited, there is a substantial literature examining other environmental agreements that can provide insights of relevance to climate treaties (e.g., Biermann et al., 2009b; Mitchell, 2003; Young, 2002a,b, 2008, 2009). Drawing on this evidence, the IPCC (Gupta et al., 2007) concludes that there is, as with national-level policy instruments, "high agreement" and "much evidence" to support a number of conclusions about international treaties. The Kyoto Protocol has stimulated national policies and the creation of carbon markets, but its economic impacts are not clear, and its overall ambition with regard to emissions reduction has been limited. There is broad agreement in the literature that, to be successful, a successor agreement to Kyoto will have to be both environmentally effective and cost effective, take account of distributional and equity considerations, and be institutionally feasible (Aldy and Stavins, 2007). These goals are most likely to be achieved if the agreement incorporates goals, specific actions and timetables, rules for participation, and institutional arrangements and provisions for reporting and compliance. Of particular importance are the extent of engagement by national governments and the stringency and timing of the goals (Gupta et al., 2007).

TABLE 17.1 National Environmental Policy Instruments and Evaluative Criteria

	Criteria			
Instrument	Environmental effectiveness	Cost-effectiveness	Meets distributional considerations	Institutional feasibility
Regulations and Standards	Emission levels set directly, though subject to exceptions Depends on deferrals and compliance	Depends on design; uniform application often leads to higher overall compliance costs	Depends on level playing field; small/new actors may be disadvantaged	Depends on technical capacity; popular with regulators, in countries with weak functioning markets
Taxes and charges	Depends on ability to set tax at a level that induces behavioral change	Better with broad application; higher administrative costs where institutions are weak	Regressive; can be improved with revenue recycling	Often politically unpopular; may be difficult to enforce with underdeveloped institutions
Tradable permits	Depends on emissions cap, participation and compliance	Decreases with limited participation and fewer sectors	Depends on initial permit allocation, may pose difficulties for small emitters	Requires well-functioning markets and complementary institutions

Voluntary agreements	Depends on program design, including clear targets, a baseline scenario, third-party involvement in design and review, and monitoring provisions	Depends on flexibility and extent of government incentives, rewards and penalties	Benefits accrue only to participants	Often politically unpopular; requires significant number of administrative staff
Subsidies and other incentives	Depends on program design; less certain than regulations/standards.	Depends on level and program design; can be market-distorting	Benefits selected participants; possibly some that do not need it	Popular with recipients; potential resistance from vested interests. Can be difficult to phase out
Research and development	Depends on consistent funding, when technologies are developed, and policies for diffusion. May have high benefits in long term	Depends on program design and the degree of risk	Initially benefits selected participants; potentially easy for funds to be misallocated	Requires many separate decisions; depends on research capacity and long-term funding

NOTE: Evaluations are predicated on assumptions that instrument are representative of best practice rather than theoretically perfect. This assessment is based primarily on experiences and literature from developed countries, since peer-reviewed articles on the effectiveness of instruments in other counties were limited. Applicability in specific counties, sectors, and circumstances—particularly developing counties and economies in transition—may differ greatly. Environmental and cost effectiveness may be enhanced when instruments are strategically combined and adapted to local circumstances.
SOURCE: Gupta et al. (2007).

The IPCC (Gupta et al., 2007) notes that a large number of actions are being undertaken to reduce emissions by corporations, by local and regional governments (including U.S. states), and by nongovernmental organizations. It concludes that there is "high agreement" and "much evidence" that these actions have some effect on emissions and stimulate innovative policies and technologies but generally have limited impact in the absence of national policies.

On the adaptation side, as mentioned earlier, the UNFCCC and the Kyoto Protocol support adaptation planning and action through the National Adaptation Programmes of Action for 49 Least Developed Countries and have created the Adaptation Fund "to finance concrete adaptation projects and programmes in developing country Parties to the Kyoto Protocol that are particularly vulnerable to the adverse effects of climate change."[1] The Adaptation Fund is financed from a share of proceeds from the CDM and from other sources of funding.

There are many challenges to effective adaptation policy. These include: (1) cross-scale integration of decision making; (2) removal of legal and institutional barriers at higher levels of governance that may inhibit policy decisions at lower levels of governance; (3) unfunded mandates, lack of clarity about authority, and lack of mechanisms for cross-scale and cross-sector coordination and collaboration; (4) effective linking of science and decision making across levels; (5) identification of efficiencies, co-benefits and potential negative feedbacks among adaptation options and between mitigation and adaptation efforts in various sectors and across levels; and (6) the monitoring and evaluation of implementation of policies occurring (and depending on actions) at multiple levels (e.g., Adger et al., 2009b). The *Adapting* panel report (NRC, 2010a) discusses many of these issues in detail.

RESEARCH CHALLENGES ASSOCIATED WITH POLICY DESIGN AND IMPLEMENTATION

The need for future climate policies that are broader in scope, more flexible, and more ambitious than current policies will also require that policy makers employ iterative decision making and adaptive risk management (see Box 3.1). This poses new and expanded research challenges. Among the most important are (1) monitoring compliance with treaties, (2) assessing the benefits and costs of climate targets, and (3) examining complex and interacting policies.

[1] *http://unfccc.int/cooperation_and_support/financial_mechanism/adaptation_fund/items/3659.php.*

Monitoring Compliance with Treaties Intended to Reduce Climate Change

Research has shown that improving the effectiveness of international agreements will require a variety of mechanisms to verify compliance (Mitchell, 2003; Winkler, 2008; see also Chapter 16). Scholarship in this area has pointed out many of the constraints to monitoring and implementation, including establishing baselines, measuring GHGs, documenting additionality (that is, what countries and other actors are doing in addition to what would have been implemented in the absence of climate agreements) and "leakage" (emissions reductions in areas with strong policies being offset by increases in areas with weaker policies; Laurance, 2007; Santilli et al., 2005). Substantial improvements in technical capabilities will be required to meet these needs.

Numerous methods for performing more direct GHG measurements exist or have been proposed. For example, CO_2 could be measured directly at large concentrated sources, to supplement indirect measurements calculated from fuel inputs (Ackerman and Sundquist, 2008). An expanded network of ground-based, tall-tower, aircraft and satellite measurements of atmospheric CO_2 (including its isotopic signature) could be combined with atmospheric circulation models to infer regional anthropogenic CO_2 signals among natural sources and sinks of CO_2. In particular, a high-precision, high-resolution satellite system such as the Orbiting Carbon Observatory (which crashed at launch in February 2009) could provide the critical baseline CO_2 information against which decadal CO_2 trends can be verified following a climate treaty (NRC, 2009h). A recent NRC study (2010k) examined a number of these approaches, including their potential use in treaty monitoring and verification. The technology for monitoring changes in land use has also been an active area of research for decades and continues to grow in sophistication (Asner, 2009; GOFC-GOLD, 2008; Moran, 2009).

Verification of climate treaties will also require enhanced institutional arrangements (Winkler, 2008). At present, most work on reporting GHG emissions and removal due to human activities follows the UNFCCC protocol for activities in four sectors: energy; industrial processes and product use; agriculture, forestry, and other land use; and waste. In the United States, the Environmental Protection Agency (EPA) is responsible for the annual national summary report of GHG emissions and sinks, and the Department of Energy's Energy Information Administration provides energy statistics in greater detail. The EPA also recently issued a Mandatory Reporting of Greenhouse Gases Rule (EPA, 2009b), which sets up procedures for reporting from large sources and suppliers in the United States. Monitoring GHG concentrations is primarily the responsibility of NOAA, as part of the Global Atmosphere Watch of the World Meteorological Organization (WMO, 2009a).

The defense, intelligence, and diplomatic communities have considerable experience with designing both technology and institutional arrangements to effectively monitor treaty compliance, and in particular to deploy remote sensing for fine-scale local observations. Expanded engagement of these communities might substantially advance the pace at which the science of monitoring and institutional design evolves and thus provides enhanced support for decision making around international treaties. National and international law enforcement agencies as well as traditional treaty enforcement institutions may also need to be involved since most proposed policies have the potential for fraud and falsification (Gibbs et al., 2009; INECE, 2009; Yang, 2006). As in prior national security agreements, effective verification mechanisms may require surmounting discomfort, in the United States as elsewhere, over provisions allowing access for international inspectors.

Finally, establishing standards and certification mechanisms will be extremely important to reduce emissions. Standards and certification are sets of rules and procedures that are intended to ensure that sellers of credits are following steps that ensure that carbon is actually being sequestered and thus are closely related to monitoring. Proposals are appearing in the literature on how to develop and implement such standards (Oldenburg et al., 2009). These could be informed by the existing literature on how standards and certifications are used to shape the use of technology, including how such standards are negotiated, implemented, and enforced with varying degrees of effectiveness (Bingen and Busch, 2006; Eden, 2009; Hatanaka et al., 2005). These issues are also closely connected with the discussion of monitoring and observation discussed above.

Assessing the Costs and Benefits of Climate Targets

One of the most critical issues in policy design is comparing and assessing different trajectories to achieve GHG emissions reductions and evaluating the consequences and implications of those trajectories for human and environmental systems. A recent NRC study (NRC, 2010j) examined the implications for a range of climate stabilization targets. In contrast, this subsection provides a high-level overview of the social science research needs associated with analytic methods to evaluate targets, focusing on the two major alternative approaches: benefit-cost analysis and cost-effectiveness analysis.

Benefit-cost analysis is a method of systematic evaluation of the total social consequences of any decision or strategy. Applied to climate change, it has been used to assess alternative GHG emissions trajectories, typically by comparing a few simple alternative trajectories—each with associated projections of GHG concentrations, global average temperature, climate change impacts, and their valuation—with projections of the cost and effort required to achieve the trajectory relative to some baseline. Benefit-cost analysis requires expressing climate change impacts and lost services in an overall monetary metric so they can be compared to estimates of the costs associated with policies to limit the magnitude of future climate change. Ancillary costs and co-benefits of climate policies, which refer to costs and benefits in other areas (such as changes in local air pollution) resulting from climate policies, are sometimes included in the calculations as well. Such analysis can be conducted for a specific region or nation, or for the world.

Benefit-cost analysis can examine the expected benefits and costs of a particular target or policy or, by looking across targets, can identify an optimum target that maximizes net social benefits. If costs and benefits can be systematically and reliably projected and compared (discussed below are five of the major challenges that must be met to accomplish this objective), the socially optimal level of GHG emissions will be the level where the marginal benefit of reducing GHG emissions further will be equal to the marginal cost of making further GHG emissions reductions. If these calculations can be done credibly, decision makers can use this information to (1) set a limit on GHG emissions, (2) set a price on GHG emissions (whether implemented through market mechanisms or full costing of regulatory programs like emissions standards), and (3) get some sense of how important the climate problem is relative to other major societal problems.

An alternative approach, cost-effectiveness analysis, stipulates some limit on climate change (e.g., a future limit on human-caused radiative forcing or global-average temperature change) as a fixed goal, without evaluating the climate damages associated with the goal, then compares the costs of alternative emissions and policy trajectories to achieve that goal. For example, cost-effectiveness analysis has been used to compare alternative trajectories by which global emissions slow their growth and then decline to meet specified limits on atmospheric CO_2 concentration or human-caused radiative forcing in the 22nd century (e.g., starting the decline immediately versus growing for a decade or two and then declining faster [CCSP, 2007c; Richels et al., 2007; van Vuuren et al., 2006; Wigley et al., 1996]). A variant of cost-effectiveness analysis that has been used for climate change, called "safe landing" or "tolerable windows" analysis, defines two such constraints, one on the amount (and sometimes the rate) of climate change and another on the maximum rate of global emissions reduction. It then examines the cost and feasibility of alternative trajectories that stay within those boundaries (Füssel et al., 2003).

Cost-effectiveness approaches are often used when the costs and benefits of some action differ greatly in character, and the benefits are subject to greater uncertainty or controversy. In this circumstance, cost-effectiveness analysis allows analytically based comparisons of decisions without requiring that all impacts—in this case, damages from climate change and costs of emissions reduction—be reduced to a single metric. However, the implicit value this imputes to GHG emissions reductions is still equal to the marginal cost of GHG emission reductions that results from hitting the target or staying within the tolerable window. Of course, this implicit value can then be used to adjust the target if that value is felt to be lower or higher than the aggregated marginal value of the climate change impacts avoided. Such an iterative approach to GHG target setting allows multiple metrics to be used in evaluating the impacts of climate changes without completely abandoning the discipline provided by the strict application of cost-benefit analysis.

Formal policy-analytic methods such as benefit-cost and cost-effectiveness analysis can be powerful tools for informing decisions and illuminating structural issues underlying them and have had significant influence in climate policy debates. However, in practice, several major challenges must be met to provide reliable guidance to policy (e.g., Adler and Posner, 2006; Atkinson and Mourato, 2008; Dietz, 1994; Graves, 2007). The modeling community has made significant progress in addressing each of these challenges (see references within each section), but further progress in each area would greatly improve the usefulness of the results produced.

Five challenges are particularly difficult and influential in contributing to differences in cost-benefit valuations between alternative studies. These challenges, discussed in the paragraphs below, have to do with being able to systematically and comprehensively evaluate the benefits of GHG emissions reductions, being able to consistently and comprehensively project the costs of GHG emissions reductions, or being able to compare costs and benefits over time, under uncertainty, and across different socioeconomic groups.

1. *Estimating the social value of goods and services, particularly for impacts on ecosystems, climate-related amenities, or other resources and values for which market prices do not exist.* If formal policy-analytic methods are to be used to inform the choice of climate targets, rather than merely the choice of alternative means to meet a specified target, then all consequences of climate change and of efforts to limit it must be made comparable and valued. Economic theory argues that prices in well-functioning markets reflect the full social value of the goods and services that are exchanged, so market prices can be used to value changes in those goods caused by climate change. For impact sectors where markets exist like agriculture, this allows structural models

calibrated with market data to be used to value changes in activity levels that result from climate change. However, many important things that will be affected by climate change, such as environmental amenities, ecosystem services, and human health effects, are not exchanged in markets and so have no market price to provide guidance on their social valuation. This problem is pervasive in many areas of environmental assessment, and various methods, including contingent valuation and hedonic pricing approaches, have been developed to infer people's valuations for nonmarketed goods from their choices in related markets or suitably disciplined surveys (Arrow et al., 1993; Atkinson and Mourato, 2008; Carson, 1997; Mendelsohn and Olmstead, 2009). However, the range of estimates from the various studies is large and there is no consensus on the best approaches.

2. *Valuing uncertain outcomes, particularly high-consequence events whose probability is believed (but not known) to be low at low levels of warming, but increases with greater climate forcing, often called the problem of "fat tails."* Outcomes like these could plausibly result from dramatic irreversibilities in the climate or climate-impacted systems (e.g., a large ice sheet like Greenland melts very rapidly, increasing sea levels and reducing the reflection of sunlight from it, or large amounts of GHGs are released from warming permafrost). In principle, uncertain outcomes can be given a probability weight so that more likely outcomes are given greater weight and less likely—but much worse—outcomes are given lesser weight. "Fat tails" then generally refers to the case where the probability of very-high-consequence outcomes is still high enough that the product of that probability times the valuation of climate damages that would result from that outcome is large (i.e., does not approach zero because the probability of the outcome goes to zero more slowly than the impact valuation of that outcome increases). At present, it is difficult to estimate the probabilities of uncertain climate outcomes, but when these uncertainties are included in an analysis, the results can be very sensitive to assumptions that are made about the probability distribution associated with these low-probability/high-consequence events, and result in quite different conclusions (Nordhaus, 2009; Stern, 2007; Weitzman, 2007b, 2009; Yohe and Tol, 2007). Equally or more important here can be assessing how people perceive and act on the different risks that they face.

3. *Comparing costs, damages, and impacts in the near and long term (setting a social discount rate).* The rationale for discounting future costs and benefits (i.e., assigning them a lower value than immediate ones) has been discussed for more than 80 years (see Portney and Weyant [1999] for an overview). Discounting has both an ethical and a scientific component, and when these are correctly distinguished, the case for some form of discounting is compelling (Arrow et al., 2004; Heal, 1997; Nordhaus, 2008; Weitzman, 2007a). There are substantial disagreements, however, over the appropriate functional

form and quantitative magnitude of discount factors, and whether it is appropriate to apply the same approach to monetary costs and environmental damages (Atkinson and Mourato, 2008; Dasgupta and Ramsey, 2008; Dietz and Stern, 2008; Graves, 2007; Heal, 2009; Yohe, 2006). In addition, since many of the people who will be affected by climate change the most have not yet been born, an equitable way of factoring their preferences, which cannot be directly measured, into the calculations needs to be developed (Portney and Weyant, 1999).

Because many costs of reducing climate change occur in the near term while the most serious of the climate impacts avoided would be further in the future, socially optimal levels of climate change limitation in a benefit-cost framework can be quite sensitive to choices about discounting; lower discount rates usually imply stronger and earlier action to limit climate change than higher discount rates.

4. *Estimating how policy will influence technological change.* It is possible that technological innovation will create opportunities to reduce GHG emissions at lower than present costs, but the rate of such innovation and the relative influence and mechanisms of various possible ways to stimulate it are subject to substantial uncertainties. Alternative models of induced technological change highlight the influence of policies to raise the effective price of emissions, learning-by-doing, public versus private investments in research and development structured in various ways, basic science versus specifically targeted research, and overall investment driven by aggregate economic growth. These alternative models can imply substantial differences in preferred policies, but available data appear to not discriminate strongly between them (Goulder, 2004; Grübler, et al., 2002).

5. *Incorporating equity considerations into the analysis.* The costs and benefits of climate change adaptation and limitation will be unevenly distributed across space, time, and social and economic groups. There will be substantial differences across regions within the United States and across the globe (USGCRP, 2009a; World Bank, 2009). Although costs and benefits could in principle be weighted to incorporate equity concerns (Atkinson and Mourato, 2008; Kverndokk and Rose, 2008), in practice this poses significant challenges of observing and projecting disaggregated costs and benefits and, if aggregation is required, identifying defensible equity-based weights. Moreover, formal analyses of climate change responses have examined only aggregate effects at the level of the jurisdiction considered. International aggregations of climate change impacts are often valued in terms of losses in income, which tends to bias the weighting toward richer and away from poorer people who have less to lose but will feel percentage losses in income more.

That there are significant research challenges that remain regarding the major elements of cost-benefit analysis as currently applied to climate change policy evaluation means that great care needs to be exercised in communicating the results of these calculations to decision makers. Consumers of these studies need to know what factors are included in the analysis and how, and which ones are left out or only partially represented. They can then add their own assessments of the missing elements and perspectives to the numbers they get from the cost-benefit calculations in order to provide a more complete picture of the value of different policy alternatives. At the same time, the raw numbers themselves, if interpreted correctly, can often help decision makers set bounds on appropriate actions, especially if we are far away from the optimum.

Examining Complex and Interacting Policies

While many policy analyses, such as benefit-cost, assume rather generic policy instruments (e.g., a single tax on GHG emissions or a single cap-and-trade policy that applies to all fossil fuel consumption in the nation uniformly), actual policies are much more complex. They also interact with other climate and nonclimate policies at different scales and jurisdictions and their institutional design and implementation critically shapes their effectiveness (Young, 2002a). Previous National Research Council reports and the international community have detailed the research agenda in this area (see in particular Biermann et al., 2009b; NRC, 2005a, 2008h).

Three key topics emerge from these analyses and our own assessment of the challenges faced by policy analysis in supporting climate change decision making: introducing realistic complexity into analyses of climate policy, coordinating across levels of government, and equity and distributional issues. These topics are explored in the paragraphs that follow.

Introducing Realistic Complexity into Analyses of Climate Policy

In the United States as of December 2009, 32 states and the District of Columbia had adopted mandatory standards that, over the next 10 to 20 years, will require that between about 10 and 15 percent[2] of the energy supplied by utilities come from alternative and renewable sources (Pew Center on Global Climate Change, 2009). Four other states have voluntary standards. State and local governments as well as the federal government have a variety of programs, including labeling, appliance standards, and

[2] States vary in how the standards are defined so comparisons of goals can only be approximate.

investment in technology development and adoption, that are intended to promote energy efficiency (North Carolina State University, 2010). It is not completely clear how existing policies will be affected by, or will affect, future ones, especially across governance scales. If policy analysis is to inform policy decisions, then it has to find ways to understand and model these complex interactions (Selin and VanDeveer, 2007).

These interactions can have substantial influence on the effectiveness of polices. As the IPCC has concluded, programs intended to inform and influence behavior can multiply the effects of other policies. For example, home weatherization programs offering identical financial incentives differed in impact by more than an order of magnitude, depending on how they were implemented (Stern et al., 1986). Legislation and regulations also involve political compromises that add complexity and cause actual policies to deviate from their original goals (Pressman and Widalvsky, 1973). Moreover, domestic climate policies could enhance or retard the U.S. balance of trade depending on how they are structured (Houser et al., 2008).

Nor are these interactions restricted to the U.S. context. International climate policy will interact with many other international agreements and laws. For example, trade agreements may either contradict or complement mechanisms for enforcing emissions limitations (Weber and Peters, 2009). Efforts to encourage transfer of technologies to reduce emissions may be facilitated or inhibited by intellectual property agreements (Brewer, 2008). Development funding can enhance or retard efforts to reduce and adapt to climate change (Klein et al., 2007; World Bank, 2009). A substantial literature has identified the possibility of these complex interactions, but their implications are only just beginning to be explored in depth. There may be advantages to such complex policies, if they can be designed taking into account both political reality and the implications of the complexity involved—complexity that may lead to more robust policy (Anderies et al., 2004; Andersson and Ostrom, 2008; Ostrom, 2007; Pinto and De Oliveira, 2008). Interactions can be also positive as policy instruments try to reap co-benefits across policy goals. A few mechanisms, such as the CDM and the United Nations Collaborative Programme on Reducing Emissions from Deforestation and Forest Degradation in Developing Countries, seek to reap co-benefits both across mitigation and adaptation and across climate policy and development. For example, the CDM allows Kyoto Annex 1 countries to offset their carbon emissions by generating carbon credits (through the creation and implementation of projects) in Annex 2 countries. Besides generating carbon credits, CDMs are also required to produce a "development dividend" by creating jobs, promoting sustainable development, and other methods (the definition of what constitutes sustainable development requirements varies substantially across countries). However, empirical research has found that success in promoting sustainable development has been mixed (Olsen, 2007; Pearson, 2007; Sutter and Parreño, 2007).

Coordinating Across Levels of Government

Many levels of government are already engaged in adapting to and limiting the magnitude of climate change (Betsill and Bulkeley, 2006; Betstill and Rabe, 2009; Paterson, 2009; Rabe, 2008; Schreurs, 2008; Selin and VanDeveer, 2007). Public-private partnerships and public-social partnerships (between business and communities) add to the complexity of emerging policy (Lemos and Agrawal, 2006). This complexity raises important questions about the legitimacy of climate policy, including the scope and means of representation of stakeholder interests (Falkner, 2003; Ford, 2003). It also raises questions about the distribution of resources, and about how negative externalities from the policies can be avoided or corrected (Bäckstrand, 2006; Cashore, 2002; Lemos and Agrawal, 2006).

The multilevel governance system of climate policy presents both opportunities and challenges for policy makers. The federal government can learn from and build on policy "experiments" enacted at the state level and capitalize on existing networks to expand political coalitions (Peterson and Rose, 2006). However, the capacity of decision makers operating at any one level can be enhanced or (more frequently) constrained by the policies at other levels (Adger et al., 2007; Betsill and Bulkeley, 2006; Moser, 2009b). The appropriate mode of governing depends on the character of the problem and available resources (including knowledge), the dynamics of the sector involved, the availability of policy options for other policy actors, and the constellation of political interests around a policy (Dietz and Henry, 2008; Dietz et al., 2003; Ostrom, 2005, 2007; Selin and VanDeveer, 2007).

Recent literature also suggests that polycentric policy (i.e., policy that does not originate from and is not implemented in just one, central decision-making unit but is carried out by multiple, linked centers of authority) may be more robust and adaptable than policies implemented by a single unit of government (Andersson and Ostrom, 2008; Ostrom, 2007, 2010; Pinto and De Oliveira, 2008).

In the case of adaptation policy, the implications of multilevel, hybrid forms and polycentric governance both domestically and internationally are many and varied. Processes at the global and national levels will influence local adaptation decisions and vice versa; in the United States and around the world, a great variety of actors and institutions including local, regional, state, federal, and tribal authorities will influence those decisions (e.g., Agrawal, 2008; Armitage et al., 2007; Bulkeley, 2005; Cash et al.,

2006b; Moser, 2009b; Rabe, 2008; Urwin and Jordan, 2008). For example, in less developed regions, adaptation policy critically intersects with development and decentralization of government authority (Agrawala, 2004; Burton et al., 2007; Eakin and Lemos, 2006; Klein et al., 2007; Kok et al., 2008; World Bank, 2009).

Equity and Distributional Issues

Climate change is a quintessential equity problem since those who have been historically least responsible for causing it will be disproportionally negatively affected by it (Adger et al., 2006; Baer et al., 2000; O'Brien and Leichenko, 2003; Roberts and Parks, 2007; World Bank, 2009). There are usually three main sources of inequality shaping climate policy dialogues: historical responsibility for the problem, who will likely bear the brunt of its negative impacts, and who will be responsible to solve it (O'Brien and Leichenko, 2003; Parks and Roberts, 2010). In addition, other distributional and equity factors need to be considered in the design of adaptation policy. For example, many of those most severely affected by climate change are often those least able to engage effectively in policy decision-making processes. Many policies have the potential for indirect or secondary impacts that may be inequitably distributed (Kates, 2000).

While a rich literature explores different equity aspects of climate change as a problem, including exploring the three aspects of inequality mentioned above in greater detail (e.g., Dow et al., 2006; Roberts and Parks, 2007; Schneider and Lane, 2006; World Bank, 2009), there has been less empirical research carried out about specific ways in which equity issues have or can shape policy design and implementation, especially from the point of view of developing countries, who have pointed out global inequality as a main impediment for international cooperation (Parks and Roberts, 2010).

With regard to adaptation policy in less-developed regions, equity and distribution of costs and benefits of climate change is intrinsically related to the structural inequality and multiplicity of stressors that shape vulnerability to climate impact including poverty, lack of education and access to health care, and war and conflict (see Chapters 11 and 16). Equity issues will also greatly drive political debates in the domestic climate policy context (see *Limiting the Magnitude of Future Climate Change* [NRC, 2010c]). For example, the fear of job losses is already prevalent in fossil fuel-dependent industries and regions of the United States (Peterson and Rose, 2006). Policies that place a price on carbon will affect various industries and regions of the country differently and will differentially affect socioeconomic groups within regions (Oladosu and Rose, 2007).

RESEARCH NEEDS

As scientific knowledge and public awareness about the potential consequences of climate change have grown, the nation and the world have moved from trying to decide whether or not to have climate policy to trying to decide what policies will most effectively limit the magnitude of future climate change and help populations and their infrastructure adapt to its impacts. Scientific analysis can inform this discussion by elucidating the factors that influence the adoption, implementation, and effectiveness of both domestic and international agreements. This research includes improving methods for quantifying and comparing benefits, costs, and risks associated with climate change and climate policies; developing methods for analyzing complex policies and combinations of policies; learning how to design policies that work at multiple levels of governance; and examining climate policies in a broad context, including overall sustainability goals, concerns with equity, and relationships with an array of nonclimate policies. The challenges are substantial but so are the opportunities for both advancing science and gaining scientific knowledge that contributes to effective policy making. Some specific research needs include the following.

Continue to improve understanding of what leads to the adoption and implementation of international agreements on climate and other environmental issues and on what forms of these agreements are most effective at achieving their goals. Given the current state of our understanding of international agreements on climate and sustainability, the International Human Dimensions Programme (Biermann et al., 2009a, 2010), previous NRC reports (NRC, 2005a), and others (Young et al., 2008) have considered research needs. Drawing on these, we identify five key research questions in this area:

- How do multiple international agreements interact with each other and with public and private policies at the national, regional, and local levels?
- How are international agreements and their implementation influenced by and how do they influence nongovernmental actors, such as private firms and nongovernmental organizations?
- How can international agreements utilize an adaptive risk-management approach to responding to climate change?
- How can accountability and legitimacy of international agreements and the mechanisms that implement them be ensured?
- How can international agreements best take account of fairness and equity concerns?

Develop protocols, institutions, and technologies for monitoring and verifying compliance with international agreements. In addition to research on international agreements, decisions about participating in such agreements need to be made with full awareness of the institutional and technological capabilities for monitoring and verifying that participants to the agreement are fulfilling their commitments, and for informing future policy changes or refinements in an adaptive risk-management framework. Observations of GHG emissions and concentrations are an especially pressing need and are the subject of a recent NRC study (see NRC, 2010k). A system of observations and measurements designed to support international agreements or financial transactions (for example, a cap-and-trade system for carbon emissions) will likely have to meet a higher level of scrutiny than systems used purely for scientific research. In particular, systems to support treaty compliance need to pay special attention to data surety, authentication, reliability, accuracy, and transparency. For use in verification, GHG measurements must be accurate and have sufficient spatial and temporal coverage to distinguish human emissions from natural background variations. An additional and related concern relates to the need to monitor what different actors, including states and private organizations, are doing and how it interplays and affects overall policy goals. For example, a single country or even a private organization may attempt to implement large-scale solar radiation management (see Chapter 15). Observing systems designed to monitor and verify treaty compliance could also be used to monitor and support evaluations of the direct impacts and unintended consequences of such approaches. Better measurements will need to be integrated with a better understanding of how standards and certification can be used effectively to encourage compliance. Research is also needed on the links among measurements, effective enforcement strategies, and standards and certification.

Continue to improve methods for estimating costs, benefits, and cost effectiveness. Research on valuation is advancing in part through improved methodologies for eliciting stated preferences, especially through methods that draw on approaches from decision sciences (such as making valuation a problem for public deliberation) and in part by the accumulation of more valuation studies that make integration and cross-study comparisons (or meta-analyses) feasible. Finding appropriate discount rates and identifying appropriate ways to handle equity effects of climate policies are in part public choices, but a program of scientific analysis can both identify better ways to handle these issues in benefit-cost and cost-effectiveness analyses and develop tools for better assessing the appropriate values to use, including valid and reliable methods of eliciting preferences. Better characterization of uncertainty across all aspects of climate change science and better integration of uncertainty into analytical tools are also extremely important for improving policy design. Finally, better understanding and modeling of technological innovation could lead to more realistic estimation of costs.

Develop methods for analyzing complex, hybrid policies. Real policies include many complexities that address the needs and concerns of diverse sectors, regions, and interests. As a result, nearly all policies are hybrids that involve some elements of regulation and standards, some aspect of market incentives, and some degree of voluntary action on the part of individuals, firms, governments, and communities. Most existing policy analysis tools, such as benefit-cost analysis, were developed to examine simple policies that use a single modality to influence behavior, what might be termed an "ideal type" or "pure form" policy. However, in reality, many policies are complex. To provide useful analysis to decision makers shaping policy, scientists need to analyze the costs, benefits, and risks associated with complex hybrid forms of policy, which can sometimes be done with extensions of existing quantitative tools but may often require new, more advanced tools. Chapter 4 of the report discusses some of the tools currently available and under development that can be applied to this task.

Further understanding of how institutions interact in the context of multilevel governance and adaptive management. Recent research suggests that polycentric approaches to policy may be well suited to adaptive risk management. However, the interaction of multiple policies with multiple goals and approaches to affecting change, adopted and implemented in multiple contexts, can also lead to less-than-ideal results, and even situations where the outcome is wholly ineffective or even harmful. For example, policies that will change or affect water resource distributions across a multistate watershed (such as the Colorado River) will require enormous coordination among the affected states but will also have important implications for different water-dependent sectors (including agriculture, energy, flood management, industry and urban water users, and ecosystem managers) within any one state. Research is needed to characterize how policies interact across scales and intentions and to diagnose forms of policies that are most and least effective when implemented in the context of other policies. In addition, the problem of effectively linking scientific analysis to public deliberation becomes much more complex when there are multiple stakeholder groups involved, and especially when some of them reflect primarily local interests and perspectives while others are national or even global interests and perspectives (NRC, 2008h).

Develop analytical approaches that examine and evaluate climate policy taking into account its full range of effects including those on human well-being and ecosystems integrity, unintended consequences and equity effects. Policies rarely if ever do only one thing and rarely if ever affect everyone equally. Climate policy must be considered in the larger context of sustainable development (e.g., World Bank, 2009), and doing so will require assessment of the full range of effects of climate policy on human well-being and ecosystem health. Also needed are better metrics for comparing different outcomes, as noted above, and especially because regions, sectors, regions, local communities, and even different groups within communities will be differentially vulnerable to climate change and efforts to limit and adapt to it (see Chapter 16). It is also important to consider equity across social groups and time, so that current efforts to limit or adapt to climate change do not have major negative effects on human well-being and ecosystem health in several decades or centuries. Since equity effects are major elements in both the domestic and international debates about climate policy, a sounder understanding of these issues would aid in both the design of policy and in moving toward adoption and implementation.

References

Ackerman, K. V., and E. T. Sundquist. 2008. Comparison of two U.S. power-plant carbon dioxide emissions data sets. *Environmental Science & Technology* 42(15):5688-5693.

Adams, P. N., and D. L. Inman. 2009. *Climate Change and Potential Hotspots of Coastal Erosion Along the Southern California Coast—Final Report*. CEC-500-2009-022-F, Sacramento, California Energy Commission.

Adger, W. N., J. Paavola, S. Huq, and M. J. Mace, eds. 2006. *Fairness in Adaptation to Climate Change*. Cambridge, MA: MIT Press.

Adger, W. N., S. Agrawala, M. M. Q. Mirza, C. Conde, K. L. O'Brien, J. Pulhin, R. Pulwarty, B. Smit, and K. Takahashi. 2007. Assessment of adaptation practices, options, constraints and capacity. In *Climate Change 2007: Impacts, Adaptation and Vulnerability: Contribution of Working Group II to the Fourth Assessment Report of the Intergovernmental Panel on Climate Change*. M. L. Parry, O. F. Canziani, J. P. Palutikof, C. E. Hanson, and P. J. Van Der Linden, eds. Cambridge: Cambridge University Press.

Adger, W. N., I. Lorenzoni, and K. O'Brien. 2009a. Adaptation now. In *Adapting to Climate Change: Thresholds, Values, Governance*. W. N. Adger, I. Lorenzoni, and K. L. O'Brien, eds. Cambridge: Cambridge University Press.

Adger, W. N., S. Dessai, M. Goulden, M. Hulme, I. Lorenzoni, D. R. Nelson, L. O. Naess, J. Wolf, and A. Wreford. 2009b. Are there social limits to adaptation to climate change? *Climatic Change* 93(3-4):335-354.

Adler, M. D., and E. A. Posner. 2006. *New Foundations of Cost-Benefit Analysis*. Cambridge, MA: Harvard University Press.

Agrawala, S. 2004. Adaptation, development assistance and planning: Challenges and opportunities. *IDS Bulletin—Institute of Development Studies* 35:50-54.

Agrawal, A. 2008. The role of local institutions in adaptation to climate change. In *Social Dimensions of Climate Change Workshop*. Washington, DC: Social Development Department, The World Bank.

Agrawal, A., and N. Perrin. 2008. *Climate Adaptation, Local Institutions, and Rural Livelihoods*. Ann Arbor, MI: International Forestry Resources and Institutions Program, University of Michigan.

AGU (American Geophysical Union). 2009. *Geoengineering the Climate System. A Position Statement of the American Geophysical Union* (adopted by the AGU Council on December 13, 2009). Washington, DC: AGU.

Airame, S., J. E. Dugan, K. D. Lafferty, H. Leslie, D. A. Mcardle, and R. R. Warner. 2003. Applying ecological criteria to marine reserve design: A case study from the California Channel Islands. *Ecological Applications* 13 (1):S170-S184.

Akbari, H., M. Pomerantz, and H. Taha. 2001. Cool surfaces and shade trees to reduce energy use and improve air quality in urban areas. *Solar Energy* 70(3):295-310.

Akbari, H., S. Menon, and A. Rosenfeld. 2009. Global cooling: Increasing world-wide urban albedos to offset CO_2. *Climatic Change* 94(3-4):275-286.

Albrecht, A., D. Schindler, K. Grebhan, U. Kohnle, and H. Mayer. 2009. Storminess over the North-Atlantic European region under climate change—a review. *Allgemeine Forst Und Jagdzeitung* 180(5-6):109-118.

Aldy, J. E., and R. N. Stavins. 2007. *Architectures for Agreement*. Cambridge, MA: Cambridge University Press.

Alheit, J., and E. Hagen. 1997. Long-term climate forcing of European herring and sardine populations. *Fisheries Oceanography* 6:130-139.

Allan, J. D., M. A. Palmer, and N. L. Poff. 2005. Climate change and freshwater ecosystems. Pp. 272-290 in *Climate Change and Biodiversity*. T. E. Lovejoy and L. Hannah, eds. New Haven, CT: Yale University Press.

Alley, R. B., J. Marotzke, W. D. Nordhaus, J. T. Overpeck, D. M. Peteet, R. A. Pielke, Jr., R. T. Pierrehumbert, P. B. Rhines, T. F. Stocker, L. D. Talley, and J. M. Wallace. 2003. Abrupt climate change. *Science* 299(5615):2005-2010.

Alley, W. M. 1984. The Palmer Drought Severity Index—Limitations and assumptions. *Journal of Climate and Applied Meteorology* 23:1100-1109.

Alley, W. M., T. E. Reilly, and F. O. Lehn. 1999. Sustainability of ground-water resources. In *U.S. Geological Survey Circular*. Washington, DC: U.S. Geological Survey, Department of the Interior. Available at *http://purl.access.gpo.gov/GPO/LPS22429*. Accessed December 3, 2009.

AMS (American Meteorological Society). 2009. Geoengineering the Climate System. A Policy Statement of the American Meteorological Society (adopted by the AMS Council on July 20, 2009). Washington, DC: AMS.

Anderies, J. M., M. A. Janssen, and E. Ostrom. 2004. A framework to analyze the robustness of social-ecological systems from an institutional perspective. *Ecology and Society* 9(1):18. Available at *http://www.ecologyandsociety.org/vol9/iss1/art18/print.pdf*. Accessed May 10, 2010.

Anderson, P. J., and J. F. Piatt. 1999. Community reorganization in the Gulf of Alaska following ocean climate regime shift. *Marine Ecology-Progress Series* 189:117-123.

Andersson, K., and E. Ostrom. 2008. Analyzing decentralized resource regimes from a polycentric perspective. *Policy Sciences* 41(1):71-93.

Andreadis, K. M., and D. P. Lettenmaier. 2006. Trends in 20th century drought over the continental United States. *Geophysical Research Letters* 33(10).

Andreae, M. O., and P. Merlet. 2001. Emission of trace gases and aerosols from biomass burning. *Global Biogeochemical Cycles* 15(4):955-966.

Angel, S., S. Sheppard, and D. Civco. 2005. *The Dynamics of Global Urban Expansion*. Washington, DC: Department of Transport and Urban Development, The World Bank.

Angert, A., S. Biraud, C. Bonfils, C. C. Henning, W. Buermann, J. Pinzon, C. J. Tucker, and I. Fung. 2005. Drier summers cancel out the CO_2 uptake enhancement induced by warmer springs. *Proceedings of the National Academy of Sciences of the United States of America* 102:10823-10827.

Annan, J. D., J. C. Hargreaves, R. Ohgaito, A. Abe-Ouchi, and S. Emori. 2005. Efficiently constraining climate sensitivity with paleoclimate simulations. *Scientific Online Letters on the Atmosphere* 1:181-184.

Ansolabehere, S., and D. M. Koninsky. 2009. Public attitudes toward construction of new power plants. *Public Opinion Quarterly* 73(3):566-577.

Anthoff, D., R. S. J. Tol, and G.W. Yohe. 2009. Risk aversion, time preference, and the social cost of carbon. *Environmental Research Letters* 4(2):024002.

Anthoff, D., R. Nicholls, and R. Tol. 2010. The economic impact of substantial sea-level rise. *Mitigation and Adaptation Strategies for Global Change* 15(4):321-335.

Antle, J. M. 2009. *Agriculture and the Food System*. Washington, DC: Resources for the Future.

Archer, C. L., and M. Z. Jacobson. 2005. Evaluation of global wind power. *Journal of Geophysical Research* 110:D12110.

Archer, D., and B. Buffett. 2005. Time-dependent response of the global ocean clathrate reservoir to climatic and anthropogenic forcing. *Geochemistry, Geophysics, Geosystems* 6.

Armitage, D., F. Berkes, and N. Doubleday, eds. 2007. *Adaptive Co-Management: Collaboration, Learning, and Multi-Level Governance*. Seattle, WA: University of Washington Press.

Arneth, A., N. Unger, M. Kulmala, and M. O. Andreae. 2009. Clean the air, heat the planet? *Science* 326(5953):672-673.

Arriaga, L., A. E. Castellanos, E. Moreno, and J. Alarcon. 2004. Potential ecological distribution of alien invasive species and risk assessment: A case study of buffel grass in arid regions of Mexico. *Conservation Biology* 18(6):1504-1514.

Arrigo, K. R., G. Van Dijken, and S. Pabi. 2008. Impact of a shrinking Arctic ice cover on marine primary production. *Geophysical Research Letters* 35 (19).

Arrow, K., R. Solow, P. R. Portney, E. E. Leamer, R. Radner, and H. Schuman. 1993. Report of the NOAA Panel on Contingent Valuation. *Federal Register* 58(1993):4601-4614.

Arrow, K., P. Dasgupta, L. Goulder, G. Daily, P. Ehrlich, G. Heal, S. Levin, K.-G. Maler, S. Schneider, D. Starrett, and B. Walker. 2004. Are we consuming too much? *Journal of Economic Perspectives* 18(3):147-172.

Arunachalam, V. S., and E. L. Fleischer. 2008. The global energy landscape and materials innovation. *MRS Bulletin* 33(4):264-276.

Arvai, J., G Bridge, N. Dolsak, R. Franzese, T. Koontz, A. Luginbuhl, P. Robbins, K. Richards, K. Smith Korfmacher, B. Sohngen, J. Tansey, and A. Thompson. 2006. Adaptive management of the global climate problem: Bridging the gap between climate research and climate policy. *Climatic Change* 78:217-225.

ASCE (American Society of Civil Engineers). 2009. *Report Card for America's Infastructure*. Available at *http://www.infrastructurereportcard.org/*. Accessed April 29, 2010.

Asilomar Scientific Organizing Committee. 2010. Statement from the Asilomare International Conference on Climate Intervention Technologies, March 26, 2010. Climate Response Fund, *www.climateresponsefund.org*.

Asner, G. P. 2009. Tropical forest carbon assessment: Integrating satellite and airborne mapping approaches. *Environmental Research Letters* 4(3).

Atkinson, G., and S. Mourato. 2008. Environmental cost-benefit analysis. *Annual Review of Environment and Resources* 33(1):317-344.

Auffhammer, M., V. Ramanathan, and J. R. Vincent. 2006. Integrated model shows that atmospheric brown clouds and greenhouse gases have reduced rice harvests in India. *Proceedings of the National Academy of Sciences of the United States of America* 103(52):19668-19672.

Auld, G., L. H. Gulbrandsen, and C. L. McDermott. 2008. Certification schemes and the impacts on forests and forestry. *Annual Review of Environment and Resources* 33:187-211.

Backlund, P., A. Janetos, and D. Schimel. 2008. *Effects of Climate Change on Agriculture, Land Resources, Water Resources and Biodiversity in the United States.* Synthesis and Assessment Product 4.3. Washington, DC: U.S. Climate Change Science Program.

Bäckstrand, K. 2006. Democratizing global environmental governance? Stakeholder democracy after the World Summit on Sustainable Development. *European Journal of International Relations* 12(4):467-498.

Baer, P., J. Harte, B. Haya, A. V. Herzog, J. Holdren, N. E. Hultman, D. M. Kammen, R. B. Norgaard, and L. Raymond. 2000. Climate change: Equity and greenhouse gas responsibility. *Science* 289:2287.

Baker, D. J., R. W. Schmitt, and C. Wunsch. 2007. Endowments and new institutions for long-term observations. *Oceanography* 20(4):10-14.

Bakun, A. 1990. Global climate change and intensification of coastal ocean upwelling *Science* 247 (4939):198-201.

Bakun, A., and S. J. Weeks. 2004. Greenhouse gas buildup, sardines, submarine eruptions and the possibility of abrupt degradation of intense marine upwelling ecosystems. Ecology Letters 7 (11):1015-1023.

Bala, G., K. Caldeira, M. Wickett, T. J. Phillips, D. B. Lobell, C. Delire, and A. Mirin. 2007. Combined climate and carbon-cycle effects of large-scale deforestation. *Proceedings of the National Academy of Sciences of the United States of America* 104(16):6550-6555.

Baldocchi, D., and S. Wong. 2008. Accumulated winter chill is decreasing in the fruit growing regions of California. *Climatic Change* 87:S153-S166.

Bamber, J. L., R. L. Layberry, and S. P. Gogineni. 2001. A new ice thickness and bedrock data set for the Greenland ice sheet, measurement, data reduction, and errors. *Journal of Geophysical Research* 106(D24):33773-33733, 33780.

Bamber, J. L., R. E. M. Riva, B. L. A. Vermeersen, and A. M. Lebrocq. 2009. Reassessment of the potential sea-level rise from a collapse of the West Antarctic Ice Sheet. *Science* 324(5929):901-903.

Barbier, E. B., E. W. Koch, B. R. Silliman, S. D. Hacker, E. Wolanski, J. Primavera, E. F. Granek, S. Polasky, S. Aswani, L. A. Cramer, D. M. Stoms, C. J. Kennedy, D. Bael, C. V. Kappel, G. M. E. Perillo, and D. J. Reed. 2008. Coastal ecosystem-based management with nonlinear ecological functions and values. *Science* 319(5861):321-323.

Barker, T., I. Bashmakov, A. Alharthi, M. Amann, L. Cifuentes, J. Drexhage, M. Duan, O. Edenhofer, B. Flannery, M. Grubb, M. Hoogwijk, F. I. Ibitoye, C. J. Jepma, W. A. Pizer, and K. Yamaji. 2007a. Mitigation from a cross-sectoral perspective. In *Climate Change 2007: Mitigation. Contribution of Working Group III to the Fourth Assessment Report of the Intergovernmental Panel on Climate Change*. B. Metz, O. R. Davidson, P. R. Bosch, R. Dave, and L. A. Meyer, eds. Cambridge, U.K. and New York: Cambridge University Press.

Barnett, J. 2001. *The Meaning of Environmental Security: Ecological Politics and Policy in the New Security Era.* London: Zed Books.

Barnett, J. 2003. Security and climate change. *Global Environmental Change-Human and Policy Dimensions* 13(1):7-17.

Barnett, J. 2009. The prize of peace (is eternal vigilance): A cautionary editorial essay on climate geopolitics. *Climatic Change* 96(1-2):1-6.

Barnett, T. P., D. W. Pierce, K. M. Achuta Rao, P. J. Gleckler, B. D. Santer, J. M. Gregory, and W. M. Washington. 2005a. Penetration of a warming signal in the world's oceans: Human impacts. *Science* 309:284-287.

Barnett, T. P., J. C. Adam, and D. P. Lettenmaier. 2005b. Potential impacts of a warming climate on water availability in snow-dominated regions. *Nature* 438(7066):303-309.

Barnett, T. P., D. W. Pierce, H. G. Hidalgo, C. Bonfils, B. D. Santer, T. Das, G. Bala, A. W. Wood, T. Nozawa, A. A. Mirin, D. R. Cayan, and M. D. Dettinger. 2008. Human-induced changes in the hydrology of the western United States. *Science* 319(5866):1080-1083.

Bartlett, S. 2008. Climate change and urban children: Impacts and implications for adaptation in low- and middle-income countries. *Environment and Urbanization* 20(2):501-519.

Baskett, M. L., S. A. Levin, S. D. Gaines, and J. Dushoff. 2005. Marine reserve design and the evolution of size at maturation in harvested fish. *Ecological Applications* 15 (3):882-901.

Bates, B., and Z. Kundzewicz. 2008. *Climate Change and Water, IPCC Technical Paper 6*. Geneva: IPCC Secretariat.

Battin, J., M. Wiley, M. Ruckelshaus, R. Palmer, E. Korb, K. Bartz, and H. Imaki. 2007. Projected impacts of climate change on salmon habitat restoration. *Proceedings of the National Academy of Sciences of the United States of America* 104(16):6720-6725.

Batty, M. 2008. The size, scale, and shape of cities. *Science* 319(5864):769-771.

Bayley, P. B. 1981. Fish yield from the Amazon in Brazil—comparison with African river yields and management possibilities. *Transactions of the American Fisheries Society* 110(3):351-359.

Becker, G. S. 2010 (March 17). *U.S. Food and Agricultural Imports: Safeguards and Selected Issues* (RL34198). CRS (Congressional Research Service). Text available at LexisNexis Congressional Research Digital Collection. *http://www.nationalaglawcenter.org/assets/crs/RS22734.pdf*. Accessed July 13, 2010.

Beckley, B. D., F. G. Lemoine, S. B. Luthcke, R. D. Ray, and N. P. Zelensky. 2007. A reassessment of global and regional mean sea level trends from TOPEX and Jason-1 altimetry based on revised reference frame and orbits. *Geophysical Research Letters* 34(14).

Behrenfeld, M. J., R. T. O'malley, D. A. Siegel, C. R. Mcclain, J. L. Sarmiento, G. C. Feldman, A. J. Milligan, P. G. Falkowski, R. M. Letelier, and E. S. Boss. 2006. Climate-driven trends in contemporary ocean productivity. *Nature* 444 (7120):752-755.

Bell, M. L., F. Dominici, and J. M. Samet. 2005. A meta-analysis of time-series studies of ozone and mortality with comparison to the national morbidity, mortality, and air pollution study. *Epidemiology* 16(4):436-445.

Bell, M. L., R. D. Peng, and F. Dominici. 2006. The exposure-response curve for ozone and risk of mortality and the adequacy of current ozone regulations. *Environmental Health Perspectives* 114(4):532-536.

Bell, M. L., R. Goldberg, C. Hogrefe, P. L. Kinney, K. Knowlton, B. Lynn, J. Rosenthal, C. Rosenzweig, and J. A. Patz. 2007. Climate change, ambient ozone, and health in 50 US cities. *Climatic Change* 82(1-2):61-76.

Beller-Simms, N., H. Ingram, D. Feldman, N. Mantua, K. Jacobs, and A. Waples, eds. 2008. *Decision-Support Experiments and Evaluations Using Seasonal to Interannual Forecasts and Observational Data—A Focus on Water Resources*. Synthesis and Assessment Product 5.3. Washington, DC: U.S. Climate Change Science Program, 192 pp.

Bender, M. A., T. R. Knutson, R. E. Tuleya, J. J. Sirutis, G. A. Vecchi, S. T. Garner, and I. M. Held. 2010. Modeled impact of anthropogenic warming on the frequency of intense Atlantic hurricanes. *Science* 327(5964):454-458.

Benestad, R. E. 2005. A review of the solar cycle length estimates. *Geophysical Research Letters* 32:L15714, doi:10.1029/2005GL023621.

Beniston, M. 2004. The 2003 heat wave in Europe: A shape of things to come? An analysis based on Swiss climatological data and model simulations. *Geophysical Research Letters* 31(2).

Bernstein, P. L. 1998. *Against the Gods: The Remarkable Story of Risk*. New York: John Wiley.

Betsill, M. M. 2001. Mitigating climate change in US cities: Opportunities and obstacles. *Local Environment* 6:393-406.

Betsill, M. M., and H. Bulkeley. 2006. Cities and the multilevel governance of global climate change. *Global Governance* 12(2):141-159.

Betsill, M. M., and B. G. Rabe. 2009. Climate change and multi-level governance: The emerging state and local roles. In *Towards Sustainable Communities*, 2nd edition. D. A. Mazmanian and M. E. Kraft, eds. Cambridge, MA: MIT Press.

Betts, R. A., O. Boucher, M. Collins, P. M. Cox, P. D. Falloon, N. Gedney, D. L. Hemming, C. Huntingford, C. D. Jones, D. M. H. Sexton, and M. J. Webb. 2007. Projected increase in continental runoff due to plant responses to increasing carbon dioxide. *Nature* 448(7157):1037-1041.

Biermann, F., M. M. Bestill, J. Gupta, N. Kanie, L. Lebel, D. Liverman, H. Schroeder, and B. Siebenhüner. 2009a. *Earth System Governance: People, Places and the Planet. Science and Implementation Plan of the Earth System Governance Project.* Bonn: Earth Systems Governance Project.

Biermann, F., P. Pattberg, H. van Asselt, and F. Zelli. 2009b. The fragmentation of global governance architectures: A framework for analysis. *Global Environmental Politics* 9(4):14-40.

Biermann, F., M. M. Betsill, S. C. Vieira, J. Gupta, N. Kanie, L. Lebel, D. Liverman, H. Schroeder, B. Siebenhüner, P. Z. Yanda, and R. Zondervan. 2010. Navigating the anthropocene: The Earth System Governance Project strategy paper. *Current Opinion in Environmental Sustainability* 2(3): 202-208.

Bishr, M., and L. Mantelas. 2008. A trust and reputation model for filtering and classifying knowledge about urban growth. *GeoJournal* 72(3):229-237.

Bindoff, N. L., J. Willebrand, V. Artale, A. Cazenave, J. Gregory, S. Gulev, K. Hanawa, C. Le Quéré, S. Levitus, Y. Nojiri, C. K. Shum, L. D. Talley, and A. Unnikrishnan. 2007. Observations: Oceanic climate change and sea level. In *Climate Change 2007: The Physical Science Basis. Contribution of Working Group I to the Fourth Assessment Report of the Intergovernmental Panel on Climate Change* S. Solomon, D. Qin, M. Manning, Z. Chen, M. Marquis, K. B. Averyt, M.Tignor, and H. L. Miller, eds. Cambridge, U.K.: Cambridge University Press.

Bingen, J., and L. Busch, eds. 2006. *Agricultural Standards: The Shape of the Global Food and Fiber System.* Dordrecht, Netherlands: Springer.

Black, J. S., P. C. Stern, and J. T. Elworth. 1985. Personal and contextual influences on household energy adaptations. *Journal of Applied Psychology* 70(1):3-21.

Blackford, J. C., and F. J. Gilbert. 2007. pH variability and CO2 induced acidification in the North Sea. *Journal of Marine Systems* 64 (1-4):229-241.

Blasing, T. J. 2008. *Recent Greenhouse Gas Concentrations.* Carbon Dioxide Information Analysis Center, Oak Ridge National Laboratory. Available at *http://cdiac.ornl.gov/pns/current_ghg.html.* Accessed December 3, 2009.

Boden, T. A., G. Marland, and R. J. Andres. 2009. *Global, Regional, and National Fossil-Fuel CO_2 Emissions.* Carbon Dioxide Information Analysis Center, Oak Ridge National Laboratory, U.S. Department of Energy, Oak Ridge, TN, doi:10.3334/CDIAC/00001.

Boesch. D., J. C. Field, and D. Scavia. 2000. *The Potential Consequences of Climate Variability and Change on Coastal Areas and Marine Resources. Report of the Coastal Areas and Marine Resources Sector Team.* U.S. National Assessment of the Potential Consequences of Climate Variability and Change. Washintgon, DC: U.S. Global Change Research Program.

Bonan, G. B. 1999. Frost followed the plow: Impacts of deforestation on the climate of the United States. *Ecological Applications* 9(4):1305-1315.

Bonan, G. B. 2008. Forests and climate change: Forcings, feedbacks, and the climate benefits of forests. *Science* 320(5882):1444-1449.

Borgerson, S. G. 2008. Arctic meltdown—The economic and security implications of global warming. *Foreign Affairs* 87(2):63-77.

Borlaug, N. 2007. Feeding a hungry world. *Science* 318:359-359.

Bosello, F., R. Roson, and R. S. J. Tol. 2007. Economy-wide estimates of the implications of climate change: Sea level rise. *Environmental & Resource Economics* 37(3):549-571.

Bostrom, A., and D. Lashof. 2007. Weather it's climate change? Pp. 31-43 in *Creating a Climate for Change: Communicating Climate Change and Facilitating Social Change.* S. C. Moser and L. Dilling, eds. Cambridge, U.K.: Cambridge University Press.

Bostrom, A., M. G. Morgan, B. Fischhoff, and D. Read. 1994. What do people know about global climate change? 1. Mental models. *Risk Analysis* 14(6):959-970.

Bounoua, L., R. Defries, G. J. Collatz, P. Sellers, and H. Khan. 2002. Effects of land cover conversion on surface climate. *Climatic Change* 52(1-2):29-64.

Bower, K., T. Choularton, J. Latham, J. Sahraei, and S. Salter. 2006. Computational assessment of a proposed technique for global warming mitigation via albedo-enhancement of marine stratocumulus clouds. *Atmospheric Research* 82(1-2):328-336.

Boyer, T. P., S. Levitus, J. I. Antonov, R. A. Locarnini, and H. E. Garcia. 2005. Linear trends in salinity for the World Ocean, 1955-1998. *Geophysical Research Letters* 32:L01604, doi:10.1029/2004GL021791.

Boyle, E. A., and L. Keigwin. 1987. North Atlantic thermohaline circulation during the past 20,000 years linked to high-latitude surface temperature. *Nature* 330:35-40.

Braden, J. B., D. G. Brown, J. Dozier, P. Gober, S. M. Hughes, D. R. Maidment, S. L. Schneider, P. W. Schultz, J. S. Shortle, S. K. Swallow, and C. M. Werner. 2009. Social science in a water observing system. *Water Resources Research* 45.

Bradley, R. S., M. Vuille, H. F. Diaz, and W. Vergara. 2006. Threats to water supplies in the tropical Andes. *Science* 312(5781):1755-1756.

Brauch, H. G. 2005. *Threats, Challenges, Vulnerabilities and Risks in Environmental and Human Security*. Bonn, Germany: Institute for Environment and Human Security.

Brazel, A., N. Selover, R. Vose, and G. Heisler. 2000. The tale of two climates—Baltimore and Phoenix urban LTER sites. *Climate Research* 15(2):123-135.

Brechin, S. R. 2003. Comparative public opinion and knowledge on global climatic change and the Kyoto Protocol: The US versus the rest of the world? *International Journal of Sociology and Social Policy* 23(10):106-134.

Brewer, T. L. 2008. Climate change technology transfer: A new paradigm and policy agenda. *Climate Policy* 8(5):516-526.

Briffa, K. R., P. D. Jones, F. H. Schweingruber, and T. J. Osborn. 1998. Influence of volcanic eruptions on Northern Hemisphere summer temperature over the past 600 years. *Nature* 393(6684):450-455.

Broad, K., A. S. P. Pfaff, and M. H. Glantz. 2002. Effective and equitable dissemination of seasonal-to-interannual climate forecasts: Policy implications from the Peruvian fishery during El Nino 1997-98. *Climatic Change* 54(4):415-438.

Brody, S. D., S. Zahran, H. Grover, and A. Vedlitz. 2008. A spatial analysis of local climate change policy in the United States: Risk, stress, and opportunity. *Landscape and Urban Planning* 87(1):33-41.

Broecker, W. S. 1987. Unpleasant surprises in the greenhouse? *Nature* 328:123-126.

Broecker, W. S. 1997. Thermohaline circulation, the Achilles heel of our climate system: Will man-made CO_2 upset the current balance? *Science* 278(5343):1582-1588.

Broecker, W. S. 2002. *The Glacial World According to Wally*, 3rd edition. Palisades, NY: Eldigio Press.

Brooke, C. 2008. Conservation and adaptation to climate change. *Conservation Biology* 22:1471-1476.

Brüker, G. 2005. Vulnerable populations: Lessons learnt from the summer 2003 heat waves in Europe. *Eurosurveillance* 10(7):147.

Bruinsma, J. 2003. *World Agriculture: Towards 2015/2030: An FAO Perspective*. Rome: Food and Agriculture Organization of the United Nations.

Brundtland, G. H. 1987. *Report of the World Commission on Environment and Development: "Our Common Future."* New York: United Nations.

Brutsaert, W., and M. B. Parlange. 1998. Hydrologic cycle explains the evaporation paradox. *Nature* 396(6706):30.

Bryden, H. L., H. R. Longworth, and S. A. Cunningham. 2005. Slowing of the Atlantic meridional overturning circulation at 25 degrees North. *Nature* 438(7068):655-657.

BTS (Bureau of Transportation Statistics)/Federal Highway Administration. 2001. National Household Travel Survey (NHTS) 2001—National Data and Analysis Tool CD. Washington, DC.

Buesseler, K. O., S. C. Doney, D. M. Karl, P. W. Boyd, K. Caldeira, F. Chai, K. H. Coale, H. J. W. De Baar, P. G. Falkowski, K. S. Johnson, R. S. Lampitt, A. F. Michaels, S. W. A. Naqvi, V. Smetacek, S. Takeda, and A. J. Watson. 2008. Environment - Ocean iron fertilization - Moving forward in a sea of uncertainty. *Science* 319 (5860):162-162.

Bulkeley, H. 2001. Governing climate change: The politics of risk society. *Transactions of the Institute of British Geographers* 26(4):430-447.

Bulkeley, H. 2005. Reconfiguring environmental governance: Towards a politics of scales and networks. *Political Geography* 24(8):875-902.

Burby, R. J. 1998. *Cooperating with Nature: Confronting Natural Hazards with Land Use Planning for Sustainable Communities*. Washington, DC: Joseph Henry Press.

Burger, N., L. Ecola, T. Light, and M. Toman. 2009. *Evaluating Options for U.S. Greenhouse-Gas Mitigation Using Multiple Criteria*. Santa Monica, CA: RAND Corporation.

Burke, M. B., D. B. Lobell, and L. Guarino. 2009. Shifts in African crop climates by 2050, and the implications for crop improvement and genetic resources conservation. *Global Environmental Change* 19(3):317-325.

Burroughs, W. 2003. *Climate into the 21st Century*. World Meteorological Organization. Cambridge: Cambridge University Press, 240 pp.

Burton I, L. Bizikova, T. Dickinson, and Y. Howard. 2007. Integrating adaptation into policy: Upscaling evidence from local to global. *Climate Policy* 7:371-376.

Busby, J. 2007. *Climate Change and National Security: An Agenda for Action*. New York: Council on Foreign Relations.

CAA (Clean Air Act). 1990. P. L. 101-549.

Caldeira, K., and L. Wood. 2008. Global and Arctic climate engineering: Numerical model studies. *Philosophical Transactions of the Royal Society A: Mathematical, Physical and Engineering Sciences* 366(1882):4039-4056.

Camp, C. D., and K. K. Tung. 2007. Surface warming by the solar cycle as revealed by the composite mean difference projection. *Geophysical Research Letters* 34(14).

Campbell, K. M., J. Gulledge, J. R. Mcneill, J. Podesta, P. Ogden, L. Fuerth, R. J. Woolsey, A. T. Lennon, J. Smith, and R. Weitz. 2007. *The Age of Consequences: The Foreign Policy and National Security Implications of Global Climate Change*. Ft. Belvoir, VA: Defense Technical Information Center.

Campbell-Lendrum, D., and C. Corvalán 2007. Climate change and developing-country cities: Implications for environmental health and equity. *Journal of Urban Health* 84:109-117.

Campbell, J. E., D. B. Lobell, and C. B. Field. 2009. Greater transportation energy and GHG offsets from bioelectricity than ethanol. *Science* 324(5930):1055-1057.

Canadell, J. G., and M. R. Raupach. 2008. Managing forests for climate change mitigation. *Science* 320(5882):1456-1457.

Canadell, J. G., C. Le Quere, M. R. Raupach, C. B. Field, E. T. Buitenhuis, P. Ciais, T. J. Conway, N. P. Gillett, R. A. Houghton, and G. Marland. 2007. Contributions to accelerating atmospheric CO_2 growth from economic activity, carbon intensity, and efficiency of natural sinks. *Proceedings of the National Academy of Sciences of the United States of America* 104(47):18866-18870.

Cao, L., and K. Caldeira. 2008. Atmospheric CO2 stabilization and ocean acidification. *Geophysical Research Letters* 35 (19):5.

Carrico, A., M. P. Vandenbergh, P. C. Stern, G. T. Gardner, T. Dietz, and J. M. Gilligan. 2010. Energy and climate change: Key lessons for implementing the behavioral wedge. *Journal of Energy and Environmental Law* 1(10-24).

Carroll, A., and C. Somerville. 2009. Cellulosic biofuels. *Annual Review of Plant Biology* 60(1):165-182.

Carson, R. T. 1997. Contingent valuation: Theoretical advances and empirical tests since the NOAA panel. *American Journal of Agricultural Economics* 79(5):1501-1507.

Carson, R. T. 2010. The environmental Kuznets curve: Seeking empirical regularity and theoretical structure. *Review of Environmental Economics and Policy* 4:3-23.

Cash, D. W. 2001. "In order to aid in diffusing useful and practical information": Agricultural extension and boundary organizations. *Science, Technology & Human Values* 26(4):431-453.

Cash, D. W., and J. Buizer. 2005. *Knowledge-Action Systems for Seasonal to Interannual Climate Forecasting: Summary of a Workshop*. Report to the Roundtable on Science and Technology for Sustainability, Policy and Global Affairs. Washington, DC: The National Academies Press.

Cash, D. W., W. C. Clark, F. Alcock, N. M. Dickson, N. Eckley, D. H. Guston, J. Jäger, and R. B. Mitchell. 2003. Knowledge systems for sustainable development. *Proceedings of the National Academy of Sciences of the United States of America* 100(14):8086-8091.

Cash, D. W., J. C. Borck, and A. G. Patt. 2006a. Countering the loading-dock approach to linking science and decision making—comparative analysis of El Nino/Southern Oscillation (ENSO) forecasting systems. *Science, Technology & Human Values* 31(4):465-494.

Cash, D. W., N. W. Adger, F. Berkets, P. Garden, L. Lebel, P. Olsson, L. Pritchard, and O. Young. 2006b. Scale and cross-scale dynamics: Governance and information in a multilevel world. *Ecology and Society* 11(2).

Cashore, B. 2002. Legitimacy and the privatization of environmental governance: How non-state market-driven (NSMD) governance systems gain rule-making authority. *Governance* 15(4):503-529.

Catarious, D. M., Jr., and R. H. Espach. 2009. *Impacts of Climate Change on Colombia's National and Regional Security*. CNA Corporation. Available at *http://www.dtic.mil/cgi-bin/GetTRDoc?AD=ADA509086&Location=U2&doc=GetTRDoc.pdf*. Accessed May 13, 2010.

Cavlovic, T., K. H. Baker, R. P. Berrens, and K. Gawande. 2000. A meta-analysis of Kuznets curve studies. *Agriculture and Resource Economics Review* 29:32-42.

Cayan, D., M. Tyree, M. Dettinger, H. Hidalgo, T. Das, E. Maurer, P. Bromirski, N. Graham, and R. Flick. 2009. *Climate Change Scenarios and Sea Level Rise Estimates for the California 2008 Climate Change Scenarios Assessment*. PIER Research Report CEC-500-2009-014. Sacramento, CA: California Energy Commission.

Cazenave, A., and R. S. Nerem. 2004. Present-day sea level change: Observations and causes. *Reviews of Geophysics* 42(3).

Cazenave, A., D. P. Chambers, P. Cipollini, L. L. Fu, J. W. Hurell, M. Merrifield, S. Nerem, H. P. Plag, C. K. Shum, and J. Willis. 2010. The Challenge for Measuring Sea Level Rise and Regional and Global Trends. Plenary Paper presented at OceanObs09, September 21-25, 2009, Venice, Italy.

CCSP (Climate Change Science Program and the Subcommittee on Global Change Research). 2003. Strategic plan for the U.S. Climate Change Science Program Washington, D.C: U.S. Climate Change Science Program.

CCSP (Climate Change Science Program). 2007a. *The First State of the Carbon Cycle Report (SOCCR): The North American Carbon Budget and Implications for the Global Carbon Cycle*. Synthesis and Assessment Product 2.2. A Report by the U.S. Climate Change Science Program and the Subcommittee on Global Change Research. A. W. King, L. Dilling, G. P. Zimmerman, D. M. Fairman, R. A. Houghton, G. Marland, A. Z. Rose, and T. J. Wilbanks, eds. Asheville, NC: National Climatic Data Center, National Oceanic and Atmospheric Administration, 242 pp.

CCSP. 2007b. *Global Change Scenarios: Their Development and Use*. Sub-report 2.1B of Synthesis and Assessment Product 2.1 by the U.S. Climate Change Science Program and the Subcommittee on Global Change Research, E. A. Parson, V. R. Burkett, K. Fisher-Vanden, D. W. Keith, L. O. Mearns, H. M. Pitcher, C. E. Rosenzweig, and M. D. Webster, eds. Washington, DC: Office of Biological and Environmental Research, Department of Energy.

CCSP. 2007c. *Scenarios of Greenhouse Gas Emissions and Atmospheric Concentrations*. Sub-report 2.1A of Synthesis and Assessment Product 2.1 by the U.S. Climate Change Science Program and the Subcommittee on Global Change Research, L. Clarke, J. Edmonds, H. Jacoby, H. Pitcher, J. Reilly, and R. Richels, eds. Washington, DC: Office of Biological and Environmental Research, Department of Energy.

CCSP. 2008a. *Analyses of the Effects of Global Change on Human Health and Welfare and Human Systems*. Synthesis and Assessment Product 4.6 by the U.S. Climate Change Science Program and the Subcommittee on Global Change Research, J. L. Gamble, ed., and K. L. Ebi, F. G. Sussman, and T. J. Wilbanks, authors. Washington, DC: U.S. Environmental Protection Agency.

CCSP. 2008b. *The Effects of Climate Change on Agriculture, Land Resources, Water Resources, and Biodiversity in the United States*. Synthesis and Assessment Product 4.3 by the U.S. Climate Change Science Program and the Subcommittee on Global Change Research, P. Backlund, A. Janetos, D. Schimel, J. Hatfield, K. Boote, P. Fay, L. Hahn, C. Izaurralde, B. A. Kimball, T. Mader, J. Morgan, D. Ort, W. Polley, A. Thomson, D. Wolfe, M. G. Ryan, S. R. Archer, R. Birdsey, C. Dahm, L. Heath, J. Hicke, D. Hollinger, T. Huxman, G. Okin, R. Oren, J. Randerson, W. Schlesinger, D. Lettenmaier, D. Major, L. Poff, S. Running, L. Hansen, D. Inouye, B. P. Kelly, L. Meyerson, B. Peterson, and R. Shaw, eds. Washington, DC: U.S. Department of Agriculture, 362 pp.

CCSP. 2008c. *Climate Models: An Assessment of Strengths and Limitations*. Synthesis and Assessment Product 3.1 by the U.S. Climate Change Science Program, D. C. Bader, C. Covey, W. J. Gutowski, I. M. Held, K. E. Kunkel, R. L. Miller, R. T. Tokmakian, and M. H. Zhang, authors. Washington, DC: U.S. Department of Energy.

CCSP. 2008d. Preliminary Review of Adaptation Options for Climate-Sensitive Ecosystems and Resources. Synthesis and Assessment Product 4.4 by the U.S. Climate Change Science Program and the Subcommittee on Global Change Research, S. H. Julius and J. M. West, eds.; J. S. Baron, L. A. Joyce, P. Kareiva, B. D. Keller, M. A. Palmer, C. H. Peterson, and J. M. Scott, authors. Washington, DC: U.S. Environmental Protection Agency, 873 pp.

CCSP. 2008e. *Trends in Emissions of Ozone-Depleting Substances, Ozone Layer Recovery, and Implications for Ultraviolet Radiation Exposure*. Synthesis and Assessment Product 2.4 by the U.S. Climate Change Science Program and the Subcommittee on Global Change Research, A. R. Ravishankara, M. J. Kurylo, and C. A. Ennis, eds. Asheville, NC: National Climatic Data Center, NOAA, Department of Commerce, 240 pp.

CCSP. 2008f. *Weather and Climate Extremes in a Changing Climate. Regions of Focus: North America, Hawaii, Caribbean, and U.S. Pacific Islands*. Synthesis and Assessment Product 3.3 by the U.S. Climate Change Science Program and the Subcommittee on Global Change Research, T. R. Karl, G. A. Meehl, C. D. Miller, S. J. Hassol, A. M. Waple, and W. L. Murray, eds. Washington, DC: National Climatic Data Center, NOAA, Department of Commerce, 164 pp.

CCSP, 2009a. *Coastal Sensitivity to Sea-Level Rise: A Focus on the Mid-Atlantic Region*. Synthesis and Assessment Product 4.1 by the U.S. Climate Change Science Program and the Subcommittee on Global Change Research, J. G. Titus (coordinating lead author), K. E. Anderson, D. R. Cahoon, D. B. Gesch, S. K. Gill, B. T. Gutierrez, E. R. Thieler, and S. J. Williams (lead authors). Washington, DC: U.S. Environmental Protection Agency, 320 pp.

CCSP. 2009b. *Thresholds of Climate Change in Ecosystems*. Synthesis and Assessment Product 4.2 by the U.S. Climate Change Science Program and the Subcommittee on Global Change Research, D. B. Fagre, C. W. Charles, C. D. Allen, C. Birkeland, F. S. Chapin III, P. M. Groffman, G. R. Guntenspergen, A. K. Knapp, A. D. McGuire, P. J. MulhollandD. P. C. Peters, D. D. Roby, and G. Sugihara, eds. Washington, DC: U.S. Geological Survey, Department of the Interior.

CDC (Centers for Disease Control and Prevention). 2006. Heat-related deaths—United States, 1999-2003. *Morbidity and Mortality Weekly Report* 55(29):796-798. Available at *http://www.cdc.gov/mmwr/preview/mmwrhtml/mm5529a2.htm*. Accessed on October 26, 2010.

CDWR (California Deptartment of Water Resources). 2008. *Managing an Uncertain Future: Climate Change Adaptation Strategies for California's Water*. Sacremento, CA: CDWR. Available at *http://www.water.ca.gov/climatechange/docs/ClimateChangeWhitePaper.pdf*. Accessed on October 26, 2010..

CNA Corporation. 2007. *National Security and the Threat of Climate Change*. Available at *http://securityandclimate.cna.org/report/*. Accessed May 13, 2010.

CNA Corporation. 2009. *Socioeconomic and Security Implications of Climate Change in China*. Conference Paper. Alexandria, VA: CNA Corporation, 24 pp.

Chameides, W. L., P. S. Kasibhatla, J. Yienger, and H. Levy. 1994. Growth of continental-scale metro-agro-plexes, regional ozone pollution and world food-production. *Science* 264(5155):74-77.

Chan, F., J. A. Barth, J. Lubchenco, A. Kirincich, H. Weeks, W. T. Peterson, and B. A. Menge. 2008. Emergence of Anoxia in the California Current Large Marine Ecosystem. *Science* 319 (5865):920.

Changnon, S. 1969. Recent studies of urban effects on precipitation in the U.S. *Bulletin of the American Meteorological Society* 50(6):411-421.

Changnon, S. A., K. E. Kunkel, and B. C. Reinke. 1996. Impacts and responses to the 1995 heat wave: A call to action. *Bulletin of the American Meteorological Society* 77(7):1497-1506.

Chao, B. F., Y. H. Wu, and Y. S. Li. 2008. Impact of artificial reservoir water impoundment on global sea level. *Science* 320(5873):212-214.

Chapin, F. S. I., P. A. Matson, and H. A. Mooney. 2002. *Principles of Terrestrial Ecosystem Ecology*. New York: Springer.

Chapin, F. S., T. S. Rupp, A. M. Starfield, L. O. Dewilde, E. S. Zavaleta, N. Fresco, J. Henkelman, and A. D. Mcguire. 2003. Planning for resilience: Modeling change in human-fire interactions in the Alaskan boreal forest. *Frontiers in Ecology and the Environment* 1(5):255-261.

Chavez, F. P., J. Ryan, S. E. Lluch-Cota, and M. Niquen C. 2003. From anchovies to sardines and back: Multidecadal change in the Pacific Ocean. *Science* 299(5604):217-221.

Cheung, W. W. L., V. W. Y. Lam, J. L. Sarmiento, K. Kearney, R. Watson, and D. Pauly. 2009. Projecting global marine biodiversity impacts under climate change scenarios. *Fish and Fisheries* 10(3):235-251.

Chhatre, A., and A. Agrawal. 2008. Forest commons and local enforcement. *Proceedings of the National Academy of Sciences of the United States of America* 105(36):13286-13291.

Christensen, J. H., B. Hewitson, A. Busuioc, A. Chen, X. Gao, I. Held, R. Jones, R. K. Kolli, W.-T. Kwon, R. Laprise, V. Magaña Rueda, L. Mearns, C. G. Menéndez, J. Räisänen, A. Rinke, A. Sarr, and P. Whetton. 2007. Regional climate projections. In *Climate Change 2007: The Physical Science Basis. Contribution of Working Group I to the Fourth Assessment Report of the Intergovernmental Panel on Climate Change*, S. Solomon, D. Qin, M. Manning, Z. Chen, M. Marquis, K. B. Averyt, M. Tignor, and H. L. Miller, eds. Cambridge, U.K. and New York: Cambridge University Press.

Christensen, N. S., A. W. Wood, N. Voisin, D. P. Lettenmaier, and R. N. Palmer. 2004 The effects of climate change on the hydrology and water resources of the Colorado River basin. *Climatic Change* 62(1-3):337-363.

Christy, J. R., R. W. Spencer, and W. D. Braswell. 2000. MSU tropospheric temperatures: Dataset construction and radiosonde comparisons. *Journal of Atmospheric and Oceanic Technology* 17:1153-1170.

Christy, J. R., R. W. Spencer, W. B. Norris, and W. D. Braswell. 2003. Error estimates of version 5.0 of MSU-AMSU bulk atmospheric temperatures. *Journal of Atmospheric and Oceanic Technology* 20:613-629.

Church, J. A., and N. J. White. 2006. A 20th century acceleration in global sea-level rise. *Geophysical Research Letters* 33(1).

Church, J. A., N. J. White, T. Aarup, W. S. Wilson, P. L. Woodworth, C. M. Domingues, J. R. Hunter, and K. Lambeck. 2008. Understanding global sea levels: Past, present and future. *Sustainability Science* 3(1):9-22, doi:10.1007/s11625-008-0042-4.

Cifuentes, L., V. H. Borja-Aburto, N. Gouveia, G. Thurston, and D. L. Davis. 2001a. Assessing the health benefits of urban air pollution reductions associated with climate change mitigation (2000-2020): Santiago, Sao Paulo, Mexico City, and New York City. *Environmental Health Perspectives* 109(Suppl 3):419-425.

Cifuentes, L., V. H. Borja-Aburto, N. Gouveia, G. Thurston, and D. Lee Davis. 2001b. Hidden health benefits of greenhouse gas mitigation. *Science* 293(5533):1257-1259.

Clark, B., A. Jorgenson, and J. Kentor. 2010. Militarization and energy consumption: A test of treadmill of destruction theory in comparative perspective. *International Journal of Sociology* 40(2):23-43.

Clark, G., S. C. Moser, S. J. Ratick, K. Dow, W. B. Meyer, S. Emani, W. Jin, J. X. Kasperson, R. E. Kasperson, and H. E. Schwarz. 1998. Assessing the vulnerability of coastal communities to extreme storms: The case of Revere, MA, USA. *Mitigation and Adaptation Strategies for Global Change* 3:59-82.

Clarke, L., J. Edmonds, V. Krey, R. Richels, S. Rose, and M. Tavoni. 2009. International climate policy architectures: Overview of the EMF 22 international scenarios. *Energy Economics* 31(2):S64-S81.

Cohen, B. 2006. Urbanization in developing countries: Current trends, future projections, and key challenges for sustainability. *Technology in Society* 28:63-80.

Cole, C. V., J. Duxbury, J. Freney, O. Heinemeyer, K. Minami, A. Mosier, K. Paustian, N. Rosenberg, N. Sampson, D. Sauerbeck, and Q. Zhao. 1997. Global estimates of potential mitigation of greenhouse gas emissions by agriculture. *Nutrient Cycling in Agroecosystems* 49(1-3):221-228.

Cole, M. A., and E. Neumayer. 2004. Examining the impact of demographic factors on air pollution. *Population and Environment* 26(1):5-21.

Collins, T. W. 2005. Households, forests, and fire hazard vulnerability in the American West: A case study of a California community. *Global Environmental Change Part B: Environmental Hazards* 6(1):23-37.

Confalonieri, U. 2007. Specific vulnerabilities. *Bulletin of the World Health Organization* 85:827-828.

Confalonieri, U., B. Menne, R. Akhtar, K. L. Ebi, M. Hauengue, R. S. Kovats, B. Revich, and A. Woodward. 2007. Human health. Pp. 391-431 in *Climate Change 2007: Impacts, Adaptation and Vulnerability. Contribution of Working Group II to the Fourth Assessment Report of the Intergovernmental Panel on Climate Change.* M. L. Parry, O. F. Canziani, J. P. Palutikof, P. J. van der Linden, and C. E. Hanson, eds. Cambridge, U.K.: Cambridge University Press.

Conroy, M. S. 2006. *Branded: How the "Certification Revolution" Is Transforming Global Corporations.* Gabriola Island, BC: New Society, Xvi, 335 pp.

Cook, B. I., R. L. Miller, and R. Seager. 2009. Amplification of the North American "Dust Bowl" drought through human-induced land degradation. *Proceedings of the National Academy of Sciences of the United States of America* 106(13):4997-5001.

Cooper, T. F., G. De'Ath, K. E. Fabricius, and J. M. Lough. 2008. Declining coral calcification in massive Porites in two nearshore regions of the northern Great Barrier Reef. *Global Change Biology* 14 (3):529-538.

Cooper, O. R., D. D. Parrish, A. Stohl, M. Trainer, P. Nedelec, V. Thouret, J. P. Cammas, S. J. Oltmans, B. J. Johnson, D. Tarasick, T. Leblanc, I. S. Mcdermid, D. Jaffe, R. Gao, J. Stith, T. Ryerson, K. Aikin, T. Campos, A. Weinheimer, and M. A. Avery. 2010. Increasing springtime ozone mixing ratios in the free troposphere over western North America. *Nature* 463(7279):344-348.

Costanza, R., W. J. Mitsch, and J. W. Day. 2006. A new vision for New Orleans and the Mississippi delta: Applying ecological economics and ecological engineering. *Frontiers in Ecology and the Environment* 4(9):465-472.

Crabtree, G. W., and N. S. Lewis. 2007. Solar energy conversion. *Physics Today* 60(3):37-42.

Cramer, J. C., B. Hackett, P. P. Craig, E. Vine, M. Levine, T. M. Dietz, and D. Kowalczyk. 1984. Structural behavioral determinants of residential energy use: Summer electricity use in Davis. *Energy* 9(3):207-216.

Critchley, C. R. 2008. Public opinion and trust in scientists: The role of the research context, and the perceived motivation of stem cell researchers. *Public Understanding of Science* 17(3):309-327.

Crossett, K. M., T. J. Culliton, P. C. Wiley, and T. R. Goodspeed. 2005. *Population Trends Along the Coastal United States: 1980-2008*. Silver Spring, MD: National Ocean Service, NOAA, U.S. Department of Commerce.

Crowell, M., S. Edelman, K. Coulton, and S. McAfee. 2007. How many people live in coastal areas? *Journal of Coastal Research* 23(5):iii-vi.

Crowell, M., K. Coulton, C. Johnson, J. Westcott, D. Bellomo, S. Edelman, and E. Hirsch. 2010. An Estimate of the U.S. Population Living in 100-Year Coastal Flood Hazard Areas. *Journal of Coastal Research* 26(2): 201-211.

Crutzen, P. 2006. Albedo enhancement by stratospheric sulfur injections: A contribution to resolve a policy dilemma? *Climatic Change* 77(3):211-220.

Crutzen, P. J., A. R. Mosier, K. A. Smith, and W. Winiwarter. 2008. N_2O release from agro-biofuel production negates global warming reduction by replacing fossil fuels. *Atmospheric Chemistry and Physics* 8(2):389-395.

Csatho, B. M., J. F. Bolzan, C. J. Van Der Veen, A. F. Schenk, and D.-C. Lee. 1999. Surface velocities of a Greenland outlet glacier from high-resolution visible satellite imagery. *Polar Geography* 23(1):71-82.

Curriero, F. C., K. S. Heiner, J. M. Samet, S. L. Zeger, L. Strug, and J. A. Patz. 2002. Temperature and mortality in 11 cities of the eastern United States. *American Journal of Epidemiology* 155:80-87.

Curry, R., B. Dickson, and I. Yashayaev. 2003. A change in the freshwater balance of the Atlantic Ocean over the past four decades. *Nature* 426:826-829.

Cutter, S. L., J. T. Mitchell, and M. S. Scott. 2000. Revealing the vulnerability of people and places: A case study of Georgetown County, South Carolina. *Annals of the Association of American Geographers* 90(4):713-737.

Cvetkovich, G., and R. E. Loefstedt, eds. 1999. *Social Trust and the Management of Risk*. London: Earthscan.

Dabelko, G. D. 2009. Planning for climate change: The security community's precautionary principle. *Climatic Change* 96(1-2):13-21.

Dai, A. G., K. E. Trenberth, and T. T. Qian. 2004. A global dataset of Palmer Drought Severity Index for 1870-2002: Relationship with soil moisture and effects of surface warming. *Journal of Hydrometeorology* 5(6):1117-1130.

Dai, A. G., T. T. Qian, K. E. Trenberth, and J. D. Milliman. 2009. Changes in continental freshwater discharge from 1948 to 2004. *Journal of Climate* 22(10):2773-2792.

Daily, G. C., and P. A. Matson. 2008. Ecosystem services: From theory to implementation. *Proceedings of the National Academy of Sciences of the United States of America* 105(28):9455-9456.

Daily, G. C., S. Polasky, J. Goldstein, P. M. Kareiva, H. A. Mooney, L. Pejchar, T. H. Ricketts, J. Salzman, and R. Shallenberger. 2009. Ecosystem services in decision making: Time to deliver. *Frontiers in Ecology and the Environment* 7(1):21-28.

Dailey, P. S., G. Zuba, G. Ljung, I. M. Dima, and J. Guin. 2009. On the relationship between North Atlantic sea surface temperatures and US hurricane landfall risk. *Journal of Applied Meteorology and Climatology* 48(1):111-129.

Dalby, S. 2009. *Security and Environmental Change*. Cambridge: Polity Press.

Daley, R. 1991. *Atmospheric Data Analysis*. Cambridge Atmospheric and Space Science Series. New York: Cambridge University Press, 457 pp.

Darwin, R., M. Tsigas, J. Lewandrowski, and A. Raneses. 1995. *World Agriculture and Climate Change: Economic Adaptations*. Agricultural Economics Reports 33933, Economic Research Service, U.S. Department of Agriculture.

Darwin, R. F., and R. S. J. Tol. 2001. Estimates of the economic effects of sea level rise. *Environmental and Resource Economics* 19(2):113-129.

Das, S., and J. R. Vincent. 2009. Mangroves protected villages and reduced death toll during Indian super cyclone. *Proceedings of the National Academy of Sciences of the United States of America* 106(18):7357-7360.

Dasgupta, P., and F. Ramsey. 2008. *Comments on the Stern Review's Economics of Climate Change*. Cambridge, U.K.: Cambridge University Press. Available at *http://www.econ.cam.ac.uk/faculty/dasgupta/STERN.pdf*. Accessed May 7, 2010.

Dasgupta, S., B. Laplante, H. Wang, and D. Wheeler. 2002. Confronting the environmental Kuznets curve. *Journal of Economic Perspectives* 16(1):147-168.

Davis, B., and D. Belkin. 2008. Food inflation, riots spark worries for world leaders: IMF, World Bank push for solutions; Turmoil in Haiti. *The Wall Street Journal*, April 14.

Davis, R. E., P. C. Knappenberger, P. J. Michaels, and W. M. Novicoff. 2004. Seasonality of climate-human mortality relationships in US cities and impacts of climate change. *Climate Research* 26(1):61-76.

Davis, S. C., S. W. Diegel, and R. G. Boundy. 2008. *Transportation Energy Data Book*, 27th ed. ORNL-6981. Washington, DC: U.S. Department of Energy.

Davis, S. C., K. J. Anderson-Teixeira, and E. H. DeLucia. 2009. Life-cycle analysis and the ecology of biofuels. *Trends in Plant Science* 14(3):140-146.

Day, J. W., D. F. Boesch, E. J. Clairain, G. P. Kemp, S. B. Laska, W. J. Mitsch, K. Orth, H. Mashriqui, D. J. Reed, L. Shabman, C. A. Simenstad, B. J. Streever, R. R. Twilley, C. C. Watson, J. T. Wells, and D. F. Whigham. 2007. Restoration of the Mississippi Delta: Lessons from Hurricanes Katrina and Rita. *Science* 315(5819):1679-1684.

Delta Vision Blue Ribbon Task Force. 2007. *Delta Vision: Our Vision for the California Delta*. Sacramento, CA: Resources Agency.

Defries, R., and L. Bounoua. 2004. Consequences of land use change for ecosystem services: A future unlike the past. *GeoJournal* 61(4):345-351.

Defries, R., F. Achard, S. Brown, M. Herold, D. Murdiyarso, B. Schlamadinger, and C. De Souza. 2007. Earth observations for estimating greenhouse gas emissions from deforestation in developing countries. *Environmental Science and Policy* 10(4):385-394.

DeFries, R. S., T. Rudel, M. Uriarte, and M. Hansen. 2010. Deforestation driven by urban population growth and agricultural trade in the twenty-first century. *Nature Geoscience* 3:178-181, doi:10.1038/NGEO756.

Denholm, P., E. Ela, B. Kirby, and M. Milligan. 2010. *Role of Energy Storage with Renewable Electricity Generation*. NREL Report TP-6A2-47187. Washington, DC: U.S. Department of Energy, 61 pp.

Department of Commerce. 1975. Bureau of the Census, Historical Statistics of the United States.

Dessai, S., and M. Hulme. 2007. Assessing the robustness of adaptation decisions to climate change uncertainties: A case study on water resources management in the east of England. *Global Environmental Change—Human and Policy Dimensions* 17(1):59-72.

Dessai, S., W. N. Adger, M. Hulme, J. Turnpenny, J. Kohler, and R. Warren. 2004. Defining and experiencing dangerous climate change—an editorial essay. *Climatic Change* 64(1-2):11-25.

Diamond, J. M. 2005. *Collapse: How Societies Choose to Fail or Succeed*. New York: Viking.

Dickson, B., I. Yashayaev, J. Meincke, B. Turrell, S. Dye, and J. Holfort. 2002. Rapid freshening of the deep North Atlantic Ocean over the past four decades. *Nature* 416:832-837.

Dietz, S., and N. Stern. 2008. Why economic analysis supports strong action on climate change: A response to the Stern Review's critics. *Review of Environmental Economics and Policy 2*(1):94-113.

Dietz, T. 1994. What should we do? Human ecology and collective decision making. *Human Ecology Review* 1:301-309.

Dietz, T., and A. D. Henry. 2008. Context and the commons. *Proceedings of the National Academy of Sciences of the United States of America* 105:13189-13190.

Dietz, T., E. Ostrom, and P. C. Stern. 2003. The struggle to govern the commons. *Science* 302(5652):1907-1912.

Dietz, T., E. A. Rosa, and R. York. 2007. Driving the human ecological footprint. *Frontiers in Ecology and the Environment* 5(1):13-18.

Dietz, T., E. A. Rosa, and R. York. 2009a. Environmentally efficient well-being: Rethinking sustainability as the relationship between human well-being and environmental impacts. *Human Ecology Review* 16(1):114-123.

Dietz, T., G. T. Gardner, J. Gilligan, P. C. Stern, and M. P. Vandenbergh. 2009b. Household actions can provide a behavioral wedge to rapidly reduce US carbon emissions. *Proceedings of the National Academy of Sciences of the United States of America* 106(44):18452-18456.

Dietz, T., E. A. Rosa, and R. York. 2009c. Human driving forces of global change: Dominant perspectives. In *Human Footprints on the Global Environment: Threats to Sustainability*, E. A. Rosa, A. Diekmann, T. Dietz, and C. C. Jaeger, eds. Cambridge, MA: MIT Press.

Diffenbaugh, N. S., C. H. Krupke, M. A. White, and C. E. Alexander. 2008. Global warming presents new challenges for maize pest management. *Environmental Research Letters* 3(4).

Dilling, L., and B. Farhar. 2007. Making it easy: Establishing energy efficiency and renewable energy as routine best practice. In *Creating a Climate for Change: Communicating Climate Change and Facilitating Social Change*, S. Moser and L. Dilling, eds. Cambridge, U.K.: Cambridge University Press.

Dlugokencky, E. J., S. Houweling, L. Bruhwiler, K. A. Masarie, P. M. Lang, J. B. Miller, and P. P. Tans. 2003. Atmospheric methane levels off: Temporary pause or a new steady-state?, *Geophysical Research Letters* 30(19):1992, doi:10.1029/2003GL018126.

Dlugokencky, E. J., L. Bruhwiler, J. W. C. White, L. K. Emmons, P. C. Novelli, S. A. Montzka, K. A. Masarie, P. M. Lang, A. M. Crotwell, J. B. Miller, and L. V. Gatti. 2009. Observational constraints on recent increases in the atmospheric CH4 burden. *Geophys. Res. Lett.* 36 (18):L18803.

DOD (U.S. Department of Defense). 2010 (February 1). *The Quadrennial Defense Review*. Washington, DC: DOD.

Dodman, D. 2009. Blaming cities for climate change? An analysis of urban greenhouse gas emissions inventories. *Environment and Urbanization* 21(1):185-201.

DOE (U.S. Department of Energy). 2004. *Water Energy Resources of the United States with Emphasis on Low Head/Low Power Resources, Energy Efficiency and Renewable Energy: Wind and Hydropower Technologies*. Washington, DC: DOE.

DOE. 2006. *Feasibility Assessment of the Water Energy Resources of the United States for New Low Power and Small Hydro Classes of Hydroelectric Plants, Energy Efficiency and Renewable Energy: Wind and Hydropower Technologies*. Washington, DC: DOE.

DOE. 2008a. *20% Wind Energy by 2030: Increasing Wind Energy's Contribution to U.S. Electricity Supply*. Washington, DC: DOE. Available at *http://www.20percentwind.org/20percent_wind_energy_report_revOct08.pdf*. Accessed May 8, 2010.

DOE. 2008b. *Scientific Grand Challenges: Challenges in Climate Change Science and the Role of Computing at the Extreme Scale*. Report from the Workshop Held November 6-7. Washington, DC: DOE.

DOE. 2009a. *Modern Shale Gas Development in the United States: A Primer*. Washington, DC: DOE. Available at *http://www.netl.doe.gov/technologies/oil-gas/publications/EPreports/Shale_Gas_Primer_2009.pdf*. Accessed May 7, 2010.

DOE. 2009b. *Science Challenges and Future Directions: Climate Change Integrated Assessment Research*. Report from the U.S. Department of Energy, Office of Science, Office of Biological and Environmental Research Workshop on Integrated Assessment, November 2008.

DOE. 2009c. *Strategies for the Commercialization and Deployment of Greenhouse Gas Intensity-Reducing Technologies and Practices*. Committee on Climate Change Science and Technology Integration. Washington, DC: DOE.

DOE. 2010. *Fossil Energy Techline: DOE Awards Cooperative Agreement for Post-Combustion Carbon Capture Project*. Washington, DC: DOE. Available at *http://fossil.energy.gov/news/techlines/2010/10007-DOE_Awards_Cooperative_Agreement.html*. Accessed March 12, 2010.

DOI (Department of the Interior). 2009. *Salazar Launches DOI Climate Change Response Strategy*. Available at *http://www.doi.gov/archive/news/09_News_Releases/091409.html*. Accessed May 3, 2010.

Doney, S. C., V. J. Fabry, R. A. Feely, and J. A. Kleypas. 2009. Ocean acidification: The other CO_2 problem. *Annual Review of Marine Science* 1:169-192.

Donner, L. J., and W. G. Large. 2008. Climate modeling. *Annual Review of Environment and Resources* 33(1):1-17.

Donoghue, E. R., M. Nelson, G. Rudis, R. I. Sabogal, J. T. Watson, G. Huhn, and G. Luber. 2003. Heat-related deaths—Chicago, Illinois, 1996-2001, and United States, 1979-1999. *Morbidity and Mortality Weekly Report*. 52(26);610-613

DOT (Department of Transportation). 2008. *Research and Innovative Technology Administration. National Transportation Statistics*. Bureau of Transportation Statistics. Available at *http://www.bts.gov/publications/national_transportation_statistics*. Accessed May 3, 2010.

Douglas, M. 1985. *Risk Acceptability According to the Social Sciences. Social Research Perspectives: Occasional Reports on Current Topics*. New York: Russell Sage Foundation.

Dow, K., R. Kasperson, and M. Bohn. 2006. Exploring the social justice implications of adaptation and vulnerability. Pp. 79-96 in *Fairness in Adaptation to Climatic Change*, N. Adger, J. Paavola, S. Huq, and M. J. Mace, eds. Cambridge, MA: MIT Press.

Dueck, T. A., R. D. Visser, H. Poorter, S. Persijn, A. Gorissen, W. D. Visser, A. Schapendonk, J. Verhagen, J. Snel, F. J. M. Harren, A. K. Y. Ngai, F. Verstappen, H. Bouwmeester, L. A. C. J. Voesenek, and A. V. D. Werf. 2007. No evidence for substantial aerobic methane emission by terrestrial plants: A ^{13}C-labelling approach. *New Phytologist* 175(1):29-35.

Durfee, J. L. 2006. "Social change" and "status quo" framing effects on risk perception: An exploratory experiment. *Science Communication* 27(4) :459-495.

Durieux, L., L. A. T. Machado, and H. Laurent. 2003. The impact of deforestation on cloud cover over the Amazon arc of deforestation. *Remote Sensing of Environment* 86(1):132-140.

Dutta, K., E. A. G. Schuur, J. C. Neff, and S. A. Zimov. 2006. Potential carbon release from permafrost soils of Northeastern Siberia. *Global Change Biology* 12(12):2336-2351.

Dyurgerov, M., and M. Meier. 2005. Glaciers and the changing earth system: A 2004 snapshot. *Occasional Paper, Institute of Arctic and Alpine Research, University of Colorado*. Boulder, CO: Institute of Arctic and Alpine Research, University of Colorado. Available at *http://instaar.colorado.edu/other/download/OP58%5Fdyurgerov%5Fmeier.pdf.*

Eagle, N., A. Pentland, and D. Lazer. 2009. Inferring friendship network structure by using mobile phone data. *Proceedings of the National Academy of Sciences of the United States of America* 106(36):15274-15278.

Eakin, H., and J. Conley. 2002. Climate variability and the vulnerability of ranching in southeastern Arizona: A pilot study. *Climate Research* 21(3):271-281.

Eakin, H., and M. C. Lemos. 2006. Adaptation and the state: Latin America and the challenge of capacity-building under globalization. *Global Environmental Change* 16:7-18.

Eakin, H., and A. L. Luers. 2006. Assessing the vulnerability of social-environmental systems. *Annual Review of Environment and Resources* 31:365-394.

Eakin, H., E. L. Tompkins, D. R. Nelson, and J. M. Anderies. 2009. Hidden costs and disparate uncertainties: Trade-offs in approaches to climate policy. Pp. 212-226 in *Living with Climate Change: Are There Limits to Adaptation?* W. N. Adger et al., eds. Cambridge, U.K.: Cambridge University Press.

Early, J. T. 1989. Space-based solar shield to offset greenhouse effect. *Journal of the British Interplanetary Society* 42:567-569.

Easterling, D. R., and M. F. Wehner. 2009. Is the climate warming or cooling? *Geophysical Research Letters* 36:L08706, doi:10.1029/2009GL037810.

Easterling, W. E., P. K. Aggarwal, P. Batima, K. M. Brander, L. Erda, S. M. Howden, A. Kirilenko, J. Morton, J.-F. Soussana, J. Schmidhuber, and F. N. Tubiello. 2007. Food, fibre and forest products. Pp. 273-313 in *Climate Change 2007: Impacts, Adaptation and Vulnerability. Contribution of Working Group II to the Fourth Assessment Report of the Intergovernmental Panel on Climate Change*, M. L. Parry, O. F. Canziani, J. P. Palutikof, P. J. van der Linden, and C. E. Hanson, eds. Cambridge, U.K.: Cambridge University Press.

Ebi, K. L., and G. Mcgregor. 2008. Climate change, tropospheric ozone and particulate matter, and health impacts. *Environmental Health Perspectives* 116(11):1449-1455.

Ebi, K. L., J. Balbus, P. L. Kinney, E. Lipp, D. Mills, M. S. Oneill, and M. L. Wilson. 2009. US funding is insufficient to address the human health impacts of and public health responses to climate variability and change. *Environmental Health Perspectives* 117(6):857-862.

Eddy, J. A. 1976. Maunder minimum. *Science* 192(4245):1189-1202.

Eden, S. 2009. The work of environmental governance networks: Traceability, credibility and certification by the Forest Stewardship Council. *Geoforum* 40:383-394.

EIA (Energy Information Agency). 2007. *Renewable Energy Annual, 2005.* Washington, DC: DOE.

EIA (Energy Information Administration). 2009. *Annual Energy Review 2008.* Washington, DC: EIA. Available at *http://www.eia.doe.gov/aer/pdf/aer.pdf.* Accessed May 5, 2010.

EISA (Energy Independence and Security Act) of 2007. P. L. 110-140, December 19, 2007.

Eliasson, I. 2000. The use of climate knowledge in urban planning. *Landscape and Urban Planning* 48(1-2):31-44.

Elliott, D. L., L. L. Wendell, and G. L. Gower. 1991. *An Assessment of the Available Windy Land Area and Wind Energy Potential in the Contiguous United States.* Report PNL-7789. Richland, WA: Pacific Northwest Laboratory.

Ellis, E. C., and N. Ramankutty. 2008. Putting people in the map: Anthropogenic biomes of the world. *Frontiers in Ecology and the Environment* 6(8):439-447.

Emanuel, K. 2005. Increasing destructiveness of tropical cyclones over the past 30 years. *Nature* 436(7051):686-688.

Enfield, D. B., A. M. Mestas-Nuñez, and P. J. Trimble. 2001. The Atlantic multidecadal oscillation and its relation to rainfall and river flows in the continental U.S. *Geophysical Research Letters* 28:2077-2080.

Engle, N. L., and M. C. Lemos. 2010. Unpacking governance: Building adaptive capacity to climate change of river basins in Brazil. *Global Environmental Change* 20(1):4-13.

EPA (Environmental Protection Agency). 2002 (August). *Estimated Per Capita Fish Consumption in the United States*. Report EPA-821-C-02-003, Washington, DC.

EPA. 2006. *Global Anthropogenic Non-CO_2 Greenhouse Gas Emissions: 1990-2020*. Washington, DC: Office of Atmospheric Programs, Climate Change Division.

EPA. 2008. *A Screening Assessment of the Potential Impacts of Climate Change on Combined Sewer Overflow (CSO) Mitigation in the Great Lakes and New England Regions*. Global Change Research Program, National Center for Environmental Assessment, Report EPA/600/R-07/033F, Washington, DC. Available at *http://www.epa.gov/ncea*. Accessed on October 26, 2010. Springfield, VA: National Technical Information Service.

EPA. 2009a. *ENERGY STAR Programmable Thermostat Suspension Memo*. Available at *http://www.energystar.gov/ia/partners/prod_development/revisions/downloads/thermostats/Spec_Suspension_Memo_May2009.pdf*. Accessed on October 26, 2010.

EPA. 2009b. *Inventory of U.S. Greenhouse Gas Emissions and Sinks: 1990-2007*. EPA Report 430-R-09-004, Washington, DC.

EPA. 2009c. *Light-Duty Automotive Technology, Carbon Dioxide Emissions, and Fuel Economy Trends: 1975 Through 2009*. Washington, DC: EPA.

EPA. 2010a. *LCA Resources*. Available at *http://www.epa.gov/nrmrl/lcaccess/resources.html*. Accessed September 10, 2010.

EPA. 2010b. *EPA and NHTSA to Propose Greenhouse Gas and Fuel Efficiency Standards for Heavy-Duty Trucks; Begin Process for Further Light-Duty Standards: Regulatory Announcement*. Available at *http://www.epa.gov/oms/climate/regulations/420f10038.htm*. Accessed July 13, 2010.

EPICA community members. 2004. Eight glacial cycles from an Antarctic ice core. *Nature* 429(6992):623-628.

EPRI (Electric Power Research Institute). 2007a. *Assessment of Waterpower Potential and Development Needs*. Palo Alto, CA: EPRI. Available at *http://www.aaas.org/spp/cstc/docs/07_06_1ERPI_report.pdf*. Accessed May 7, 2010.

Epstein, P. R. 2005. Climate change and human health. *New England Journal of Medicine* 353(14):1433-1436.

Erikson, K. 1994. *A New Species of Trouble: Explorations in Disaster, Trauma, and Community*, 1st ed. New York: W.W. Norton.

Etheridge, D. M., L. P. Steele, R. L. Langenfelds, R. J. Francey, J.-M. Barnola, and V. I. Morgan. 1996. Natural and anthropogenic changes in atmospheric CO_2 over the last 1000 years from air in Antarctic ice and firn. *Journal of Geophysical Research* 101:4115-4128.

Etheridge, D. M., L. P. Steele, R. J. Francey, and R. L. Langenfelds. 2002. Historical CH_4 records since about 1000 A.D. from ice core data. In *Trends: A Compendium of Data on Global Change*. Oak Ridge, TN: Carbon Dioxide Information Analysis Center, Oak Ridge National Laboratory, DOE.

European Commission Joint Research Centre. 2010. *LCA Tools, Services, and Data*. Available at *http://lca.jrc.ec.europa.eu/lcainfohub/datasetArea.vm*. Accessed September 10, 2010.

Evans, P. B. 2002. *Livable Cities? Urban Struggles for Livelihood and Sustainability*. Berkelely, CA: University of California Press.

Eviner, V. T., and F. S. Chapin. 2003. Functional matrix: A conceptual framework for predicting multiple plant effects on ecosystem processes. *Annual Review of Ecology Evolution and Systematics* 34:455-485.

Ewing, R., and F. Rong. 2008. The impact of urban form on US residential energy use. *Housing Policy Debate* 19(1):1-30.

Ewing, R., K. Bartholomew, S. Winkelman, J. Walters, and D. Chen. 2007. *Growing Cooler: Evidence on Urban Development and Climate Change*. Washington, DC: Urban Land Institute.

Fabry, V. J., B. A. Seibel, R. A. Feely, and J. C. Orr. 2008. Impacts of ocean acidification on marine fauna and ecosystem processes. *ICES Journal of Marine Science* 65(3):414-432.

Fairbanks, R. G. 1989. A 17,000 year glacial euststic sea level record: Influence of glacial melting rates on the Younger Dryas event and deep ocean circulation. *Nature* 342:637-641.

Falkner, R. 2003. Private environmental governance and international relations: Exploring the links. *Global Environmental Politics* 3(2):72-87.

FAO (Food and Agriculture Organization of the United Nations). 2009. *The State of the World Fisheries and Aquaculture 2008*. Rome: FAO Fisheries and Aquaculture Department.

Farber, S. C., R. Costanza, and M. A. Wilson. 2002. Economic and ecological concepts for valuing ecosystem services. *Ecological Economics* 41(3):375-392.

Fargione, J., J. Hill, D. Tilman, S. Polasky, and P. Hawthorne. 2008. Land clearing and the biofuel carbon debt. *Science* 319(5867):1235-1238.

Farrell, A. E., and J. Jäger, eds. 2005. *Assessments of Regional and Global Environmental Risks: Designing Processes for the Effective Use of Science in Decisionmaking.* Washington, DC: RFF Press.

Farrell, A. E., R. J. Plevin, B. T. Turner, A. D. Jones, M. O'Hare, and D. M. Kammen. 2006. Ethanol can contribute to energy and environmental goals. *Science* 311(5760):506-508.

Fawcett, A., K. Calvin, P. Delachesnaye, J. Reilly, and J. Weyant. 2009. U.S. climate policy architectures: Overview of the EMF 22 U.S. scenarios. *Energy Economics* 31:S198-S211.

Fawcett, R. 2007. Has the world cooled since 1998? *Bulletin of the Australian Meteorological and Oceanographic Society* 20:141-148.

Fernandes, L., J. Day, A. Lewis, S. Slegers, B. Kerrigan, D. Breen, D. Cameron, B. Jago, J. Hall, D. Lowe, J. Innes, J. Tanzer, V. Chadwick, L. Thompson, K. Gorman, M. Simmons, B. Barnett, K. Sampson, G. De'ath, B. Mapstone, H. Marsh, H. Possingham, I. Ball, T. Ward, K. Dobbs, J. Aumend, D. Slater, and K. Stapleton. 2005. Establishing representative no-take areas in the Great Barrier Reef: Large-scale implementation of theory on marine protected areas. *Conservation Biology* 19 (6):1733-1744.

FHA (Federal Highway Administration). 2008. *Highway Statistics 2008.* Washington, DC: U.S. Department of Transportation. Available at *http://www.fhwa.dot.gov/policyinformation/statistics/2008/*. Accessed May 7, 2010.

Fiala, N. 2008. Meeting the demand: An estimation of potential future greenhouse gas emissions from meat production. *Ecological Economics* 67(3):412-419.

Ficke, A. D., C. A. Myrick, and L. J. Hansen. 2007. Potential impacts of global climate change on freshwater fisheries. *Reviews in Fish Biology and Fisheries* 17(4):581-613.

Field, C. B., M. J. Behrenfeld, J. T. Randerson, and P. Falkowski. 1998. Primary production of the biosphere: Integrating terrestrial and oceanic components. *Science* 281 (5374):237-240.

Field, C. B., G. C. Daily, F. W. Davis, S. Gaines, P. A. Matson, J. Melack, and N. L. Miller. 1999. *Confronting Climate Change in California: Ecological Impacts on the Golden State.* Cambridge, MA: Union of Concerned Scientists and the Ecological Society of America.

Field, C. B., D. B. Lobell, H. A. Peters, and N. R. Chiariello. 2007a. Feedbacks of terrestrial ecosystems to climate change. *Annual Review of Environment and Resources* 32:1-29.

Field, C. B., L. D. Mortsch, M. Brklacich, D. L. Forbes, P. Kovacs, J. A. Patz, S. W. Running, and M. J. Scott. 2007b. North America. Pp. 617-652 in *Climate Change 2007: Impacts, Adaptation and Vulnerability. Contribution of Working Group II to the Fourth Assessment Report of the Intergovernmental Panel on Climate Change*, M. L. Parry, O. F. Canziani, J. P. Palutikof, P. J. van der Linden, and C. E. Hanson, eds. Cambridge, U.K.: Cambridge University Press.

Filion, Y. R. 2008. Impact of urban form on energy use in water distribution systems. *Journal of Infrastructure Systems* 14(4):337-346.

Filleul, L., S. Cassadou, S. Medina, P. Fabres, A. Lefranc, D. Eilstein, A. Le Tertre, L. Pascal, B. Chardon, M. Blanchard, C. Declercq, J. F. Jusot, H. Prouvost, and M. Ledrans. 2006. The relation between temperature, ozone, and mortality in nine French cities during the heat wave of 2003. *Environmental Health Perspectives* 114(9):1344-1347.

Finnveden, G., M. Z. Hauschild, T. Ekvall, J. Guinée, R. Heijungs, S. Hellweg, A. Koehler, D. Pennington, and S. Suh. 2009. Recent developments in life cycle assessment. *Journal of Environmental Management* 91:1-21.

Firestone, J., W. Kempton, and A. B. Krueger. 2009. Public acceptance of offshore wind power projects in the United States. *Wind Energy* 12(2):183-202.

Fischhoff, B. 1981. *Acceptable Risk.* New York: Cambridge University Press.

Fischhoff, B. 2007. Nonpersuasive communication about matters of greatest urgency: Climate change. *Environmental Science & Technology* 41:7204-7208.

Fischlin, A., G. F. Midgley, J. T. Price, R. Leemans, B. Gopal, C. Turley, M. D. A. Rounsevell, O. P. Dube, J. Tarazona, and A. A. Velichko. 2007. Ecosystems, their properties, goods, and services. Pp. 211-272 in *Climate Change 2007: Impacts, Adaptation and Vulnerability. Contribution of Working Group II to the Fourth Assessment Report of the Intergovernmental Panel on Climate Change*, M. L. Parry, O. F. Canziani, J. P. Palutikof, P. J. van der Linden, and C. E. Hanson, eds. Cambridge, U.K.: Cambridge University Press.

Flanner, M. G., C. S. Zender, J. T. Randerson, and P. J. Rasch. 2007. Present-day climate forcing and response from black carbon in snow. *Journal of Geophysical Research* 112(D11).

Fletcher, E. A. 2001. Solar thermal processing: A review. *Journal of Solar Energy Engineering* 123(2):63-74.

Flick, R. E. 1998. Comparison of California tides, storm surges, and mean sea level during the El Niño winters of 1982-83 and 1997-98. *Shore & Beach* 6(3):7-11.

Flick, R. E., J. F. Murray, and L. C. Ewing. 1999. *Trends in U.S. Tidal Datum Statistics and Tide Range: A Data Report Atlas* (SIO Reference Series 99-20). La Jolla, CA: Scripps Institution of Oceanography.

Flick, R. E., J. F. Murray, and L. C. Ewing. 2003. Trends in U.S. tidal datum statistics and tide range. *Journal of Waterway, Port, Coastal and Ocean Engineering* 129(4):155-164.

Fofonoff, N. P. 1985. Physical-properties of seawater—a new salinity scale and equation of state for seawater. *Journal of Geophysical Research* 90(NC2):3332-3342.

Foley, J. A., R. Defries, G. P. Asner, C. Barford, G. Bonan, S. R. Carpenter, F. S. Chapin, M. T. Coe, G. C. Daily, H. K. Gibbs, J. H. Helkowski, T. Holloway, E. A. Howard, C. J. Kucharik, C. Monfreda, J. A. Patz, I. C. Prentice, N. Ramankutty, and P. K. Snyder. 2005. Global consequences of land use. *Science* 309(5734):570-574.

Folke, C., A. Jansson, J. Larsson, and R. Costanza. 1997. Ecosystem appropriation by cities. *Ambio* 26(3):167-172.

Ford, L. H. 2003. Challenging global environmental governance: Social movement agency and global civil society. *Global Environmental Politics* 3:120-134.

Forster, P., V. Ramaswamy, P. Artaxo, T. Berntsen, R. Betts, D. W. Fahey, J. Haywood, J. Lean, D. C. Lowe, G. Myhre, J. Nganga, R. Prinn, G. Raga, M. Schulz, R. Van Dorland, G. Bodeker, O. Boucher, W. D. Collins, T. J. Conway, E. Dlugokencky, J. W. Elkins, D. Etheridge, P. Foukal, P. Fraser, M. Geller, F. Joos, C. D. Keeling, S. Kinne, K. Lassey, U. Lohmann, A. C. Manning, S. Montzka, D. Oram, K. O'shaughnessy, S. Piper, G. K. Plattner, M. Ponater, N. Ramankutty, G. Reid, D. Rind, K. Rosenlof, R. Sausen, D. Schwarzkopf, S. K. Solanki, G. Stenchikov, N. Stuber, T. Takemura, C. Textor, R. Wang, R. Weiss, and T. Whorf. 2007. *Changes in Atmospheric Constituents and in Radiative Forcing*. Cambridge, U.K. and New York: Cambridge University Press.

FRA (Federal Railroad Administration). 2009. *Vision for High Speed Rail in America*. Washington, DC: U.S. Department of Transportation. Available at *http://www.fra.dot.gov/downloads/rrdev/hsrstrategicplan.pdf*. Accessed May 7, 2010.

Freudenburg, W. R., and S. K. Pastor. 1992. Public responses to technological risks. *Sociological Quarterly* 33(3):389-412.

Frey, B. S. 2008. *Happiness: A Revolution in Economics*. Cambridge, MA: MIT Press.

Friedlingstein, P., P. Cox, R. Betts, L. Bopp, W. Von Bloh, V. Brovkin, P. Cadule, S. Doney, M. Eby, I. Fung, G. Bala, J. John, C. Jones, F. Joos, T. Kato, M. Kawamiya, W. Knorr, K. Lindsay, H. D. Matthews, T. Raddatz, P. Rayner, C. Reick, E. Roeckner, K. G. Schnitzler, R. Schnur, K. Strassmann, A. J. Weaver, C. Yoshikawa, and N. Zeng. 2006. Climate-carbon cycle feedback analysis: Results from the (CMIP)-M-4 model intercomparison. *Journal of Climate* 19(14):3337-3353.

Frische, M., K. Garofalo, T. H. Hansteen, R. Borchers, and J. Harnisch. 2006. The origin of stable halogenated compounds in volcanic gases. *Environmental Science and Pollution Research* 13:406-413.

Frumhoff, P. C. 2007. *Confronting Climate Change in the U.S. Northeast: Science, Impacts, and Solutions*. Cambridge, MA: UCS Publications.

Fu, Q., and C. M. Johanson. 2005. Satellite-derived vertical dependence of tropical tropospheric temperature trends. *Geophysical Research Letters* 32:L10703, doi:10.1029/2004GL022266.

Fu, Q, C. M. Johanson, S. G. Warren, and D. J. Seidel. 2004. Contribution of stratospheric cooling to satellite-inferred tropospheric temperature trends. *Nature* 429:55-58.

Fu, Q., C. M. Johanson, J. M. Wallace, and T. Reichler. 2006. Enhanced mid-latitude tropospheric warming in satellite measurements. *Science* 312(5777):1179, doi:10.1126/science.1125566.

Fu, G. B, S. P. Charles, and J. J. Yu 2009. A critical overview of pan evaporation trends over the last 50 years. *Climatic Change* 97:193-214.

Füssel, H. M., and R. J. T. Klein. 2006. Climate change vulnerability assessments: An evolution of conceptual thinking. *Climatic Change* 75(3):301-329.

Füssel, H.-M., F. L. Toth, J. G. van Minnen, and F. Kaspar. 2003. Climate impact response functions as impact tools in the tolerable windows approach. *Climatic Change* 56(1-2):91-117.

Gaedke, U., D. Ollinger, E. Bauerle, and D. Straile. 1998. The Impact of the Interannual Variability in Hydrodynamic Conditions on the Plankton Development in Lake Constance in Spring and Summer. Universitat Konstanz.

Gage, K. L., T. R. Burkot, R. J. Eisen, and E. B. Hayes. 2008. Climate and vectorborne diseases. *American Journal of Preventive Medicine* 35(5):436-450.

Gardiner, S. 2010. Is "arming the future" with geoengineering really the lesser evil? Some doubts about the ethics of intentionally manipulating the climate system. In Climate Ethics: Essential Readings, S. M. Gardiner, S. Caney, D. Jamieson, and H. Shue, eds. New York: Oxford University Press.

Gardner, G. T., and P. C. Stern. 1996. *Environmental Problems and Human Behavior*. Boston: Allyn and Bacon.

Garrett, K. A., S. P. Dendy, E. E. Frank, M. N. Rouse, and S. E. Travers. 2006. Climate change effects on plant disease: Genomes to ecosystems. *Annual Review of Phytopathology* 44:489-509.

Gazeau, F., C. Quiblier, J. M. Jansen, J. P. Gattuso, J. J. Middelburg, and C. H. R. Heip. 2007. Impact of elevated CO_2 on shellfish calcification. *Geophysical Research Letters* 34(7).

GCRA (Global Change Research Act). 1990. P. L. 101-606, Title 15, Chapter 56A: Global Change Research.

Gedney, N., P. M. Cox, R. A. Betts, O. Boucher, C. Huntingford, and P. A. Stott. 2006. Continental runoff: A quality-controlled global runoff data set (Reply). *Nature* 444(7120):E14-E15.

Georgakakos, K. P., N. E. Graham, E. M. Carpenter, A. P. Georgakakos, and H. Yao. 2005. Integrating climate-hydrology forecasts and multi-objective reservoir management for northern California. *EOS* 86(12):4.

Gerten, D., S. Rost, W. Von Bloh, and W. Lucht. 2008. Causes of change in 20th century global river discharge. *Geophysical Research Letters* 35(20):5.

Gibbard, S., K. Caldeira, G. Bala, T. J. Phillips, and M. Wickett. 2005. Climate effects of global land cover change. *Geophysical Research Letters* 32:L23705.

Gibbs, C., J. E. Fry, C. Oliver, and E. F. McGarrell. 2009. Carbon trading markets: An overview. Presentation to the Interpol Environmental Crimes Committee Pollution Working Group, Lyon, France.

Gigerenzer, G. 2008. *Rationality for Mortals: How People Cope with Uncertainty*. Oxford: Oxford University Press.

Gille, S. T. 2002. Warming of the Southern Ocean since the 1950s. *Science* 295 (5558):1275-1277.

Gilles, J. L., and C. Valdivia. 2009. Local forecast communcation in the altiplano. *Bulletin of the American Meteorological Society* 90(1):85-91.

Gillett, N. P., A. J. Weaver, F. W. Zwiers, and M. F. Wehner. 2004. Detection of volcanic influence on global precipitation. *Geophysical Research Letters* 31.

Gine, X., R. Townsend, and J. Vickery. 2008. Patterns of rainfall insurance participation in rural India. *World Bank Economic Review* 22(3):539.

Ginley, D., M. A. Green, and R. Collins. 2008. Solar energy conversion toward 1 terawatt. *MRS Bulletin* 33(4):355-364.

Gleick, P. H. 2000. *Water—The Potential Consequences of Climate Variability and Change for the Water Resources of the United States*. Oakland, CA: Pacific Institute for Studies in Development, Environment, and Security. Available at *http://www.gcrio.org/NationalAssessment/water/water.pdf*.

Gleick, P. H. 2003a. Global freshwater resources: Soft-path solutions for the 21st century. *Science* 302(5650):1524-1528.

Gleick, P. H. 2003b. Water use. *Annual Review of Environment and Resources* 28:275-314.

Glynn, P. W. 1991. Coral-Reef Bleaching in the 1980s and Possible Connections with Global Warming. *Trends in Ecology & Evolution* 6 (6):175-179.

GOFC-GOLD. 2008. *Reducing Greenhouse Gas Emissions from Deforestation and Degradation in Developing Countries: A Sourcebook of Methods and Procedures for Monitoring, Measuring and Reporting*. GOFC-GOLD Project Office, Report COP13-2, Alberta, Canada, hosted by Natural Resources Canada.

GOFC-GOLD. 2009. *Reducing Greenhouse Gas Emissions from Deforestation and Degradation in Developing Countries: A Sourcebook of Methods and Procedures for Monitoring, Measuring and Reporting*. GOFC-GOLD Project Office, Report COP14-2, Alberta, Canada, hosted by Natural Resources Canada.

Golubev, V. S., J. H. Lawrimore, P. Y. Groisman, N. A. Speranskaya, S. A. Zhuravin, M. J. Menne, T. C. Peterson, and R. W. Malone. 2001. Evaporation changes over the contiguous United States and the former USSR: A reassessment. *Geophysical Research Letters* 28(13):2665-2668.

Gornitz, V., 2009. Sea level change, post-glacial. Pp. 887-893 in *Encyclopedia of Paleoclimatology & Ancient Environments*, V. Gornitz, ed. Dordrecht, Netherlands: Springer.

Gornitz, V., S. Lebedeff, and J. Hansen. 1982. Global sea level trend in the past century. *Science* 215(4540):1611-1614.

Gough, C. M., C. S. Vogel, H. P. Schmid, and P. S. Curtis. 2008. Controls on annual forest carbon storage: Lessons from the past and predictions for the future. *Bioscience* 58(7):609-622.

Goulder, L. H. 2004. *Induced Technological Change and Climate Policy*. Arlington, VA: Pew Center on Global Climate Change.

Govindasamy, B., and K. Caldeira. 2000. Geoengineering Earth's radiation balance to mitigate CO_2-induced climate change. *Geophysical Research Letters* 27(14): 2141-2144.

Grantham, B. A., F. Chan, K. J. Nielsen, D. S. Fox, J. A. Barth, A. Huyer, J. Lubchenco, and B. A. Menge. 2004. Upwelling-driven near-shore hypoxia signals ecosystem and oceanographic changes in the northeast Pacific. *Nature* 429 (6993):749-754.

Graves, P. E. 2007. *Environmental Economics: A Critique of Benefit-Cost Analysis*. Lanham, MD: Rowman & Littlefield.

Gray, L. J., J. D. Haigh and R. G. Harrison, 2005. The influence of solar changes on the Earth's climate, Hadley Centre technical note 62, Publisher: MET Office, pp. 1-81.

Grebmeier, J. M., J. E. Overland, S. E. Moore, E. V. Farley, E. C. Carmack, L. W. Cooper, K. E. Frey, J. H. Helle, F. A. Mclaughlin, and S. L. Mcnutt. 2006. A major ecosystem shift in the northern Bering Sea. *Science* 311(5766):1461-1464.

Green, M. A. 2003. Crystalline and thin-film silicon solar cells: State of the art and future potential. *Solar Energy* 74(3):181-192.

Green, R. 2005. Electricity and markets. *Oxford Review of Economic Policy* 21(1):67-87.

Gregory, J. M., P. Huybrecht, and S. Raper. 2004. Threatened loss of the Greenland ice sheet. *Nature* 428:616.

Gregory, P. J., S. N. Johnson, A. C. Newton, and J. S. Ingram. 2009. Integrating pests and pathogens into the climate change/food security debate. *Journal of Experimental Botany* 60(10):2827-2838.

Grimm, N. B., S. H. Faeth, N. E. Golubiewski, C. L. Redman, J. G. Wu, X. M. Bai, and J. M. Briggs. 2008. Global change and the ecology of cities. *Science* 319(5864):756-760.

Grimmond, S. 2007. Urbanization and global environmental change: Local effects of urban warming. *Geographical Journal* 173:83-88.

Grinsted, A., J. Moore, and S. Jevrejeva. 2009. Reconstructing sea level from paleo and projected temperatures 200 to 2100 Ad. *Climate Dynamics* 34(4): 461-472,.

Groisman, P. Y., and R. W. Knight. 2008. Prolonged dry episodes over the conterminous United States: New tendencies emerging during the last 40 years. *Journal of Climate* 21:1850-1862.

Grossman, G. M., and A. B. Krueger. 1995. Economic growth and the environment. *Quarterly Journal of Economics* 110(2):353-377.

Grove, J. M. 2009. Cities: Managing densely settled biological systems. In *Principles of Ecosystem Stewardship: Resilience-Based Natural Resource Management in a Changing World*, F. S. Chapin III, G. P. Kofinas, and C. Folke, eds. New York: Springer.

Grübler, A., N. Nakicenovic, and W. D. Nordhaus, eds. 2002. *Technological Change and the Environment*. Washington, DC: Resources for the Future.

Gryparis, A., B. Forsberg, K. Katsouyanni, A. Analitis, G. Touloumi, J. Schwartz, E. Samoli, S. Medina, H. R. Anderson, E. M. Niciu, H. E. Wichmann, B. Kriz, M. Kosnik, J. Skorkovsky, J. M. Vonk, and Z. Dortbudak. 2004. Acute effects of ozone on mortality from the "Air Pollution and Health: A European Approach" project. *American Journal of Respiratory and Critical Care Medicine* 170(10):1080-1087.

Gu, L., D. D. Baldocchi, S. C. Wofsy, J. W. Munger, J. J. Michalsky, S. P. Urbanski, and T. A. Boden. 2003. Response of a deciduous forest to the Mount Pinatubo eruption: Enhanced photosynthesis. *Science* 299(5615):2035-2038.

Guan, B., and S. Nigam. 2008. Pacific sea surface temperatures in the twentieth century: An evolution-centric analysis of variability and trend. *Journal of Climate* 21:2790-2809.

Guan, B., and S. Nigam. 2009. Analysis of Atlantic SST variability factoring inter-basin links and the secular trend: Clarified structure of the Atlantic Multidecadal Oscillation. *Journal of Climate* 22(15):4228-4240.

Guinotte, J. M., and V. J. Fabry. 2008. Ocean acidification and its potential effects on marine ecosystems. *Annals of the New York Academy of Sciences* 1134(2008):320-342.

Gullison, R. E., P. C. Frumhoff, J. G. Canadell, C. B. Field, D. C. Nepstad, K. Hayhoe, R. Avissar, L. M. Curran, P. Friedlingstein, C. D. Jones, and C. Nobre. 2007. Environment: Tropical forests and climate policy. *Science* 316(5827):985-986.

Guneralp, B., and K. C. Seto. 2008. Environmental impacts of urban growth from an integrated dynamic perspective: A case study of Shenzhen, South China. *Global Environmental Change-Human and Policy Dimensions* 18(4):720-735.

Gupta, J., and H. Van Asselt. 2006. Helping operationalise Article 2: A transdisciplinary methodological tool for evaluating when climate change is dangerous. *Global Environmental Change-Human and Policy Dimensions* 16(1):83-94.

Gupta, S., D. A. Tirpak, N. Burger, J. Gupta, N. Höhne, A. I. Boncheva, G. M. Kanoan, C. Kolstad, J. A. Kruger, A. Michaelowa, S. Murase, J. Pershing, T. Saijo, and A. Sari. 2007. Policies, instruments, and co-operative arrangements. In *Climate Change 2007: Mitigation. Contribution of Working Group III to the Fourth Assessment Report of the Intergovernmental Panel on Climate Change*, B. Metz, O. R. Davidson, P. R. Bosch, R. Dave, and L. A. Meyer, eds. Cambridge, U.K. and New York: Cambridge University Press.

Gurgel, A., J. M. Reilly, and S. Paltsev. 2007. Potential land use implications of a global biofuels industry. *Journal of Agricultural and Food Industrial Organization* 5(2).

Gutowski, W. J., G. C. Hegerl, G. J. Holland, T. R. Knutson, L. O. Mearns, R. J. Stouffer, P. J. Webster, M. F. Wehner, and F. W. Zwiers. 2008. *Causes of Observed Changes in Extremes and Projections of Future Changes in Weather and Climate Extremes in a Changing Climate: Regions of Focus: North America, Hawaii, Caribbean, and U.S. Pacific Islands.* Synthesis and Assessment Product 3.3. T. R. Karl, G. A. Meehl, C. D. Miller, S. J. Hassol, A. M. Waple, and W. L. Murray, eds. Washington, DC: Climate Change Science Program.

Guttman, N. B. 1989. Statistical descriptors of climate. *Bulletin of the American Meteorological Society* 70:602-607.

Haines, A., and J. A. Patz. 2004. Health effects of climate change. *Journal of the American Medical Association* 291(1):99-103.

Hall, M. H. P., and D. B. Fagre. 2003. Modeled climate-induced glacier change in Glacier National Park, 1850-2100. *Bioscience* 53(2):131-140.

Halpern, B. S., S. Walbridge, K. A. Selkoe, C. V. Kappel, F. Micheli, C. D'Agrosa, J. F. Bruno, K. S. Casey, C. Ebert, H. E. Fox, R. Fujita, D. Heinemann, H. S. Lenihan, E. M. P. Madin, M. T. Perry, E. R. Selig, M. Spalding, R. Steneck, and R. Watson. 2008. A global map of human impact on marine ecosystems. *Science* 319(5865):948-952.

Hamilton, L. C., and B. D. Keim. 2009. Regional variation in perceptions about climate change. *International Journal of Climatology* 29(15):2348-2352.

Hamin, E. M., and N. Gurran. 2009. Urban form and climate change: Balancing adaptation and mitigation in the U.S. and Australia. *Habitat International* 33(3):238-245.

Hamlet, A. F., P. W. Mote, M. P. Clark, and D. P. Lettenmaier. 2007. Twentieth-century trends in runoff, evapotranspiration, and soil moisture in the western United States. *Journal of Climate* 20(8):1468-1486.

Hanna, E., J. Cappelen, X. Fettweis, P. Huybrechts, A. Luckman, and M. H. Ribergaard. 2009. Hydrologic response of the Greenland ice sheet: The role of oceanographic warming. *Hydrological Processes* 23(1):7-30.

Hansen, J., and S. Lebedev. 1987. Global trends of measured surface air temperatures. *Journal of Geophysical Research* 92(D11):13345-13372.

Hansen, J., and M. Sato. 2004. Greenhouse gas growth rates. *Proceedings of the National Academy of Sciences of the United States of America* 101(46):16109-16114.

Hansen, J., R. Ruedy, J. Glascoe, and M. Sato. 1999. GISS analysis of surface temperature change. *Journal of Geophysical Research* 104(D24):30997-31022.

Hansen, J., R. Ruedy, M. Sato, M. Imhoff, W. Lawrence, D. Easterling, T. Peterson, and T. Karl. 2001. A closer look at United States and global surface temperature change. *Journal of Geophysical Research* 106:23947-23963, doi:10.1029/2001JD000354.

Hansen, J., M. Sato, R. Ruedy, L. Nazarenko, A. Lacis, G. A. Schmidt, G. Russell, I. Aleinov, M. Bauer, S. Bauer, N. Bell, B. Cairns, V. Canuto, M. Chandler, Y. Cheng, A. Del Genio, G. Faluvegi, E. Fleming, A. Friend, T. Hall, C. Jackman, M. Kelley, N. Kiang, D. Koch, J. Lean, J. Lerner, K. Lo, S. Menon, R. Miller, P. Minnis, T. Novakov, V. Oinas, J. Perlwitz, J. Perlwitz, D. Rind, A. Romanou, D. Shindell, P. Stone, S. Sun, N. Tausnev, D. Thresher, B. Wielicki, T. Wong, M. Yao, and S. Zhang. 2005. Efficacy of climate forcings. *Journal of Geophysical Research* 110(D18):D18104.

Hansen, J., M. Sato, R. Ruedy, K. Lo, D. W. Lea, and M. Medina-Elizade. 2006. Global temperature change. *Proceedings of the National Academy of Sciences of the United States of America* 103(39):14288-14293.

Hansen, J., M. Sato, P. Kharecha, D. Beerling, R. Berner, V. Masson-Delmotte, M. Pagani, M. Raymo, D. L. Royer, and J. C. Zachos. 2008. Target atmospheric CO_2: Where should humanity aim? *The Open Atmospheric Science Journal* 2:217-231.

Hansen, J. E. 2005. A slippery slope: How much global warming constitutes "dangerous anthropogenic interference"? *Climatic Change* 68(3):269-279.

Hardoy, J., and G. Pandiella. 2009. Urban poverty and vulnerability to climate change in Latin America. *Environment and Urbanization* 21(1):203-224.

Hare, S. R., N. J. Mantua, and R. C. Francis. 1999. Inverse production regimes: Alaska and West Coast Pacific salmon. *Fisheries* 24(1):6-14.

Harries, J. E., and J. M. Futyan. 2006. On the stability of the Earth's radiative energy balance: Response to the Mt. Pinatubo eruption. *Geophysical Research Letters* 33:L23814, doi:10.1029/2006GL027457.

Harris, J., and C. Blumstein, eds. 1984. *What Works: Documenting Energy Conservation in Buildings*. Washington, DC: American Council for an Energy-Efficient Economy.

Hatanaka, M., C. Bain, and L. Busch. 2005. Third-party certification in the global agrifood system. *Food Policy* 30(3):354-369.

Haupt, B. J., and D. Seidov. 2007. Strengths and weaknesses of the global ocean conveyor: Inter-basin freshwater disparities as the major control. *Progress in Oceanography* 73(3-4):358-369.

Hay, S. I., J. Cox, D. J. Rogers, S. E. Randolph, D. I. Stern, G. D. Shanks, M. F. Myers, and R. W. Snow. 2002. Climate change and the resurgence of malaria in the East African highlands. *Nature* 415(6874):905-909.

Hayden, B. P. 1999. Climate change and extratropical storminess in the United States: An assessment. *Journal of the American Water Resources Association* 35(6):1387-1397.

Hayhoe, K., D. Cayan, C. B. Field, P. C. Frumhoff, E. P. Maurer, N. L. Miller, S. C. Moser, S. H. Schneider, K. N. Cahill, E. E. Cleland, L. Dale, R. Drapek, R. M. Hanemann, L. S. Kalkstein, J. Lenihan, C. K. Lunch, R. P. Neilson, S. C. Sheridan, and J. H. Verville. 2004. Emissions pathways, climate change, and impacts on California. *Proceedings of the National Academy of Sciences of the United States of America* 101(34):12422-12427.

Hayhoe, K., C. P. Wake, T. G. Huntington, L. F. Luo, M. D. Schwartz, J. Sheffield, E. Wood, B. Anderson, J. Bradbury, A. Degaetano, T. J. Troy, and D. Wolfe. 2007. Past and future changes in climate and hydrological indicators in the US Northeast. *Climate Dynamics* 28(4):381-407.

Hayhoe, K., S. Sheridan, L. Kalkstein, and S. Greene. 2010. Climate change, heat waves, and mortality projections for Chicago. *Journal of Great Lakes Research* 36(2):65-73, doi:10.1016/j.jglr.2009.12.009.

Heal, G. 1997. Discounting and climate change: An editorial comment. *Climatic Change* 37(2):335-343.

Heal, G. 2009. The economics of climate change: A post-Stern perspective. *Climatic Change* 96(3):275-297.

Heberger, M., H. Cooley, P. Herrera, and P. H. Gleick. 2009. The Impacts of Sea-Level Rise on the California Coast—Final Report. Sacramento: California Energy Commission Report CEC-500-2009-024-F.

Hecht, J. E. 2005. *National Environmental Accounting: Bridging the Gap Between Ecology and Economy*. Washington, DC: Resources for the Future.

Hegerl, G. C., and S. Solomon. 2009. Risks of climate engineering. *Science* 325(5943):955-956.

Hegerl, G. C., T. R. Karl, M. Allen, N. L. Bindoff, N. Gillett, D. Karoly, X. Zhang, and F. Zwiers. 2006. Climate change detection and attribution: Beyond mean temperature signals. *Journal of Climate* 19:5058-5077.

Hegerl, G. C., F. W. Zwiers, P. Braconnot, N. P. Gillett, Y. Luo, J. A. Marengo Orsini, N. Nicholls, J. E. Penner, and P. A. Stott. 2007. Understanding and attributing climate change. In *Climate Change 2007: The Physical Science Basis. Contribution of Working Group I to the Fourth Assessment Report of the Intergovernmental Panel on Climate Change*, S. Solomon, D. Qin, M. Manning, Z. Chen, M. Marquis, K. B. Averyt, M. Tignor, and H. L. Miller, eds. Cambridge, U.K. and New York: Cambridge University Press.

Heinz Center. 2008a. *The State of the Nation's Ecosystems 2008*. Washington, DC: Island Press and Heinz III Center for Science, Economics and the Environment.

Heinz Center. 2008b. *A Survey of Climate Change Adaptation Planning*. Washington, DC: H. John Heinz III Center for Science, Economics and the Environment.

Held, I. M., and B. J. Soden. 2006. Robust responses of the hydrological cycle to global warming. *Journal of Climate* 19(21):5686-5699.

Helly, J. J., and L. A. Levin. 2004. Global distribution of naturally occurring marine hypoxia on continental margins. *Deep-Sea Research Part I-Oceanographic Research Papers* 51 (9):1159-1168.

Henry, A. D. 2009. The challenge of learning for sustainability: A prolegomenon to theory. *Human Ecology Review* 16(2):131-140.

Hermann, W. A. 2006. Quantifying global exergy resources. *Energy* 31(12):1685-1702.

Hess, J. J., J. N. Malilay, and A. J. Parkinson. 2008. Climate change: The importance of place. *American Journal of Preventive Medicine* 35(5):468-478.

Hesselbjerg, J., and B. Hewitson. 2007. Regional climate projections. In *Climate Change 2007: The Physical Science Basis. Contribution of Working Group I to the Fourth Assessment Report of the Intergovernmental Panel on Climate Change*, S. Solomon, D. Qin, M. Manning, Z. Chen, M. Marquis, K. B. Averyt, M. Tignor, and H. L. Miller, eds. Cambridge, U.K.: Cambridge University Press.

Hill, M., and P. Hupe. 2009. *Implementing Public Policy: An Introduction to the Study of Operational Governance*. Los Angeles, CA: Sage.

Hiltermann, T. J. N., J. Stolk, S. C. Van Der Zee, B. Brunekreef, C. R. De Bruijne, P. H. Fischer, C. B. Ameling, P. J. Sterk, P. S. Hiemstra, and L. Van Bree. 1998. Asthma severity and susceptibility to air pollution. *European Respiratory Journal* 11(3):686-693.

Hoegh-Guldberg, O. 1999. Climate change, coral bleaching and the future of the world's coral reefs. *Marine and Freshwater Research* 50 (8):839-866.

Hofmann, G. E., M. J. O'Donnell, and A. E. Todgham. 2008. Using functional genomics to explore the effects of ocean acidification on calcifying marine organisms. *Marine Ecology-Progress Series* 373:219-225.

Holbrook, S. J., R. J. Schmitt, and J. S. Stephens. 1997. Changes in an assemblage of temperate reef fishes associated with a climate shift. *Ecological Applications* 7(4):1299-1310.

Holgate, S. J., and P. L. Woodworth. 2004. Evidence for enhanced coastal sea level rise during the 1990s. *Geophysical Research Letters* 31(7):L07305.

Holland, D. M., R. H. Thomas, B. De Young, M. H. Ribergaard, and B. Lyberth. 2008. Acceleration of Jakobshavn Isbrae triggered by warm subsurface ocean waters. *Nature Geoscience* 1(10):659-664.

Horne, R., T. Grant, and K. Verghese. 2009. *Life Cycle Assessment: Principles, Practices and Prospects*. Colingwood, Victoria, Australia: CSIRO Publishing.

Horton, R., C. Herweijer, C. Rosenzweig, J. Liu, V. Gornitz, and A. C. Ruane. 2008. Sea level rise projections for current generation CGCMs based on the semi-empirical method. *Geophysical Research Letters* 35: L02715, doi:10.1029/2007GL032486.

Houghton, R. A. 2005. Aboveground forest biomass and the global carbon balance. *Global Change Biology* 11(6):945-958.

Houser, T., R. Bradley, B. Childs Staley, J. Werksman, and R. Heilmayr. 2008. *Leveling the Carbon Playing Field: International Competition and U.S. Climate Policy Design*. Washington, DC: World Resouces Institute.

Houweling, S., T. Rockmann, I. Aben, F. Keppler, M. Krol, J. F. Meirink, E. J. Dlugokencky, and C. Frankenberg. 2006. Atmospheric constraints on global emissions of methane from plants. *Geophysical Research Letters* 33:L15821, doi:10.1029/2006GL026162.

Howat, I. M., I. Joughin, and T. A. Scambos. 2007. Rapid changes in ice discharge from Greenland outlet glaciers. *Science* 315(5818):1559-1561.

Hu, P. S., and T. R. Reuscher. 2005. *Summary of Travel Trends: 2001 National Household Travel Survey*, Report FHWA-PL-07-010. Washington, DC: Federal Highway Administration, U.S. Department of Transportation.

Hu, W., K. Mengersen, A. Mcmichael, and S. Tong. 2008. Temperature, air pollution and total mortality during summers in Sydney, 1994-2004. *International Journal of Biometeorology* 52(7):689-696.

Huang, R. X. 1999. Mixing and energetics of the oceanic thermohaline circulation. *Journal of Physical Oceanography* 29(4):727-746.

Hughes, T. P., A. H. Baird, D. R. Bellwood, M. Card, S. R. Connolly, C. Folke, R. Grosberg, O. Hoegh-Guldberg, J. B. C. Jackson, J. Kleypas, J. M. Lough, P. Marshall, M. Nystrom, S. R. Palumbi, J. M. Pandolfi, B. Rosen, and J. Roughgarden. 2003. Climate Change, Human Impacts, and the Resilience of Coral Reefs. *Science* 301 (5635):929-933.

Huijbregts, M. A. J., S. Hellweg, R. Frischknecht, K. Hungerbühler, and A. J. Hendriks. 2008. Ecological footprint accounting in the life cycle assessment of products. *Ecological Economics* 64:798-807.

Huitema, D., E. Mostert, W. Egas, S. Moellenkamp, C. Pahl-Wostl, and R. Yalcin. 2009. Adaptive water governance: Assessing the institutional prescriptions of adaptive (co-)management from a governance perspective and defining a research agenda. *Ecology and Society* 14(1).

Hunt, G. L., and P. J. Stabeno. 2002. Climate change and the control of energy flow in the southeastern Bering Sea. *Progress in Oceanography* 55 (1-2):5-22.

Hunt, R., D. W. Hand, M. A. Hannah, and A. M. Neal. 1991. Response to CO_2 enrichment in 27 herbaceous species. *Functional Ecology* 5(3):410-421.

Huntington, T. G. 2004. Climate change, growing season length, and transpiration: Plant response could alter hydrologic regime. *Plant Biology* 6(6):651-653.

ICSU-IGFA (International Council for Science—International Group of Funding Agencies for Global Change Research). 2009. *Review of the International Geosphere-Biosphere Programme (IGBP)*. Paris: International Council for Science, 57 pp. Avaliable at *http://www.icsu.org*.

IHDP (International Human Dimensions Programme). 2005. *Science Plan: Urbanization and Global Environmental Change*. International Human Dimensions Programme on Global Environmental Change Report 15. Bonn, Germany: IHDP Secretariat.

IHDP. 2009. *International Human Dimensions Programme on Global Environmental Change Annual Report for 2009*. Available at *http://www.ihdp.unu.edu/*.

IEA (International Energy Agency). 2009. *Key World Energy Statistics 2009*. Paris. Available at *http://www.iea.org/textbase/nppdf/free/2009/key_stats_2009.pdf*. Accessed on October 26, 2010.

INECE (International Network for Environmental Compliance and Enforcement). 2009. *INECE Special Report on Climate Compliance: United Nations Climate Change Conference in Copenhagen—COP 15*. Washington, DC: INECE.

IPCC (Intergovernmental Panel on Cliamte Change). 2005. *Carbon Dioxide Capture and Storage*. B. Metz, O. Davidson, H. de Coninck, M. Loos, and L. Meyer, eds. Cambridge, U.K.: Cambridge University Press, 431 pp.

IPCC. 2007a. *Climate Change 2007: The Physical Science Basis. Contribution of Working Group I to the Fourth Assessment Report of the IPCC*, S. Solomon, D. Qin, M. Manning, Z. Chen, M. Marquis, K. B. Averyt, M. Tignor, and H. L. Miller, eds. Cambridge, U.K.: Cambridge University Press, 996 pp.

IPCC. 2007b. *Climate Change 2007: Synthesis Report. Contribution of Working Groups I, II and III to the Fourth Assessment Report of the Intergovernmental Panel on Climate Change*, Core Writing Team, R. K. Pachauri, and A. Reisinger, eds. Geneva, 104 pp.

IPCC. 2007c. *Climate Change 2007: Impacts, Adaptation and Vulnerability. Contribution of Working Group II to the Fourth Assessment Report of the Intergovernmental Panel on Climate Change*, M. L. Parry, O. F. Canziani, J. P. Palutikof, P. J. van der Linden, and C. E. Hanson, eds. Cambridge, U.K.: Cambridge University Press, 976 pp.

IPCC. 2007d. *Climate Change 2007: Mitigation. Contribution of Working Group III to the Fourth Assessment Report of the Intergovernmental Panel on Climate Change*, B. Metz, O. R. Davidson, P. R. Bosch, R. Dave, and L. A. Meyer, eds. Cambridge, U.K.: Cambridge University Press, 851 pp.

IPCC. 2009. Chapter Outline of the Working Group II Contribution to the IPCC Fifth Assessment Report (AR5), IPCC-XXXI/Doc. 20, Rev. 1 (28.X.2009), Bali, October 26-29. Available at *http://www.ipcc.ch/meeting_documentation/meeting_documentation_31th_session.htm*. Accessed on October 26, 2010.

Ito, K., S. F. De Leon, and M. Lippmann. 2005. Associations between ozone and daily mortality: Analysis and meta-analysis. *Epidemiology* 16(4):446-457.

Jack, B. K., C. Kousky, and K. R. E. Sims. 2008. Designing payments for ecosystem services: Lessons from previous experience with incentive-based mechanisms. *Proceedings of the National Academy of Sciences of the United States of America* 105(28):9465-9470.

Jackson, R. B., J. T. Randerson, J. G. Canadell, R. G. Anderson, R. Avissar, D. D. Baldocchi, G. B. Bonan, K. Caldeira, N. S. Diffenbaugh, C. B. Field, B. A. Hungate, E. G. Jobbagy, L. M. Kueppers, M. D. Nosetto, and D. E. Pataki. 2008. Protecting climate with forests. *Environmental Research Letters* 3(4).

Jacob, D. J., and D. A. Winner. 2009. Effect of climate change on air quality. *Atmospheric Environment* 43(1):51-63.

Jacques, P. 2006. Downscaling climate models and environmental policy: From global to regional politics. *Journal of Environmental Planning and Management* 49(2):301-307.

Jaeger, C., O. Renn, E. A. Rosa, and T. Webler. 2001. *Risk, Uncertainty and Rational Action*. London: Earthscan.

Jaffe, A. B., and R. N. Stavins. 1994. The energy efficienct gap—What does it mean? *Energy Policy* 22(10):804-810.

Jaffe, A. B., R. G. Newell, and R.N. Stavins. 2005. A tale of two market failures: Technology and environmental policy. *Ecological Economics* 54(2-3):164-174.

Jamieson, D. 1996. Ethics and intentional climate change. *Climatic Change* 33(3):323-336.

Janetos, A., L. Hansen, D. Inouye, B. P. Kelly, L. Meyerson, B. Peterson, and R. Shaw. 2008. Biodiversity. In *The Effects of Climate Change on Agriculture, Land Resources, Water Resources, and Biodiversity in the United States*. Washington, DC: Climate Change Science Program.

Jansen, E., J. Overpeck, K. R. Briffa, J.-C. Duplessy, F. Joos, V. Masson-Delmotte, D. Olago, B. Otto-Bliesner, W. R. Peltier, S. Rahmstorf, R. Ramesh, D. Raynaud, D. Rind, O. Solomina, R. Villalba, and D. Zhang. 2007. Palaeoclimate. In *Climate Change 2007: The Physical Science Basis. Contribution of Working Group I to the Fourth Assessment Report of the Intergovernmental Panel on Climate Change*, S. Solomon, D. Qin, M. Manning, Z. Chen, M. Marquis, K. B. Averyt, M. Tignor, and H. L. Miller, eds. Cambridge, U.K.: Cambridge University Press.

Janzen, H. H. 2004. Carbon cycling in earth systems—a soil science perspective. *Agriculture Ecosystems & Environment* 104(3):399-417.

Jaramillo, P., C. Samaras, H. Wakeley, and K. Meisterling. 2009. Greenhouse gas implications of using coal for transportation: Life cycle assessment of coal-to-liquids, plug-in hybrids, and hydrogen pathways. *Energy Policy* 37(7):2689-2695.

Jauregui, E., and E. Romales. 1996. Urban effects on convection precipitation in Mexico City. *Atmospheric Environment* 30:3382-3389.

Jenkins, A., and D. Holland. 2007. Melting of floating ice and sea level rise. *Geophysical Research Letters* 34(16).

Jin, M. L., R. E. Dickinson, and D. L. Zhang. 2005. The footprint of urban areas on global climate as characterized by MODIS. *Journal of Climate* 18(10):1551-1565.

Johnston, P. A., E. R. M. Archer, C. H. Vogel, C. N. Bezuidenhout, W. J. Tennant, and R. Kuschke. 2004. Review of seasonal forecasting in South Africa: Producer to end-user. *Climate Research* 28(1):67-82.

Joireman, J., H. Truelove, and B. Deull. In press. Effect of outdoor temperature, heat primes and anchoring on belief in global warming. *Journal of Environmental Psychology*.

Jones, P. G., and P. K. Thornton. 2003. The potential impacts of climate change on maize production in Africa and Latin America in 2055. *Global Environmental Change-Human and Policy Dimensions* 13(1):51-59.

Jones, T. S., A. P. Liang, E. M. Kilbourne, M. R. Griffin, P. A. Patriarca, S. G. F. Wassilak, R. J. Mullan, R. F. Herrick, H. D. Donnell, K. W. Choi, and S. B. Thacker. 1982. MORBIDITY AND MORTALITY ASSOCIATED WITH THE JULY 1980 HEAT-WAVE IN ST-LOUIS AND KANSAS-CITY, MO. *Jama-Journal of the American Medical Association* 247 (24):3327-3331.

Jorgenson, A. K. 2007. Does foreign investment harm the air we breathe and the water we drink? A cross-national study of carbon dioxide emissions and organic water pollution in less-developed countries, 1975 to 2000. *Organization & Environment* 20(2):137-156.

Jorgenson, A. K. 2009. The transnational organization of production, the scale of degradation, and ecoefficiency: A study of carbon dioxide emissions in less-developed countries. *Human Ecology Review* 16(1):64-74.

Joughin, I., S. Tulaczyk, R. Bindschadler, and S. F. Price. 2002. Changes in west Antarctic ice stream velocities: Observation and analysis. *Journal of Geophysical Research-Solid Earth* 107(B11).

Joughin, I., S. B. Das, M. A. King, B. E. Smith, I. M. Howat, and T. Moon. 2008. Seasonal speedup along the western flank of the Greenland ice sheet. *Science* 320(5877):781-783.

Jouzel, J., V. Masson-Delmotte, O. Cattani, G. Dreyfus, S. Falourd, G. Hoffmann, B. Minster, J. Nouet, J. M. Barnola, J. Chappellaz, H. Fischer, J. C. Gallet, S. Johnsen, M. Leuenberger, L. Loulergue, D. Luethi, H. Oerter, F. Parrenin, G. Raisbeck, D. Raynaud, A. Schilt, J. Schwander, E. Selmo, R. Souchez, R. Spahni, B. Stauffer, J. P. Steffensen, B. Stenni, T. F. Stocker, J. L. Tison, M. Werner, and E.W. Wolff. 2007. Orbital and millennial Antarctic climate variability over the past 800,000 years. *Science* 317(5839):793-797.

JSOST (Joint Subcommittee on Ocean Science and Technology). 2007. *Charting the Course for Ocean Science for the United States for the Next Decade: An Ocean Research Priorities Plan and Implementation Strategy*. Joint Subcommittee on Ocean Science and Technology, 84 pp. Available at *http://www.rsmas.miami.edu/groups/niehs/JSOST_briefings_general.pdf*. Accessed on October 26, 2010..

Kalff, J. 2002. *Limnology*. Upper Saddle River, NJ: Prentice Hall.

Kalkstein, L. S., J. S. Greene, D. M. Mills, A. D. Perrin, J. P. Samenow, and J.-C. Cohen. 2008. Analog European heat waves for U.S. cities to analyze impacts on heat-related mortality. *Bulletin of the American Meteorological Society* 89(1):75-85.

Kalnay, E. 2002. *Atmospheric Modeling, Data Assimilation and Predictability*. New York: Cambridge University Press, 350 pp.

Kalnay, E., M. Kanamitsu, R. Kistler, W. Collins, D. Deaven, L. Gandin, M. Iredell, S. Saha, G. White, J. Woollen, Y. Zhu, A. Leetmaa, B. Reynolds, M. Chelliah, W. Ebisuzaki, W. Higgins, J. Janowiak, K. C. Mo, C. Ropelewski, J. Wang, R. Jenne, and D. Joseph. 1996. The NCEP/NCAR 40-year reanalysis project. *Bulletin of the American Meteorological Society* 77(3):437-471.

Kapnick, S., and A. Hall. 2009. *Observed Changes in the Sierra Nevada Snowpack: Potential Causes and Concerns*. Sacramento: California Energy Commission. Available at *http://www.energy.ca.gov/2009publications/CEC-500-2009-016/CEC-500-2009-016-F.PDF*. Accessed June 16, 2010.

Karl, T. R., and C. N. Williams, Jr. 1987. An approach to adjusting climatological time series for discontinuous inhomogeneities. *Journal of Climate and Applied Meteorology* 20:1744-1763.

Kasperson, J. X., R. E. Kasperson, and B. L. Turner. 2009. Vulnerability of coupled human-ecological systems to global environmental change. Pp. 231-294 in *Human Footprints on the Global Environment: Threats to Sustainability*, E. A. Rosa, A. Diekmann, T. Dietz, and C. Jaeger, eds. Cambridge, MA: MIT Press.

Kates, R., W. Hohenemser, and C. Kasperson. 1984. *Perilous Progress: Managing the Hazards of Technology. A Westview Special Studies in Science, Technology and Public Policy*. Boulder, CO: Westview Press.

Kates, R. W. 2000. Cautionary tales: Adaptation and the global poor. *Climatic Change* 45(1):5-17.

Kates, R. W., W. C. Clark, R. Corell, J. M. Hall, C. C. Jaeger, I. Lowe, J. J. Mccarthy, H. J. Schellnhuber, B. Bolin, N. M. Dickson, S. Faucheux, G. C. Gallopin, A. Grübler, B. Huntley, J. Jager, N. S. Jodha, R. E. Kasperson, A. Mabogunje, P. Matson, H. Mooney, B. Moore, T. O'riordan, and U. Svedin. 2001. Sustainability science. *Science* 292(5517):641-642.

Kates, R. W., C. E. Colten, S. Laska, and S. P. Leatherman. 2006. Reconstruction of New Orleans after Hurricane Katrina: A research perspective. *Proceedings of the National Academy of Sciences of the United States of America* 103(40):14653-14660.

Katsman, C., W. Hazeleger, S. Drijfhout, G. Oldenborgh, and G. Burgers. 2008. Climate scenarios of sea level rise for the northeast Atlantic Ocean: A study including the effects of ocean dynamics and gravity changes induced by ice melt. *Climatic Change* 91(3-4):351-374.

Katsouyanni, K., G. Touloumi, C. Spix, J. Schwarte, F. Balducci, S. Medina, G. Rossi, B. Wojtyniak, J. Sunyer, L. Bacharova, J. P. Schouten, A. Ponka, and H. R. Anderson. 1997. Short term effects of ambient sulphur dioxide and particulate matter on mortality in 12 European cities: Results from time series data from the APHEA project. *British Medical Journal* 314(7095):1658-1663.

Kaufman, D. S., D. P. Schneider, N. P. Mckay, C. M. Ammann, R. S. Bradley, K. R. Briffa, G. H. Miller, B. L. Otto-Bliesner, J. T. Overpeck, B. M. Vinther, and Arctic Lakes 2k Project Members. 2009. Recent warming reverses long-term Arctic cooling. *Science* 325(5945):1236-1239.

Keeling, C. D., S. C. Piper, R. B. Bacastow, M. Wahlen, T. P. Whorf, M. Heimann, and H. A. Meijer. 2005. Atmospheric CO_2 and $^{13}CO_2$ exchange with the terrestrial biosphere and oceans from 1978 to 2000: Observations and carbon cycle implications. Pp. 83-113 in *A History of Atmospheric CO_2 and Its Effects on Plants, Animals, and Ecosystems*, J. R. Ehleringer, T. E. Cerling, and M. D. Dearing, eds. New York: Springer-Verlag.

Keeling, R. F., S.C. Piper, A.F. Bollenbacher, and J.S. Walker. 2009. Atmospheric CO_2 records from sites in the SIO air sampling network. In Trends: A Compendium of Data on Global Change. Carbon Dioxide Information Analysis Center, Oak Ridge National Laboratory, U.S. Department of Energy, Oak Ridge, TN, doi:10.3334/CDIAC/atg.035.

Kehrwald, N. M., L. G. Thompson, T. D. Yao, E. Mosley-Thompson, U. Schotterer, V. Alfimov, J. Beer, J. Eikenberg, and M. E. Davis. 2008. Mass loss on Himalayan glacier endangers water resources. *Geophysical Research Letters* 35(22).

Keith, D. W. 2000. Geoengineering the climate: History and prospect. *Annual Review of Energy and the Environment* 25:245-284.

Keith, D. W., J. F. Decarolis, D. C. Denkenberger, D. H. Lenschow, S. L. Malyshev, S. Pacala, and P. J. Rasch. 2004. The influence of large-scale wind power on global climate. *Proceedings of the National Academy of Sciences of the United States of America* 101(46):16115-16120.

Kelly, P. M., and W. N. Adger. 2000. Theory and practice in assessing vulnerability to climate change and facilitating adaptation. *Climatic Change* 47(4):325-352.

Kempton, W. 1991. Lay perspectives on global climate change. Pp. 29-69 in *Energy Efficiency and the Environment: Forging the Link*, E. L. Vine, D. Crawley, and P. Centolella, eds. Washington, DC: American Council for an Energy Efficient Economy.

Keppler, F., J. T. G. Hamilton, M. Braäÿ, and T. Rockmann. 2006. Methane emissions from terrestrial plants under aerobic conditions. *Nature* 439(7073):187-191.

Keutartz, J. 1999. Engineering the environment: The politics of nature development. Pp. 83-102 in Living with Nature: Environmental Politics as Cultural Discourse, F. Fischer and M. Hajer, eds. Oxford: Oxford University Press.

Khagram, S. 2004. *Dams and Development: Transnational Struggles for Water and Power*. Ithaca, NY: Cornell University Press.

Khagram, S., and S. Ali. 2006. Environment and security. *Annual Review of Environment and Resources* 31:395-411.

Khatiwala, S., F. Primeau, and T. Hall. 2009. Reconstruction of the history of anthropogenic CO_2 concentrations in the ocean. *Nature* 462(7271):346-349.

Khvorostyanov, D. V., P. Ciais, and G. Krinner. 2008. Vulnerability of east Siberia's frozen carbon stores to future warming. *Geophysical Research Letters* 35(10):L10703, doi:10710.11029/12008GL033639.

Kildow, J., and C. S. Colgan. 2005. *California's Ocean Economy*. Monterey, CA: National Ocean Economics Program.

Kim, H., S. Kim, and B. E. Dale. 2009. Biofuels, land use change, and greenhouse gas emissions: Some unexplored variables. *Environmental Science and Technology* 43:961-967.

Kim, J., and D. Waliser. 2009. A Projection of the Cold Season Hydroclimate in California in Mid-Twenty-First Century Under the SRES-A1B Emission Scenario. Draft Paper. Sacramento, CA: California Energy Commission.

King County. 2008. *Vulnerability of Major Wastewater Facilities to Flooding from Sea-Level Rise*. Seattle, WA: King County, Department of Natural Resources and Parks, Wastewater Treatment Division.

Kingston, D. G., M. C. Todd, R. G. Taylor, J. R. Thompson, and N. W. Arnell. 2009. Uncertainty in the estimation of potential evapotranspiration under climate change. *Geophysical Research Letters* 36:L20403.

Kinlan, B. P., and S. D. Gaines. 2003. Propagule dispersal in marine and terrestrial environments: A community perspective. *Ecology* 84 (8):2007-2020.

Kirllenko, A. and R. Sedjo. 2007. Climate change impacts on forestry. *Proceedings of the National Academy of Sciences of the United States of America* 104(50):19697-19702.

Kirshen, P., C. Watson, E. Douglas, A. Gontz, J. Lee, and Y. Tian. 2008. Coastal flooding in the northeastern United States due to climate change. *Mitigation and Adaptation Strategies for Global Change* 13(5):437-451.

Klein, R. J. T., S. E. H. Eriksen, L. O. Naess, A. Hammill, T. M. Tanner, C. Robledo, K. L. O'Brien. 2007. Portfolio screening to support the mainstreaming of adaptation to climate change into development assistance. *Climatic Change*, 84:23-44.

Klopper, E., C. H. Vogel, and W. A. Landman. 2006. Seasonal climate forecasts—Potential agricultural-risk management tools? *Climatic Change* 76(1-2):73-90.

Knight, J., J. J. Kennedy, C. Folland, G. Harris, G. S. Jones, M. Palmer, D. Parker, A. Scaife, and P. Stott. 2009. Do global temperature trends over the last decade falsify climate predictions? In "State of the Climate in 2008," T. C. Peterson and M. O. Baringer, eds. Special Supplement to the *Bulletin of the American Meteorological Society* 90.

Knowles, N., M. D. Dettinger, and D. R. Cayan. 2006. Trends in snowfall versus rainfall in the western United States. *Journal of Climate* 19(18):4545-4559.

Knutson, T. R., J. L. Mcbride, J. Chan, K. Emanuel, G. Holland, C. Landsea, I. Held, J. P. Kossin, A. K. Srivastava, and M. Sugi. 2010. Tropical cyclones and climate change. *Nature Geoscience* 3(3):157-163.

Knutti, R., G. A. Meehl, M. R. Allen, and D. A. Stainforth. 2006. Constraining climate sensitivity from the seasonal cycle in surface temperature. *Journal of Climate* 19:4224-4233.

Kok, M., B. Metz, J. Verhagen, and S. Van Rooijen. 2008. Integrating development and climate policies: National and international benefits. *Climate Policy* 8:103-118.

Kopp, R. E., F. J. Simons, J. X. Mitrovica, A. C. Maloof, and M. Oppenheimer. 2009. Probabilistic assessment of sea level during the last interglacial stage. *Nature* 462:863-868, doi:10.1038/nature08686.

Krueger, A. B. 2009. *Measuring the Subjective Well-Being of Nations*. Chicago: University of Chicago Press.

Kubiszewski, I., C. J. Cleveland, and P. K. Endres. 2010. Meta-analysis of net energy return for wind power systems. *Renewable Energy* 35:218-225.

Kundzewicz, Z. W., U. Ulbrich, T. Brucher, D. Graczyk, A. Kruger, G. C. Leckebusch, L. Menzel, I. Pinskwar, M. Radziejewski, and M. Szwed. 2005. Summer floods in central Europe: Climate change track? *Natural Hazards* 36(1-2):165-189.

Kundzewicz, Z. W., L. J. Mata, N. W. Arnell, P. Döll, P. Kabat, B. Jiménez, K. A. Miller, T. Oki, Z. Sen, and I. A. Shiklomanov. 2007. Freshwater resources and their management. Pp. 173-210 in *Climate Change 2007: Impacts, Adaptation and Vulnerability. Contribution of Working Group II to the Fourth Assessment Report of the Intergovernmental Panel on Climate Change*, M. L. Parry, O. F. Canziani, J. P. Palutikof, P. J. van der Linden, and C. E. Hanson, eds. Cambridge, U.K.: Cambridge University Press.

Kundzewicz, Z. W., L. J. Mata, N. W. Arnell, P. Dall, B. Jimenez, K. Miller, T. Oki, and Z. Åžen. 2009. Reply to "Climate, hydrology and freshwater: Towards an interactive incorporation of hydrological experience into climate research": Water and climate projections. *Hydrological Sciences Journal* 54(2):406-415.

Kunreuther, H. 2008. *Reducing Losses from Catastrophic Risks Through Long-Term Insurance and Mitigation*. Philadelphia: Risk Management and Decision Processes Center, The Wharton School, University of Pennsylvania.

Kverndokk, S., and A. Rose. 2008. *Equity and Justice in Global Warming Policy*. Fondazione Eni Enrico Mattei (FEEM) Working Paper 80.2008. Milan, Italy: FEEM.

Kwok, R., and D. A. Rothrock. 2009. Decline in Arctic sea ice thickness from submarine and ICESat records: 1958-2008. *Geophysical Research Letters* 36:L15501, doi:10.1029/2009GL039035.

Lackner, K. S., P. Grimes, and H.-J. Ziock. 1999. The case for carbon dioxide extraction from air. *SourceBook—The Energy Industry's Journal of Issues* 57(9):6-10.

Lake, J. A., and R. N. Wade. 2009. Plant-pathogen interactions and elevated CO_2: Morphological changes in favour of pathogens. *Journal of Experimental Botany* 60(11):3123-3131.

Lambeck, K., and J. Chappell. 2001. Sea level change through the last glacial cycle. *Science* 292(5517):679-686, doi:10.1126/science.1059549.

Lambert, F. H., N. P. Gillett, D. I. A. Stone, and C. Huntingford. 2005. Attribution studies of observed land precipitation changes with nine coupled models. *Geophysical Research Letters* 32:L18704.

Lane, L., K. Caldeira, R. Chatfield, and S. Langhoff, eds. 2007. *Workshop Report on Managing Solar Radiation*. NASA/CP-2007-214558. Moffett Field, CA: NASA Ames Research Center.

Larsen, P. H., S. Goldsmith, O. Smith, M. L. Wilson, K. Strzepek, P. Chinowsky, and B. Saylor. 2008. Estimating future costs for Alaska public infrastructure at risk from climate change. *Global Environmental Change-Human and Policy Dimensions* 18(3):442-457.

Larson, E. D. 2007. *Sustainable Bioenergy: A Framework for Decision Makers: UN-Energy*. Available at *http://esa.un.org/un-energy/pdf/susdev.Biofuels.FAO.pdf*. Accessed November 1, 2010.

Latham, J. 2002. Amelioration of global warming by controlled enhancement of the albedo and longevity of low-level maritime clouds. *Atmospheric Science Letters* 3(2-4):52-58.

Laurance, W. F. 2007. A new initiative to use carbon trading for tropical forest conservation. *Biotropica* 39(1):20-24.

Laurenti, G. 2007. *Fish and Fishery Products. World Apparent Consumption Statistics Based on Food Balance Sheets (1961-2003)*. FAO Fisheries Circular 821, Revison 8. Rome: FAO.

Lawrence, D. M., and A. G. Slater. 2005. A projection of severe near-surface permafrost degradation during the 21st century. *Geophysical Research Letters* 32(24).

Le Quéré, C., M. R. Raupach, J. G. Canadell, G. Marland, L. Bopp, P. Ciais, T. J. Conway, S. C. Doney, R. A. Feely, P. Foster, P. Friedlingstein, K. Gurney, R. A. Houghton, J. I. House, C. Huntingford, P. E. Levy, M. R. Lomas, J. Majkut, N. Metzl, J. P. Ometto, G. P. Peters, I. C. Prentice, J. T. Randerson, S. W. Running, J. L. Sarmiento, U. Schuster, S. Sitch, T. Takahashi, N. Viovy, G. R. van der Werf, and F. I. Woodward. 2009. Trends in the sources and sinks of carbon dioxide. Nature Geoscience 2:831-836, doi:10.1038/ngeo689.

Leakey, A. D. B., E. A. Ainsworth, C. J. Bernacchi, A. Rogers, S. P. Long, and D. R. Ort. 2009. Elevated CO_2 effects on plant carbon, nitrogen, and water relations: Six important lessons from FACE. *Journal of Experimental Botany* (published online April 28), doi:10.1093/jxb/erp096.

Lean, J. L. and T. N. Woods. 2010. Solar total and spectral irradiance: Measurements and models. In Evolving Solar Physics and the Climates of Earth and Space, K. Schrijver and G. Siscoe, eds. Cambridge, U.K.: Cambridge University Press.

Lehmann, J., and S. Joseph, eds. 2009. *Biochar for Environmental Management: Science and Technology*. London: Earthscan.

Lehmann, J., and S. Sohi. 2008. Comment on "Fire-derived charcoal causes loss of forest humus." *Science* 321(5894).

Lehman, S. J., and L. D. Keigwin. 1992. Sudden changes in North Atlantic circulation during the last deglaciation. *Nature* 356:757-762.

Leiserowitz, A. 2007. *International Public Opinion, Perception, and Understanding of Global Climate Change*. Human Development Report Office Occasional Paper 2007/31. New York: United Nations.

Leiserowitz, A. 2010. Climate change risk perceptions and behavior in the United States. In *Climate Change Science and Policy*, S. Schneider, A. Rosencranz, and M. Mastrandrea, eds. Washington, DC: Island Press.

Lemke, P., J. Ren, R. B. Alley, I. Allison, J. Carrasco, G. Flato, Y. Fujii, G. Kaser, P. Mote, R. H. Thomas, and T. Zhang. 2007. Observations: Changes in snow, ice and frozen ground. In *Climate Change 2007: The Physical Science Basis. Contribution of Working Group I to the Fourth Assessment Report of the Intergovernmental Panel on Climate Change*, S. Solomon, D. Qin, M. Manning, Z. Chen, M. Marquis, K. B. Averyt, M. Tignor, and H. L. Miller, eds. Cambridge, U.K.: Cambridge University Press.

Lemos, M. C. 2008. What influences innovation adoption by water managers? Climate information use in Brazil and the United States. *Journal of the American Water Resources Association* 44(6):1388-1396.

Lemos, M. C., and A. Agrawal. 2006. Environmental governance. *Annual Review of Environment and Resources* 31(3):1-329.

Lemos, M. C., and L. Dilling. 2007. Equity in forecasting climate: Can science save the world's poor? *Science and Public Policy* 34:109-116.

Lempert, R. J. 2002. A new decision sciences for complex systems. *Proceedings of the National Academy of Sciences of the United States of America* 99:7309-7313.

Lempert, R. J., and M. T. Collins. 2007. Managing the risk of uncertain threshold responses: Comparison of robust, optimum, and precautionary approaches. *Risk Analysis* 27(4):1009-1026.

Lempert, R. J., S. W. Popper, and S. C. Bankes. 2003. Shaping the next one hundred years: New methods for quantitative, long-term policy analysis. Santa Monica, CA: RAND. Available at *http://www.rand.org/publications/MR/MR1626/*. Accessed on October 26, 2010.

Lenton, T. M., and N. E. Vaughan. 2009. The radiative forcing potential of different climate geoengineering options. *Atmospheric Chemistry and Physics* 9(15):5539-5561.

Lenton, T. M., H. Held, E. Kriegler, J. W. Hall, W. Lucht, S. Rahmstorf, and H. J. Schellnhuber. 2008. Tipping elements in the Earth's climate system. *Proceedings of the National Academy of Sciences of the United States of America* 105(6):1786-1793.

Lenzen, M. 2008. Life cycle energy and greenhouse gas emissions of nuclear energy: A review. *Energy Conversion and Management* 49: 2178-2199.

Lester, S. E., B. S. Halpern, K. Grorud-Colvert, J. Lubchenco, B. I. Ruttenberg, S. D. Gaines, S. Airame, and R. R. Warner. 2009. Biological effects within no-take marine reserves: a global synthesis. *Marine Ecology-Progress Series* 384:33-46.

Leuliette, E. W., R. S. Nerem, and G. T. Mitchum. 2004. Calibration of TOPEX/Poseidon and Jason altimeter data to construct a continuous record of mean sea level change. *Marine Geodesy* 27(1-2):79-94.

Leung, L. R., L. O. Mearns, F. Giorgi, and R. L. Wilby. 2003. Regional climate research: Needs and opportunities. *Bulletin of the American Meteorological Society* 84(1):89-95.

Levitus, S. 1989. Interpentadal variability of salinity in the upper 150 m of the North Atlantic Ocean, 1970-74 versus 1955-59, *Journal of Geophysical Research* 94:9679-9685.

Levitus, S., J. I. Antonov, J. L. Wang, T. L. Delworth, K. W. Dixon, and A. J. Broccoli. 2001. Anthropogenic warming of Earth's climate system. *Science* 292(5515):267-270.

Levitus, S., J. Antonov, and T. Boyer. 2005. Warming of the world ocean, 1955-2003, *Geophysical Research Letters* 32:L02604, doi:10.1029/2004GL021592.

Levitus, S., J. I. Antonov, T. P. Boyer, R. A. Locarnini, H. E. Garcia, and A. V. Mishonov. 2009. Global ocean heat content 1955-2008 in light of recently revealed instrumentation problems. *Geophysical Research Letters* 36:L07608.

Levy, J. I., S. M. Chemerynski, and J. A. Sarnat. 2005. Ozone exposure and mortality: An empiric Bayes metaregression analysis. *Epidemiology* 16(4):458-468.

Lewis, N. S. 2007. Toward cost-effective solar energy use. *Science* 315(5813):798-801.

Lewis, C. L., and M. A. Coffroth. 2004. The acquisition of exogenous algal symbionts by an octocoral after bleaching. *Science* 304 (5676):1490-1492.

Li, W. H., R. Fu, and R. E. Dickinson. 2006. Rainfall and its seasonality over the Amazon in the 21st century as assessed by the coupled models for the IPCC AR4. *Journal of Geophysical Research* 111(D2).

Lima, F. P., N. Queiroz, P. A. Ribeiro, S. J. Hawkins, and A. M. Santos. 2006. Recent changes in the distribution of a marine gastropod, *Patella rustica Linnaeus*, 1758, and their relationship to unusual climatic events. *Journal of Biogeography* 33(5):812-822.

Lin, C. Y., F. Chen, J. C. Huang, W. C. Chen, Y. A. Liou, W. N. Chen, and S. C. Liu. 2008. Urban heat island effect and its impact on boundary layer development and land-sea circulation over northern Taiwan. *Atmospheric Environment* 42(22):5635-5649.

Lindseth, G. 2004. The Cities for Climate Protection Campaign (CCPC) and the framing of local climate policy. *Local Environment* 9(4):325-336.

Lins, H. F., and J. R. Slack. 2005. Seasonal and regional characteristics of US streamflow trends in the United States from 1940 to 1999. *Physical Geography* 26(6):489-501.

Littell, J.S., M. McGuire Elsner, L.C. Whitely Binder, and A.K. Snover, eds. 2009. *The Washington Climate Change Impacts Assessment: Evaluating Washington's Future in a Changing Climate*. Seattle, WA: Climate Impacts Group, University of Washington.

Liu, J. G., G. C. Daily, P. R. Ehrlich, and G. W. Luck. 2003. Effects of household dynamics on resource consumption and biodiversity. *Nature* 421(6922):530-533.

Liverman, D. 2009. The geopolitics of climate change: Avoiding determinism, fostering sustainable development. *Climatic Change* 96(1-2):7-11.

Lobell, D. B., K. N. Cahill, and C. B. Field. 2007. Historical effects of temperature and precipitation on California crop yields. *Climatic Change* 81(2):187-203.

Lobell, D. B., M. B. Burke, C. Tebaldi, M. D. Mastrandrea, W. P. Falcon, and R. L. Naylor. 2008. Prioritizing climate change adaptation needs for food security in 2030. *Science* 319(5863):607-610.

Logan, J. A., J. Regniere, and J. A. Powell. 2003. Assessing the impacts of global warming on forest pest dynamics. *Frontiers in Ecology and the Environment* 1(3):130-137.

Lohmann, U., and J. Feichter. 2005. Global indirect aerosol effects: A review. *Atmospheric Chemistry and Physics* 5:715-737.

Long, S. P., E. A. Ainsworth, A. D. B. Leakey, J. Nosberger, and D. R. Ort. 2006. Food for Thought: Lower-Than-Expected Crop Yield Stimulation with Rising CO2 Concentrations. *Science* 312 (5782):1918-1921.

Lonsdale, K., T. Downing, R. Nicholls, D. Parker, A. Vafeidis, R. Dawson, et al. 2008. Plausible responses to the threat of rapid sea-level rise in the Thames Estuary. *Climatic Change* 91(1):145-169.

Loreau, M., S. Naeem, P. Inchausti, J. Bengtsson, J. P. Grime, A. Hector, D. U. Hooper, M. A. Huston, D. Raffaelli, B. Schmid, D. Tilman, and D. A. Wardle. 2001. Biodiversity and ecosystem functioning: Current knowledge and future challenges. *Science* 294(5543):804-808.

Lorenzoni, I., and N. Pidgeon. 2006. Public views on climate change: European and USA perspectives. *Climatic Change* 77(1):73-95.

Lorenzoni, I., N. F. Pidgeon, and R. E. O'Connor. 2005. Dangerous climate change: The role for risk research. *Risk Analysis* 25(6):1387-1398.

Lovelock, J. 2008. A geophysiologist's thoughts on geoengineering. *Philosophical Transactions of the Royal Society A* 366(1882):3883-3890.

Lowe, D. C. 2006. Vicarious Experiences vs. Scientific Information in Climate Change Risk Perception and Behaviour: A Case Study of Undergraduate Students in Norwich, UK. Norwich, UK: Tyndall Centre for Climate Change Research.

Lozier, M. S., S. Leadbetter, R. G. Williams, V. Roussenov, M. S. C. Reed, and N. J. Moore. 2008. The spatial pattern and mechanisms of heat-content change in the North Atlantic. *Science* 319(5864):800-803, doi:10.1126/science.1146436.

Lubell, M., S. Zahran, and A. Vedlitz. 2007. Collective action and citizen responses to global warming. *Political Behavior* 29(3):391-413.

Luedeling, E., M. Zhang, and E. H. Givetz. 2009. Climatic changes lead to declining winter chill for fruit and nut trees in California during 1950-2099. *PLoS ONE* 4(7): e6166.

Luers, A. L., D. B. Lobell, L. S. Sklar, C. L. Addams, and P. A. Matson. 2003. A method for quantifying vulnerability, applied to the agricultural system of the Yaqui Valley, Mexico. *Global Environmental Change-Human and Policy Dimensions* 13(4):255-267.

Lüthi, D., M. Le Floch, B. Bereiter, T. Blunier, J.-M. Barnola, U. Siegenthaler, D. Raynaud, J. Jouzel, H. Fischer, K. Kawamura, and T. F. Stocker. 2008. High-resolution carbon dioxide concentration record 650,000-800,000 years before present. *Nature* 453(7193):379-382, doi:10.1038/nature06949.

Lythe, M. B., and D. G. Vaughan. 2001. BEDMAP: A new ice thickness and subglacial topographic model of Antarctica. *Journal of Geophysical Research* 106(B6):11335-11351.

Mackenbach, J. P., A. E. Kunst, and C. W. N. Looman. 1992. Seasonal-variation in mortality in the Netherlands. *Journal of Epidemiology and Community Health* 46(3):261-265.

Mage, D., G. Ozolins, P. Peterson, A. Webster, R. Orthofer, V. Vandeweerd, and M. Gwynne. 1996. Urban air pollution in megacities of the world. *Atmospheric Environment* 30(5):681-686.

Maibach, E., K. Wilson, and J. Witte. 2010. A *National Survey of Television Meteorologists About Climate Change: Preliminary Findings*. Fairfax, VA: George Mason University, Center for Climate Change Communication.

Malhi, Y., J. T. Roberts, R. A. Betts, T. J. Killeen, W. H. Li, and C. A. Nobre. 2008. Climate change, deforestation, and the fate of the Amazon. *Science* 319(5860):169-172.

Manning, M., M. Petit, D. Easterling, J. Murphy, A. Patwardhan, H.-H. Rogner, R. Swart, and G. Yohe, eds. 2004. *Describing Scientific Uncertainties in Climate Change to Support Analysis of Risk and of Options*. May 2004 IPCC Workshop Report. Boulder, CO: IPCC Working Group I Technical Support Unit, 138 pp. Available at *http://www.ipcc.ch/*. Accessed on October 26, 2010.

Mantua, N. J., and S. R. Hare. 2002. The Pacific decadal oscillation. *Journal of Oceanography* 58(1):35-44.

Mantua, N. J., S. R. Hare, Y. Zhang, J. M. Wallace, and R. C. Francis. 1997. A Pacific interdecadal climate oscillation with impacts on salmon production. *Bulletin of the American Meteorological Society* 78:1069-1079.

Marenco, A., H. Gouget, P. Nedelec, J. P. Pages, and F. Karcher. 1994. Evidence of a long-term increase in tropospheric ozone from Pic Du Midi data series - consequences - positive radiative forcing. *Journal of Geophysical Research* 99(D8):16617-16632.

Margreta, M., C. Ford, and M. A. Dipo. 2009. *U.S. Freight on the Move: Highlights from the 2007 Commodity Flow Survey Preliminary Data*. Washington, DC: U.S. Department of Transportation. Available at *http://www.bts.gov/publications/bts_special_report/2009_09_30/*. Accessed May 7, 2010.

Mari, X. 2008. Does ocean acidification induce an upward flux of marine aggregates? *Biogeosciences* 5(4):1023-1031.

Martin, J. H., and S. E. Fitzwater. 1988. Iron deficiency limits phytoplankton growth in the north-east Pacific subarctic. *Nature* 331 (6154):341-343.

Marx, S. M., and E. U. Weber. 2009. Decision making under climate uncertainty: The power of understanding judgment and decision processes. In *Climate Change in the Great Lakes Region: Decision Making under Uncertainty*. East Lansing, MI: MSU Press.

Marx, S. M., E. U. Weber, B. S. Orlove, A. Leiserowitz, D. H. Krantz, C. Roncoli, and J. Phillips. 2007. Communication and mental processes: Experiential and analytic processing of uncertain climate information. *Global Environmental Change-Human and Policy Dimensions* 17(1):47-58.

Matson, P. A., R. Naylor, and I. Ortiz-Monasterio. 1998. Integration of environmental, agronomic, and economic aspects of fertilizer management. *Science* 280(5360):112-115.

Matthes, F. E. 1939. Report of the Committee on Glaciers. *Transactions of the American Geophysical Union* 518-523.

Matthews, H. D., and K. Caldeira. 2007. Transient climate-carbon simulations of planetary geoengineering. *Proceedings of the National Academy of Sciences of the United States of America* 104(24):9949-9954.

Mauget, S. A. 2003. Multidecadal regime shifts in US streamflow, precipitation, and temperature at the end of the twentieth century. *Journal of Climate* 16(23):3905-3916.

Maurer, E. P., and P. B. Duffy. 2005. Uncertainty in projections of streamflow changes due to climate change in California. *Geophysical Research Letters* 32(3):5.

Mazmanian, D., and P. A. Sabatier. 1981. *Effective Policy Implementation*. Lexington, MA: D. C. Heath.

McCarl, B. A. 2008. *U.S. Agriculture in the Climate Change Squeeze: Part I: Sectoral Sensitivity and Vulnerability*. Report to the National Environmental Trust.

McCarthy, M. P., H. A. Titchner, P. W. Thorne, S. F. B. Tett, L. Haimberger, and D. E. Parker. 2008. Assessing bias and uncertainty in the HadAT-adjusted radiosonde climate record. *Journal of Climate* 21(4):817-832.

McCay, B. J., and S. Jentoft. 2009. Uncommon ground: Critical perspectives on common property. In *Human Footprints on the Global Environment: Threats to Sustainability*, E. A. Rosa, A. Diekmann, T. Dietz, and C. C. Jaeger, eds. Cambridge, MA: MIT Press.

McConnell, J. R., R. Edwards, G. L. Kok, M. G. Flanner, C. S. Zender, E. S. Saltzman, J. R. Banta, D. R. Pasteris, M. M. Carter, and J. D. W. Kahl. 2007. 20th-century industrial black carbon emissions altered Arctic climate forcing. *Science* 317(5843):1381-1384.

McDonald A, Riha S, DiTommaso A, and A. DeGaetano. 2009. Climate change and the geography of weed damage: Analysis of US maize systems suggests the potential for significant range transformations. *Agriculture Ecosystems & Environment* 130(3-4):131-140.

McDowall, R. M. 1992. Particular problems for the conservation of diadromous fish. *Aquatic Conservation-Marine and Freshwater Ecosystems* 2(4):351-355.

McGowan, J. A., D. R. Cayan, and L. M. Dorman. 1998. Climate-ocean variability and ecosystem response in the northeast Pacific. *Science* 281(5374):210-217.

McGranahan, G., D. Balk, and B. Anderson. 2007. The rising tide: Assessing the risks of climate change and human settlements in low elevation coastal zones. *Environment and Urbanization* 19(1):17-37.

McGuire, A. D., F. S. Chapin, J. E. Walsh, and C. Wirth. 2006. Integrated regional changes in Arctic climate feedbacks: Implications for the global climate system. *Annual Review of Environment and Resources* 31:61-91.

McLaughlin, S. B., D. Ugarte, C. T. Garten, L. R. Lynd, M. A. Sanderson, V. R. Tolbert, and D. D. Wolf. 2002. High-value renewable energy from prairie grasses. *Environmental Science & Technology* 36(10):2122-2129.

McManus, J. F., R. Francois, J.-M. Gherardi, L. D. Keigwin, and S. Brown-Leger. 2004. Collapse and rapid resumption of Atlantic meridional circulation linked to deglacial climate changes. *Nature* 428:834-837, doi:10.1038/nature02494.

McMichael, A. J., R. E. Woodruff, and S. Hales. 2006. Climate change and human health: present and future risks. *Lancet* 367(9513):859-869.

MEA (Millennium Ecosystem Assessment). 2005. *Ecosystems and Human Well-Being. The Millennium Ecosystem Assessment series*. Washington, DC: Island Press, 4 vols.

Mears, C. A. and F. J. Wentz. 2005. The Effect of Drifting Measurement Time on Satellite-Derived Lower Tropospheric Temperature *Science*, 309: 1548-1551

Meehl, G. A., T. F. Stocker, W. D. Collins, P. Friedlingstien, A. T. Gaye, et al. 2007a. Global climate projections. In *Climate Change 2007a: The Physical Science Basis. Contribution of Working Group I to the Fourth Assessment Report of the Intergovernmental Panel on Climate Change*, S. Solomon, D. Qin, M. Manning, Z. Chen, M. Marquis, K. B. Averyt, M. Tignor, and H. L. Miller, eds. Cambridge, U.K.: Cambridge University Press.

Meehl, G. A., C. Covey, T. Delworth, M. Latif, B. McAvaney, J. F. B. Mitchell, R. J. Stouffer, and K. E. Taylor. 2007b. The WCRP CMIP3 multi-model dataset: A new era in climate change research. *Bulletin of the American Meteorological Society* 88:1383-1394.

Meehl, G. A., J. M. Arblaster, K. Matthes, F. Sassi, and H. van Loon. 2009a. Amplifying the Pacific climate system response to a small 11-year solar cycle forcing. *Science* 325(5944):1114-1118.

Meehl, G. A., L. Goddard, J. Murphy, R. J. Stouffer, G. Boer, G. Danabasoglu, K. Dixon, M. A. Giorgetta, A. M. Greene, E. Hawkins, G. Hegerl, D. Karoly, N. Keenlyside, M. Kimoto, B. Kirtman, A. Navarra, R. Pulwarty, D. Smith, D. Stammer, and T. Stockdale. 2009b. Decadal prediction. *Bulletin of the American Meteorological Society* 90(10):1467-1485.

Meehl, G. A., C. Tebaldi, G. Walton, D. Easterling, and L. Mcdaniel. 2009c. Relative increase of record high maximum temperatures compared to record low minimum temperatures in the U.S. *Geophysical Research Letters* 36(23):L23701.

Meinke, H., S. M. Howden, P. C. Struik, R. Nelson, D. Rodriguez, and S. C. Chapman. 2009. Adaptation science for agriculture and natural resource management—urgency and theoretical basis. *Current Opinion in Environmental Sustainability* 1(1):69-76.

Mendelsohn, R., and S. Olmstead. 2009. The economic valuation of environmental amenities and disamenities: Methods and applications. *Annual Review of Environment and Resources* 34:325-347, doi:10.1146/annurev-environ-011509-135201.

Menne, M. J., and C. N. Williams. 2009. Homogenization of Temperature Series via Pairwise Comparisons. *Journal of Climate* 22 (7):1700-1717.

Menne, M. J., C. N. Williams, Jr., and R. S. Vose. 2009. The U.S. historical climatology network monthly temperature data, version 2. *Bulletin of the American Meteorological Society* 90:993-1007, doi:10.1175/2008BAMS2613.1.

Menon, S., J. Hansen, L. Nazarenko, and Y. F. Luo. 2002. Climate effects of black carbon aerosols in China and India. *Science* 297(5590):2250-2253.

Mileti, D. S. 1999. Disasters by design: A reassessment of natural hazards in the United States. In *Natural Hazards and Disasters*. Washington, DC: Joseph Henry Press.

Miller, A. W., A. C. Reynolds, C. Sobrino, and G. F. Riedel. 2009. Shellfish face uncertain future in high CO_2 world: Influence of acidification on oyster larvae calcification and growth in estuaries. *PLoS One* 4(5):e5661.

Miller, L., and B. C. Douglas. 2004. Mass and volume contributions to twentieth-century global sea level rise. *Nature* 428(6981):406-409.

Milly, P. C. D., R. T. Wetherald, K. A. Dunne, and T. L. Delworth. 2002. Increasing risk of great floods in a changing climate. *Nature* 415(6871):514-517.

Milner, A. M., L. E. Brown, and D. M. Hannah. 2009. Hydroecological response of river systems to shrinking glaciers. *Hydrological Processes* 23(1):62-77.

Mirza, M. M. Q. 2003. Climate change and extreme weather events: Can developing countries adapt? *Climate Policy* 3(3):233-248.

MIT (Massachusetts Institute of Technology). 2003. *The Future of Nuclear Power*. Cambridge, MA: MIT Press.

MIT. 2006. *The Future of Geothermal Energy in the 21st Century: Impact of Enhanced Geothermal Systems (EGS) on the United States*. Cambridge: MIT Press.

Mitchell, D. L., and W. Finnegan. 2009. Modification of cirrus clouds to reduce global warming. *Environmental Research Letters* 4(4):045102.

Mitchell, R. B. 2003. International environmental agreements: A survey of their features, formation, and effects. *Annual Review of Environment and Resources* 28:429-461.

Mitrovica, J. X., M. E. Tamisiea, J. L. Davis, and G. A. Milne. 2001. Recent mass balance of polar ice sheets inferred from patterns of global sea-level change. *Nature* 409(6823):1026-1029.

MMS (Minerals Management Service). 2006. *Technology White Paper on Wave Energy Potential on the U.S. Outer Continental Shelf. Renewable Energy and Alternate Use Program*. U.S. Department of the Interior. Available at *http://ocsenergy.anl.gov*. Accessed July 13, 2010.

Mohan, J. E., L. H. Ziska, W. H. Schlesinger, R. B. Thomas, R. C. Sicher, K. George, and J. S. Clark. 2006. Biomass and toxicity responses of poison ivy (*Toxicodendron radicans*) to elevated atmospheric CO_2. *Proceedings of the National Academy of Sciences of the United States of America* 103(24):9086-9089.

Moore, S. E., and H. P. Huntington. 2008. Arctic marine mammals and climate change: Impacts and resilience. *Ecological Applications* 18 (2):S157-S165.

Mora, C., R. A. Myers, M. Coll, S. Libralato, T. J. Pitcher, R. U. Sumaila, D. Zeller, R. Watson, K. J. Gaston, and B. Worm. 2009. Management effectiveness of the world's marine fisheries. *PLoS Biology* 7(6):e1000131.

Moran, E. 2009. Progress in the last ten years in the study of land use/cover change and the outlook for the next decade. In *Human Footprints on the Global Environment: Threats to Sustainability*, E. A. Rosa, A. Diekmann, T. Dietz, and C. C. Jaeger, eds. Cambridge, MA: MIT Press.

Moran, E. F., and E. Ostrom. 2005. *Seeing the Forest and the Trees: Human-Environment Interactions in Forest Ecosystems*. Cambridge, MA: MIT Press.

Morgan, M. G., B. Fischhoff, A. Bostrom, and C. J. Atman. 2001. *Risk Communication: A Mental Models Approach*. New York: Cambridge University Press.

Moser, S. 2009b. Whether our levers are long enough and the fulcrum strong? Exploring the soft underbelly of adaptation decisions and actions. In *Living with Climate Change*, N. W. Adger, eds. Cambridge: Cambridge University Press.

Moser, S., and A. Luers. 2008. Managing climate risks in California: The need to engage resource managers for successful adaptation to change. *Climatic Change* 87(0):309-322.

Moser, S., R. Kasperson, G. Yohe, and J. Agyeman. 2008. Adaptation to climate change in the Northeast United States: Opportunities, processes, constraints. *Mitigation and Adaptation Strategies for Global Change* 13(5):643-659.

Moser, S. C. 2009a. Good Morning, America! The Explosive U.S. Awakening to the Need for Adaptation. Charleston, SC: Coastal Services Center, National Ocean Service, NOAA.

Moser, S. C. 2010. Communicating climate change: History, challenges, process and future directions. *Wiley Interdisciplinary Reviews: Climate Change* 1(1):31-53.

Moser, S. C., and L. Dilling. 2007. *Creating a Climate for Change: Communicating Climate Change and Facilitating Social Change*. Cambridge: Cambridge University Press.

Moser, S. C., and J. Tribbia. 2006. Vulnerability to inundation and climate change impacts in California: Coastal managers' attitudes and perceptions. *Marine Technology Society Journal* 40(4):35-44.

Moser, S. C., and J. Tribbia. 2007. *Vulnerability to Coastal Impacts of Climate Change: Coastal Managers' Attitudes, Knowledge, Perceptions, and Actions*. PIER Project Report CEC-500-2007-082 prepared for the California Energy Commission.

Mosier, A., and C. Kroeze. 2000. Potential impact on the global atmospheric N_2O budget of the increased nitrogen input required to meet future global food demands. *Chemosphere—Global Change Science* 2(3-4):465-473.

Mosier, A. R., J. M. Duxbury, J. R. Freney, O. Heinemeyer, K. Minami, and D. E. Johnson. 1998. Mitigating agricultural emissions of methane. *Climatic Change* 40(1):39-80.

Moss, R. H., and S. H. Schneider. 2000. Uncertainties in the IPCC TAR: Recommendations to lead authors for more consistent assessment and reporting. Pp. 33-51 in *Guidance Papers on the Cross Cutting Issues of the Third Assessment Report of the IPCC*, R. Pachauri, T. Taniguchi, and K. Tanaka, eds. Geneva: World Meteorological Organization.

Moss, R. H., B. Lim, E. L. Malone, and A. L. Brenkert. 2002. *Developing Socio-Economic Scenarios for Use in Vulnerability and Adaptation Assessments*. New York: United Nations Development Programme.

Moss, R. H., J. A. Edmonds, K. A. Hibbard, M. R. Manning, S. K. Rose, D. P. Van Vuuren, T. R. Carter, S. Emori, M. Kainuma, T. Kram, G. A. Meehl, J. F. B. Mitchell, N. Nakicenovic, K. Riahi, S. J. Smith, R. J. Stouffer, A. M. Thomson, J. P. Weyant, and T. J. Wilbanks. 2010. The next generation of scenarios for climate change research and assessment. *Nature* 463(7282):747-756.

Mote, P. W., A. F. Hamlet, M. P. Clark, and D. P. Lettenmaier. 2005. Declining mountain snowpack in western North America. *Bulletin of the American Meteorological Society* 86(1):39-49..

Mote, T. L. 2007. Greenland surface melt trends 1973-2007: Evidence of a large increase in 2007. *Geophysical Research Letters* 34(22).

Mudway, I. S., and F. J. Kelly. 2000. Ozone and the lung: A sensitive issue. *Molecular Aspects of Medicine* 21(1-2):1-48.

Mueter, F. J., and M. A. Litzow. 2008. Sea ice retreat alters the biogeography of the Bering Sea Continental Shelf. *Ecological Applications* 18(2):309-320.

Murphy, D. M., S. Solomon, R. W. Portmann, K. H. Rosenlof, P. M. Forster, and T. Wong. 2009. An observationally based energy balance for the Earth since 1950. *Journal of Geophysical Research* 114: D17107.

Murphy, J. M., D. M. H. Sexton, D. N. Barnett, G. S. Jones, M. J. Webb, M. Collins, and D. A. Stainforth. 2004. Quantification of modelling uncertainties in a large ensemble of climate change simulations. *Nature* 430:768-772.

Miyamoto, K. 1997. *Renewable Biological Systems for Alternative Sustainable Energy Production*, FAO Agricultural Services Bulletin 128. Rome: FAO.

Najjar, R., L. Patterson, and S. Graham. 2009. Climate simulations of major estuarine watersheds in the Mid-Atlantic region of the US. *Climatic Change* 95(1):139-168.

Nakicenovic, N. 2000. *Special Report on Emissions Scenarios: A Special Report of Working Group III of the Intergovernmental Panel on Climate Change*. New York: Cambridge University Press.

NASA (National Aeronautics and Space Administration). 2001. *Goddard Space Flight Center Scientific Visualization Studio, "Earth at Night 2001."* Available at *http://visibleearth.nasa.gov/view_rec.php?id=11795*. Access on October 26, 2010.

NASA. 2008. Advanced Global Atmospheric Gases Experiment (AGAGE). Sponsored by NASA's Atmospheric Composition Focus Area in Earth Science. Available at *http://agage.eas.gatech.edu/index.htm*. Accessed May 13, 2010.

NASA GISS (Goddard Institute for Space Studies). 2010. *Datasets and Images. Global Annual Mean Surface Air Temperature Change*. Available at *http://data.giss.nasa.gov/gistemp/graphs/*. Accessed May 13, 2010.

Naughton, M. P., A. Henderson, M. C. Mirabelli, R. Kaiser, J. L. Wilhelm, S. M. Kieszak, C. H. Rubin, and M. A. Mcgeehin. 2002. Heat-related mortality during a 1999 heat wave in Chicago. *American Journal of Preventive Medicine* 22(4):221-227.

NCDC (National Climatic Data Center). 2006. *Federal Climate Complex Global Surface Summary of Day Data. Version 7*. Available at ftp://ftp.ncdc.noaa.gov/pub/data/gsod/readme.txt. Accessed May 13, 2010.

Neftel, A., H. Friedli, E. Moor, H. Lötscher, H. Oeschger, U. Siegenthaler, and B. Stauffer. 1994. Historical CO_2 record from the Siple Station ice core. In *Trends: A Compendium of Data on Global Change*. Oak Ridge, TN: Carbon Dioxide Information Analysis Center, Oak Ridge National Laboratory, DOE.

Nelson, E., G. Mendoza, J. Regetz, S. Polasky, H. Tallis, D. R. Cameron, K. M. A. Chan, G. C. Daily, J. Goldstein, P. M. Kareiva, E. Lonsdorf, R. Naidoo, T. H. Ricketts, and M. R. Shaw. 2009. Modeling multiple ecosystem services, biodiversity conservation, commodity production, and tradeoffs at landscape scales. *Frontiers in Ecology and the Environment* 7(1):4-11.

Nepstad, D. C., C. M. Stickler, and O. T. Almeida. 2006. Globalization of the Amazon soy and beef industries: Opportunities for conservation. *Conservation Biology* 20(6):1595-1603.

NIC (National Intelligence Council). 2009a. *China: The Impact of Climate Change to 2030—Geopolitical Implications: Conference Report*. NIC-CR 2009-09, June. Prepared jointly by CENTRA Technology, Inc. and Scitor Corporation. Available at *http://www.dni.gov/nic/PDF_GIF_otherprod/climate_change/cr200909_china_climate_change.pdf*. Accessed May 13, 2010.

NIC. 2009b. *China: The Impact of Climate Change to 2030: Special Report*. NIC-2009-02D. Prepared by Joint Global Change Research Institute and Battelle Memorial Institute, Pacific Northwest Division. Available at *http://www.dni.gov/nic/PDF_GIF_otherprod/climate_change/climate2030_china.pdf*. Accessed May 13, 2010.

NIC. 2009c. *India: The Impact of Climate Change to 2030—Geopolitical Implications: Conference Report*. NIC-CR 2009-07, May. Prepared jointly by CENTRA Technology, Inc. and Scitor Corporation. Available at *http://www.dni.gov/nic/PDF_GIF_otherprod/climate_change/cr200907_india_climate_change.pdf*. Accessed May 13, 2010.

NIC. 2009d. *India: The Impact of Climate Change to 2030: Special Report*. NIC-2009-03D. Prepared By Joint Global Change Research Institute and Battelle Memorial Institute, Pacific Northwest Division. Available at *http://www.dni.gov/nic/PDF_GIF_otherprod/climate_change/climate2030_india.pdf*. Accessed May 13, 2010.

NIC. 2009e. *North Africa: Special Report*. NIC-2009-07D. Prepared By Joint Global Change Research Institute and Battelle Memorial Institute, Pacific Northwest Division. Available at *http://www.dni.gov/nic/PDF_GIF_otherprod/climate_change/climate2030_north_africa.pdf*. Accessed May 13, 2010.

NIC. 2009f. *Russia: The Impact of Climate Change to 2030—Geopolitical Implications: Conference Report*. NIC-CR 2009-16, September. Prepared jointly by CENTRA Technology, Inc. and Scitor Corporation. Available at *http://www.dni.gov/nic/PDF_GIF_otherprod/climate_change/cr200916_russia_climate_change.pdf*. Accessed May 13, 2010.

NIC. 2009g. *Russia: The Impact of Climate Change to 2030: Special Report*. NIC-2009-04D. Prepared by Joint Global Change Research Institute and Battelle Memorial Institute, Pacific Northwest Division. Available at *http://www.dni.gov/nic/PDF_GIF_otherprod/climate_change/climate2030_russia.pdf*. Accessed May 13, 2010.

NIC. 2009h. *Southeast Asia and Pacific Islands: Special Report*. NIC-2009-06D. Prepared by Joint Global Change Research Institute and Battelle Memorial Institute, Pacific Northwest Division. Available at *http://www.dni.gov/nic/PDF_GIF_otherprod/climate_change/climate2030_southeast_asia_pacific_islands.pdf*. Accessed May 13, 2010.

Nichols, N. 1999. Cognitive illusions, heuristics, and climate prediction. *Bulletin of the American Meteorological Society* 80(7):1385-1396.

Nicholls, R. J. 2004. Coastal flooding and wetland loss in the 21st century: Changes under the SRES climate and socio-economic scenarios. *Global Environmental Change-Human and Policy Dimensions* 14(1):69-86.

Nicholls, R. J., and R. S. J. Tol. 2006. Impacts and responses to sea-level rise: A global analysis of the SRES scenarios over the twenty-first century. *Philosophical Transactions of the Royal Society A* 364(1841):1073-1095.

Nicholls, R. J., F. M. J. Hoozemans, and M. Marchand. 1999. Increasing flood risk and wetland losses due to global sea-level rise: Regional and global analyses. *Global Environmental Change-Human and Policy Dimensions* 9:S69-S87.

Nicholls, R. J., P. P. Wong, V. R. Burkett, J. O. Codignotto, J. E. Hay, R. F. McLean, S. Ragoonaden, and C. D. Woodroffe. 2007. Coastal systems and low-lying areas. Pp. 315-356 in *Climate Change 2007: Impacts, Adaptation and Vulnerability. Contribution of Working Group II to the Fourth Assessment Report of the Intergovernmental Panel on Climate Change*, M. L. Parry, O. F. Canziani, J. P. Palutikof, P. J. van der Linden, and C. E. Hanson, eds. Cambridge, U.K.: Cambridge University Press.

Nick, F. M., A. Vieli, I. M. Howat, and I. Joughin. 2009. Large-scale changes in Greenland outlet glacier dynamics triggered at the terminus. *Nature Geoscience* 2(2):110-114.

NIFC (National Interagency Fire Center). 2008. *Total Wildland Fires and Acres(1960-2007)*. National Interagency Coordination Center 2008. Available at *http://www.nifc.gov/fire_info/fires_acres.htm*. Accessed on October 26, 2010.

Nilsson, J., G. Brostrom, and G. Walin. 2003. The thermohaline circulation and vertical mixing: Does weaker density stratification give stronger overturning? *Journal of Physical Oceanography* 33(12):2781-2795.

Nisbet, M. C., and C. Mooney. 2007. Framing science. *Science* 316(5821) :56.

NOAA (National Oceanic and Atmospheric Administration). 2010. *Commerce Department Proposes Establishment of NOAA Climate Service*. Available at *http://www.noaanews.noaa.gov/stories2010/20100208_climate.html*. Accessed May 3, 2010.

NOAA/ESRL (Earth System Research Laboratory). 2009. *The NOAA Annual Greenhouse Gas Index (AGGI)*. Available at *http://www.esrl.noaa.gov/gmd/aggi/*. Accessed on October 26, 2010.

NOAA/NCDC. 2008. *The USHCN Version 2 Serial Monthly Dataset*. Available at *http://www.ncdc.noaa.gov/oa/climate/research/ushcn/*. Accessed May 13, 2010.

Noerdlinger, P. D., and K. R. Brower. 2007. The melting of floating ice raises the ocean level. *Geophysical Journal International* 170(1):145-150.

Norby, R. J., E. H. DeLucia, B. Gielen, C. Calfapietra, C. P. Giardina, J. S. King, J. Ledford, H. R. McCarthy, D. J. P. Moore, R. Ceulemans, P. De Angelis, A. C. Finzi, D. F. Karnosky, M. E. Kubiske, M. Lukac, K. S. Pregitzer, G. E. Scarascia-Mugnozza, W. H. Schlesinger, and R. Oren. 2005. Forest response to elevated CO_2 is conserved across a broad range of productivity. *Proceedings of the National Academy of Sciences of the United States of America* 102(50):18052-18056.

Nordhaus, W. 2008. *A Question of Balance: Weighing the Options on Global Warming Policies*. New Haven, CT: Yale University Press.

Norman, E., and K. Bakker. 2009. Transgressing scales: Water governance across the Canada-U.S. borderland. *Annals of the Association of American Geographers* 99:99-117.

NRC (National Research Council). 1980a. *Improving Risk Communication*. Washington, DC: National Academy Press.

NRC. 1980b. *Energy in Transition, 1985-2010: Final Report of the Committee on Nuclear and Alternative Energy Systems*. Washington, DC: National Academies Press.

NRC. 1980c. *Strategy for the National Climate Program*. Washington, DC: National Academy Press.

NRC. 1986. The National Climate Program: Early Achievements and Future Directions. Washington, DC: National Academies Press.

NRC. 1988. Toward an Understanding of Global Change: Initial Priorities for U.S. Contributions to the International Geosphere - Biosphere Program. Washington, DC: National Academies Press.

NRC. 1990a. *Research Strategies for the U.S. Global Change Research Program*. Washington, DC: National Academy Press.

NRC. 1990b. *Sea-Level Change*. Washington, DC: National Academy Press.

NRC. 1992a. *Global Environmental Change: Understanding the Human Dimensions.* Washington, DC: National Academy Press.

NRC. 1992b. *Policy Implications of Greenhouse Warming: Mitigation, Adaptation, and the Science Base.* Washington, DC: National Academy Press.

NRC. 1993. *Research to Protect, Restore, and Manage the Environment.* Committee on Environmental Research. Washington, DC: National Academy Press.

NRC. 1996. *Understanding Risk: Informing Decisions in a Democratic Society.* Washington, DC: National Academy Press.

NRC. 1997a. *Environmentally Significant Consumption: Research Directions. Human Dimensions of Global Environmental Change: Research Pathways for the Next Decade.* Washington, DC: National Academy Press.

NRC. 1997b. *Nature and Human Society: The Quest for a Sustainable World.* Washington, DC: National Academy Press.

NRC. 1999a. *Hydrologic Science Priorities for the U.S. Global Change Research Program: An Initial Assessment.* Washington, DC: National Academy Press.

NRC. 1999b. *Making Climate Forecasts Matter.* Washington, DC: National Academy Press.

NRC. 1999c. *Our Common Journey: A Transition Toward Sustainability.* Washington, DC: National Academy Press.

NRC. 2000. Clean Coastal Waters Understanding and Reducing the Effects of Nutrient Pollution. Washington, DC: National Academy Press.

NRC. 2001. *Grand Challenges in Environmental Sciences.* Washington, DC: National Academy Press.

NRC. 2002a. *Abrupt Climate Change: Inevitable Surprises.* Washington, DC: The National Academies Press.

NRC. 2002b. *The Drama of the Commons.* Washington, DC: The National Academies Press.

NRC. 2003a. *Planning Climate and Global Change Research: A Review of the Draft U.S. Climate Change Science Program Strategic Plan.* Washington, DC: The National Academies Press.

NRC. 2003b. *Understanding Climate Change Feedbacks.* Washington, DC: The National Academies Press.

NRC. 2004a. *Climate Data Records from Environmental Satellites.* Washington, DC: The National Academies Press.

NRC. 2004b. *Implementing Climate and Global Change Research: A Review of the Final U.S. Climate Change Science Program Strategic Plan.* Washington, DC: The National Academies Press.

NRC. 2005a. *Decision Making for the Environment: Social and Behavioral Science Research Priorities.* Washington, DC: The National Academies Press.

NRC. 2005b. *Earth Science and Applications from Space: Urgent Needs and Opportunities to Serve the Nation.* Washington, DC: The National Academies Press.

NRC. 2005c. *Population, Land Use, and Environment: Research Directions.* Washington, DC: The National Academies Press.

NRC. 2005d. *Radiative Forcing of Climate Change.* Washington, DC: The National Academies Press.

NRC. 2005e. *Review of NOAA's Plan for the Scientific Stewardship Program.* Washington, DC: The National Academies Press.

NRC. 2005f. Thinking Strategically: The Appropriate Use of Metrics for the Climate Change Science Program. Washington, DC: National Academies Press.

NRC. 2006a. *Commuting in America III.* Washington, DC: The National Academies Press.

NRC. 2006b. *Surface Temperature Reconstructions for the Last 2,000 Years.* Washington, DC: The National Academies Press.

NRC. 2007a. *Analysis of Global Change Assessments: Lessons Learned.* Washington, DC: The National Academies Press.

NRC. 2007b. *Colorado River Basin Water Management: Evaluating and Adjusting to Hydroclimatic Variability.* Washington, DC: The National Academies Press.

NRC. 2007c. *Earth Science and Applications from Space: National Imperatives for the Next Decade and Beyond.* Washington, DC: The National Academies Press.

NRC. 2007d. *Environmental Data Management at NOAA: Archiving, Stewardship, and Access.* Washington, DC: The National Academies Press.

NRC. 2007e. *Environmental Impacts of Wind-Energy Projects.* Washington, DC: The National Academies Press.

NRC. 2007f. *Evaluating Progress of the U.S. Climate Change Science Program: Methods and Preliminary Results.* Washington, DC: The National Academies Press.

NRC. 2007g. *Polar Icebreakers in a Changing World: An Assessment of U.S. Needs.* Washington, DC: The National Academies Press.

NRC. 2007h. *Research and Networks for Decision Support in the NOAA Sectoral Applications Research Program*. Washington DC: The National Academies Press.

NRC. 2007i. *Putting People on the Map: Protecting Confidentiality with Linked Social-Spatial Data*. Washington, DC: The National Academies Press.

NRC. 2008a. *Bioinspired Chemistry for Energy: A Workshop Summary to the Chemical Sciences Roundtable*. Washington, DC: The National Academies Press.

NRC. 2008b. *Ecological Impacts of Climate Change*. Washington, DC: The National Academies Press.

NRC. 2008c. *Earth Observations from Space: The First 50 Years of Scientific Achievement*. Washington, DC: The National Academies Press

NRC. 2008d. *Ensuring the Climate Record from the NPOESS and GOES-R Spacecraft: Elements of a Strategy to Recover Measurement Capabilities Lost in Program Restructuring*. Washington, DC: The National Academies Press.

NRC. 2008e. *Estimating Mortality Risk Reduction and Economic Benefits from Controlling Ozone Air Pollution*. Washington, DC: The National Academies Press.

NRC. 2008f. *Minerals, Critical Minerals, and the U.S. Economy*. Washington, DC: The National Academies Press.

NRC. 2008g. *Potential Impacts of Climate Change on US Transportation*. Washington, DC: The National Academies Press.

NRC. 2008h. *Public Participation in Environmental Assessment and Decision Making*. Washington, DC: The National Academies Press.

NRC. 2008i. *Transitioning to Sustainability Through Research and Development on Ecosystem Services and Biofuels: Workshop Summary*. Washington, DC: The National Academies Press.

NRC. 2009a. *America's Energy Future: Electricity from Renewable Resources: Status, Prospects, and Impediments*. Washington, DC: The National Academies Press.

NRC. 2009b. *America's Energy Future: Liquid Transportation Fuels from Coal and Biomass*. Washington, DC: The National Academies Press.

NRC. 2009c. *America's Energy Future: Real Prospects for Energy Efficiency in the United States*. Washington, DC: The National Academies Press.

NRC. 2009d. *America's Energy Future: Technology and Transformation*. Washington, DC: The National Academies Press.

NRC. 2009e. *Driving and the Built Environment: The Effects of Compact Development on Motorized Travel, Energy Use, and CO_2 Emissions*. Washington, DC: The National Academies Press.

NRC. 2009f. *Hidden Costs of Energy: Unpriced Consequences of Energy Production and Use*. Washington, DC: The National Academies Press.

NRC. 2009g. *Informing Decisions in a Changing Climate*. Washington, DC: The National Academies Press.

NRC. 2009h. *Letter Report on the Orbiting Carbon Observatory*. Washington, DC: The National Academies Press.

NRC. 2009i. *A New Biology for the 21st Century: Ensuring the United States Leads the Coming Biology Revolution*. Washington, DC: The National Academies Press.

NRC. 2009j. *Observing Weather and Climate from the Ground Up: A Nationwide Network of Networks*. Washington, DC: The National Academies Press.

NRC. 2009k. *Restructuring Federal Climate Research to Meet the Challenges of Climate Change*. Washington, DC: The National Academies Press.

NRC. 2009l. *Scientific Value of Arctic Sea Ice Imagery Derived Products*. Washington, DC: The National Academies Press.

NRC. 2010a. *America's Climate Choices: Adapting to the Impacts of Climate Change*. Washington, DC: The National Academies Press.

NRC. 2010b. *America's Climate Choices: Informing an Effective Response to Climate Change*. Washington, DC: The National Academies Press.

NRC. 2010c. *America's Climate Choices: Limiting the Magnitude of Future Climate Change*. Washington, DC: The National Academies Press.

NRC. 2010d. *Certifiably Sustainable?: The Role of Third-Party Certification Systems: Report of a Workshop*. Washington, DC: The National Academies Press.

NRC. 2010e. *National Security Implications of Climate Change for U.S. Naval Forces: Letter Report*. Washington, DC: The National Academies Press

NRC. 2010f. *Ocean Acidification: A National Strategy to Meet the Challenges of a Changing Ocean.* Washington, DC: The National Academies Press.

NRC. 2010g (in press). *Potential Energy Savings and Greenhouse Gas Reductions from Transportation.* Washington, DC: The National Academies Press.

NRC. 2010h. *Realizing the Energy Potential of Methane Hydrate for the United States.* Washington, DC: The National Academies Press.

NRC. 2010i. *Technologies and Approaches to Reducing the Fuel Consumption of Medium- and Heavy-Duty Vehicles.* Washington, DC: The National Academies Press.

NRC. 2010j. *Climate Stabilization Targets: Emissions, Concentrations, and Impacts over Decades to Millennia.* Washington, DC: The National Academies Press.

NRC. 2010k. *Verifying Greenhouse Gas Emissions: Methods to Support International Climate Agreements.* Washington, DC: The National Academy Press.

NSIDC (National Snow and Ice Data Center). 2010. *Sea Ice Extent.* Available at *http://nsidc.org/data/seaice_index/.* Accessed May 13, 2010.

NYCDEP (New York City Department of Environmental Protection). 2008. *Assessment and Action Plan. Report 1.* New York: NYCDEP.

O'Brien, K. L., and R. M. Leichenko. 2000. Double exposure: Assessing the impacts of climate change within the context of globalization. *Global Environmental Change* 10:221-232.

O'Brien, K., and R. M. Leichenko. 2003. Winners and losers in the context of global change. *Annals of the Association of American Geographers* 93:89-103.

O'Brien, K., R. Leichenko, U. Kelkar, H. Venema, G. Aandahl, H. Tompkins, A. Javed, S. Bhadwal, S. Barg, L. Nygaard, and J. West. 2004. Mapping vulnerability to multiple stressors: Climate change and globalization in India. *Global Environmental Change-Human and Policy Dimensions* 14(4):303-313.

O'Brien, K., S. Eriksen, L. Sygna, and L. O. Naess. 2006. Questioning complacency: Climate change impacts, vulnerability, and adaptation in Norway. *Ambio* 35(2):50-56.

O'Brien, K., Et Al. 2008. Disaster Risk Reduction, Climate Change Adaptation and Human Security. Oslo, Norway: Report prepared for the Royal Norwegian Ministry of Foreign Affairs by the Global Environmental Change and Human Security (GECHS) Project.

O'Brien, K., J. Wolf, and L. Sygna. 2009. Global environmental change and human security (GECHS) in a decade that matters. *IHDP Update* June:2.

Oladosu, G., and A. Rose. 2007. Income distribution impacts of climate change mitigation policy in the Susquehann River Basin Economy. *Energy Economics* 29:520-544.

Oldenburg, C. M., J.-P. Nicot, and S. L. Bryant. 2009. Case studies of the application of the Certification Framework to two geologic carbon sequestration sites. *Energy Procedia* 1:63-70.

Ollinger, S. V., A. D. Richardson, M. E. Martin, D. Y. Hollinger, S. E. Frolking, P. B. Reich, L. C. Plourde, G. G. Katul, J. W. Munger, R. Oren, M.-L. Smith, K. T. Paw, P. V. Bolstad, B. D. Cook, M. C. Day, T. A. Martin, R. K. Monson, and P. Schmid. 2008. *Proceedings of the National Academy of Sciences of the United States of America* 105(49):19336-19341.

Olsen, K. A. 2007. The clean development mechanism's contribution to sustainable development: A review of the literature. *Climatic Change* (84):59-73.

Olsthoorn, X., P. van der Werff, L. Bouwer, and D. Huitema. 2008. Neo-Atlantis: The Netherlands under a 5-m sea level rise. *Climatic Change* 91(1):103-122.

O'Neill, S., and S. Nicholson-Cole. 2009. "Fear won't do it": Promoting positive engagement with climate change through visual and iconic representations. *Science Communication* 30(3):355-379.

Oppenheimer, M. 2005. Defining dangerous anthropogenic interference: The role of science, the limits of science. *Risk Analysis* 25(6):1399-1407.

Orians, G. H., and D. Policansky. 2009. Scientific bases of macroenvironmental indicators. *Annual Review of Environment and Resources* 34:375-404.

Orr, J. C., V. J. Fabry, O. Aumont, L. Bopp, S. C. Doney, R. A. Feely, A. Gnanadesikan, N. Gruber, A. Ishida, F. Joos, R. M. Key, K. Lindsay, E. Maier-Reimer, R. Matear, P. Monfray, A. Mouchet, R. G. Najjar, G. K. Plattner, K. B. Rodgers, C. L. Sabine, J. L. Sarmiento, R. Schlitzer, R. D. Slater, I. J. Totterdell, M. F. Weirig, Y. Yamanaka, and A. Yool. 2005. Anthropogenic ocean acidification over the twenty-first century and its impact on calcifying organisms. *Nature* 437(7059):681-686.

Ostrom, E. 2005. *Understanding Institutional Diversity*. Princeton, NJ: Princeton University Press.

Ostrom, E. 2007. A diagnostic approach for going beyond panaceas. *Proceedings of the National Academy of Sciences of the United States of America* 104:15181-15187.

Ostrom, E. 2010. A multi-scale approach to coping with climate change and other collective action problems. *Solutions* 1(2):27-36. Available at *http://thesolutionsjournal.com/node/565*. Accessed May 10, 2010.

Ostrom, E., and H. Nagendra. 2006. Insights on linking forests, trees, and people from the air, on the ground, and in the laboratory. *Proceedings of the National Academy of Sciences of the United States of America* 103(51):19224-19231.

Overland, J. E., and P. J. Stabeno. 2004, Is the climate of the Bering Sea warming and affecting the ecosystem?, *Eos Transactions*. AGU, 85:309-316.

Overpeck, J. T., and J. E. Cole. 2006. Abrupt change in Earth's climate system. *Annual Review of Environment and Resources* 31(1):1-31.

Overpeck, J. T., and J. L. Weiss. 2009. Projections of future sea level becoming more dire. *Proceedings of the National Academy of Sciences of the United States of America* 106(51):21461-21462.

Overpeck, J. T., B. L. Otto-Bliesner, G. H. Miller, D. R. Muhs, R. B. Alley, and J. T. Kiehl. 2006. Paleoclimatic evidence for future ice-sheet instability and rapid sea-level rise. *Science* 311(5768):1747-1750.

Paaijmans, K. P., A. F. Read, and M. B. Thomas. 2009. Understanding the link between malaria risk and climate. *Proceedings of the National Academy of Sciences of the United States of America* 106(33):13844-13849.

Pabi, S., G. L. Van Dijken, and K. R. Arrigo. 2008. Primary production in the Arctic Ocean, 1998-2006. *Journal of Geophysical Research-Oceans* 113 (C8).

Pagano, T. C., H. C. Hartmann, and S. Sorooshian. 2002. Factors affecting seasonal forecast use in Arizona water management: A case study of the 1997-98 El Nino. *Climate Research* 21(3):259-269.

Page, E. A. 2008. Distributing the burdens of climate change. *Environmental Politics* 17(4):556-575.

Palecki, M. A., S. A. Changnon, and K. E. Kunkel. 2001. The nature and impacts of the July 1999 heat wave in the midwestern United States: Learning from the lessons of 1995. *Bulletin of the American Meteorological Society* 82(7):1353-1367.

Pardey, P. G., and N. M. Beintema. 2002. *Slow Magic: Agricultural R&D a Century After Mendel*. University of Minnesota, Center for International Food and Agricultural Policy.

Parks, B.C., and J.T. Roberts, 2010. Climate change, social theory and justice. *Theory, Culture & Society* 27(2-3):1-32.

Parmesan, C., and H. Galbraith. 2004. *Observed Impacts of Global Climate Change in the U.S.* Arlington, VA: Pew Center on Global Climate Change.

Parmesan, C., and G. Yohe. 2003. A globally coherent fingerprint of climate change impacts across natural systems. *Nature* 421(6918):37-42.

Parola, P., C. Socolovschi, L. Jeanjean, I. Bitam, P. E. Fournier, A. Sotto, P. Labauge, and D. Raoult. 2008. Warmer weather linked to tick attack and emergence of severe Rickettsioses. *PLoS Neglected Tropical Diseases* 2(11).

Parris, T. M., and R. W. Kates. 2003. Characterizing and measuring sustainable development. *Annual Review of Environment and Resources* 28:559-586.

Parry, M. L., C. Rosenzweig, A. Iglesias, M. Livermore, and G. Fischer. 2004. Effects of climate change on global food production under SRES emissions and socio-economic scenarios. *Global Environmental Change-Human and Policy Dimensions* 14(1):53-67.

Parson, E. A. 2008. Useful global-change scenarios: current issues and challenges. *Environmental Research Letters* 3(4).

Pascual, M., J. A. Ahumada, L. F. Chaves, X. Rodo, and M. Bouma. 2006. Malaria resurgence in the East African highlands: Temperature trends revisited. *Proceedings of the National Academy of Sciences of the United States of America* 103(15):5829-5834.

Paskal, C. 2007. How climate change is pushing the boundaries of security and foreign policy. London, UK: Chatham House.

Paterson, M. 2009. Post-hegemonic climate politics? *British Journal of Politics & International Relations* 11(1):140-158.

Patz, J. A., D. Campbell-Lendrum, T. Holloway, and J. A. Foley. 2005. Impacts of regional climate change on human health. *Nature* 438:310-317.

Patz, J. A., S. J. Vavrus, C. K. Uejio, and S. L. Mclellan. 2008. Climate change and waterborne disease risk in the Great Lakes region of the US. *American Journal of Preventive Medicine* 35(5):451-458.

Paustian, K., et al. 2004. *Agricultural Mitigation of Greenhouse Gases: Science and Policy Options*. Council on Agricultural Science and Technology(CAST) Report R141 2004, 120 pp.

Pearson, B. 2007. Market failure: Why the clean development mechanism won't promote clean development. *Journal of Clean Development* 15:247-252.

Pederson, G. T., S. T. Gray, D. B. Fagre, and L. J. Graumlich. 2006. Long-duration drought variability and impacts on ecosystem services: A case study from Glacier National Park, Montana. *Earth Interactions* 10.

Pelling, M. 2003. *The Vulnerability of Cities: Natural Disasters and Social Resilience*. London: Earthscan.

Pellow, D. N., and R. J. Brulle. 2007. Poisoning the planet: The struggle for environmental justice. *Contexts* 6(1):37-41.

Peltier, W. R., and R. G. Fairbanks. 2006. Global glacial ice volume and Last Glacial Maximum duration from an extended Barbados sea level record. *Quaternary Science Reviews* 25:3322.

Pendleton, L., P. King, C. Mohn, D. G. Webster, R. K. Vaughn, and P. Adams. 2009. *Estimating the Potential Economic Impacts of Climate Change on Southern California Beaches—Final Report*, CEC-500-2009-033-F. California Energy Commission.

Perkins, C. and A.W. Weimer. 2004. Likely near-term solar-thermal water splitting technologies. *International Journal of Hydrogen Energy* 29:1587-1599.

Perkins, C., and A.W. Weimer. 2009. Solar-thermal production of renewable hydrogen. *AIChE Journal* 55(2):286-293.

Perry, A. L., P. J. Low, J. R. Ellis, and J. D. Reynolds. 2005. Climate change and distribution shifts in marine fishes. *Science* 308(5730):1912-1915.

Peterson, T., and A. Rose. 2006. Reducing the conflict between climate policy and energy policy in the U.S.: The important role of states. *Energy Policy* 34:619-631.

Pew Oceans Commission. 2003 (May). *America's Living Oceans: Charting a Course for Sea Change. A Report to the Nation.* Arlington, VA: Pew Oceans Commission.

Pfeffer, W. T., J. T. Harper, and S. O'Neel. 2008. Kinematic constraints on glacier contributions to 21st-century sea-level rise. *Science* 321(5894):1340-1343.

Piao, S., P. Friedlingstein, P. Ciais, N. De Noblet-Ducoudra, D. Labat, and S. Zaehle. 2007. Changes in climate and land use have a larger direct impact than rising CO_2 on global river runoff trends. *Proceedings of the National Academy of Sciences of the Untied States of America* 104(39):15242-15247.

Pickett, S. T. A., M. L. Cadenasso, J. M. Grove, P. M. Groffman, L. E. Band, C. G. Boone, W. R. Burch, C. S. B. Grimmond, J. Hom, J. C. Jenkins, N. L. Law, C. H. Nilon, R. V. Pouyat, K. Szlavecz, P. S. Warren, and M. A. Wilson. 2008. Beyond urban legends: An emerging framework of urban ecology, as illustrated by the Baltimore Ecosystem Study. *Bioscience* 58(2):139-150.

Pidgeon, N., R. E. Kasperson, and P. Slovic, eds. 2003. *The Social Amplification of Risk*. Cambridge: Cambridge University Press.

Pidgeon, N. F., I. Lorenzoni, and W. Poortinga. 2008. Climate change or nuclear power—No thanks! A quantitative study of public perceptions and risk framing in Britain. *Global Environmental Change* 18(1):69-85.

Pielke, R. A., R. Avissar, M. Raupach, A. J. Dolman, X. B. Zeng, and A. S. Denning. 1998. Interactions between the atmosphere and terrestrial ecosystems: Influence on weather and climate. *Global Change Biology* 4(5):461-475.

Pierce, D. W., T. P. Barnett, H. G. Hidalgo, T. Das, C. Bonfils, B. D. Santer, G. Bala, M. D. Dettinger, D. R. Cayan, A. Mirin, A. W. Wood, and T. Nozawa. 2008. Attribution of declining western US snowpack to human effects. *Journal of Climate* 21(23):6425-6444.

Pinter, N., A. A. Jemberie, J. W. F. Remo, R. A. Heine, and B. S. Ickes. 2008. Flood trends and river engineering on the Mississippi River system. *Geophysical Research Letters* 35:L23404.

Pinto, R. F., and J. A. P. De Oliveira. 2008. Implementation challenges in protecting the global environmental commons: The case of climate change policies in Brazil. *Public Administration and Development* 28(5):340-350.

Pirard, P., S. Vandentorren, M. Pascal, K. Laaidi, A. Le Tertre, S. Cassadou, and M. Ledrans. 2005. Summary of the mortality impact assessment of the 2003 heat wave in France. *Eurosurveillance* 10(7):153-156.

Platt, R. H. 1996. *Land Use and Society: Geography, Law, and Public Policy*. Washington, DC: Island Press.

Platt, R. H. 1999. *Disasters and Democracy: The Politics of Extreme Natural Events*. Washington, DC: Island Press.

PNWA (Pacific Northwest Waterways Association). 2009. *The Role of Hydropower in the Northwest*. Available at *http://www.pnwa.net/new/Articles/Hydropower.pdf*. Accessed December 2, 2009.

Poff, N. L., J. D. Olden, D. M. Merritt, and D. M. Pepin. 2007. Homogenization of regional river dynamics by dams and global biodiversity implications. *Proceedings of the National Academy of Sciences of the United States of America* 104(14):5732-5737.

Pohl, C. 2008. From science to policy through transdisciplinary research. *Environmental Science & Policy* 11:45-63.

Polovina, J. J., G. T. Mitchum, and G. T. Evans. 1995. Decadal and basin-scale variation in mixed layer depth and the impact on biological production in the Central and North Pacific, 1960-88. *Deep-Sea Research Part I-Oceanographic Research Papers* 42 (10):1701-1716.

Polsky, C., R. Neff, and B. Yarnal. 2007. Building comparable global change vulnerability assessments: The vulnerability scoping diagram. *Global Environmental Change* 17(3-4):472-485.

Porfiriev, B. 2009. Community resilience and vulnerability to disasters: Qualitative models and megacities—a comparison with small towns. *Environmental Hazards* 8:23-27.

Portmann, R. W., S. Solomon, and G. C. Hegerl. 2009. Spatial and seasonal patterns in climate change, temperatures, and precipitation across the United States. *Proceedings of the National Academy of Sciences of the United States of America* 106(18):7324-7329.

Portney, P., and J. P. Weyant, eds. 1999. *Discounting and Intergenerational Equity*. Washington, DC: RFF Press.

Poumadère, M., C. Mays, G. Pfeifle, and A. T. Vafeidis. 2008. Worst case scenario as stakeholder decision support: A 5- to 6-m sea level rise in the Rhone delta, France. *Climatic Change* 91(1-2):123-143.

Prakash, A., and M. Potoski. 2006. *The Voluntary Environmentalists: Green Clubs, ISO 14001, and Voluntary Environmental Regulations*. Cambridge, UK: Cambridge University Press.

Pressman, J. L., and A. Wildavsky. 1984. *Implementation: How Great Expectations in Washington Are Dashed in Oakland; or, Why It's Amazing that Federal Programs Work at All, This Being a Saga of the Economic Development Administration as Told by Two Sympathetic Observers Who Seek to Build Morals on a Foundation of Ruined Hopes*, revised 3rd ed. The Oakland Project series. Berkeley, CA: University of California Press.

Pritchard, H. D., R. J. Arthern, D. G. Vaughan, and L. A. Edwards. 2009. Extensive dynamic thinning on the margins of the Greenland and Antarctic ice sheets. *Nature* 461:971-975.

PSAC (President's Science Advisory Committee). 1965. *Restoring the Quality of Our Environment: Report of the Environmental Pollution Panel*. Washington, DC: The White House, 317 pp.

Rabalais, N. N. and R. E. Turner (eds.). 2001. Coastal Hypoxia: Consequences for Living Resources and Ecosystems. Coastal and Estuarine Studies 58, American Geophysical Union, Washington, D.C.

Rabe, B.G. 2008. States on steroids: The intergovernmental odyssey of American climate policy. *Review of Policy Research* 25(2):105-128.

Raento, M., A. Oulasvirta, and N. Eagle. 2009. Smartphones: An emerging tool for social scientists. *Sociological Methods and Research* 37:426-454.

Rahmstorf, S. 1995. Bifurcations of the Atlantic thermohaline circulation in response to changes in the hydrological cycle. *Nature* 378:145-149.

Rahmstorf, S. 2007. A semi-empirical approach to projecting future sea-level rise. *Science* 315(5810):368-370.

Rahmstorf, S. 2010. A new view on sea level rise. *Nature Reports Climate Change* (1004):44-45.

Rahmstorf, S., A. Cazenave, J. A. Church, J. E. Hansen, R. F. Keeling, D. E. Parker, and R. C. J. Somerville. 2007. Recent climate observations compared to projections. *Science* 316(5825):709-709.

Ramaswami, A., T. Hillman, B. Janson, M. Reiner, and G. Thomas. 2008. A demand-centered, hybrid life-cycle methodology for city-scale greenhouse gas inventories. *Environmental Science and Technology* 42:6455-6461.

Randall, D. A., R. A. Wood, S. Bony, R. Colman, T. Fichefet, J. Fyfe, V. Kattsov, A. Pitman, J. Shukla, J. Srinivasan, R. J. Stouffer, A. Sumi, and K. E. Taylor. 2007. Climate models and their evaluation. In *Climate Change 2007: The Physical Science Basis. Contribution of Working Group I to the Fourth Assessment Report of the Intergovernmental Panel on Climate Change*, S. Solomon, D. Qin, M. Manning, Z. Chen, M. Marquis, K. B. Averyt, M.Tignor, and H. L. Miller, eds. Cambridge, U.K.: Cambridge University Press.

Rasch, P. J., S. Tilmes, R. P. Turco, A. Robock, L. Oman, C. C. Chen, G. L. Stenchikov, and R. R. Garcia. 2008. An overview of geoengineering of climate using stratospheric sulphate aerosols. *Philosophical Transactions of the Royal Society A* 366(1882):4007-4037.

Rau, G. H. 2009. Electrochemical CO_2 capture and storage with hydrogen generation. *Energy Procedia* 1(1):823-828.

Rau, G. H., K. G. Knauss, W. H. Langer, and K. Caldeira. 2007. Reducing energy-related CO_2 emissions using accelerated weathering of limestone. *Energy* 32(8):1471-1477.

Ravishankara, A. R., S. Solomon, A. A. Turnipseed, and R. F. Warren. 1993. Atmospheric lifetimes of long-lived halogenated species. *Science* 259(5092):194-199.

Ravishankara, A. R., J. S. Daniel, and R. W. Portmann. 2009. Nitrous oxide (N_2O): The dominant ozone-depleting substance emitted in the 21st century. *Science* 326(5949):123-125.

Rayner, S., D. Lach, and H. Ingram. 2005. Weather forecasts are for wimps: Why water resource managers do not use climate forecasts. *Climatic Change* 69(2):197-227.

Rees, W., and M. Wackernagel. 2008. Urban ecological footprints: Why cities cannot be sustainable—and why they are a key to sustainability. Pp. 537-555 in *Urban Ecology. An International Perspective on the Interaction Between Humans and Nature*, J. M. Marzluff, E. Shulenberger, W. Endlicher, M. Alberti, G. Bradley, C. Ryan, U. Simon, and C. ZumBrunnen. New York: Springer.

Reich, P. B., B. A. Hungate, and Y. Q. Luo. 2006. Carbon-nitrogen interactions in terrestrial ecosystems in response to rising atmospheric carbon dioxide. *Annual Review of Ecology Evolution and Systematics* 37:611-636.

Reilly, J., F. Tubiello, B. Mccarl, D. Abler, R. Darwin, K. Fuglie, S. Hollinger, C. Izaurralde, S. Jagtap, J. Jones, L. Mearns, D. Ojima, E. Paul, K. Paustian, S. Riha, N. Rosenberg, and C. Rosenzweig. 2003. US agriculture and climate change: New results. *Climatic Change* 57(1-2):43-69.

Renn, O. 2005. *Risk Governance: Towards an Integrative Approach*. Geneva: International Risk Governance Council.

Renn, O. 2008. *Risk Governance: Coping with Uncertainty in a Complex World*. Sterling, VA: Earthscan.

Repetto, R. C. 2008. *The Climate Crisis and the Adaptation Myth*. New Haven, CT: Yale School of Forestry & Environmental Studies.

Revelle, R., and H. E. Suess. 1957. Carbon dioxide exchange between atmosphere and ocean and the question of an increase of atmospheric CO_2 during the past decades. *Tellus* 9:18-27.

Rhemtulla, J. M., D. J. Mladenoff, and M. K. Clayton. 2009. Historical forest baselines reveal potential for continued carbon sequestration. *Proceedings of the National Academy of Sciences of the United States of America* 106(15):6082-6087.

Richels, R.G., T. F. Rutherford, G. J. Blanford, and L. Clarke. 2007. Managing the transition to climate stabilization. *Climate Policy* 7:409-428.

Ricke, K., M. G. Morgan, J. Apt, D. Victor, and J. Steinbruner. 2008. Unilateral Geoengineering: Non-technical Briefing Notes for a Workshop at the Council on Foreign Relations. Washington D.C., May 05, 2008. Available at *http://www.cfr.org/content/thinktank/GeoEng_041209.pdf*. Accessed on November 16, 2010.

Ridgwell, A., J. S. Singarayer, A. M. Hetherington, and P. J. Valdes. 2009. Tackling regional climate change by leaf albedo bio-geoengineering. *Current Biology* 19(2):146-150.

Riebesell, U., K. G. Schulz, R. G. J. Bellerby, M. Botros, P. Fritsche, M. Meyerhofer, C. Neill, G. Nondal, A. Oschlies, J. Wohlers, and E. Zollner. 2007. Enhanced biological carbon consumption in a high CO_2 ocean. *Nature* 450(7169):545-548.

Rignot, E., and P. Kanagaratnam. 2006. Changes in the velocity structure of the Greenland ice sheet. *Science* 311(5763):986-990.

Rignot, E., G. Casassa, P. Gogineni, W. Krabill, A. Rivera, and R. Thomas. 2004. Accelerated ice discharge from the Antarctic Peninsula following the collapse of Larsen B ice shelf. *Geophysical Research Letters* 31(18).

Rignot, E., J. E. Box, E. Burgess, and E. Hanna. 2008. Mass balance of the Greenland ice sheet from 1958 to 2007. *Geophysical Research Letters* 35(20).

Rignot, E., M. Koppes, and I. Velicogna, 2010. Rapid submarine melting of the calving faces of West Greenland glaciers. *Nature Geoscience* 3:187-191, doi:10.1038/ngeo765.

Roberts, J.T., and B. C. Parks. 2007. *A Climate of Injustice: Global Inequality, North-South Politics, and Climate Policy*. Cambridge, MA: MIT Press.

Robertson, G. P., and S. M. Swinton. 2005. Reconciling agricultural productivity and environmental integrity: A grand challenge for agriculture. *Frontiers in Ecology and the Environment* 3(1):38-46.

Robertson, G. P., and P. M. Vitousek. 2009. Nitrogen in agriculture: Balancing the cost of an essential resource. *Annual Review of Environment and Resources* 34(1).

Robertson, G. P., E. A. Paul, and R. R. Harwood. 2000. Greenhouse gases in intensive agriculture: Contributions of individual gases to the radiative forcing of the atmosphere. *Science* 289(5486):1922-1925.

Robertson, G. P., V. H. Dale, O. C. Doering, S. P. Hamburg, J. M. Melillo, M. M. Wander, W. J. Parton, P. R. Adler, J. N. Barney, R. M. Cruse, C. S. Duke, P. M. Fearnside, R. F. Follett, H. K. Gibbs, J. Goldemberg, D. J. Mladenoff, D. Ojima, M. W. Palmer, A. Sharpley, L. Wallace, K. C. Weathers, J. A. Wiens, and W. W. Wilhelm. 2008. Agriculture: Sustainable biofuels redux. *Science* 322:49-50.

Roble, R. G., and R. E. Dickinson. 1989. How will changes in carbon dioxide and methane modify the mean structure of the mesosphere and thermosphere? *Geophysical Research Letters* 16:1441-1444.

Robock, A., M. Q. Mu, K. Vinnikov, I. V. Trofimova, and T. I. Adamenko. 2005. Forty-five years of observed soil moisture in the Ukraine: No summer desiccation (yet). *Geophysical Research Letters* 32(3).

Robock, A., L. Oman, and G. L. Stenchikov. 2008. Regional climate responses to geoengineering with tropical and Arctic SO_2 injections. *Journal of Geophysical Research* 113(D16):15.

Robock, A., A. Marquardt, B. Kravitz, and G. Stenchikov. 2009. Benefits, risks, and costs of stratospheric geoengineering. *Geophysical Research Letters* 36.

Roderick, M. L., G. D. Farquhar, S. L. Berry, and I. R. Noble. 2001. On the direct effect of clouds and atmospheric particles on the productivity and structure of vegetation. *Oecologia* 129(1):21-30.

Roe, G. H., and M. B. Baker. 2007. Why is climate sensitivity so unpredictable? *Science* 318(5850):629-632.

Rogers, W., and J. Gulledge. 2010. *Lost in Translation: Closing the Gap Between Climate Science and National Security Policy*. Center for New American Security. Available at *http://www.cnas.org/files/documents/publications/Lost%20in%20Translation_Code406_Web_0.pdf*. Accessed May 13, 2010.

Rohling, E. J., K. Grant, C. Hemleben, M. Siddall, B. A. A. Hoogakker, M. Bolshaw, and M. Kucera. 2008. High rates of sea-level rise during the last interglacial period. *Nature Geoscience* 1(1):38-42.

Rosa, E. A. 2007. *The Public Climate for Nuclear Power: The Changing of Seasons. The Role of Nuclear Power in Global and Domestic Energy Policy: Recent Developments and Future Expectations*. Howard H. Baker, Jr. Center for Public Policy.

Rosa, E. A., and D. L. Clark, Jr. 1999. Historical routes to technological gridlock: Nuclear technology as prototypical vehicle. *Research in Social Problems and Public Policy* 7:21-57.

Rosenberg, S., A. Vedlitz, D. F. Cowman, and S. Zahran. 2010. Climate change: A profile of US climate scientists' perspectives. *Climatic Change* 101:311-329.

Rosenzweig, C., and D. Hillel. 1998. *Climate Change and the Global Harvest: Potential Impacts of the Greenhouse Effect on Agriculture*. New York: Oxford University Press.

Rosenzweig, C., A. Iglesias, X. B. Yang, P. R. Epstein, and E. Chivian. 2001. Climate change and extreme weather events: Implications for food production, plant diseases, and pests. *Global Change & Human Health* 2(2):90-104.

Rosenzweig, C., W. D. Solecki, L. Parshall, M. Chopping, G. Pope, and R. Goldberg. 2005. Characterizing the urban heat island in current and future climates in New Jersey. *Global Environmental Change Part B: Environmental Hazards* 6(1):51-62.

Rosenzweig, C., G. Casassa, D. J. Karoly, A. Imeson, C. Liu, A. Menzel, S. Rawlins, T. L. Root, B. Seguin, and P. Tryjanowski. 2007. Assessment of observed changes and responses in natural and managed systems. Pp. 79-131 in *Climate Change 2007: Impacts, Adaptation and Vulnerability. Contribution of Working Group II to the Fourth Assessment Report of the Intergovernmental Panel on Climate Change*, M. L. Parry, O. F. Canziani, J. P. Palutikof, P. J. van der Linden, and C. E. Hanson, eds. Cambridge, U.K.: Cambridge University Press.

Roy, S. B., S. W. Pacala, and R. L. Walko. 2004. Can large wind farms affect local meteorology? *Journal of Geophysical Research* 109:D19101, doi:10.1029/2004JD004763.

The Royal Society. 2009 (September). *Geoengineering the Climate: Science, Governance and Uncertainty*. London: The Royal Society.

Rudel, T. K. 2005. *Tropical Forests: Regional Paths of Destruction and Regeneration in the Late Twentieth Century*. New York: Columbia University Press.

RUSI (Royal United Services Institute). 2009. *Climate-Related Impacts on National Security in Mexico and Central America, Interim Report*. Prepared by S. Fetzek. London, Great Britain: Stephen Austin & Sons.

Sabatier, P. A. 1986. Top-down and bottom-up approaches to implementation research: A critical analysis and suggested synthesis. *Journal of Public Policy* 6:21-48.

Sabine, C., and R. Feely. 2005. The oceanic sink for carbon dioxide. In *Greenhouse Gas Sinks*, edited by N. H. D Reay, J Grace, K Smith. Oxfordshire: CABI Publishing.

Sabine, C. L., R. A. Feely, N. Gruber, R. M. Key, K. Lee, J. L. Bullister, R. Wanninkhof, C. S. Wong, D. W. R. Wallace, B. Tilbrook, F. J. Millero, T. H. Peng, A. Kozyr, T. Ono, and A. F. Rios. 2004. The oceanic sink for anthropogenic CO2. *Science* 305 (5682):367-371.

Sagarin, R. D., J. P. Barry, S. E. Gilman, and C. H. Baxter. 1999. Climate-related change in an intertidal community over short and long time scales. *Ecological Monographs* 69(4):465-490.

Sailor, D. J., and H. Fan. 2002. Modeling the diurnal variability of effective albedo for cities. *Atmospheric Environment* 36(4):713-725.

Salehyan, I., 2008. From climate change to conflict? No consensus yet. *Journal of Peace Research* 45(3):315-326.

Sallenger, A. H., W. Krabill, J. Brock, R. Swift, S. Manizade, and H. Stockdon. 2002. Sea-cliff erosion as a function of beach changes and extreme wave runup during the 1997-1998 El Nino. *Marine Geology* 187(3-4):279-297.

Salter, S., G. Sortino, and J. Latham. 2008. Sea-going hardware for the cloud albedo method of reversing global warming. *Philosophical Transactions of the Royal Society A* 366(1882):3989-4006.

Saltman, T., R. Cook, M. Fenn, R. Haeuber, B. Bloomer, C. Eagar, T. Huntington, B. J. Cosby, A. Watkins, S. McLaughlin, and Executive Office of the President. 2005. *National Acid Precipitation Assessment Program Report to Congress: An Integrated Assessment*. Washington, DC: National Science and Technology Council.

Samaras, C., and K. Meisterling. 2008. Life cycle assessment of greenhouse gas emissions from plug-in hybrid vehicles: Implications for policy. *Environmental Science and Technology* 42(9):3170-3176.

Santilli, M., P. Moutinho, S. Schwartzman, D. Nepstad, L. Curran, and C. Nobre. 2005. Tropical deforestation and the Kyoto Protocol. *Climatic Change* 71(3):267-276.

Satterthwaite, D. 2008. Cities' contribution to global warming: Notes on the allocation of greenhouse gas emissions. *Environment and Urbanization* 20(2):539-549.

Scambos, T. A., J. A. Bohlander, C. A. Shuman, and P. Skvarca. 2004. Glacier acceleration and thinning after ice shelf collapse in the Larsen B embayment, Antarctica. *Geophysical Research Letters* 31(18).

Scheberle, D. 2004. *Federalism and Environmental Policy: Trust and the Politics of Implementation*, 2nd ed. Washington, DC: Georgetown University Press.

Scherr, S. J., and S. Sthapit. 2009. Farming and land use to cool the planet. In *State of the World 2009: Into a Warming World: A Worldwatch Institute Report on Progress Toward a Sustainable Society*, L. Starke, ed. New York: W.W. Norton.

Schimel, D., J. Melillo, H. Tian, A. D. Mcguire, D. Kicklighter, T. Kittel, N. Rosenbloom, S. Running, P. Thornton, D. Ojima, W. Patton, R. Kelly, M. Sykes, R. Neilson, and B. Rizzo. 2000. Contribution of increasing CO_2 and climate to carbon storage by ecosystems in the United States. *Science* 287(5460):2004.

Schimel, D. S., J. I. House, K. A. Hibbard, P. Bousquet, P. Ciais, P. Peylin, B. H. Braswell, M. J. Apps, D. Baker, A. Bondeau, J. Canadell, G. Churkina, W. Cramer, A. S. Denning, C. B. Field, P. Friedlingstein, C. Goodale, M. Heimann, R. A. Houghton, J. M. Melillo, B. Moore, D. Murdiyarso, I. Noble, S. W. Pacala, I. C. Prentice, M. R. Raupach, P. J. Rayner, R. J. Scholes, W. L. Steffen, and C. Wirth. 2001. Recent patterns and mechanisms of carbon exchange by terrestrial ecosystems. *Nature* 414(6860):169-172.

Schlesinger, M. E., and N. Ramankutty. 1994. An oscillation in the global climate system of period 65-70 years. *Nature* 367(6465):723-726.

Schmer, M. R., K. P. Vogel, R. B. Mitchell, and R. K. Perrin. 2008. Net energy of cellulosic ethanol from switchgrass. *Proceedings of the National Academy of Sciences of the United States of America* 105(2):464-469.

Schmidhuber, J., and F. N. Tubiello. 2007. Global food security under climate change. *Proceedings of the National Academy of Sciences of the United States of America* 104(50):19703-19708.

Schmidtlein, M. C., R. Deutsch, W. W. Piegorsch, and S. L. Cutter. 2008. A sensitivity analysis of the social vulnerability index. *Risk Analysis* 28(4):1099-1114.

Schneider, S., and K. Kuntz-Duriseti. 2002. Uncertainty and climate change policy. In *Climate Change Policy: A Survey*, S. Schneider, A. Rosencranz, and J. O. Niles, eds. Washington, DC: Island Press.

Schneider, S., and J. Lane. 2006. Dangers and thresholds in climate change and implications for justice. Pp. 23-52 in *Fairness in Adaptation to Climate Change*, N. Adger, J. Paavola, S. Huq, and M. J. Mace, eds. Cambridge, MA: MIT Press.

Schneider, S. H. 1996. Geoengineering: Could or should we do it? *Climatic Change* 33(3):291-302.

Schneider, S. H. 2008. Geoengineering: Could we or should we make it work? *Philosophical Transactions of the Royal Society A* 366(1882):3843-3862.

Schneider, S. H., and R. S. Chen. 1980. Carbon-dioxide warming and coastline flooding—Physical factors and climatic impact. *Annual Review of Energy* 5:107-140.

Schneider, T., and I. M. Held. 2001. Discriminants of twentieth-century changes in Earth surface temperatures. *Journal of Climate* 14:249-254.

Schreurs, M. A. 2008. From the bottom up: Local and subnational climate change politics. *Journal of Environment & Development* 17(4):343-355.

Schuur, E. A. G., J. G. Vogel, K. G. Crummer, H. Lee, J. O. Sickman, and T. E. Osterkamp. 2009. The effect of permafrost thaw on old carbon release and net carbon exchange from tundra. *Nature* 459(7246):556-559.

Seager, R., M. F. Ting, I. Held, Y. Kushnir, J. Lu, G. Vecchi, H. P. Huang, N. Harnik, A. Leetmaa, N. C. Lau, C. H. Li, J. Velez, and N. Naik. 2007. Model projections of an imminent transition to a more arid climate in southwestern North America. *Science* 316(5828):1181-1184.

Seager, R., A. Tzanova, and J. Nakamura. 2009. Drought in the Southeastern United States: Causes, variability over the last millennium and the potential for future hydroclimate change. *Journal of Climate* 22:5021-5045.

Searchinger, T., R. Heimlich, R. A. Houghton, F. X. Dong, A. Elobeid, J. Fabiosa, S. Tokgoz, D. Hayes, and T. H. Yu. 2008. Use of US croplands for biofuels increases greenhouse gases through emissions from land-use change. *Science* 319(5867):1238-1240.

Selden, T. M., and D. Song. 1994. Environmental quality and development: Is there a Kuznets curve for air pollution emissions? *Journal of Environmental Economics and Management* 27(2):147-162.

Selin, H., and S. VanDeveer. 2007. Political science and prediction: What's next for U.S. climate change policy? *Review of Policy Research* 24(1):1-27.

Semenza, J. C., C. H. Rubin, K. H. Falter, J. D. Selanikio, W. D. Flanders, H. L. Howe, and J. L. Wilhelm. 1996. Heat-related deaths during the July 1995 heat wave in Chicago. *New England Journal of Medicine* 335(2):84-90.

Seto, K. C., and J. M. Shepherd. 2009. Global urban land-use trends and climate impacts. *Current Opinion in Environmental Sustainability* 1(1):89-95.

Shakhova, N., I. Semiletov, A. Salyuk, V. Yusupov, D. Kosmach, and O. Gustafsson, 2010. Extensive methane venting to the atmosphere from sediments of the East Siberian Arctic shelf. *Science* 327(5970):1246-1250.

Shapiro, M. A., J. Shukla, G. Brunet, C. Nobre, M. Béland, R. Dole, K. Trenberth, R. Anthes, G. Asrar, L. Barrie, P. Bougeault, G. Brasseur, D. Burridge, A. Busalacchi, J. Caughey, D. Chen, J. Church, T. Enomoto, B. Hoskins, Ø. Hov, A. Laing, H. Le Treut, J. Marotzke, G. McBean, G. Meehl, M. Miller, B. Mills, J. Mitchell, M. Moncrieff, T. Nakazawa, H. Olafsson, T. Palmer, D. Parsons, D. Rogers, A. Simmons, A. Troccoli, Z. Toth, L. Uccellini, C. Velden, and J. M. Wallace. 2010. An Earth-system prediction initiative for the 21st century. *Bulletin of the American Meteorological Society*, doi:10.1175/2010BAMS2944.1.

Shaviv, N.J.. 2002. Cosmic ray diffusion from the galactic spiral arms, iron meteorites, and a possible climatic connection. *Physical Review Letters* 89:051102

Sheffield, J., and E. F. Wood. 2008. Global trends and variability in soil moisture and drought characteristics, 1950-2000, from observation-driven simulations of the terrestrial hydrologic cycle. *Journal of Climate* 21(3):432-458.

Shem, W., and M. Shepherd. 2009. On the impact of urbanization on summertime thunderstorms in Atlanta: Two numerical model case studies. *Atmospheric Research* 92(2):172-189.

Sheppard, S. R. J. 2005. Landscape visualisation and climate change: The potential for influencing perceptions and behaviour. *Environmental Science & Policy* 8(6):637-654.

Sheppard, S. R. J., and M. Meitner. 2005. Using multi-criteria analysis and visualisation for sustainable forest management planning with stakeholder groups. *Forest Ecology and Management* 207(1-2):171-187.

Sherman, K., I. Belkin, S. Seitzinger, P. Hoagland, D. Jin, M.-C. Aquarone, and S. Adams. 2009. Indicators of changing states of large marine ecosystems. In *Sustaining the World's Large Marine Ecosystems*, K. Sherman, M. C. Aquarone, and S. Adams, eds. Narragansett, RI: NOAA, National Marine Fisheries Service.

Shi, A. Q. 2003. The impact of population pressure on global carbon dioxide emissions, 1975-1996: Evidence from pooled cross-country data. *Ecological Economics* 44(1):29-42.

Shindell, D., D. Rind, N. Balachandran, J. Lean, J. Lonergan. 1999. Solar cycle variability, ozone, and climate. *Science* 284:305-308.

Shindell, D. T., G. A. Schmidt, M. E. Mann, D. Rind, and A. Waple. 2001. Solar forcing of regional climate change during the maunder minimum. *Science* 294(5549):2149-2152.

Siddall, M., T. F. Stocker, and P. U. Clark. 2009. Constraints on future sea-level rise from past sea-level change. *Nature Geoscience* 2(8):571-575.

Silverman, J., B. Lazar, L. Cao, K. Caldeira, and J. Erez. 2009. Coral reefs may start dissolving when atmospheric CO2 doubles. *Geophysical Research Letters* 36.

Sitch, S., P. M. Cox, W. J. Collins, and C. Huntingford. 2007, Indirect radiative forcing of climate change through ozone effects on the land-carbon sink. *Nature* 448:791-794.

Sivan, D., K. Lambeck, R. Toueg, A. Raban, Y. Porath, and B. Shirman. 2004. Ancient coastal wells of Caesarea Maritima, Israel, an indicator for relative sea level changes during the last 2000 years. *Earth and Planetary Science Letters* 222(1):315-330.

Slaymaker, O., and R. E. J. Kelly. 2007. *The Cryosphere and Global Environmental Change*. Environmental Systems and Global Change Series. Malden, MA: Blackwell.

Slovic, P. 2000. What does it mean to know a cumulative risk? Adolescents' perceptions of short-term and long-term consequences of smoking. *Journal of Behavioral Decision Making* 13:259-266.

Smetacek, V., and S. Nicol. 2005. Polar ocean ecosystems in a changing world. *Nature* 437(7057):362-368.

Smith, J. B., S. H. Schneider, M. Oppenheimer, G. W. Yohe, W. Hare, M. D. Mastrandrea, A. Patwardhan, I. Burton, J. Corfee-Morlot, C. H. D. Magadza, H. M. Fuessel, A. B. Pittock, A. Rahman, A. Suarez, and J. P. Van Ypersele. 2009. Assessing dangerous climate change through an update of the Intergovernmental Panel on Climate Change (IPCC) "reasons for concern." *Proceedings of the National Academy of Sciences of the United States of America* 106(11):4133-4137.

Smith, K. A., and F. Conen. 2004. Impacts of land management on fluxes of trace greenhouse gases. *Soil Use and Management* 20:255-263.

Smith, P. 2004. Engineered biological sinks on land. In *The Global Carbon Cycle: Integrating Humans, Climate, and the Natural World*, C. B. Field and M. R. Raupach, eds. Washington, DC: Island Press.

Smith, P., D. Martino, Z. Cai, D. Gwary, H. Janzen, P. Kumar, B. McCarl, S. Ogle, F. O'Mara, C. Rice, B. Scholes, and O. Sirotenko. 2007. Agriculture. In *Climate Change 2007: Mitigation. Contribution of Working Group III to the Fourth Assessment Report of the Intergovernmental Panel on Climate Change*, B. Metz, O. R. Davidson, P. R. Bosch, R. Dave, and L. A. Meyer, eds. Cambridge, U.K., and New York: Cambridge University Press.

Snyder, M. A., L. C. Sloan, N. S. Diffenbaugh, and J. L. Bell. 2003. Future climate change and upwelling in the California Current. *Geophysical Research Letters* 30 (15).

Socolow, R. H. 1978. *Saving Energy in the Home: Princeton's Experiment at Twin Rivers*. Cambridge, MA: Ballinger.

Socolow, R. H., and A. Glaser. 2009. Balancing risks: Nuclear energy & climate change. *Daedalus* 138(4):31-44.

Soden, B. J., R. T. Wetherald, G. L. Stenchikov, and A. Robock. 2002. Global cooling after the eruption of Mount Pinatubo: A test of climate feedback by water vapor. *Science* 296(5568):727-730.

Sokolov, A. P., P. H. Stone, C. E. Forest, R. Prinn, M. C. Sarofim, M. Webster, S. Paltsev, C. A. Schlosser, D. Kicklighter, S. Dutkiewicz, J. Reilly, C. Wang, B. Felzer, J. M. Melillo, and H. D. Jacoby. 2009. Probabilistic forecast for twenty-first-century climate based on uncertainties in emissions(without policy) and climate parameters. *Journal of Climate* 22(19):5175-5204.

Solecki, W. D., C. Rosenzweig, L. Parshall, G. Pope, M. Clark, J. Cox, and M. Wiencke. 2005. Mitigation of the heat island effect in urban New Jersey. *Global Environmental Change Part B: Environmental Hazards* 6(1):39-49.

Solomon, S., D. Qin, M. Manning, R. B. Alley, T. Berntsen, N. L. Bindoff, Z. Chen, A. Chidthaisong, J. M. Gregory, G. C. Hegerl, M. Heimann, B. Hewitson, B. J. Hoskins, F. Joos, J. Jouzel, V. Kattsov, U. Lohmann, T. Matsuno, M. Molina, N. Nicholls, J. Overpeck, G. Raga, V. Ramaswamy, J. Ren, M. Rusticucci, R. Somerville, T. F. Stocker, P. Whetton, R. A. Wood, and D. Wratt. 2007. Technical summary. In *Climate Change 2007: The Physical Science Basis. Contribution of Working Group I to the Fourth Assessment Report of the Intergovernmental Panel on Climate Change*, S. Solomon, D. Qin, M. Manning, Z. Chen, M. Marquis, K. B. Averyt, M. Tignor, and H. L. Miller, eds. Cambridge, U.K., and New York: Cambridge University Press.

Solomon, S., G.-K. Plattner, R. Knutti, and P. Friedlingstein. 2009. Irreversible climate change due to carbon dioxide emissions. *Proceedings of the National Academy of Sciences of the United States of America* 106(6):1704-1709.

Solomon, S., K. H. Rosenlof, R. W. Portmann, J. S. Daniel, S. M. Davis, T. J. Sanford, and G.-K. Plattner. 2010. Contributions of stratospheric water vapor to decadal changes in the rate of global warming. *Science* 327(5970):1219-1223.

Sorensen, T. C. 1990. Rethinking national security. *Foreign Affairs* 69(3):1-18.

Southward, A. J., S. J. Hawkins, and M. T. Burrows. 1995. 70 Years observations of changes in distribution and abundance on zooplankton and intertidal orgnanisms in the western English Channel in relation to rising sea temperature. *Journal of Thermal Biology* 20 (1-2):127-155.

Spittlehouse, D. L., and R. B. Stewart. 2003. Adaptation to climate change in forest management. *BC Journal of Ecosystems and Management* 4(1): 1-11.

Steele, J. H. 1998. From carbon flux to regime shift. *Fisheries Oceanography* 7(3-4):176-181.

Steffen, W., A. Sanderson, P. Tyson, J. Jäger, P. Matson, B. Moore, Iii, F. Oldfield, K. Richardson, H. J. Schellnhuber, B. L. Turner II, and R. J. Wasson, eds. 2004. *Global Change and the Earth System: A Planet Under Pressure*. Berlin: Springer.

Stehfest, E., L. Bouwman, D. P. van Vuuren, M. G. J. den Elzen, B. Eickhout, and P. Kabat. 2009. Climate benefits of changing diet. *Climatic Change* 95(1-2):83-102.

Steinfeld, A. 2005. Solar thermochemical production of hydrogen: A review. *Solar Energy* 78(5):603-615.

Stenseth, N. C., A. Mysterud, G. Ottersen, J. W. Hurrell, K. S. Chan, and M. Lima. 2002. Ecological effects of climate fluctuations. *Science* 297(5585):1292-1296.

Sterman, J., and L. Booth Sweeney. 2007. Understanding public complacency about climate change: Adults' mental models of climate change violate conservation of matter. *Climatic Change* 80(3):213-238.

Sterman, J. D. 2008. Risk communication on climate: Mental models and mass balance. *Science* 322(5901):532-533.

Stern, D. I. 2004. The rise and fall of the environmental Kuznets curve. *World Development* 32(8):1419-1439.

Stern, N. 2007. *The Economics of Climate Change: The Stern Review*. Cambridge, U.K.: Cambridge University Press.

Stern, P. C., E. Aronson, J. M. Darley, D. H. Hill, E. Hirst, W. Kempton, and T. J. Wilbanks. 1986. The effectiveness of incentives for residential energy conservation. *Evaluation Review* 10(2):147-176.

Stern, P.C., T. J. Wilbanks, S. Cozzens, and W. Rosa. 2009. *Generic Lessons Learned about Societal Responses to Emerging Technologies Perceived as Involving Risks*. Oak Ridge National Laboratory Report ORNL/TM-2009/114, Oak Ridge, TN.

Stern, P. C., G. T. Gardner, M. P. Vandenbergh, T. Dietz, and J. M. Gilligan. 2010. Design principles for carbon emissions reduction programs. *Environmental Science and Technology*44 (13): 4847-4848 .

Stewart, I. T., D. R. Cayan, and M. D. Dettinger. 2005. Changes toward earlier streamflow timing across western North America. *Journal of Climate* 18(8):1136-1155.

Stocker, T. F. 2000. Past and future reorganizations in the climate system. *Quaternary Science Reviews* 19(1-5):301-319.

Stocker, T. F., and A. Schmittner. 1997. Influence of CO_2 emission rates on the stability of the thermohaline circulation. *Nature* 388(6645):862-865.

Stokes, D. 1997. *Pasteur's Quadrant: Basic Science and Technological Innovation*. Washington, DC: Brookings Press.

Stokes, C. R., S. D. Gurney, M. Shahgedanova, and V. Popovnin. 2006. Late-20th-century changes in glacier extent in the Caucasus Mountains, Russia/Georgia. *Journal of Glaciology* 52(176):99-109.

Stolaroff, J., D. Keith, and G. Lowry. 2006. A pilot-scale prototype contactor for CO_2 capture from ambient air: Cost and energy requirements. Presented at the 8th International Conference on Greenhouse Gas Control Technologies (GHGT-8), Trondheim, Norway.

Stott, P. A., D. A. Stone, and M. R. Allen. 2004. Human contribution to the European heatwave of 2003. *Nature* 432(7017):610-614.

Straneo, F., G. S. Hamilton, D. A. Sutherland, L. A. Stearns, F. Davidson, M. O. Hammill, G. B. Stenson, and A. Rosing-Asvid. 2010. Rapid circulation of warm subtropical waters in a major glacial fjord in East Greenland. *Nature Geoscience* 3:182-186, doi:10.1038/NGEO764.

Sun, Y., S. Solomon, A. G. Dai, and R. W. Portmann. 2007. How often will it rain? *Journal of Climate* 20:4801-4818.

Sutter, C., and J. C. Parreño. 2007. Does the current Clean Development Mechanism (CDM) deliver its sustainable development claim? *Climactic Change* (84):75-90.

Svensmark, H. 1998. Influence of cosmic rays on Earth's climate. *Physical Review Letters* 81:5027.

Svensmark, H. 2006. Imprint of galactic dynamics on Earth's climate. *Astronomische Nachrichten* 327(9):866-870.

Swann, A. L., I. Y. Fung, S. Levis, G. B. Bonan, and S. C. Doney. 2010. Changes in Arctic vegetation amplify high-latitude warming through the greenhouse effect. *Proceedings of the National Academy of Sciences of the United States of America* 107:1295-1300.

Swanson, K. L., G. Sugihara, and A. A. Tsonis. 2009. Long-term natural variability and 20th century climate change. *Proceedings of the National Academy of Sciences of the United States of America* 106(38):16120-16123.

Swinton, S. M., F. Lupi, G. P. Robertson, and D. A. Landis. 2006. Ecosystem services from agriculture: Looking beyond the usual suspects. *American Journal of Agricultural Economics* 88(5):1160-1166.

Swofford, J., and M. Slattery. 2010. Public attitudes of wind energy in Texas: Local communities in close proximity to wind farms and their effect on decision-making. *Energy Policy* 38(5):2508-2519.

Taha, H., S. Douglas, and J. Haney. 1997. Mesoscale meteorological and air quality impacts of increased urban albedo and vegetation. *Energy and Buildings* 25(2):169-177.

Takahashi, T., S. C. Sutherland, R. A. Feely, and R. Wanninkhof. 2006. Decadal change of the surface water pCO(2) in the North Pacific: A synthesis of 35 years of observations. *Journal of Geophysical Research-Oceans* 111 (C7).

Tallis, H. M., and P. Kareiva. 2006. Shaping global environmental decisions using socio-ecological models. *Trends in Ecology & Evolution* 21(10):562-568.

Tans, P. 2010. *NOAA/ESRL (NOAA Earth System Research Laboratory). Trends in Atmospheric Carbon Dioxide*. Available at *http://www.esrl.noaa.gov/gmd/ccgg/trends/*. Accessed on October 26, 2010.

Tebaldi, C., and R. Knutti. 2007. The use of the multi-model ensemble in probabilistic climate projections. *Philosophical Transactions of the Royal Society A* 365(1857):2053-2075.

Teodoro, M. P. 2009. Contingent professionalism: Bureaucratic mobility and the adoption of water conservation rates. *Journal of Public Administration Research and Theory* 20(2):437-459, doi:10.1093/jopart/mup012.

Thomalla, F., T. Downing, E. Spanger-Siegfried, G. Y. Han, and J. Rockstrom. 2006. Reducing hazard vulnerability: Towards a common approach between disaster risk reduction and climate adaptation. *Disasters* 30(1):39-48.

Thomas, R., E. Frederick, W. Krabill, S. Manizade, and C. Martin. 2006. Progressive increase in ice loss from Greenland. *Geophysical Research Letters* 33(10).

Thompson, D. W. J., and J. M. Wallace. 2000. Annular modes in the extratropical circulation. Part I: Month-to-month variability. *Journal of Climate* 13:1000-1016.

Thompson, D. W. J., and J. M. Wallace. 2001. Regional climate impacts of the Northern Hemisphere annular mode. *Science* 3(5527):85-89.

Thompson, D. W. J., J. J. Kennedy, J. M. Wallace, and P. D. Jones. 2008. A large discontinuity in the mid-twentieth century in observed global-mean surface temperature. *Nature* 453(7195):646-649.

Thompson, D. W. J., J. M. Wallace, P. D. Jones, and J. J. Kennedy. 2009. Identifying signatures of natural climate variability in time series of global-mean surface temperature: Methodology and insights. *Journal of Climate* 22(22):6120-6141.

Thresher, R., M. Robinson, and P. Veers. 2007. To capture the wind. *IEEE Power & Energy Magazine* 5(6):34-46.

Tietenberg, T. 2002. The tradable permits approach to protecting the commons: What have we learned? Pp. 197-232 in The Drama of the Commons. Committee on the Human Dimensions of Global Change, E. Ostrom et al., eds. Washington, DC: National Academy Press.

Tilman, D., J. Knops, D. Wedin, P. Reich, M. Ritchie, and E. Siemann. 1997. The influence of functional diversity and composition on ecosystem processes. *Science* 277(5330):1300-1302.

Tilman, D., J. Fargione, B. Wolff, C. D'Antonio, A. Dobson, R. Howarth, D. Schindler, W. H. Schlesinger, D. Simberloff, and D. Swackhamer. 2001. Forecasting agriculturally driven global environmental change. *Science* 292(5515):281-284.

Tilman, D., R. Socolow, J. A. Foley, J. Hill, E. Larson, L. Lynd, S. Pacala, J. Reilly, T. Searchinger, C. Somerville, and R. Williams. 2009. Beneficial biofuels—the food, energy, and environment trilemma. *Science* 325:270-271.

Tilmes, S., R. Muller, and R. Salawitch. 2008. The sensitivity of polar ozone depletion to proposed geoengineering schemes. *Science* 320(5880):1201-1204.

Toth, F., and E. Hizsnyik. 2008. Managing the inconceivable: Participatory assessments of impacts and responses to extreme climate change. *Climatic Change* 91(1):81-101.

Travis, D. J., A. Carleton, and R. G. Lauritsen. 2002. Contrails reduce daily temperature range. *Nature* 418:601, doi:10.1038/418601.

Trenberth, K. E., and A. Dai. 2007. Effects of Mount Pinatubo volcanic eruption on the hydrological cycle as an analog of geoengineering. *Geophysical Research Letters* 34.

Trenberth, K. E., L. Smith, T. Qian, A. Dai, J. Fasullo, 2007: Estimates of the global water budget and its annual cycle using observational and model data. *J. Hydrometeorol.*, 8, 758-769.

Trenberth, K. E., and J. Fasullo. 2008. Energy budgets of Atlantic hurricanes and changes from 1970. *Geochemistry Geophysics Geosystems* 9.

Trenberth, K. E., and J. T. Fasullo. 2010. Tracking Earth's energy. *Science* 328(5976):316-317.

Trenberth, K. E., P. D. Jones, P. Ambenje, R. Bojariu, D. Easterling, A. Klein Tank, D. Parker, F. Rahimzadeh, J. A. Renwick, M. Rusticucci, B. Soden, and P. Zhai. 2007. Observations: Surface and atmospheric climate change. In *Climate Change 2007: The Physical Science Basis. Contribution of Working Group I to the Fourth Assessment Report of the Intergovernmental Panel on Climate Change*, S. Solomon, D. Qin, M. Manning, Z. Chen, M. Marquis, K. B. Averyt, M. Tignor, and H. L. Miller, eds. Cambridge, U.K.: Cambridge University Press.

Tribbia, J., and S. C. Moser. 2008. More than information: What coastal managers need to plan for climate change. *Environmental Science & Policy* 11(4):315-328.

Turner, B. L. 2009. Sustainability and forest transitions in the southern Yucatán: The land architecture approach. *Land Use Policy* 27(2):170-180.

Turner, B. L., R. E. Kasperson, P. A. Matson, J. J. Mccarthy, R. W. Corell, L. Christensen, N. Eckley, J. X. Kasperson, A. Luers, M. L. Martello, C. Polsky, A. Pulsipher, and A. Schiller. 2003a. A framework for vulnerability analysis in sustainability science. *Proceedings of the National Academy of Sciences of the United States of America* 100(14):8074-8079.

Turner, B. L., P. A. Matson, J. J. Mccarthy, R. W. Corell, L. Christensen, N. Eckley, G. K. Hovelsrud-Broda, J. X. Kasperson, R. E. Kasperson, A. Luers, M. L. Martello, S. Mathiesen, R. Naylor, C. Polsky, A. Pulsipher, A. Schiller, H. Selin, and N. Tyler. 2003b. Illustrating the coupled human-environment system for vulnerability analysis: Three case studies. *Proceedings of the National Academy of Sciences of the United States of America* 100(14):8080-8085.

Turner, R. K., W. N. Adger, and R. Brouwer. 1998. Ecosystem services value, research needs, and policy relevance: A commentary. *Ecological Economics* 25(1):61-65.

UCAR (University Corporation for Atmospheric Research). 2007. *North American Regional Climate Change Assessment Program*. Available at *http://www.narccap.ucar.edu/about/index.html*. Accessed May 13, 2010.

UN (United Nations). 2007. *World Urbanization Prospects: The 2007 Revision*. New York: UN.

UN. 2009. *Millennium Development Goals Report*. United Nations Publications.

UNDP (United Nations Development Programme). 1994. *Human Development Report*. New York: Oxford University Press.

UNEP, WMO, NOAA, and NASA. 1994. *Scientific Assessment of Ozone Depletion: Executive Summary*. World Meteorological Organization Global Ozone Research and Monitoring Project Report 37.

UNFCCC (United Nations Framework Convention on Climate Change). 1992. *United Nations Conference on Environment and Development (UNCED). Rio de Janeiro*. Available at *http://unfccc.int/resource/docs/convkp/conveng.pdf*. Accessed on October 26, 2010.

UNFCCC. 2009. *National Adaptation Programmes of Action*. *http://unfccc.int/national_reports/napa/items/2719.php*. Accessed December 3, 2009.

Unger, J. 2004. Intra-urban relationship between surface geometry and urban heat island: Review and new approach. *Climate Research* 27(3):253-264.

University of Oxford, Tyndall Centre, and Hadley Centre Met Office. 2009. Four Degrees and Beyond. International Climate Conference. September 28-30, 2009. Oxford, UK. Available at *http://www.eci.ox.ac.uk/4degrees/index.php*. Accessed on October 26, 2010.

Urwin, K., and A. Jordan. 2008. Does public policy support or undermine climate change adaptation? Exploring policy interplay across different scales of governance. *Global Environmental Change-Human and Policy Dimensions* 18(1):180-191.

U.S. Commission on Ocean Policy. 2004. *An Ocean Blueprint for the 21st Century. Final Report.* Washington, DC: U.S. Commission on Ocean Policy.

USGCRP (U.S. Global Change Research Program.) 2001. *Climate Change Impacts on the United States: The Potential Consequences of Climate Variability and Change.* New York: Cambridge University Press.

USGCRP. 2009a. *Global Climate Change Impacts in the United States.* New York: Cambridge University Press.

USGCRP. 2009b. *Our Changing Planet: The U.S. Climate Change Science Program for Fiscal Year 2009.* A Report by the Climate Change Science Program and the Subcommittee on Global Change Research. A Supplement to the President's Fiscal Year 2009 Budget. New York: Cambridge University Press.

USGS (U.S. Geological Survey). 2003. *Ground-Water Depletion Across the Nation*, U.S Geological Survey Fact Sheet 103-0.3. Available at *http://pubs.usgs.gov/fs/fs-103-03/*. Accessed on October 26, 2010.

USGS. 2004. *Estimated Use of Water in the United States in 2000.* Available at *http://pubs.usgs.gov/circ/2004/circ1268/htdocs/text-total.html*. Accessed December 3, 2009.

USGS. 2008. *USGS Repeat Photography Project Documents Retreating Glaciers in Glacier National Park.* Available at *http://www.nrmsc.usgs.gov/repeatphoto*. Accessed May 3, 2010.

USGS. 2009. *Hydroelectric Power Water Use*. Available at *http://ga.water.usgs.gov/edu/wuhy.html*. Accessed December 2, 2009.

Vanderheiden, S. 2008. *Atmospheric Justice: A Political Theory of Climate Change.* New York: Oxford University Press.

Van Der Werf, G. R., J. Dempewolf, S. N. Trigg, J. T. Randerson, P. S. Kasibhatla, L. Gigliof, D. Murdiyarso, W. Peters, D. C. Morton, G. J. Collatz, A. J. Dolman, and R. S. Defries. 2008. Climate regulation of fire emissions and deforestation in equatorial Asia. *Proceedings of the National Academy of Sciences of the United States of America* 105(51):20350-20355.

Van Vuuren, D. P., J. Weyant, and F. de la Chesnaye. 2006. Multi-gas scenarios to stabilize radiative forcing. *Energy Economics* 28(1):102-120.

Van Zalinge, N., N. Thuok, T. S. Tana, D. Loeung. 2000. Where there is water, there is fish? Cambodian fisheries issues in a Mekong River Basin perspective. Pp. 37-48 in *Common Property in the Mekong: Issues of Sustainability and Subsistence*, M. Ahmed and P. Hirsch, eds. Working Papers,

The WorldFish Center, number 13972 6.

Vayda, A. P. 1988. Actions and consequences as objects of explanation in human ecology. In *Human Ecology: Research and Applications*, R. J. Borden, J. Jacobs, and G. L. Young, eds. College Park, MD: Society for Human Ecology.

Velders, G. J. M., S. O. Andersen, J. S. Daniel, D. W. Fahey, and M. McFarland. 2007. The importance of the Montreal Protocol in protecting climate. *Proceedings of the National Academy of Sciences of the United States of America* 104:4814-4819.

Velicogna, I. 2009. Increasing rates of ice mass loss from the Greenland and Antarctic ice sheets revealed by GRACE. *Geophysical Research Letters* 36.

Vellinga, M., and R. A. Wood. 2002. Global climatic impacts of a collapse of the Atlantic thermohaline circulation. *Climatic Change* 54(3):251-267.

Vermeer, M., and S. Rahmstorf. 2009. Global sea level linked to global temperature. *Proceedings of the National Academies of Sciences of the United States of America* 106(51):21527-21532.

Victor, D. G., K. Raustiala, and E. G. Skolnikoff, eds. 1998. *The Implementation and Effectiveness of International Environmental Commitments: Theory and Practice.* Cambridge, MA: MIT Press.

Vine, E. L, J. A. Sathaye, and W. R. Makundi. 2001. An overview of guidelines and issues for the monitoring, evaluation, reporting, verification, and certification of forestry projects for climate change mitigation. *Global Environmental Change* 11:203-216.

Virgoe, J. 2009. International governance of a possible geoengineering intervention to combat climate change. *Climatic Change* 95(1-2):103-119.

Vitousek, P. M., H. A. Mooney, J. Lubchenco, and J. M. Melillo. 1997. Human domination of Earth's ecosystems. *Science* 277(5325):494-499.

Vogel, C., and K. O'Brien. 2006. Who can eat information? Examining the effectiveness of seasonal climate forecasts and regional climate-risk management strategies. *Climate Research* 33(1):111-122.

Vogel, C., S. C. Moser, R. E. Kasperson, and G. D. Dabelko. 2007. Linking vulnerability, adaptation, and resilience science to practice: Pathways, players, and partnerships. *Global Environmental Change* 17(3-4):349-364.

Vogler, J., and C. Bretherton. 2006. The European Union as a protagonist to the United States on climate change. *International Studies Perspectives* 7(1):1-22.

Vorosmarty, C. J., P. Green, J. Salisbury, and R. B. Lammers. 2000. Global water resources: Vulnerability from climate change and population growth. *Science* 289(5477):284-288.

Wackernagel, M., N. B. Schultz, D. Deumling, A. C. Linares, M. Jenkins, V. Kapos, C. Monfreda, J. Loh, N. Myers, R. B. Norgaard, and J. Randers. 2002. Tracking the ecological overshoot of the human economy. *Proceedings of the National Academy of Sciencesof the United States of America* 99:9266-9271.

Walters, C., and A. M. Parma. 1996. Fixed exploitation rate strategies for coping with effects of climate change. *Canadian Journal of Fisheries and Aquatic Sciences* 53(1):148-158.

Walther, G.-R., E. Post, P. Convey, A. Menzel, C. Parmesan, T. J. C. Beebee, J.-M. Fromentin, O. Hoegh-Guldberg, and F. Bairlein. 2002. Ecological responses to recent climate change. *Nature* 416(6879):389-395.

Wang, B., H. U. Neue, and H. P. Samonte. 1997. Effect of cultivar difference ("IR72,""IR65598," and "Dular") on methane emission. *Agriculture Ecosystems & Environment* 62(1):31-40.

Wang, C. 2007. Impact of direct radiative forcing of black carbon aerosols on tropical convective precipitation. *Geophysical Research Letters* 34(5).

Wang, Y. H., and D. J. Jacob. 1998. Anthropogenic forcing on tropospheric ozone and OH since preindustrial times. *Journal of Geophysical Research* 103(D23):31123-31135.

Wardle, D. A., M. C. Nilsson, and O. Zackrisson. 2008. Fire-derived charcoal causes loss of forest humus. *Science* 320(5876):629-629.

Watson, R., and D. Pauly. 2001. Systematic distortions in world fisheries catch trends. *Nature* 414(6863):534-536.

Watson, R. T., M. C. Zinyowera, and R. H. Moss. 1997. *The Regional Impacts of Climate Change: An Assessment of Vulnerability*. Cambridge, U.K.: Cambridge University Press.

WBCSD (World Business Council for Sustainable Development). 2004. *Mobility 2030: Meeting the Challenges to Sustainability*. Geneva: WBCSD.

Weart, S. R. 2008. *The Discovery of Global Warming: Revised and Expanded Edition*. Cambridge, MA: Harvard University Press.

Weber, C. L. 2009. Measuring structural change and energy use: Decomposition of the US economy from 1997 to 2002. *Energy Policy* 37(4):1561-1570.

Weber, C. L., and G.P. Peters. 2009. Climate change policy and international trade: Policy considerations in the U.S. *Energy Policy 37*(2):432-440.

Weber, E. U. 2006. Experience-based and description-based perceptions of long-term risk: Why global warming does not scare us (yet). *Climatic Change* 77(1-2):103-120.

Weber, S. L., T. J. Crowley, and G. Der Schrier. 2004. Solar irradiance forcing of centennial climate variability during the Holocene. *Climate Dynamics* 22(5):539-553.

Weitzman, M. L. 2007a. A review of the Stern Review on the economics of climate change. *Journal of Economic Literature* 47:703-724.

Weitzman, M. L. 2007b. *Structural Uncertainty and the Value of Statistical Life in the Economics of Catastrophic Climate Change*. NBER Working Paper 13490. Washington, DC: National Bureau of Economic Research.

Weitzman, M. L. 2009. On modeling and interpreting the economics of catastrophic climate change. *Review of Economics and Statistics* 91(1).

West, J. J., and H. Dowlatabadi. 1999. On assessing the economic impacts of sea-level rise on developed coasts. In *Climate, Change and Risk*, T. Downing, A. Olsthoorn, and R. S. J. Tol, eds. London: Routledge.

Westerling, A. L. 2009. Climate change, growth, and California wildfire final paper. Sacramento, CA: California Energy Commission.

Westerling, A. L., and B. P. Bryant. 2008. Climate change and wildfire in California. *Climatic Change* 87:S231-S249.

Westerling, A. L., H. G. Hidalgo, D. R. Cayan, and T. W. Swetnam. 2006. Warming and earlier spring increase western US forest wildfire activity. *Science* 313(5789):940-943.

Weyant, J. P. 2009. A perspective on integrated assessment. *Climatic Change* 95(3-4):317-323.

WGA (Western Governors' Association). 2006. *Clean and Diversified Energy Initiative: Geothermal Task Force Report*. Washington, DC: WGA. Available at *http://www.westgov.org/wga/initiatives/cdeac/Geothermal-full.pdf*. Accessed May 7, 2010.

Whitfield, S., E. A Rosa, T. Dietz, and A. Dan. 2009. The future of nuclear power: Value orientations and risk perceptions. *Risk Analysis* 29:425-437.

Wigley, T. M. L., R. Richels, and J. A. Edmonds. 1996. Economic and environmental choices in the stabilization of atmospheric CO_2 concentrations. *Nature* 379:240-243.

Wigley, T. M. L., C. M. Ammann, B. D. Santer, and S. C. B. Raper. 2005. Effect of climate sensitivity on the response to volcanic forcing. *Journal of Geophysical Research* 110:D09107, doi:10.1029/2004JD005557.

Wilbanks, T. J. 2005. Issues in developing a capacity for integrated analysis of mitigation and adaptation. *Environmental Science & Policy* 8(6):541-547.

Wilby, R. L., J. Troni, Y. Biot, L. Tedd, B. C. Hewitson, D. M. Smith, and R. T. Sutton. 2009. A review of climate risk information for adaptation and development planning. *International Journal of Climatology* 29(9):1193-1215.

Wilcox, B. P., Y. Huang, and J. W. Walker. 2008. Long-term trends in streamflow from semiarid rangelands: Uncovering drivers of change. *Global Change Biology* 14(7):1676-1689.

Wilkinson, C. R. 2000. *Status of coral reefs of the world, 2000*. Cape Ferguson, Qld.: Australian Institute of Marine Science.

Williams, J. E., A. L. Haak, N. G. Gillespie, H. M. Neville, and W. T. Colyer. 2007. *Healing Troubled Waters: Preparing Trout and Salmon Habitat for a Changing Climate*. Arlington, VA: Trout Unlimited.

Williams, M., and M. Zhang. 2008. *Challenges to Offshore Wind Development in the United States*. Cambridge, MA: MIT-USGS Science Impact Collaborative.

Wilson, C., and H. Dowlatabadi. 2007. Models of decision making and residential energy use. *Annual Review of Environment and Resources* 32:169-203.

Wilson, K. M. 2002. Forecasting the future: How television weathercasters' attitudes and beliefs about climate change affect their cognitive knowledge on the science. *Science Communication* 24(2):246-268.

Wilson, M. A., and R. B. Howarth. 2002. Discourse-based valuation of ecosystem services: Establishing fair outcomes through group deliberation. *Ecological Economics* 41(3):431-443.

Wilson, R. W., F. J. Millero, J. R. Taylor, P. J. Walsh, V. Christensen, S. Jennings, and M. Grosell. 2009. Contribution of Fish to the Marine Inorganic Carbon Cycle. *Science* 323 (5912):359-362.

Winkler, H. 2008. Measurable, reportable and verifiable—The keys to mitigation in the Copenhagen deal. *Climate Policy* 8:534-547.

Wise, M., K. Calvin, A. Thomson, L. Clarke, B. Bond-Lamberty, R. Sands, S. J. Smith, A. Janetos, and J. Edmonds. 2009b. Implications of limiting CO_2 concentrations for land use and energy. *Science* 324(5931):1183-1186.

Wise, M. A., K. V. Calvin, A. M. Thomson, L. E. Clarke, B. Bond- Lamberty, R. D. Sands, S. J. Smith, A. C. Janetos, and J. A. Edmonds. 2009a. *The Implications of Limiting CO_2 Concentrations for Agriculture, Land Use, Land-Use Change Emissions and Bioenergy*. Richland, WA: Pacific Northwest National Laboratory (PNNL).

Wittwer, E. 2006. Upper Midwest freight corridor study. In *Railroads and Freight in the Future*, T. M. Adams, ed. Midwest Regional University Transportation Center, College of Engineering, Department of Civil and Environmental Engineering, University of Wisconsin, Madison.

WMO (World Meteorological Organization). 2009a. *Global Atmosphere Watch*. Available at *http://www.wmo.ch/pages/prog/arep/gaw/gaw_home_en.html*. Accessed December 3, 2009.

WMO. 2009b. *World Climate Conference 3: High Level Declaration*. Available at *http://www.wmo.int/wcc3/declaration_en.php*. Accessed May 3, 2010.

Wolfe, D., L. Ziska, C. Petzoldt, A. Seaman, L. Chase, and K. Hayhoe. 2008. Projected change in climate thresholds in the Northeastern U.S.: Implications for crops, pests, livestock, and farmers. *Mitigation and Adaptation Strategies for Global Change* 13(5):555-575.

Wolock, D. M., and G. M. Hornberger. 1991. Hydrological effects of changes in levels of atmospherica carbron-dioxide. *Journal of Forecasting* 10(1-2):105-116.

Woodward, F. I. 2002. Potential impacts of global elevated CO_2 concentrations on plants. *Current Opinion in Plant Biology* 5(3):207-211.

World Bank. 2006. *Where is the Wealth of Nations? Measuring Capital for the 21st Century*. Washington, DC: World Bank.

World Bank. 2007. *World Development Report 2008: Agriculture for Development*. Washington, DC: World Bank.

World Bank. 2009. *World Development Report 2010 Climate Change*. Washington, DC: World Bank.

Worm, B., R. Hilborn, J. K. Baum, T. A. Branch, J. S. Collie, C. Costello, M. J. Fogarty, E. A. Fulton, J. A. Hutchings, S. Jennings, O. P. Jensen, H. K. Lotze, P. M. Mace, T. R. Mcclanahan, C. Minto, S. R. Palumbi, A. M. Parma, D. Ricard, A. A. Rosenberg, R. Watson, and D. Zeller. 2009. Rebuilding global fisheries. *Science* 325(5940):578-585.

Wu, S.-Y., B. Yarnal, and A. Fisher. 2002. Vulnerability of coastal communities to sea-level rise: A case study of Cape May County, New Jersey. *Climate Research* 22:255-270.

Wu, S.-Y., R. Najjar, and J. Siewert. 2009. Potential impacts of sea-level rise on the Mid- and Upper-Atlantic Region of the United States. *Climatic Change*, 95(1-2):121-138, doi:10.1007/s10584-008-9522-x.

Wunsch, C., R. M. Ponte, and P. Heimbach. 2007. Decadal trends in sea level patterns: 1993-2004. *Journal of Climate* 20:5889-5911.

WWEA (World Wind Energy Association). 2010. *World Wind Energy Report 2009*. Available at *http://www.wwindea.org/home/images/stories/worldwindenergyreport2009_s.pdf*. Accessed May 3, 2010.

Yalowitz, K. S., J. F. Collins, and R. A. Virginia. 2008. The Arctic Climate Change and Security Policy Conference: Final Report and Findings. Hanover, NH, December 1-3, 2008. Carnegie Engowment for Interational Peace and University of the Arctic Institute for Applied Circumpolar Policy.

Yang, T. 2006. International treaty enforcement as a public good: The role of institutional deterrent sanctions in international environmental agreements. *Michigan Journal of International Law* 27:1131.

Yates, K. K., and R. B. Halley. 2006. CO32- concentration and pCO(2) thresholds for calcification and dissolution on the Molokai reef flat, Hawaii. *Biogeosciences* 3 (3):357-369.

Ye, B. S., D. Q. Yang, and D. L. Kane. 2003. Changes in Lena River streamflow hydrology: Human impacts versus natural variations. *Water Resources Research* 39(7):14.

Yin, J. J., M. E. Schlesinger, and R. J. Stouffer. 2009. Model projections of rapid sea-level rise on the northeast coast of the United States. *Nature Geoscience* 2(4):262-266.

Yohe, G. 2006. Some thoughts on the damage estimates presented in the stern review—An editorial. *Integrated Assessment Journal* 6(3):65-72.

Yohe, G., and R. S. J. Tol. 2007. *Precaution and a Dismal Theorem: Implications for Climate Policy and Climate Research*. FNU Working Papers 145. Hamburg, Germany: University of Hamburg.

Yohe, G., J. E. Neumann, and P. Marshall. 1999. The economic damage induced by sea level rise in the United States. In *The Impact of Climate Change on the United States Economy*, R. Mendelsohn and J. E. Neumann, eds. Cambridge, U.K.: Cambridge University Press.

York, R. 2009. The challenges of measuring environmental sustainability comment on "Political and Social Foundations for Environmental Sustainability." *Political Research Quarterly* 62(1):205-208.

York, R., and M. H. Gossard. 2004. Cross-national meat and fish consumption: Exploring the effects of modernization and ecological context. *Ecological Economics* 48(3):293-302.

York, R., E. A. Rosa, and T. Dietz. 2002. Bridging environmental science with environmental policy: Plasticity of population, affluence, and technology. *Social Science Quarterly* 83(1):18-34.

York, R., E. A. Rosa, and T. Dietz. 2003. Footprints on the earth: The environmental consequences of modernity. *American Sociological Review* 68(2):279-300.

Young, O. 2008. The architecture of global environmental governance: Bringing science to bear on policy. *Global Environmental Politics* 8(1):14-32.

Young, O. 2009. The arctic in play: Governance in a time of rapid change. *The International Journal of Marine and Coastal Law* 24(2):423-442.

Young, O., H. Schroeder, and L. A. King, eds. 2008. *Institutions and Environmental Change: Principal Findings, Applications, and Research Frontiers*. Cambridge, MA: MIT Press.

Young, O. R. 2002a. *The Institutional Dimensions of Environmental Change: Fit, Interplay, and Scale, Global Environmental Accord; Variation: Global Environmental Accords*. Cambridge, MA: MIT Press.

Young, O. R. 2002b. Institutional interplay: The environmental consequences of cross-scale interactions. In *The Drama of the Commons*, E. Ostrom, T. Dietz, N. Dolsak, P. C. Stern, S. Stonich, and E. U. Weber, eds. Washington, DC: The National Academies Press.

Zacherl, D., S. D. Gaines, and S. I. Lonhart. 2003. The limits to biogeographical distributions: Insights from the northward range extension of the marine snail, *Kelletia kelletii* (Forbes, 1852). *Journal of Biogeography* 30(6):913-924.

Zahran, S., S. D. Brody, A. Vedlitz, H. Grover, and C. Miller. 2008. Vulnerability and capacity: Explaining local commitment to climate-change policy. *Environment and Planning C-Government and Policy* 26(3):544-562.

Zhang, D. D., P. Brecke, H. F. Lee, Y. Q. He, and J. Zhang. 2007b. Global climate change, war, and population decline in recent human history. *Proceedings of the National Academy of Sciences of the United States of America* 104(49):19214-19219.

Zhang, J., and K. R. Smith. 2003. Indoor air pollution: A global health concern. *British Medical Bulletin* 68:209-225.

Zhang, K. Q., B. C. Douglas, and S. P. Leatherman. 2004. Global warming and coastal erosion. *Climatic Change* 64(1-2):41-58.

Zhang, L., D. J. Jacob, K. F. Boersma, D. A. Jaffe, J. R. Olson, K. W. Bowman, J. R. Worden, A. M. Thompson, M. A. Avery, R. C. Cohen, J. E. Dibb, F. M. Flock, H. E. Fuelberg, L. G. Huey, W. W. Mcmillan, H. B. Singh, and A. J. Weinheimer. 2008. Transpacific transport of ozone pollution and the effect of recent Asian emission increases on air quality in North America: An integrated analysis using satellite, aircraft, ozonesonde, and surface observations. *Atmospheric Chemistry and Physics* 8(20):6117-6136.

Zhang, R., T. L. Delworth, and I. M. Held. 2007a. Can the Atlantic Ocean drive the observed multidecadal variability in Northern Hemisphere mean temperature? *Geophysical Research Letters* 34(2).

Zhang, X. 2010. Sensitivity of arctic summer sea ice coverage to global warming forcing: Towards reducing uncertainty in arctic climate change projections. *Tellus A* 62(3):220-227.

Zhao, M., A. J. Pitman, and T. Chase. 2001. The impact of land cover change on the atmospheric circulation. *Climate Dynamics* 17(5-6):467-477.

Zimmerman, J. B., J. R. Mihelcic, and J. Smith. 2008. Global stressors on water quality and quantity. *Environmental Science & Technology* 42(12):4247-4254.

Zimov, S. A., E. A. G. Schuur, and F. S. Chapin. 2006. Permafrost and the global carbon budget. *Science* 312(5780):1612-1613.

Ziska, L. H. 2000. The impact of elevated CO_2 on yield loss from a C3 and C4 weed in field-grown soybean. *Global Change Biology* 6(8):899-905.

Ziska, L. H. 2003. Evaluation of the growth response of six invasive species to past, present and future atmospheric carbon dioxide. *Journal of Experimental Botany* 54(381):395-404.

Ziska, L. H., J. R. Teasdale, and J. A. Bunce. 1999. Future atmospheric carbon dioxide may increase tolerance to glyphosate. *Weed Science* 47(5):608-615.

Zuwallack, R. 2009. Piloting data collection via cell phones: Results, experiences, and lessons learned. *Field Methods* 21:388-406.

Zwally, H. J., W. Abdalati, T. Herring, K. Larson, J. Saba, and K. Steffen. 2002. Surface melt-induced acceleration of Greenland ice-sheet flow. *Science* 297(5579):218-222.

APPENDIX A

America's Climate Choices: Membership Lists

COMMITTEE ON AMERICA'S CLIMATE CHOICES

ALBERT CARNESALE (Chair), University of California, Los Angeles
WILLIAM CHAMEIDES (Vice Chair), Duke University, Durham, North Carolina
DONALD F. BOESCH, University of Maryland Center for Environmental Science, Cambridge
MARILYN A. BROWN, Georgia Institute of Technology, Atlanta
JONATHAN CANNON, University of Virginia, Charlottesville
THOMAS DIETZ, Michigan State University, East Lansing
GEORGE C. EADS, Charles River Associates, Washington, D.C.
ROBERT W. FRI, Resources for the Future, Washington, D.C.
JAMES E. GERINGER, Environmental Systems Research Institute, Cheyenne, Wyoming
DENNIS L. HARTMANN, University of Washington, Seattle
CHARLES O. HOLLIDAY, JR., DuPont, Wilmington, Delaware
KATHARINE L. JACOBS,* Arizona Water Institute, Tucson
THOMAS KARL,* National Oceanic and Atmospheric Administration, Asheville, North Carolina
DIANA M. LIVERMAN, University of Arizona, Tuscon and University of Oxford, United Kingdom
PAMELA A. MATSON, Stanford University, California
PETER H. RAVEN, Missouri Botanical Garden, St. Louis
RICHARD SCHMALENSEE, Massachusetts Institute of Technology, Cambridge
PHILIP R. SHARP, Resources for the Future, Washington, D.C.
PEGGY M. SHEPARD, WE ACT for Environmental Justice, New York, New York
ROBERT H. SOCOLOW, Princeton University, New Jersey
SUSAN SOLOMON, National Oceanic and Atmospheric Administration, Boulder, Colorado
BJORN STIGSON, World Business Council for Sustainable Development, Geneva, Switzerland
THOMAS J. WILBANKS, Oak Ridge National Laboratory, Tennessee
PETER ZANDAN, Public Strategies, Inc., Austin, Texas

PANEL ON LIMITING THE MAGNITUDE OF FUTURE CLIMATE CHANGE

ROBERT W. FRI (Chair), Resources for the Future, Washington, D.C.
MARILYN A. BROWN (Vice Chair), Georgia Institute of Technology, Atlanta
DOUG ARENT, National Renewable Energy Laboratory, Golden, Colorado
ANN CARLSON, University of California, Los Angeles
MAJORA CARTER, Majora Carter Group, LLC, Bronx, New York
LEON CLARKE, Joint Global Change Research Institute (Pacific Northwest National Laboratory/University of Maryland), College Park, Maryland
FRANCISCO DE LA CHESNAYE, Electric Power Research Institute, Washington, D.C.
GEORGE C. EADS, Charles River Associates, Washington, D.C.
GENEVIEVE GIULIANO, University of Southern California, Los Angeles
ANDREW J. HOFFMAN, University of Michigan, Ann Arbor
ROBERT O. KEOHANE, Princeton University, New Jersey
LOREN LUTZENHISER, Portland State University, Oregon
BRUCE MCCARL, Texas A&M University, College Station
MACK MCFARLAND, DuPont, Wilmington, Delaware
MARY D. NICHOLS, California Air Resources Board, Sacramento
EDWARD S. RUBIN, Carnegie Mellon University, Pittsburgh, Pennsylvania
THOMAS H. TIETENBERG, Colby College (retired), Waterville, Maine
JAMES A. TRAINHAM, RTI International, Research Triangle Park, North Carolina

PANEL ON ADAPTING TO THE IMPACTS OF CLIMATE CHANGE

KATHARINE L. JACOBS* (Chair, through January 3, 2010), University of Arizona, Tucson
THOMAS J. WILBANKS (Chair), Oak Ridge National Laboratory, Tennessee
BRUCE P. BAUGHMAN, IEM, Inc., Alabaster, Alabama
ROBERT BEACHY,* Donald Danforth Plant Sciences Center, Saint Louis, Missouri
GEORGES C. BENJAMIN, American Public Health Association, Washington, D.C.
JAMES L. BUIZER, Arizona State University, Tempe
F. STUART CHAPIN III, University of Alaska, Fairbanks
W. PETER CHERRY, Science Applications International Corporation, Ann Arbor, Michigan
BRAXTON DAVIS, South Carolina Department of Health and Environmental Control, Charleston
KRISTIE L. EBI, IPCC Technical Support Unit WGII, Stanford, California
JEREMY HARRIS, Sustainable Cities Institute, Honolulu, Hawaii
ROBERT W. KATES, Independent Scholar, Bangor, Maine

HOWARD C. KUNREUTHER, University of Pennsylvania Wharton School of Business, Philadelphia
LINDA O. MEARNS, National Center for Atmospheric Research, Boulder, Colorado
PHILIP MOTE, Oregon State University, Corvallis
ANDREW A. ROSENBERG, Conservation International, Arlington, Virginia
HENRY G. SCHWARTZ, JR., Jacobs Civil (retired), Saint Louis, Missouri
JOEL B. SMITH, Stratus Consulting, Inc., Boulder, Colorado
GARY W. YOHE, Wesleyan University, Middletown, Connecticut

PANEL ON ADVANCING THE SCIENCE OF CLIMATE CHANGE

PAMELA A. MATSON (Chair), Stanford University, California
THOMAS DIETZ (Vice Chair), Michigan State University, East Lansing
WALEED ABDALATI, University of Colorado at Boulder
ANTONIO J. BUSALACCHI, JR., University of Maryland, College Park
KEN CALDEIRA, Carnegie Institution of Washington, Stanford, California
ROBERT W. CORELL, H. John Heinz III Center for Science, Economics and the Environment, Washington, D.C.
RUTH S. DEFRIES, Columbia University, New York, New York
INEZ Y. FUNG, University of California, Berkeley
STEVEN GAINES, University of California, Santa Barbara
GEORGE M. HORNBERGER, Vanderbilt University, Nashville, Tennessee
MARIA CARMEN LEMOS, University of Michigan, Ann Arbor
SUSANNE C. MOSER, Susanne Moser Research & Consulting, Santa Cruz, California
RICHARD H. MOSS, Joint Global Change Research Institute (Pacific Northwest National Laboratory/University of Maryland), College Park, Maryland
EDWARD A. PARSON, University of Michigan, Ann Arbor
A. R. RAVISHANKARA, National Oceanic and Atmospheric Administration, Boulder, Colorado
RAYMOND W. SCHMITT, Woods Hole Oceanographic Institution, Massachusetts
B. L. TURNER II, Arizona State University, Tempe
WARREN M. WASHINGTON, National Center for Atmospheric Research, Boulder, Colorado
JOHN P. WEYANT, Stanford University, California
DAVID A. WHELAN, The Boeing Company, Seal Beach, California

PANEL ON INFORMING EFFECTIVE DECISIONS AND ACTIONS RELATED TO CLIMATE CHANGE

DIANA LIVERMAN (Co-chair), University of Arizona, Tucson
PETER RAVEN (Co-chair), Missouri Botanical Garden, Saint Louis
DANIEL BARSTOW, Challenger Center for Space Science Education, Alexandria, Virginia
ROSINA M. BIERBAUM, University of Michigan, Ann Arbor
DANIEL W. BROMLEY, University of Wisconsin-Madison
ANTHONY LEISEROWITZ, Yale University
ROBERT J. LEMPERT, The RAND Corporation, Santa Monica, California
JIM LOPEZ,* King County, Washington
EDWARD L. MILES, University of Washington, Seattle
BERRIEN MOORE III, Climate Central, Princeton, New Jersey
MARK D. NEWTON, Dell, Inc., Round Rock, Texas
VENKATACHALAM RAMASWAMY, National Oceanic and Atmospheric Administration, Princeton, New Jersey
RICHARD RICHELS, Electric Power Research Institute, Inc., Washington, D.C.
DOUGLAS P. SCOTT, Illinois Environmental Protection Agency, Springfield
KATHLEEN J. TIERNEY, University of Colorado at Boulder
CHRIS WALKER, The Carbon Trust LLC, New York, New York
SHARI T. WILSON, Maryland Department of the Environment, Baltimore

Asterisks (*) denote members who resigned during the study process.

Panel on Advancing the Science of Climate Change: Statement of Task

The Panel on Advancing the Science of Climate Change will first provide a concise overview of past, present, and future climate change, including its causes and its impacts, and then recommend steps to advance our current understanding. The panel will consider both the natural climate system and the human activities responsible for driving climate change and altering the vulnerability of different regions, sectors, and populations; it will also consider the scientific advances needed to better understand the effectiveness of actions taken to limit the magnitude of future climate change and adapt to its impacts. The panel will be challenged to treat climate variables and the associated human activities and ecological processes as a single system, rather than a collection of individual elements. The panel will describe the observations, research programs, next-generation models, and other activities and tools that could improve our present understanding of climate change and its interactions with ecological and human systems, as well as the data, activities, and physical and human assets needed to support these activities. It is anticipated that the panel will convene a major workshop focusing on the research needed to better understand the potential efficacy, impacts, and risks of various "geoengineering" proposals (see Appendix E).

Ultimately, the goal of this panel is to answer the third question in the Statement of Task for the study ("What can be done to better understand climate change and its interactions with human and ecological systems?"). The panel will be challenged to produce a report that is broad and authoritative, yet concise and useful to decision makers. The costs, benefits, limitations, trade-offs, and uncertainties associated with different options and strategies should be assessed qualitatively and, to the extent practicable, quantitatively, using scenarios of future climate change and vulnerability developed in coordination with the Committee on America's Climate Choices and the other study panels. The panel should also provide policy-relevant (but not policy-prescriptive) input to the committee on the following overarching questions:

- What short-term actions can be taken to better understand climate change and its interactions with human and ecological systems?

- What promising long-term strategies, investments, and opportunities could be pursued to advance the science of climate change?
- What are the major scientific and technological advances (e.g., new observations, improved models, research priorities, etc.) needed to extend our understanding of climate change and its interactions with other systems?
- What are the major impediments (e.g., practical, institutional, economic, ethical, intergenerational, etc.) to advancing the science of climate change, and what can be done to overcome these impediments?
- What can be done to advance the science of climate change at different levels (e.g., local, state, regional, national, and in collaboration with the international community) and in different sectors (e.g., nongovernmental organizations, the business community, the research and academic communities, individuals and households, etc.)?

Panel on Advancing the Science of Climate Change: Biographical Sketches

Pamela A. Matson (Chair) (NAS) is the Chester Naramore Dean of Stanford University's School of Earth Sciences. She is also the Richard and Rhoda Goldman Professor of Environmental Studies, a senior fellow of the Woods Institute for Environment, and co-leader of Stanford's Initiative on Environment and Sustainability. Her research focuses on biogeochemical cycling and land-water interactions in tropical forests and agricultural systems, and on sustainability science. Together with hydrologists, atmospheric scientists, economists, and agronomists, she analyzes the economic drivers and environmental consequences of land use and resource use decisions in agricultural systems, with the objective of identifying practices that are economically and environmentally sustainable. She and her research team also evaluate the vulnerability of human-environment systems to climate and other global changes. Pamela joined the Stanford faculty in 1997, following positions as professor at the University of California, Berkeley and research scientist at NASA. She earned her B.S. at the University of Wisconsin-Eau Claire, her M.S. at Indiana University, and her Ph.D. at Oregon State University. She was the founding chair of the National Academies Roundtable on Science and Technology for Sustainability, is a past president of the Ecological Society of America, and served on the science committee for the International Geosphere-Biosphere Program. She currently serves on the board of trustees of the World Wildlife Fund. She was elected to the American Academy of Arts and Sciences in 1992 and to the National Academy of Sciences in 1994. In 1995, Dr. Matson was selected as a MacArthur Fellow and in 1997 was elected a Fellow of the American Association for the Advancement of Science. In 2002 she was named the Burton and Deedee McMurtry University Fellow in Undergraduate Education at Stanford.

Thomas Dietz (Vice Chair) is a professor of sociology and environmental science and policy and Assistant Vice President for Environmental Research at Michigan State University (MSU). He holds a Ph.D. in ecology from the University of California, Davis, and a bachelor of general studies from Kent State University. At MSU he was Founding Director of the Environmental Science and Policy Program and Associate Dean in the Colleges of Social Science, Agriculture and Natural Resources and Natural Science. Dr.

Dietz is a Fellow of the American Association for the Advancement of Science and has been awarded the Sustainability Science Award of the Ecological Society of America, the Distinguished Contribution Award of the American Sociological Association Section on Environment, Technology and Society, and the Outstanding Publication Award, also from the American Sociological Association Section on Environment, Technology and Society, and the Gerald R. Young Book Award from the Society for Human Ecology. At the National Research Council he has served as chair of the Committee on Human Dimensions of Global Change and the Panel on Public Participation in Environmental Assessment and Decision Making. Dr. Dietz has also served as Secretary of Section K (Social, Economic, and Political Sciences) of the American Association for the Advancement of Science and is the former president of the Society for Human Ecology. He has co-authored or co-edited 11 books and more than 100 papers and book chapters. His current research examines the human driving forces of environmental change, environmental values, and the interplay between science and democracy in environmental issues.

Waleed Abdalati is the director of the Earth Science and Observation Center in CIRES at the University of Colorado, where he is also an associate professor of geography. He conducts research on high-latitude glaciers and ice sheets using satellite and airborne instruments. From 2004 to 2008 he was head of the NASA Goddard Space Flight Center's Cryospheric Sciences Branch, supervising a group of scientists that work with satellite and aircraft instruments to understand the Earth's changing ice cover. From 2000 to 2006, he managed NASA's Cryospheric Sciences Program, overseeing NASA-funded research efforts on glaciers, ice sheets, sea ice, and polar climate. During that time, he also served as Program Scientist for NASA's Ice Cloud and land Elevation Satellite (ICESat), which has as its primary objective understanding changes in the Earth's ice cover. From 1996 through 2000, Dr. Abdalati was a research scientist at NASA's Goddard Space Flight Center, and from 1986 to 1990 he worked as a mechanical engineer in the aerospace industry. Dr. Abdalati received the Presidential Early Career Award for Scientists and Engineers from the White House in 1999 and the NASA Exceptional Service Medal in 2004. He earned his Ph.D. in geography from the University of Colorado in 1996, an M.S. degree in aerospace engineering sciences from the University of Colorado in 1991, and a B.S. in mechanical engineering from Syracuse University in 1986.

Antonio J. Busalacchi, Jr., is director of the Earth System Science Interdisciplinary Center (ESSIC) and professor in the Department of Atmospheric and Oceanic Science at the University of Maryland, College Park. His research interests include tropical ocean circulation, its role in the coupled climate system, and climate variability and predictability. Dr. Busalacchi has been involved in the activities of the World Climate Research Program (WCRP) for many years and currently is chair of the Joint Scientific

Committee that oversees the WCRP and previously was co-chair of the scientific steering group for its subprogram on Climate Variability and Predictability (CLIVAR). Dr. Busalacchi has extensive NRC experience as chair of the Board on Atmospheric Sciences and Climate, the Climate Research Committee and the Committee on a Strategy to Mitigate the Impact of Sensor Descopes and Demanifests on the NPOESS and GOES-R Spacecraft, and member of the Committee on Earth Studies, the Panel on the Tropical Ocean Global Atmosphere (TOGA) Program, and the Panel on Ocean Atmosphere Observations Supporting Short-Term Climate Predictions. He holds a Ph.D. in oceanography from Florida State University.

Ken Caldeira is a scientist at the Carnegie Institution for Science's Department of Global Ecology. His lab investigates ongoing changes to Earth's climate and carbon cycle, climate and carbon-cycle changes in the ancient past, ocean carbon cycle and biogeochemistry, ocean acidification, land cover and climate change, carbon-neutral energy for economic growth and environmental preservation, and geoengineering. Dr. Caldeira previously worked as an environmental scientist and physicist at Lawrence Livermore National Laboratory, where he researched long-term evolution of climate and geochemical cycles; ocean carbon sequestration; numerical simulation of climate, carbon, and biogeochemistry; marine biogeochemical cycles; and approaches to supplying energy services with diminished environmental footprint. Dr. Caldeira received his M.S. and Ph.D. degrees in atmospheric sciences from the New York University Department of Applied Science.

Robert W. Corell is the Global Change Director at the H. John Heinz III Center for Science, Economics, and the Environment. Prior to this he worked as a Senior Policy Fellow at the Policy Program of the American Meteorological Society and an Affiliate of the Washington Advisory Group. He recently completed an appointment as a Senior Research Fellow in the Belfer Center for Science and International Affairs at Harvard University's Kennedy School of Government. He is actively engaged in research concerned with the sciences of global change and the connection between science and public policy, particularly research activities that are focused on global and regional climate change, related environmental issues, and science to promote understanding of vulnerability and sustainable development. He was recently honored with a National Conservation Award for Science, in recognition of his more than four decades of environmental science work. He co-chairs an international strategic planning group that is developing a strategy designed to harness science, technology, and innovation for sustainable development, serves as the Chair of the Arctic Climate Impact Assessment, counsels as Senior Science Advisor to ManyOne.Net, and is Chair of the Board of the Digital Universe Foundation. He was Assistant Director for Geosciences at the National Science Foundation, where he had oversight for the Atmospheric, Earth, and

Ocean Sciences and the global change programs of the National Science Foundation (NSF). He also led the U.S. Global Change Research Program from 1987 to 2000. He was formerly a professor and academic administrator at the University of New Hampshire. He is an oceanographer and engineer by background and training. He received his Ph.D., M.S., and B.S. degrees from the Massachusetts Institute of Technology (MIT).

Ruth S. DeFries (NAS) is Denning Professor of Sustainable Development in Columbia University's Department of Ecology, Evolution and Environmental Biology. Her research investigates the relationships between human activities, the land surface, and the biophysical and biogeochemical processes that regulate the Earth's habitability. She is interested in observing land cover and land use change at regional and global scales with remotely sensed data and exploring the implications for ecological services such as climate regulation, the carbon cycle, and biodiversity. Dr. DeFries obtained a Ph.D. in 1980 from the Department of Geography and Environmental Engineering at Johns Hopkins University and a bachelor's degree in 1976 from Washington University with a major in earth science. Previously, Dr. DeFries worked at the National Research Council with the Committee on Global Change and taught at the Indian Institute of Technology in Bombay. She is a fellow of the Aldo Leopold Leadership Program.

Inez Y. Fung (NAS) is a professor of atmospheric sciences and founding co-director of the Berkeley Institute of the Environment at the University of California, Berkeley. She studies the interactions between climate change and biogeochemical cycles, particularly the processes that maintain and alter the composition of the atmosphere. Her research emphasis is on using atmospheric transport models and a coupled carbon-climate model to examine how carbon dioxide sources and sinks are changing. She is also a member of the science team for NASA's Orbiting Carbon Observatory. Dr. Fung is a recipient of the American Geophysical Union's Roger Revelle Medal, and appears in a new NAS biography series for middle-school readers, *Women's Adventure in Science*. She is a fellow of the American Meteorological Society and the American Geophysical Union, and a member of the National Academy of Sciences. She received her B.S. in applied mathematics and her Ph.D. in meteorology from MIT.

Steven Gaines is Director of the Marine Science Institute and Professor of Ecology, Evolution, and Marine Biology at the University of California at Santa Barbara (UCSB). He is a marine ecologist who studies marine conservation, the design of marine reserves, the impact of climate change on oceans, and sustainable fisheries. Dr. Gaines is a lead investigator of several groups: (1) the Partnership for Interdisciplinary Studies of Coastal Oceans (PISCO), a consortium studying marine ecosystems of the west coast of the United States, (2) the Santa Barbara Coastal Long Term Ecological Research (LTER) that studies connections between coastal watersheds and the ecology of kelp forests,

(3) the Sustainable Fisheries Group, which uses market-based approaches to enhance the sustainability of fisheries, and (4) Flow, Fish and Fishing, a biocomplexity project examining connections between ocean physics, fish, and fishing. Dr. Gaines was awarded a Pew Fellowship in 2003 to extend the conceptual framework for networks of marine reserves and uses the findings of this work to aid the ongoing efforts of the Marine Life Protection Act to establish a statewide network of marine protected areas. Steve received his Ph.D. in zoology in 1983 from Oregon State University and was a postdoctoral fellow and research scientist at Stanford University for 4 years. In 1987, he joined the faculty of Brown University and then the faculty at UCSB in 1994. Dr. Gaines became Director of the Marine Science Institute at UCSB in 1997 and has served as Acting Vice Chancellor for Research at UCSB, and Acting Dean of Science.

George M. Hornberger (NAE) is a Distinguished University Professor in the Department of Civil and Environmental Engineering and the Department of Earth and Environmental Sciences at Vanderbilt University. His research interests are catchment hydrology and hydrochemistry, as well as the transport of colloids in geological media. His work centers on the coupling of field observations with mathematical modeling, with a focus on understanding how water is routed physically through soils and rocks to streams and how hydrological processes and geochemical processes combine to produce observed stream dynamics. This modeling work allows the extension of work on individual catchments to regional scales and to the investigation of the impact of meteorological driving variables on catchment hydrology. Dr. Hornberger is a member of the American Geophysical Union, the Geological Society of America, the Society of Sigma Xi, and American Women in Science. He has served on numerous NRC studies, chaired the Board on Earth Sciences and Resources, and is currently a member of the Report Review Committee. Dr. Hornberger received his Ph.D. in hydrology from Stanford University.

Maria Carmen Lemos is an associate professor of natural resources and environment at the University of Michigan, Ann Arbor, and Senior Policy Scholar at the Udall Center for the Study of Public Policy at the University of Arizona. She currently serves as vice-chair of the Scientific Advisory Committee for the InterAmerican Institute for the Study of Climate Change (IAI) and as member of the NRC Committee on Human Dimensions of Global Change. She has M.Sc. and Ph.D. degrees in political science from MIT. During 2006-2007 she was a James Martin 21st Century School Fellow at the Environmental Change Institute at Oxford University. Her research focuses on public policy making in Latin America (Brazil, Mexico, and Bolivia) and the United States (Great Lakes region), especially related to the human dimensions of environmental change, the co-production of science and policy, and the role of technoscientific knowledge in environmental governance and in building adaptive capacity of water and disaster response systems to climate variability and change. She is a contributing author of the Intergovernmental Panel on Climate Change (IPCC) and the U.S. Climate Change Science Program Synthesis Reports.

Susanne C. Moser is Director and Principal Researcher of Susanne Moser Research and Consulting and Associate Researcher at the University of California, Santa Cruz, Institute of Marine Sciences. Previously, she was a research scientist at the Institute for the Study of Society and Environment at the National Center for Atmospheric Research and served as staff scientist at the Union of Concerned Scientists, a visiting assistant professor at Clark University, and a fellow in the Global Environmental Assessment Project at Harvard University. Her research interests include the impacts of global environmental change, especially in the coastal, public health, and forest sectors; societal responses to environmental hazards in the face of uncertainty; the use of science to support policy and decision making; and the effective communication of climate change to facilitate social change. Current work focuses on developing adaptation strategies to climate change at local and state levels, identifying ways to promote community resilience, and building decision-support systems. She is a fellow of the Aldo Leopold and Donella Meadows Leadership Programs and received a diploma in Applied Physical Geography from the University of Trier and M.A. and Ph.D. degrees in geography from Clark University.

Richard H. Moss is Senior Research Scientist with the Joint Global Change Research Institute of Pacific Northwest National Laboratory and Visiting Senior Research Scientist at the Earth Systems Science Interdisciplinary Center of the University of Maryland. He served as Director of the Office of the US Global Change Research Program/Climate Change Science Program (2000-06), Vice President and Managing Director for Climate Change at the World Wildlife Fund-US (2007-09), and Senior Director of the U.N. Foundation Energy and Climate Program (2006-2007). He also directed the Technical Support Unit of the Intergovernmental Panel on Climate Change (IPCC) impacts, adaptation, and mitigation working group (1993-1999) and served on the faculty of Princeton University (1989-91). He has been a lead author and editor of a number of assessments, reports, and research papers. Moss chairs the US National Academy of Science's standing committee on the "human dimensions" of global environmental change and serves on the editorial board of *Climatic Change*. He remains active in the IPCC and currently co-chairs the IPCC Task Group on Data and Scenario Support for Impact and Climate Analysis. He was named a Fellow of the American Association for the Advancement of Science (AAAS) in 2006, a Distinguished Associate of the U.S. Department of Energy in 2004, and a fellow of the Aldo Leopold Leadership Program in 2001. He received an M.P.A. and Ph.D. from Princeton University (Public and International Affairs) and his B.A. from Carleton College in Northfield, MN. Moss' research interests include development and use of scenarios, characterization and communication of uncertainty, and assessment of adaptive capacity and vulnerability to climate

Edward A. Parson is Joseph L. Sax Collegiate Professor of Law and Professor of Natural Resources & Environment at the University of Michigan. His research examines international environmental policy, the role of science and technology in public policy, and the political economy of regulation. Parson's recent articles have appeared in *Science*, *Nature*, *Climatic Change*, *Issues in Science and Technology*, the *Journal of Economic Literature*, and *Annual Review of Energy and the Environment*. His most recent books are *The Science and Politics of Global Climate Change* (Cambridge, 2nd ed., 2010, with Andrew Dessler) and *Protecting the Ozone Layer: Science and Strategy* (Oxford, 2003), which won the 2004 Harold and Margaret Sprout Award of the International Studies Association. Parson has chaired and served on several senior advisory committees for the National Academy of Sciences and the U.S. Global Change Research Program, including the Synthesis Team for the U.S. National Assessment of Climate Impacts. In 2005, he was appointed to the National Advisory Board of the Union of Concerned Scientists. He has worked and consulted for the International Institute for Applied Systems Analysis, the United Nations Environment Program, the Office of Technology Assessment of the U.S. Congress, the Privy Council Office of the Government of Canada, and the White House Office of Science and Technology Policy, and spent 12 years on the faculty of Harvard's Kennedy School of Government. He holds degrees in physics from the University of Toronto and in management science from the University of British Columbia, and a Ph.D. in public policy from Harvard.

Akkihebbal "Ravi" Ravishankara (NAS) is an atmospheric chemist at the National Oceanic and Atmospheric Administration (NOAA) in Boulder, Colorado. He is director of the Chemical Sciences Division of the NOAA Earth System Research Laboratory. Ravishankara's research has contributed fundamental studies of the gas-phase and surface chemistry of Earth's atmosphere. His investigations have advanced the understanding of basic chemical processes and reaction rates related to several major environmental issues, including ozone-layer depletion, climate change, and air pollution. For example, his results have led to a better understanding of the chemistry that causes the Antarctic ozone hole, identified new processes that affect ozone pollution in the lower atmosphere, and elucidated the role of aerosols and clouds in climate. In addition, he has led the evaluation of the "ozone-friendliness" and "climate friendliness" of many substances that have been proposed for use in commercial and industrial applications. Ravishankara has played leading roles in national and international reports assessing the state of the science understanding of ozone-layer depletion and other issues. He is a co-chair of the Scientific Assessment Panel of the U.N. Montreal Protocol that protects the stratospheric ozone layer. He is currently co-leading an effort within

NOAA to establish an integrated program linking climate change and air quality. A research scientist with NOAA since 1984, Ravishankara has also been an adjunct professor of chemistry at the University of Colorado in Boulder since 1989. He is a member of the National Academy of Sciences. Awards include his election as a Fellow of the American Geophysical Union, Fellow of the American Association for the Advancement of Science, Fellow of the United Kingdom Royal Society of Chemistry, recipient of the Polanyi Medal and Centenary Lectureship of the Royal Society of Chemistry, the U.S. Environmental Protection Agency's Stratospheric Ozone Protection Award, the Department of Commerce Silver Medal, and the U.S. Presidential Rank Award. Ravishankara received his doctorate in physical chemistry from the University of Florida in 1975. He has authored or co-authored over 300 scientific publications.

Raymond W. Schmitt is a senior scientist at the Woods Hole Oceanographic Institution, where he has spent most of his career. His research interests include oceanic mixing and microstructure, double-diffusive convection, the thermohaline circulation, oceanic freshwater budgets, the salinity distribution and its measurement, the use of acoustics for imaging fine structure, and the development of instrumentation. He is also interested in the intergenerational problem of sustaining long-term observations for climate. Dr. Schmitt has served on ocean sciences and polar program panels with the National Science Foundation, the Ocean Observing System Development Panel, the CLIVAR Science Steering Group, and the Ocean Studies Board. He was named a J.S. Guggenheim fellow in 1997 and has authored or co-authored over 75 publications. Dr. Schmitt earned his Ph.D. in physical oceanography from the University of Rhode Island and his B.S. in physics from Carnegie Mellon University.

B. L. (Billie Lee) Turner II (NAS) is Gilbert F. White Professor of Environment and Society in Arizona State University's School of Geographical Sciences. For most of his career (1980-2008), he taught at Clark University in Worcester, Massachusetts, where he served as the Alice C. Higgins and Milton P. Professor of Environment and Society, and Director of the Graduate School of Geography. He received his B.A. and M.A. degrees from the University of Texas at Austin in 1968 and 1969, respectively, and his Ph.D. at the University of Wisconsin, Madison, in 1974. Turner's research interests center on human-environment relationships, specifically dealing with land change science, sustainability, tropical forests, and the ancient Maya. He is currently engaged in a long-term study on deforestation and sustainability in the southern Yucatan. Dr. Turner is associated with the development of land use/cover change studies exemplified in the international programs sponsored by the IGBP and IHDP. He has also promoted the emerging field of "sustainability science," a major focus at Arizona State University. He is a former Guggenheim Fellow and Fellow of the Center for Advanced Studies in the Behavioral Sciences, and recipient of research honors from various geographical associations. He was elected to the National Academy of Sciences in 1995 and to the American Academy of Arts and Sciences in 1998.

Dr. Warren M. Washington (NAE) is a senior scientist and head of the Climate Change Research Section in the Climate and Global Dynamics Division at the National Center for Atmospheric Research (NCAR). After completing his doctorate in meteorology at Pennsylvania State University, he joined NCAR in 1963 as a research scientist. Dr. Washington's areas of expertise are atmospheric science and climate research, and he specializes in computer modeling of the Earth's climate. He serves as a consultant and advisor to a number of government officials and committees on climate system modeling. From 1978 to 1984, he served on the President's National Advisory Committee on Oceans and Atmosphere. In 1998, he was appointed to NOAA's Science Advisory Board. In 2002, he was appointed to the Science Advisory Panel of the U.S. Commission on Ocean Policy and the National Academies' Coordinating Committee on Global Change. Dr. Washington's NRC service is extensive and includes membership on the Board on Sustainable Development, the Commission on Geosciences, Environment, and Resources, the Board on Atmospheric Sciences and Climate, and the Panel on Earth and Atmospheric Sciences (chair). He is past chair of the National Science Board.

Dr. John P. Weyant is professor of management science and engineering, director of the Energy Modeling Forum (EMF), and Deputy Director of the Precourt Institute for Energy Efficiency at Stanford University. He is also a Senior Fellow of the Freeman Spogli Institute for International Studies and the Woods Institute for the Environment at Stanford. Professor Weyant earned a B.S./M.S. in aeronautical engineering and astronautics and M.S. degrees in engineering management and in operations research and statistics, all from Rensselaer Polytechnic Institute, and a Ph.D. in management science with minors in economics, operations research, and organization theory from the University of California at Berkeley. He also was also a National Science Foundation Post-Doctoral Fellow at Harvard's Kennedy School of Government. His current research focuses on analysis of global climate change policy options, energy efficiency analysis, energy technology assessment, and models for strategic planning. Weyant has been a convening lead author or lead author for the IPCC for chapters on integrated assessment, greenhouse gas mitigation, integrated climate impacts, and sustainable development, and most recently served as a review editor for the climate change mitigation working group of the IPCC's Fourth Assessment Report. He has been active in the U.S. debate on climate change policy through the Department of State, the Department of Energy, and the Environmental Protection Agency. In California, he is a member of the California Air Resources Board's Economic and Technology Advancement Advisory Committee (ETAAC), which is charged with making recommendations for technology policies to help implement AB 32, The Global Warming Solutions Act of 2006. Weyant was awarded the U.S. Association for Energy Economics' 2008 Adelmann-Frankel award for unique and enduring contributions to the field of energy economics. Weyant was honored in 2007 as a major contributor to the Nobel Peace Prize awarded to the IPCC and in 2008 by Chairman Mary Nichols for contributions to the California Air Resources Board's Economic and Technology Advancement Advisory Committee on AB 32.

Dr. David A. Whelan (NAE) is Vice President and Deputy General Manager of Boeing's Phantom Works and Chief Scientist, Integrated Defense Systems. Prior to joining Boeing in 2001, Dr. Whelan was director of the Tactical Technology Office at the Defense Advanced Research Projects Agency (DARPA), where he led the development of enabling technologies, such as unmanned vehicles and space-based radar systems. Prior to his position with DARPA, Dr. Whelan held several positions of increasing responsibility with Hughes Aircraft in the development and application of radar systems. His high-technology development experience also includes positions as research physicist for Lawrence Livermore National Laboratory (LLNL), as well as one of four lead engineers assigned for the design and development of the B-2 Stealth Bomber Program at Northrop Grumman. Dr. Whelan is vice chairman of the NRC Naval Studies Board and a member of the NRC USSOCOM Standing Committee. He is a Director of the HRL (former Hughes Research Laboratory) and serves on the LLNL Directors Review Committee for the Physics Department. Dr. Whelan is the recipient of Secretary of Defense Medal for Meritorious Civilian Service (2001), Secretary of Defense Medal for Outstanding Public Service (1998). and the U.S. Air Force Medal for Meritorious Civilian Service (2008).

Uncertainty Terminology

In assessing the state of knowledge about climate change, scientists have developed a careful terminology for expressing uncertainties around both statements of fact about a current situation (for example, "most observed warming can be attributed to human action") and statements about the likelihoods of various future outcomes (for example, "sea level could rise by several feet by 2100"). The IPCC, in particular, has devoted serious debate and discussion to appropriate ways of expressing and dealing with uncertainty around such statements (Moss and Schneider, 2000), and all recent IPCC assessments have adopted a set of terminology to describe the degree of confidence in conclusions (see, e.g., Manning et al., 2004). In estimating confidence, scientific assessment teams draw on information about "the strength and consistency of the observed evidence, the range and consistency of model projections, the reliability of particular models as tested by various methods, and, most importantly, the body of work addressed in earlier synthesis and assessment reports" (USGCRP, 2009). It is easier to employ precise uncertainty language in situations where conclusions are based on extensive quantitative data or models than in areas where data are less extensive, important research is qualitative, or models are in an earlier stage of development. Statements about the future are also generally more uncertain than statements of fact about observed changes or current trends.

Table D.1 shows the language adopted by the IPCC to describe confidence about facts or the likelihood that a statement is accurate. The U.S. Global Change Research Program's recent assessment report on *Global Climate Change Impacts on the United States*(USGCRP, 2009) uses similar language. In this report, *Advancing the Science of Climate Change*, when we draw directly on the statements of the formal national and international assessments, we adopt their terminology to describe uncertainty. However, because of the more concise nature and intent of this report, we do not attempt to quantify confidence and certainty about every statement of the science.

TABLE D.1 Language Adopted by the IPCC to Describe Confidence About Facts or the Likelihood of an Outcome

Terminology for Describing Confidence About Facts	
Very high confidence	At least 9 out of 10 chance of being correct
High confidence	About 8 out of 10 chance
Medium confidence	About 5 out of 10 chance
Low confidence	About 2 out of 10 chance
Very low confidence	Less than 1 out of 10 chance
Terminology for Describing Likelihood of an Outcome	
Virtually certain	More than 99 chances out of 100
Extremely likely	More than 95 chances out of 100
Very likely	More than 90 chances out of 100
Likely	More than 65 chances out of 100
More likely than not	More than 50 chances out of 100
About as likely as not	Between 33 and 66 chances out of 100
Unlikely	Less than 33 chances out of 100
Very unlikely	Less than 10 chances out of 100
Extremely unlikely	Less than 5 chances out of 100
Exceptionally unlikely	Less than 1 chance out of 100

SOURCE: IPCC (2007a).

APPENDIX E

The United States Global Change Research Program

The commitment of the United States to a national research program on climate began with the passage by Congress of the National Climate Program Act in 1978. The act was designed to establish a comprehensive and coordinated national climate policy and program. The following year, the National Research Council released *Strategy for the National Climate Program* (NRC, 1980c), the first of a number of reviews and advisory documents prepared by the NRC on the program.

Even though the National Climate Program Act established the National Climate Program Office as an interagency program, a subsequent review of the program by the NRC several years later (NRC, 1986) suggested that, among other problems, the Act's budget mechanism did not facilitate a coordinated and integrated program because each department and agency could and often did act independently in its budget submission. Around the same time, climate and global change issues began to rise on the scientific, political, and policy agendas. Driven by a substantial increase in the scientific literature, several high-profile discussions in Congress, and a growing recognition of the inherently interdisciplinary and interconnected nature of climate and other global changes, an NRC report in 1988, *Toward an Understanding of Global Change: Initial Priorities for U.S .Contributions to the International Geosphere-Biosphere Program* (NRC, 1988), proposed a scientific framework to improve understanding of climate and other global environmental changes.

The Global Change Research Act of 1990 as amended[1] is the currently mandated framework within which climate and global change research is implemented among U.S. federal departments and agencies. Unless altered by subsequent legislation, the Global Change Research Act of 1990 provides most of the necessary authority for a strategically integrated climate and global change research program. The Act sets the strategies and mechanisms for establishing the research and for setting priorities, stating the following inter alia:

- **A Coordinated and Integrated Research Program.** The climate and global change research program shall be coordinated and run as a national program

[1] U.S. Global Change Research Act of 1990, P.L. 101-606 (11/16/90), 104 Stat. 3096-3104 (*www.gcrio.org/gcact1990.html*).

and the agency representatives in interagency Committee[2] of the Program "shall be high ranking officials of their agency or department, wherever possible the head of the portion of that agency or department that is most relevant to the purpose of the program as describe in the Act." Further, the 1990 Act mandated that the "President should direct the Secretary of State, in cooperation with the Committee, to initiate discussions with other nations leading toward international protocols and other agreements to coordinate global change research activities ... the purpose of which is to: (i) promote international, intergovernmental cooperation on global change research; (ii) involve scientists and policymakers from developing nations in such cooperative global change research programs; and (iii) promote international efforts to provide technical and other assistance to developing nations which will facilitate improvements in their domestic standard of living while minimizing damage to the global or regional environment."

- **Priority-Setting Responsibilities of the National Research Council.** The NRC is charged in the Act with the responsibility to (i) evaluate the scientific content of the research program and plan, (ii) provide information and advice obtained from United States and international sources, and (iii) recommended priorities for future global change research. Historically, the NRC has established a variety of committees or boards to implement this responsibility.
- **Guidance for Implementing the Research Program.** The committee shall each year provide general guidance to each federal agency or department participating in the program with respect to the preparation of requests for appropriations for activities related to the program.[3] This annual guidance historically has been implemented by a "Terms of Reference" document, prepared and issued jointly by OMB and OSTP Directors, that describes the responsibilities of (i) OMB and OSTP, (ii) all participating USGCRP agencies and departments, and (iii) the federal interagency committee for developing the research program and all elements of the budget submittals. The history of the USGCRP leads to a simple conclusion: An effective program must engage the leadership at high levels of (i) OMB and OSTP and other appropriate Offices of the

[2] The Act states that "The President, through the Council (currently the NSTC and earlier the FCCSET), shall establish a Committee on Earth and Environmental Sciences (CEER). The Committee shall carry out Council functions under section 401 of the National Science and Technology Policy, Organization, and Priorities Act of 1976 (42 U.S.C. 6651) relating to global change research, for the purpose of increasing the overall effectiveness and productivity of Federal global change research efforts. The initial name of the Committee, the CEER, had its name changed to the Committee on Environment and Natural Resources (CENR) with the same charge as the CEER.

[3] This is a markedly different authority to conduct the interagency process than existed prior to 1990, resulting in implementation during the late 1980s and early 1990s that had more direct budgetary responsibility of the program's content and budget.

President, (ii) the participating agencies and departments, (iii) the committee structure under the NSTC, and (iv) the NAS through the NRC committees and boards and hence the science communities of the nation. The Act states the program of research "consider and utilize, as appropriate, reports and studies conducted by Federal agencies and departments, the NRC and other entities." Hence, the research program should be guided by NRC recommendations, and the NRC may need to reassess the means by which it provides advice on program priorities.

- **Preparing a Global Change Research Plan.** "The Chairman of the NSTC, through the Committee, shall develop a National Global Change Research Plan (including climate change) for implementation of the Program. The Plan shall contain priorities and recommendations for a national global change research. The Chairman of the Council shall submit the Plan to the Congress at least once every three years ... in developing the Plan, the Committee shall consult with academic, State, industry, and environmental groups and representatives."
- **Conducting Climate and Global Change Assessments.** "The committee shall prepare and submit to the President and the Congress an assessment every 4 years which (i) integrates, evaluates, and interprets the findings of the program and discusses the scientific uncertainties associated with such findings; (ii) analyzes the effects of global change on the natural environment, agriculture, energy production and use, land and water resources, transportation, human health and welfare, human social systems, and biological diversity; and (iii) analyzes current trends in global change, both human-induced and natural, and projects major trends for the subsequent 25 to 100 years." It was under this set of responsibilities that the United States sought to establish an international assessment process, leading to the creation of the IPCC.
- **Congressional Oversight and Appropriations.** The U.S. Congress played a critical role in the development of the 1990 Act, with senior staff working closely with OMB and OSTP to draft the 1990 Act. Thereafter, the House Committee on Science and Technology and the Senate Committee on Commerce, Science, & Transportation have provided oversight to ensure compliance with the mandates of the Act. Further, the Appropriations Committees in both the Senate and the House hold hearings on the progress being made under the Act and its mandates, as the Act requires that an annual program and budget for the USGCRP be submitted concurrent with and as a separate companion document[4] to the President's Budget.[5] Finally, the Congressional Research

[4] This is the origin of "Our Changing Planet," the program and budget document submitted in early February each year as a companion to the President's Annual Budget Submission to Congress.

[5] As a result of this set of legislatively mandated responsibilities, immediately upon the enactment of the U.S. Global Change Research Program (USGCRP), the Director of OMB and the President's Science and

Service[6] has developed a continuing oversight and review process to further document progress and accomplishments.

In the late 1990s, the USGCRP articulated a new strategic plan that focused explicitly on providing information for decision makers at regional scales. Then, in 2001, in response to a variety of inputs (including NRC, 2001), the George W. Bush Administration introduced the Climate Change Research Initiative (CCRI) to focus on key uncertainties associated with climate variability and climate change at global scales. In 2002, the administration integrated the USGCRP and the CCRI into the Climate Change Science Program (CCSP). The CCSP developed a revised strategic plan for the program that contained a focus on decision-support activities, including a series of 21 Synthesis and Assessment Products, an emphasis on adaptive management to support natural resource agencies, and comparative evaluations of response measures using integrated assessment models, scenarios, and other methods (CCSP, 2003). The National Research Council was involved in reviewing these plans (NRC, 2003a, 2004b) and later in helping to develop metrics to evaluate the progress of the program (NRC, 2005f).

With federal support ranging from $2.2 billion in 1990 (in 2008 dollars) to $1.8 billion in 2008, the USGCRP (known as the U.S. Climate Change Science Program from 2002 through 2008) has made enormous contributions to the understanding of climate change over the past two decades and provided key results and support for the IPCC. Congress, especially the Committee on Science and Technology of the House of Representatives, provided active oversight of the program in its early years, holding numerous hearings that sought to ensure compliance with the mandates of the Act. The annual budgetary guidance from the program committee to the participating agencies was a particularly important aspect of the interagency process, because it resulted in the implementation during the late 1980s and early 1990s of a process that had more direct budgetary responsibility for the program's content and budget.

Technology Advisor began issuing detailed management and budget responsibilities to the head(s) of the three components of government responsible for implementing the 1990 Act: (i) the various Offices of the President, (ii) the Secretaries and Heads of the participating Departments and Agencies, and (iii) the Committee, its Officers, and subcommittees. This document was issued several months in advance of the annual Presidential Budget development process.

[6] The Congressional Research Service has produced numerous analyses and assessments of the USGCRP and the Climate Change Science Program (CCSP). A recent example is CRS No. 98-738, *Global Climate Change: Three Policy Perspectives,* November 26, 2008, which in summary concludes that "The purpose here is not to suggest that one lens is 'better' than another, but rather to articulate the implications of the differing perspectives in order to clarify terms of debate among diverse policy communities."

Geoengineering Options to Respond to Climate Change: Steps to Establish a Research Agenda

A Workshop to Provide Input to the America's Climate Choices Study

June 15-16, 2009
Washington Court Hotel
525 New Jersey Avenue, NW, Washington, DC 20001

WORKSHOP OBJECTIVE AND SCOPE

The workshop will inform the work of the *America's Climate Choices* suite of activities by examining a number of proposed "geoengineering" approaches, or interventions in the climate system designed to diminish the amount of climate change occurring after greenhouse gases or radiatively active aerosols are released to the atmosphere. The emphasis of the workshop will be on the research needed to better understand the potential efficacy and consequences of various geoengineering approaches.

The workshop will draw on a growing body of studies and previous workshops that have examined a broad range of geoengineering issues—from the international governance of deliberate climate interventions to the feasibility of specific approaches. The particular focus of this workshop will be approaches (i) to reduce concentrations of greenhouse gases after they have been emitted to the atmosphere (e.g., CO2 capture approaches) or (ii) to limit or offset physical effects of increased greenhouse gas concentrations (e.g., Solar Radiation Management approaches). Other parts of the *America's Climate Choices* study are addressing approaches to reduce greenhouse gas emissions and adapt to climate change. Furthermore, there is already a developed research effort in CO_2 capture by conventional land-management approaches (e.g., conventional afforestation). Thus, these topics will be outside the scope of this workshop.

The workshop will be structured to bring multiple perspectives to the table—engineering, physical and environmental science, social science, policy, legal, and ethical—

to encourage an interdisciplinary dialogue and exchange of ideas, with an emphasis on the research needed to better understand the potential efficacy and consequences of various geoengineering approaches.

Workshop Agenda

Monday, June 15

8:00 Registration

8:30 Welcome — Ralph Cicerone, National Academy of Sciences

8:45 Meeting Overview—Day 1: "Getting the Issues on the Table"
Pamela Matson, Stanford University

9:00 Survey of Geoengineering Options (Including Estimates of Effectiveness, Risk, and Cost)
Ken Caldeira, Carnegie Institution

9:40 Engineering: Important Questions, State of Knowledge, and Major Uncertainties Related to Selected Geoengineering Options
David Keith, University of Calgary

10:20 Break

10:50 Physical Science: Important Questions, State of Knowledge, and Major Uncertainties Related to Selected Geoengineering Options
Daniel Schrag, Harvard University

11:30 Terrestrial Ecosystems, Complexity, and Geoengineering
Tony Janetos, University of Maryland

12:00 Working Lunch (Informal Discussion)

1:00 From Research to Field Testing and Deployment: Ethical Issues Raised By Geoengineering (Panel Discussion)

Martin Bunzl, Rutgers University (Moderator)
Stephen Gardiner, University of Washington
Dale Jamieson, New York University
William Travis, University of Colorado

2:00 Governance and Geoengineering: Who Decides and How (Panel Discussion)
Granger Morgan, Carnegie Mellon University (Moderator)
John Steinbruner, University of Maryland
Jason Blackstock, IIASA
Jay Apt, Carnegie Mellon University

3:00 Assignments/instructions to breakout groups and 20-minute break

3:30 Breakout Session 1 — All Participants

5:30 Adjourn for the day

Tuesday, June 16

8:30 Summary of Day 1/Plan for Day 2: "The Way Forward"
Pamela Matson, Stanford University

8:45 Report back from breakout groups and discussion — All Participants

10:15 Break

10:45 Reactions/Perspectives on Geoengineering (Panel Discussion)
Rob Socolow, Princeton University (Moderator)
James Fleming, Colby College
Michael Oppenheimer, Princeton University
Alan Robock, Rutgers University
Brian Toon, University of Colorado

11:45 Assignments/instructions to new breakout groups

12:00 Working Lunch

1:00 Breakout Session 2, including 20-minute break — All Participants

2:45 Report back from breakout groups following by open discussion
All Participants

4:00 Workshop Adjourns

APPENDIX G

Acronyms and Initialisms

AEF	*America's Energy Future*
AERONET	Aerosol Robotic Network
ARM	Atmosphere Radiation Measurement
CCS	carbon capture and storage
CCSP	Climate Change Science Program
CCTP	climate change technology program
CDM	Clean Development Mechanism
COPD	chronic obstructive pulmonary disease
CDR	carbon dioxide removal
C-ROADS	Climate Rapid Overview and Decision-support Simulator
DOD	U.S. Department of Defense
DOE	U.S. Department of Energy
EGS	enhanced geothermal system
ENSO	El Niño-Southern Oscillation
EOS	Earth Observing System
EPA	U.S. Environmental Protection Agency
ESSP	Earth System Science Partnership
EUMETSAT	European meteorological satellite program
FACE	free air CO_2 enrichment
GCRA	Global Change Research Act
GHG	greenhouse gas
ICESat	Ice, Cloud, and Land Elevation Satellite
ICSU	International Council of Scientific Unions
IGBP	International Geosphere-Biosphere Program
IGERT	Integrative Graduate Education and Research Traineeship
IGFA	International Group of Funding Agencies for Global Change Research
IHDP	International Human Dimensions Program

IPCC	Intergovernmental Panel on Climate Change
IUBS	International Union of Biological Science
IWRM	Integrated Water Resources Management
JI	Joint Implementation
MEA	Millennium Ecosystem Assessment
MODIS	Moderate Resolution Imaging Spectroradiometer
MPAs	marine protected areas
NAPAs	National Adaptation Programmes of Action
NASA	National Aeronautics and Space Administration
NIC	National Intelligence Council
NOAA	National Oceanic and Atmospheric Administration
NPOESS	National Polar-orbiting Operational Environmental Satellite System
NPP	net primary productivity
NRC	National Research Council
OCO	Orbiting Carbon Observatory
OMB	Office of Management and Budget
PIER	Public Interest Energy Research
PSAC	President's Science Advisory Committee
QDR	Quadrennial Defense Review
R&D	research and development
REDD	Reducing Emissions from Deforestation and Degradation
RISA	Regional Integrated Sciences and Assessments
SARP	Sector Applications Research Program
SCOPE	Scientific Committee on Problems of the Environment
SRM	solar radiation management
START	System for Analysis, Research and Training
TAO	Tropical-Atmosphere Ocean
UNFCCC	United Nations Framework Convention on Climate Change
USGCRP	U.S. Global Change Research Program

WCRP	World Climate Research Program
WMO	World Meteorological Organization
UNDP	United Nations Development Programme
UNEP	United Nations Environment Programme
UNESCO	United Nations Educational, Scientific and Cultural Organization